Lecture Notes in Computer Science 11973

Yaroslav D. Sergeyev · Dmitri E. Kvasov (Eds.)

Numerical Computations: Theory and Algorithms

Third International Conference, NUMTA 2019
Crotone, Italy, June 15–21, 2019
Revised Selected Papers, Part I

 Springer

Editors
Yaroslav D. Sergeyev (iD)
University of Calabria
Rende, Italy

Lobachevsky University of Nizhny
Novgorod
Nizhny Novgorod, Russia

Dmitri E. Kvasov (iD)
University of Calabria
Rende, Italy

Lobachevsky University of Nizhny
Novgorod
Nizhny Novgorod, Russia

ISSN 0302-9743   ISSN 1611-3349   (electronic)
Lecture Notes in Computer Science
ISBN 978-3-030-39080-8   ISBN 978-3-030-39081-5   (eBook)
https://doi.org/10.1007/978-3-030-39081-5

LNCS Sublibrary: SL1 – Theoretical Computer Science and General Issues

This Springer imprint is published by the registered company Springer Nature Switzerland AG
The registered company address is: Gewerbestrasse 11, 6330 Cham, Switzerland

Preface

This volume, edited by Yaroslav D. Sergeyev and Dmitri E. Kvasov, contains selected peer-reviewed papers from the Third Triennial International Conference and Summer School on Numerical Computations: Theory and Algorithms (NUMTA 2019) held in Le Castella – Isola Capo Rizzuto (Crotone), Italy, during June 15–21, 2019. The NUMTA 2019 conference has continued the previous successful editions of NUMTA that took place in 2013 and 2016 in Italy in the beautiful Calabria region.

NUMTA 2019 was organized by the University of Calabria, Department of Computer Engineering, Modeling, Electronics and Systems Science, Italy, in cooperation with the Society for Industrial and Applied Mathematics (SIAM), USA. This edition had the high patronage of the municipality of Crotone – the city of Pythagoras and his followers, the Pythagoreans. In fact, Pythagoras established the first Pythagorean community in this city in the 6th century B.C. It was a very special feeling for the participants of NUMTA 2019 to visit these holy, for any mathematician, places with a conference dedicated to numerical mathematics.

The goal of the NUMTA series of conferences is to create a multidisciplinary round table for an open discussion on numerical modeling nature by using traditional and emerging computational paradigms. Participants of the NUMTA 2019 conference discussed multiple aspects of numerical computations and modeling starting from foundations and philosophy of mathematics and computer science to advanced numerical techniques. New technological challenges and fundamental ideas from theoretical computer science, machine learning, linguistic, logic, set theory, and philosophy met the requirements, as well as fresh, new applications from physics, chemistry, biology, and economy.

Researchers from both theoretical and applied sciences were invited to use this excellent opportunity to exchange ideas with leading scientists from different research fields. Papers discussing new computational paradigms, relations with foundations of mathematics, and their impact on natural sciences were particularly solicited. Special attention during the conference was dedicated to numerical optimization techniques and a variety of issues related to the theory and practice of the usage of infinities and infinitesimals in numerical computations. In particular, there were a substantial number of talks dedicated to a new promising methodology allowing one to execute numerical computations with finite, infinite, and infinitesimal numbers on a new type of a computational device – the Infinity Computer patented in the EU, Russia, and the USA.

This edition of the NUMTA conference was dedicated to the 80th birthday of Professor Roman Strongin. For the past 50 years Roman Strongin has been a leader and an innovator in Global Optimization, an important field of Numerical Analysis having numerous real-life applications. His book on Global Optimization, published in 1978, was one of the first in the world on this subject. Now it is a classic and has been used by many as their first introduction and continued inspiration for Global Optimization. Since that time, Roman has published numerous books and more than 400 papers in several scientific fields and has been rewarded with many national and international honors including the President of the Russian Federation Prize. For decades Roman served as Dean, First Vice-Rector, and Rector of the famous Lobachevsky State University of Nizhny Novgorod. Since 2008 he has been President of this university. He is also Chairman of the Council of Presidents of Russian Universities, Vice-President of the Union of the Rectors of Russian Universities, and Chairman of the Public Chamber of the Nizhny Novgorod Region.

We are proud to inform you that 200 researchers from the following 30 countries participated at the NUMTA 2019 conference: Argentina, Bulgaria, Canada, China, Czech Republic, Estonia, Finland, France, Germany, Greece, India, Iran, Italy, Japan, Kazakhstan, Latvia, Lithuania, the Netherlands, Philippines, Portugal, Romania, Russia, Saudi Arabia, South Korea, Spain, Switzerland, Thailand, Ukraine, the UK, and the USA.

The following plenary lecturers shared their achievements with the NUMTA 2019 participants:

- Louis D'Alotto, USA: "Infinite games on finite graphs using Grossone"
- Renato De Leone, Italy: "Recent advances on the use of Grossone in optimization and regularization problems"
- Kalyanmoy Deb, USA: "Karush-Kuhn-Tucker proximity measure for convergence of real-parameter single and multi-criterion optimization"
- Luca Formaggia, Italy: "Numerical modeling of flow in fractured porous media and fault reactivation"
- Jan Hesthaven, Switzerland: "Precision algorithms"
- Francesca Mazzia, Italy: "Numerical differentiation on the Infinity Computer and applications for solving ODEs and approximating functions"
- Michael Vrahatis, Greece: "Generalizations of the intermediate value theorem for approximations of fixed points and zeroes of continuous functions"
- Anatoly Zhigljavsky, UK: "Uniformly distributed sequences and space-filling"

Moreover, the following tutorials were presented during the conference:

- Roberto Natalini, Italy: "Vector kinetic approximations to fluid-dynamics equations"
- Yaroslav Sergeyev, Italy and Russia: "Grossone-based Infinity Computing with numerical infinities and infinitesimals"
- Vassili Toropov, UK: "Design optimization techniques for industrial applications: Challenges and progress"

These proceedings of NUMTA 2019 consist of two volumes: Part I and Part II. The book you have in your hands is the first part containing peer-reviewed papers selected from big special streams and sessions held during the conference. The second volume contains peer-reviewed papers selected from the general stream, plenary lectures, and small special sessions of NUMTA 2019. The special streams and sessions from which the papers selected for this volume have been chosen are listed below (in the alphabetical order):

(1) Approximation: methods, algorithms and applications; organized by Alessandra De Rossi (University of Turin, Italy), Francesco Dell'Accio (University of Calabria, Italy), Elisa Francomano (University of Palermo, Italy), and Donatella Occorsio (University of Basilicata, Italy).

(2) Computational Methods for Data Analysis; organized by Rosanna Campagna, Salvatore Cuomo, and Francesco Piccialli (all from the University of Naples Federico II, Italy).

(3) First Order Methods in Optimization: Theory and Applications; organized by Simone Rebegoldi (University of Modena and Reggio Emilia, Italy) and Marco Viola (Sapienza University of Rome, Italy).

(4) High Performance Computing in Modelling and Simulation; organized by William Sparato, Donato D'Ambrosio, Rocco Rongo (all from the University of Calabria, Italy), and Andrea Giordano (ICAR–CNR, Italy).

(5) Numbers, Algorithms, and Applications; organized by Fabio Caldarola, Gianfranco D'Atri, Mario Maiolo (all from the University of Calabria, Italy), and Giuseppe Pirillo (University of Florence, Italy).

(6) Optimization and Management of Water Supply; organized by Fabio Caldarola, Mario Maiolo, Giuseppe Mendicino, and Patrizia Piro (all from the University of Calabria, Italy).

All the papers which were accepted for publication in this LNCS volume underwent a thorough peer review process (required up to three review rounds for some manuscripts) by the members of the NUMTA 2019 Program Committee and independent reviewers. The editors thank all the participants for their dedication to the success of NUMTA 2019 and are grateful to the reviewers for their valuable work. The support of the Springer LNCS editorial staff and the sponsorship of the Young Researcher Prize by Springer are greatly appreciated.

The editors express their gratitude to the institutions that offered their generous support to the international conference NUMTA 2019. This support was essential for the success of this event:

- University of Calabria (Italy)
- Department of Computer Engineering, Modeling, Electronics and Systems Science of the University of Calabria (Italy)
- Italian National Group for Scientific Computation of the National Institute for Advanced Mathematics F. Severi (Italy)
- Institute of High Performance Computing and Networking of the National Research Council (Italy)
- International Association for Mathematics and Computers in Simulation
- International Society of Global Optimization

Many thanks go to Maria Chiara Nasso from the University of Calabria, Italy, for her valuable support in the technical editing of this volume.

October 2019   Yaroslav D. Sergeyev
                                                                 Dmitri E. Kvasov

Organization

General Chair

Yaroslav Sergeyev   University of Calabria, Italy, and Lobachevsky University of Nizhny Novgorod, Russia

Scientific Committee

Lidia Aceto, Italy
Andy Adamatzky, UK
Francesco Archetti, Italy
Thomas Bäck, The Netherlands
Roberto Battiti, Italy
Roman Belavkin, UK
Giancarlo Bigi, Italy
Paola Bonizzoni, Italy
Luigi Brugnano, Italy
Sergiy Butenko, USA
Sonia Cafieri, France
Tianxin Cai, China
Cristian Calude, New Zealand
Antonio Candelieri, Italy
Mario Cannataro, Italy
Giovanni Capobianco, Italy
Domingos Cardoso, Portugal
Francesco Carrabs, Italy
Emilio Carrizosa, Spain
Leocadio Casado, Spain
Carmine Cerrone, Italy
Raffaele Cerulli, Italy
Marco Cococcioni, Italy
Salvatore Cuomo, Italy
Louis D'Alotto, USA
Oleg Davydov, Germany
Renato De Leone, Italy
Alessandra De Rossi, Italy
Kalyanmoy Deb, USA
Francesco Dell'Accio, Italy
Branko Dragovich, Serbia
Gintautas Dzemyda, Lithuania
Yalchin Efendiev, USA

Michael Emmerich, The Netherlands
Adil Erzin, Russia
Yury Evtushenko, Russia
Giovanni Fasano, Italy
Şerife Faydaoğlu, Turkey
Luca Formaggia, Italy
Elisa Francomano, Italy
Masao Fukushima, Japan
David Gao, Australia
Manlio Gaudioso, Italy
Victor Gergel, Russia
Jonathan Gillard, UK
Daniele Gregori, Italy
Vladimir Grishagin, Russia
Mario Guarracino, Italy
Nicola Guglielmi, Italy
Jan Hesthaven, Switzerland
Felice Iavernaro, Italy
Mikhail Khachay, Russia
Oleg Khamisov, Russia
Timos Kipouros, UK
Lefteris Kirousis, Greece
Yury Kochetov, Russia
Olga Kostyukova, Belarus
Vladik Kreinovich, USA
Dmitri Kvasov, Italy
Hoai An Le Thi, France
Wah June Leong, Malaysia
Øystein Linnebo, Norway
Antonio Liotta, UK
Marco Locatelli, Italy
Stefano Lucidi, Italy
Maurice Margenstern, France

Vladimir Mazalov, Russia
Francesca Mazzia, Italy
Maria Mellone, Italy
Kaisa Miettinen, Finland
Edmondo Minisci, UK
Nenad Mladenovic, Serbia
Ganesan Narayanasamy, USA
Ivo Nowak, Germany
Donatella Occorsio, Italy
Marco Panza, USA
Massimo Pappalardo, Italy
Panos Pardalos, USA
Remigijus Paulavičius, Lithuania
Hoang Xuan Phu, Vietnam
Stefan Pickl, Germany
Raffaele Pisano, France
Yuri Popkov, Russia
Mikhail Posypkin, Russia
Oleg Prokopyev, USA
Davide Rizza, UK
Massimo Roma, Italy
Valeria Ruggiero, Italy
Maria Grazia Russo, Italy
Nick Sahinidis, USA
Leonidas Sakalauskas, Lithuania
Yaroslav Sergeyev, Italy

Khodr Shamseddine, Canada
Sameer Shende, USA
Vladimir Shylo, Ukraine
Theodore Simos, Greece
Vinai Singh, India
Majid Soleimani-Damaneh, Iran
William Spataro, Italy
Maria Grazia Speranza, Italy
Giandomenico Spezzano, Italy
Rosamaria Spitaleri, Italy
Alexander Strekalovskiy, Russia
Roman Strongin, Russia
Gopal Tadepalli, India
Tatiana Tchemisova, Portugal
Gerardo Toraldo, Italy
Vassili Toropov, UK
Hiroshi Umeo, Japan
Michael Vrahatis, Greece
Song Wang, Australia
Gerhard-Wilhelm Weber, Poland
Luca Zanni, Italy
Anatoly Zhigljavsky, UK
Antanas Žilinskas, Lithuania
Julius Žilinskas, Lithuania
Joseph Zyss, France

Organizing Committee

Francesco Dall'Accio	University of Calabria, Rende (CS), Italy
Alfredo Garro	University of Calabria, Rende (CS), Italy
Vladimir Grishagin	Lobachevsky University of Nizhny Novgorod, Nizhny Novgorod, Russia
Dmitri Kvasov (Chair)	University of Calabria, Rende (CS), Italy, and Lobachevsky University of Nizhny Novgorod, Nizhny Novgorod, Russia
Marat Mukhametzhanov	University of Calabria, Rende (CS), Italy, and Lobachevsky University of Nizhny Novgorod, Nizhny Novgorod, Russia
Maria Chiara Nasso	University of Calabria, Rende (CS), Italy
Clara Pizzuti	National Research Council of Italy (CNR), Institute for High Performance Computing and Networking (ICAR), Rende (CS), Italy
Davide Rizza	University of East Anglia, Norwich, UK
Yaroslav Sergeyev	University of Calabria, Rende (CS), Italy, and Lobachevsky University of Nizhny Novgorod, Nizhny Novgorod, Russia

Sponsors

UNIVERSITÀ DELLA CALABRIA

UNIVERSITÀ DELLA CALABRIA

DIPARTIMENTO DI
INGEGNERIA INFORMATICA,
MODELLISTICA, ELETTRONICA
E SISTEMISTICA
DIMES

Dipartimento di Eccellenza 2018-2022

Consiglio Nazionale delle Ricerche
Istituto di Calcolo e Reti ad Alte Prestazioni

GNCS
iNδAM

Gruppo Nazionale
per il Calcolo Scientifico

 Springer

IMACS

In cooperation with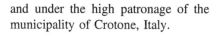

and under the high patronage of the
municipality of Crotone, Italy.

Contents – Part I

First Order Methods in Optimization: Theory and Applications

High Performance Computing in Modelling and Simulation

Numbers, Algorithms, and Applications

Optimization and Management of Water Supply

Contents – Part II

Approximation: Methods, Algorithms, and Applications

Towards an Efficient Implementation of an Accurate SPH Method

Laura Antonelli[1] , Daniela di Serafino[2] , Elisa Francomano[3](✉) ,
Francesco Gregoretti[1] , and Marta Paliaga[3]

[1] Institute for High Performance Computing and Networking (ICAR),
CNR, Naples, Italy
{laura.antonelli,francesco.gregoretti}@icar.cnr.it
[2] Department of Mathematics and Physics, University of Campania
"Luigi Vanvitelli", Caserta, Italy
daniela.diserafino@unicampania.it
[3] Department of Engineering, University of Palermo, Palermo, Italy
{elisa.francomano,marta.paliaga}@unipa.it

Abstract. A modified version of the Smoothed Particle Hydrodynamics
(SPH) method is considered in order to overcome the loss of accuracy of
the standard formulation. The summation of Gaussian kernel functions
is employed, using the Improved Fast Gauss Transform (IFGT) to reduce
the computational cost, while tuning the desired accuracy in the SPH
method. This technique, coupled with an algorithmic design for exploiting the performance of Graphics Processing Units (GPUs), makes the
method promising, as shown by numerical experiments.

Keywords: Smoothed Particle Hydrodynamics · Improved Fast Gauss
Transform · Graphics Processing Units

1 Introduction

In the last years mesh-free methods have gained much attention in many application fields [3,8]. The Smoothed Particle Hydrodynamics (SPH) is a mesh-free
technique originally introduced in astrophysics by Gingold and Monaghan [5]
and by Lucy [9], and widely used because of its ability of dealing with highly
complex geometries [1,8]. However, when irregular data distributions are considered, a loss of accuracy can occur. Many techniques have been developed to
overcome this problem. In this paper we discuss a strategy based on the Taylor
series expansion, which simultaneously improves the approximation of a function and its derivatives obtained with the SPH method. The improvement in
accuracy comes at the cost of an additional computational effort, which can be
alleviated by employing fast summations [4] in the computational scheme.

The availability of General-Purpose Graphics Processing Units (GPGPUs)
provides an opportunity to further speed up the SPH method. This work can be
also considered as a first step toward an efficient implementation of the improved

© Springer Nature Switzerland AG 2020
Y. D. Sergeyev and D. E. Kvasov (Eds.): NUMTA 2019, LNCS 11973, pp. 3–10, 2020.
https://doi.org/10.1007/978-3-030-39081-5_1

SPH version on GPGPUs. To this aim, we develop a GPU implementation of one of its most computationally intensive tasks.

The remainder of this paper is organized as follows. A brief description of the standard SPH method and its improved formulation is given in Sect. 2. A sequential implementation of the improved SPH method, to be used as a reference CPU version for the development of a GPU version, is discussed in Sect. 3 along with results of numerical experiments. A partial implementation on GPU is presented in Sect. 3.1. Conclusions are provided in Sect. 4.

2   The Improved SPH Method

The standard SPH method computes an approximation to a function $f : \Omega \subset \mathbb{R}^d \to \mathbb{R}$ starting from the *kernel approximation*

$$< f_h(\mathbf{x}) > = \int_\Omega f(\boldsymbol{\xi}) \mathsf{K}(\mathbf{x}, \boldsymbol{\xi}; h) d\Omega, \tag{1}$$

where $\mathbf{x}, \boldsymbol{\xi} \in \Omega$, $h \in \mathbb{R}^+$ is the smoothing length, and $\mathsf{K}(\mathbf{x}, \boldsymbol{\xi}; h)$ is the kernel function, which is usually normalized to unity and is required to be symmetric and sufficiently smooth. By assuming that f is at least continuously differentiable, the kernel approximation has second-order accuracy [4,5]. By considering N source points $\boldsymbol{\xi}_j \in \Omega$, each associated with a subdomain $\Omega_j \subset \Omega$, the so-called *particle approximation* of (1) is defined as

$$f_h(\mathbf{x}) = \sum_{j=1}^{N} f(\boldsymbol{\xi}_j) \mathsf{K}(\mathbf{x}, \boldsymbol{\xi}_j; h) d\Omega_j \tag{2}$$

where $d\Omega_j$ is the measure of Ω_j. Note that (2) generally does not satisfy second-order accuracy, e.g., when irregular point distributions are considered.

If f is sufficiently smooth, the k-th order Taylor expansion of $f(\boldsymbol{\xi})$ can be used in order to increase the accuracy of the approximation:

$$f(\boldsymbol{\xi}) = \sum_{|\alpha| \leq k} \frac{1}{\alpha!} (\boldsymbol{\xi} - \mathbf{x})^\alpha \mathcal{D}^\alpha f(\mathbf{x}) + O(h^{k+1}), \tag{3}$$

where $\alpha = (\alpha^{(1)}, \alpha^{(2)}, \ldots, \alpha^{(d)}) \in \mathbb{N}^d$ is a multi-index, $|\alpha| = \sum_{i=1}^{d} \alpha^{(i)}$, $\alpha! = \prod_{i=1}^{d} (\alpha^{(i)})!$, $\mathbf{y}^\alpha = y_1^{\alpha^{(1)}} \cdot y_2^{\alpha^{(2)}} \cdots \cdot y_d^{\alpha^{(d)}}$, and $\mathcal{D}^\alpha = \frac{\partial^{|\alpha|}}{(\partial x^{(1)})^{\alpha^{(1)}} \ldots (\partial x^{(d)})^{\alpha^{(d)}}}$. By multiplying (3) and its derivatives up to the k-th order by the kernel function, and integrating over Ω, we get

$$\int_\Omega f(\boldsymbol{\xi})\mathsf{K}(\mathbf{x},\boldsymbol{\xi};h)d\Omega \quad = \sum_{|\alpha|\leq k}\frac{1}{\alpha!}\int_\Omega(\boldsymbol{\xi}-\mathbf{x})^\alpha\mathcal{D}^\alpha f(\mathbf{x})\mathsf{K}(\mathbf{x},\boldsymbol{\xi};h)d\Omega$$
$$+ \int_\Omega O(h^{k+1})\mathsf{K}(\mathbf{x},\boldsymbol{\xi};h)d\Omega$$
$$\vdots \qquad\qquad (4)$$
$$\int_\Omega f(\boldsymbol{\xi})\mathcal{D}^k\mathsf{K}(\mathbf{x},\boldsymbol{\xi};h)d\Omega = \sum_{|\alpha|\leq k}\frac{1}{\alpha!}\int_\Omega(\boldsymbol{\xi}-\mathbf{x})^\alpha\mathcal{D}^\alpha f(\mathbf{x})\mathcal{D}^k\mathsf{K}(\mathbf{x},\boldsymbol{\xi};h)d\Omega$$
$$+ \int_\Omega O(h^{k+1})\mathcal{D}^k\mathsf{K}(\mathbf{x},\boldsymbol{\xi};h)d\Omega.$$

By adopting the particle approximation and neglecting the last terms in the right-hand side of (4), we get an approximation to f and its derivatives at each evaluation point $\mathbf{x}$, which can be written as a linear system of size $m = (d + k)!/(d!\,k!)$:

$$\mathbf{A}_\mathbf{x}^{(k)}\mathbf{c}_\mathbf{x}^{(k)} = \mathbf{b}_\mathbf{x}^{(k)}, \qquad\qquad (5)$$

where

$$\mathbf{A}_\mathbf{x}^{(k)} = \begin{pmatrix} \displaystyle\sum_{j=1}^N \mathsf{K}(\mathbf{x},\boldsymbol{\xi}_j;h)d\Omega_j & \cdots & \frac{1}{k!}\displaystyle\sum_{j=1}^N(\xi_j^{(d)}-x^{(d)})^k\mathsf{K}(\mathbf{x},\boldsymbol{\xi}_j;h)d\Omega_j \\ \vdots & \ddots & \vdots \\ \displaystyle\sum_{j=1}^N \mathcal{D}^k\mathsf{K}(\mathbf{x},\boldsymbol{\xi}_j;h)d\Omega_j & \cdots & \frac{1}{k!}\displaystyle\sum_{j=1}^N(\xi_j^{(d)}-x^{(d)})^k\mathcal{D}^k\mathsf{K}(\mathbf{x},\boldsymbol{\xi}_j;h)d\Omega_j \end{pmatrix}, \qquad (6)$$

$$\mathbf{c}_\mathbf{x}^{(k)} = \begin{pmatrix} f(\mathbf{x}) \\ \vdots \\ \mathcal{D}^k f(\mathbf{x}) \end{pmatrix}, \quad \mathbf{b}_\mathbf{x}^{(k)} = \begin{pmatrix} \displaystyle\sum_{j=1}^N f(\boldsymbol{\xi})\mathsf{K}(\mathbf{x},\boldsymbol{\xi}_j;h)d\Omega_j \\ \vdots \\ \displaystyle\sum_{j=1}^N f(\boldsymbol{\xi})\mathcal{D}^k\mathsf{K}(\mathbf{x},\boldsymbol{\xi}_j;h)d\Omega_j \end{pmatrix}. \qquad (7)$$

This procedure improves the accuracy of the standard method as discussed in [4], and has been successfully used in electromagnetic simulations [1].

The improved SPH method is more expensive than the standard one, because it requires, for each evaluation point $\mathbf{x}$, the construction of the matrix $\mathbf{A}_\mathbf{x}^{(k)}$ and the right-hand side $\mathbf{b}_\mathbf{x}^{(k)}$, i.e., the computation of $m^2 + m$ sums, and the solution of the linear system (5). As outlined in [4], when $\mathsf{K}(\mathbf{x},\boldsymbol{\xi};h)$ is the Gaussian kernel, the computational cost can be significantly reduced using the Fast Gauss Transform (FGT) [6]. More precisely, the FGT lowers the cost of computing M Gaussian sums using N source points from $\mathcal{O}(NM)$ to $\mathcal{O}(N+M)$, by approximating the sums with a required order of accuracy. Actually, the Improved FGT (IFGT) [10,13] is used, because it achieves higher efficiency than the original FGT, especially when the dimension d increases, by combining a suitable factorization of the exponential function in the Gauss transform with an adaptive space partitioning scheme where the N source points are grouped into clusters, and exploiting the fast decay of the Gaussian function. The description of the IFGT is beyond the scope of this paper. Here we provide only some details related to the computation of $\mathbf{A}_\mathbf{x}^{(k)}$ and $\mathbf{b}_\mathbf{x}^{(k)}$, for $d = 2$ and $k = 1$, i.e., $m = 3$. The computation of $\mathbf{A}_\mathbf{x}^{(k)}$ via IFGT requires $m(m+1)/2 = 6$ Gauss transforms using N source points, while the computation of $\mathbf{b}_\mathbf{x}^{(k)}$ requires m Gauss transforms and the evaluation of f at each source point. For any evaluation point $\mathbf{x}$, the transforms to be computed have the following form:

$$G_l(\mathbf{x}) = \sum_{j=1}^{N} w_l(\boldsymbol{\xi}_j)\mathsf{K}(\mathbf{x}, \boldsymbol{\xi}_j; h), \quad l = 1, \ldots, L,$$

where $L = m(m+3)/2 = 9$, the weights $w_l(\boldsymbol{\xi}_j)$ are defined as

$$
\begin{aligned}
w_1(\boldsymbol{\xi}_j) &= \rho_j & , \quad w_4(\boldsymbol{\xi}_j) &= \rho_j \left(\xi_j^{(1)}\right)^2, & w_7(\boldsymbol{\xi}_j) &= \rho_j f(\boldsymbol{\xi}_j), \\
w_2(\boldsymbol{\xi}_j) &= \rho_j \xi_j^{(1)}, & w_5(\boldsymbol{\xi}_j) &= \rho_j \left(\xi_j^{(2)}\right)^2, & w_8(\boldsymbol{\xi}_j) &= \rho_j \xi_j^{(1)} f(\boldsymbol{\xi}_j), \quad (8) \\
w_3(\boldsymbol{\xi}_j) &= \rho_j \xi_j^{(2)}, & w_6(\boldsymbol{\xi}_j) &= \rho_j \xi_j^{(1)} \xi_j^{(2)}, & w_9(\boldsymbol{\xi}_j) &= \rho_j \xi_j^{(2)} f(\boldsymbol{\xi}_j),
\end{aligned}
$$

$\xi_j^{(i)}$, $i = 1, 2$, is the i-th coordinate of the point $\boldsymbol{\xi}_j$, and $\rho_j = \frac{1}{\pi h^2} d\Omega_j$. The entries of $\mathbf{A}_\mathbf{x}^{(k)}$ and $\mathbf{b}_\mathbf{x}^{(k)}$ in (6) and (7) can be obtained as

$$
\begin{aligned}
A_{11}^{(k)} &= \tilde{A}_1, & A_{12}^{(k)} &= \tilde{A}_2, & A_{13}^{(k)} &= \tilde{A}_3, & b_1^{(k)} &= \tilde{A}_7, \\
A_{21}^{(k)} &= -\tilde{A}_2, & A_{22}^{(k)} &= \tilde{A}_4, & A_{23}^{(k)} &= \tilde{A}_5, & b_2^{(k)} &= -\tilde{A}_8, \\
A_{31}^{(k)} &= -\tilde{A}_3, & A_{32}^{(k)} &= \tilde{A}_5, & A_{33}^{(k)} &= -\tilde{A}_6, & b_3^{(k)} &= -\tilde{A}_9,
\end{aligned}
$$

where

$$
\begin{aligned}
\tilde{A}_1 &= G_1, & \tilde{A}_4 &= G_4 - 2x^{(1)}G_2 + (x^{(1)})^2 G_1, & \tilde{A}_7 &= G_7, \\
\tilde{A}_2 &= G_2 - x^{(1)}G_1, & \tilde{A}_5 &= G_5 - x^{(1)}G_2 - x^{(2)}G_3 + x^{(1)}x^{(2)}G_1, & \tilde{A}_8 &= G_8 - x^{(1)}G_7, \quad (9) \\
\tilde{A}_3 &= G_3 - x^{(2)}G_1, & \tilde{A}_6 &= G_6 - 2x^{(2)}G_3 + (x^{(2)})^2 G_1, & \tilde{A}_9 &= G_9 - x^{(2)}G_7,
\end{aligned}
$$

and the dependence of G_l on $\mathbf{x}$ has been removed to simplify the notations. Note that all the matrices $\mathbf{A}_\mathbf{x}^{(k)}$ and the right-hand sides $\mathbf{b}_\mathbf{x}^{(k)}$ can be computed simultaneously.

3   Implementation and Numerical Experiments

The improved SPH method can be structured into five computational tasks: (a) generate the source points $\boldsymbol{\xi}_j$ and the evaluation points $\mathbf{x}_i$; (b) compute the weights $w_l(\boldsymbol{\xi}_j)$; (c) evaluate the Gauss transforms $G_l(\mathbf{x}_i)$; (d) compute the matrices $\mathbf{A}_{\mathbf{x}_i}^{(k)}$ and the right-hand sides $\mathbf{b}_{\mathbf{x}_i}^{(k)}$; (e) solve the linear systems (5).

All the tasks were implemented in C++, with the aim of developing an efficient serial CPU version to be used as a baseline for a GPU implementation. Concerning task (a), Ω was set equal to $[0, 1]^d$ and three distributions of source points were considered: uniform d-dimensional mesh, Halton [7] and Sobol' [14] d-dimensional sequences. The Halton and Sobol' points were generated by using the C++ code available from [2]. The evaluation points $\mathbf{x}_i$ were distributed on

Table 1. Functions used in the numerical experiments.

$$f_a(x^{(1)}, x^{(2)}) = 16\, x^{(1)} x^{(2)} (1 - x^{(1)})(1 - x^{(2)})$$

$$f_b(x^{(1)}, x^{(2)}) = \tanh \frac{1}{9} (9(x^{(2)} - x^{(1)}) + 1)$$

$$f_c(x^{(1)}, x^{(2)}) = \frac{1.25 + \cos(5.4\, x^{(2)})}{6 + 6(3\, x^{(1)} - 1)^2}$$

$$f_d(x^{(1)}, x^{(2)}) = \frac{1}{3} \exp \left(-\frac{81}{16} \left(\left(x^{(1)} - \frac{1}{2} \right)^2 + \left(x^{(2)} - \frac{1}{2} \right)^2 \right) \right)$$

a uniform mesh over Ω. Task (c) exploits the IFGT method implemented in the figtree package [10], which tunes IFGT parameters to the source distribution to get tighter error bounds, according to the desired accuracy and the selected smoothing length. Task (e) was implemented by using the LAPACK routines DGETRF and DGETRS from the auto-tuning ATLAS library [15]; DGETRF computes the LU factorization of $\mathbf{A}_{\mathbf{x}_i}^{(k)}$, while DGETRS performs the corresponding triangular solves. Finally, the implementation of tasks (b) and (d) was a straightforward application of (8) and (9), storing all the weights $w_l(\boldsymbol{\xi}_j)$ and all the entries of the matrices $\mathbf{A}_{\mathbf{x}_i}^{(k)}$ and the right-hand sides $\mathbf{b}_{\mathbf{x}_i}^{(k)}$ into three arrays of lengths LN, LM and mM, respectively. Note that tasks (c) and (e) account for most of the execution time.

We performed numerical experiments with $d = 2$ and $k = 1$, using the four test functions reported in Table 1 and values of N and M much greater than the ones considered in [4]. For each distribution of source points, we set $N = (2^n + 1)^2$ and $h = 1/2^n$, with $n = 7, 8, 9, 10, 11$, and $M = (\sqrt{N} + 1)^2$. The accuracy for IFGT computations was set to 10^{-6}. The experiments were run on an Intel Xeon E5-2670 2.50 GHz CPU with 192 GB of RAM, with the Linux CentOS 6.8 operating system and the GNU 4.4.7 C++ compiler.

Figure 1 shows, for each test function f, the maximum error in approximating f with the function f_h computed by our SPH implementation,

$$\max_{i=1,\ldots,M} |f_h(\mathbf{x}_i) - f(\mathbf{x}_i)|,$$

as N varies. It also shows the maximum error for the derivative of the function with respect to $x^{(1)}$ (the error in the approximation of the derivative with respect to $x^{(2)}$ is comparable). We see that when N increases from $(2^7 + 1)^2 = 16641$ to $(2^{11} + 1)^2 = 4198401$, the maximum error in approximating f and its derivative decreases by about two orders and one order of magnitude, respectively, for each test function and each source point data set. These results confirm that increasing the number of source points can be strongly beneficial in terms of accuracy for the improved SPH method. On the other hand, the availability of high-throughput many-core processors, such as GPGPUs, encourages the development of SPH implementations able to deal with high-dimensional problems and very large sets of source and evaluation points. A first step in this direction is discussed in the next section.

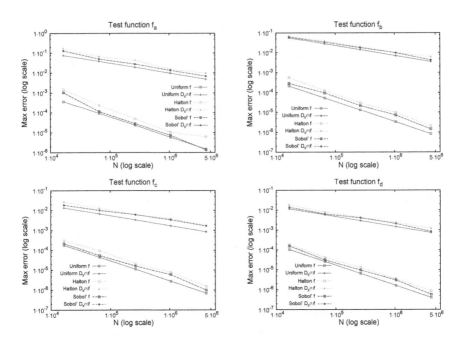

Fig. 1. Maximum error in the approximation of the test functions and their first derivatives with respect to $x^{(1)}$.

3.1   A First Step Toward a GPU Implementation

Nowadays GPUs can be found in many computers, ranging from high-end machines to laptops. In particular, an NVIDIA GPGPU is made of an array of streaming multiprocessors, each consisting of a fixed number of scalar processors, one or more instruction fetch units and on-chip fast memory. On NVIDIA GPUs, computations are performed by following the Single Instruction Multiple Threads programming model, implemented in the CUDA software framework [11], where multiple independent threads execute concurrently the same instruction. A CUDA program consists of a host program that runs on a CPU host and a kernel program that executes on a GPU device. Since thousands of cores are available in a GPU, the kernel program must expose a high level of parallelism to keep the GPU busy; balanced workload among threads, minimization of data transfers from the CPU to the GPU and vice versa, and suitable memory access patterns are also fundamental to get high performance. Our final goal is to efficiently map all the computational tasks of the improved SPH method onto GPGPUs, in order to build a parallel computational tool for SPH simulations of large-scale real-time applications. In this work we focus on task (e). The number of linear systems (5) to be solved is equal to the number, M, of evaluation points, while the size of the systems, $m = (d + k)!/(d!\,k!)$, depends on the order of the Taylor expansion, k, and the problem dimension, d. By setting $d = 2$ and $k = 1$ and letting M increase from the order of 10^4 to the order of 10^6, we are faced

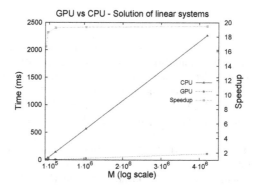

Fig. 2. GPU vs CPU performance comparison for task (e).

with the challenge of solving from thousands to millions of very small systems. In order to avoid the overhead of multiple API calls, we implemented a GPU version of task (e) that makes use of the batched dense linear algebra kernels available in the NVIDIA cuBLAS library [12]. These kernels, within a single API call, are capable of solving all the linear systems concurrently, thus increasing the exploitation of the underlying hardware. In particular, the factorization of the matrices was performed by using the cuBLAS function `cublasDgetrfBatched` and the solution of the triangular systems by using `cublasDgetrsBatched`. We executed the parallel code on an NVIDIA Tesla K20 GPU with CUDA 9.0 and compared the results with the serial version running on one core of the CPU used in the previous experiments. Figure 2 shows the run-time in milliseconds for the execution on GPU and CPU as well as the speedup achieved. For each number of evaluations, $M = (\sqrt{N} + 1)^2$, the time is the average over the runs for all the test functions and source point sets. We see that the GPU implementation of task (e) reduces the execution time by a factor up to 20, thus yielding a significant improvement in this phase of the method.

4 Conclusions

We discussed an improved SPH method based on the Taylor series expansion with the Gaussian kernel function, focusing on its efficient implementation by applying the IFGT. Numerical experiments confirmed that the accuracy of the method can significantly benefit from the use of a very large number of points. A GPU implementation of one of the SPH tasks was also developed, showing that the method can take advantage from the computational power of modern high-throughput architectures. The implementation on GPUs of the whole SPH method will be the subject of future work.

Acknowledgments. This work has been partially supported by the INdAM-GNCS research Project 2019 "Kernel-based approximation, multiresolution and subdivision methods, and related applications". This research has been also carried out within RITA (Rete ITaliana di Approssimazione).

References

1. Ala, G., Francomano, E.: A marching-on in time meshless kernel based solver for full-wave electromagnetic simulation. Numer. Algorithms **62**, 541–558 (2013)
2. Burkhardt, J.: https://github.com/cenit/jburkardt/tree/master/halton, https://github.com/cenit/jburkardt/tree/master/sobol
3. Francomano, E., Hilker, F.M., Paliaga, M., Venturino, E.: An efficient method to reconstruct invariant manifolds of saddle points. Dolomites Res. Notes Approx. **10**, 25–30 (2017)
4. Francomano, E., Paliaga, M.: Highlighting numerical insights of an efficient SPH method. Appl. Math. Comput. **339**, 899–915 (2018)
5. Gingold, R.A., Monaghan, J.J.: Smoothed particle hydrodynamics: theory and application to non-spherical stars. Mon. Not. R. Astron. Soc. **181**, 375–89 (1977)
6. Greengard, L., Strain, J.: The fast Gauss transform. SIAM J. Sci. Stat. Comput. **12**, 79–94 (1991)
7. Halton, J.H.: On the efficiency of certain quasi-random sequences of points in evaluating multi-dimensional integrals. Numer. Math. **2**, 84–90 (1960)
8. Liu, M.B., Liu, G.R.: Smoothed particle hydrodynamics (SPH): an overview and recent developments. Arch. Comput. Methods Eng. **17**, 25–76 (2010)
9. Lucy, L.B.: A numerical approach to the testing of the fission hypothesis. Astron. J. **82**, 1013–24 (1977)
10. Morariu, V.I., Srinivasan, B.V., Raykar V.C., Duraiswami, R., Davis, L.S.: Automatic online tuning for fast Gaussian summation. In: Advances in Neural Information Processing Systems, pp. 1113–1120 (2008). https://github.com/vmorariu/figtree
11. NVIDIA: CUDA C Programming Guide (2019). https://docs.nvidia.com/cuda/cuda-c-programming-guide/
12. NVIDIA: cuBLAS (2019). https://developer.nvidia.com/cublas
13. Raykar, V., Yang, C., Duraiswami, R., Gumerov, N.: Fast computation of sums of Gaussians in high dimensions. Technical report CS-TR-4767/UMIACS-TR-2005-69, University of Maryland, College Park, MD, USA (2005)
14. Sobol', I.M.: On the distribution of points in a cube and the approximate evaluation of integrals. USSR Comput. Math. Math. Phys. **7**, 86–112 (1967)
15. Whaley, R.C., Petitet, A., Dongarra, J.J.: Automated empirical optimizations of software and the ATLAS project. Parallel Comput. **27**, 3–35 (2001)

A Procedure for Laplace Transform Inversion Based on Smoothing Exponential-Polynomial Splines

Rosanna Campagna[1]([🖂]) [iD], Costanza Conti[2] [iD], and Salvatore Cuomo[3] [iD]

[1] Department of Agricultural Sciences, University of Naples Federico II,
Portici, Naples, Italy
rosanna.campagna@unina.it
[2] Department of Industrial Engineering, University of Florence, Florence, Italy
costanza.conti@unifi.it
[3] Department of Mathematics and Applications "R. Caccioppoli",
University of Naples Federico II, Naples, Italy
salvatore.cuomo@unina.it

Abstract. Multi-exponential decaying data are very frequent in applications and a continuous description of this type of data allows the use of mathematical tools for data analysis such as the Laplace Transform (LT). In this work a numerical procedure for the Laplace Transform Inversion (LTI) of multi-exponential decaying data is proposed. It is based on a new fitting model, that is a smoothing exponential-polynomial spline with segments expressed in Bernstein-like bases. A numerical experiment concerning the application of a LTI method applied to our spline model highlights that it is very promising in the LTI of exponential decay data.

Keywords: Laplace Transform Inversion · Exponential-polynomial spline · Multi-exponential data

1 Introduction

In several application fields *inverse problems* occur where, starting from observed data, the function of interest is related to them through an integral operator as follows:

Problem 1 (*Discrete data inversion*). Let be given $\{F_i\}_{i=1}^N$, s.t.:

$$F_i = F(s_i) + \varepsilon_i, \quad i = 1, ..., N, \quad \text{with} \quad F(s) = \int_a^b K(s,t)f(t)dt$$

with K suitable kernel, ε_i, $i = 1, ..., N$ unknown noises and $a, b \in \mathbb{R}$. Compute f, or some values of f.

The solution of the previous problem is related to the *Laplace transform inversion* (LTI), bearing in mind the following definition:

© Springer Nature Switzerland AG 2020
Y. D. Sergeyev and D. E. Kvasov (Eds.): NUMTA 2019, LNCS 11973, pp. 11–18, 2020.
https://doi.org/10.1007/978-3-030-39081-5_2

Definition 1 (Laplace Transform). *Let f be a given integrable function and $s_0 \in \mathbb{C}$ such that $\int_0^\infty e^{-s_0 t} f(t) dt < \infty$. Then, for $s \in \mathbb{C}$: $Re(s) > Re(s_0)$ we have $\int_0^\infty e^{-st} f(t) dt < \infty$. Defining $\mathcal{C}(f) := \{s \in \mathbb{C}, \int_0^\infty e^{-st} f(t) dt < \infty\}$ (the so called* region of convergence) *the complex valued function F:*

$$F : s \in \mathcal{C}(f) \ \to \ F(s) = \int_0^\infty e^{-st} f(t) dt \ \in \mathbb{C}$$

is the Laplace Transform (LT) *of f, generally referred to as $F(s) = \mathcal{L}[f(t)]$.*

The LT is an analytical function in its convergence region; the LTI is, in general, expressed by complex formulas; here we focus on the so-called *real* inversion and we refer to the LT main properties (see e.g. [10]):

$$\lim_{s \to \infty} F(s) = 0, \qquad \lim_{s \to \infty} s \cdot F(s) < \infty, \qquad \lim_{s \to \infty} \frac{F(s)}{s} = 0.$$

Particularly, most of the LTs have rational or exponential behaviours. The LTI consists in inverting the functional operator, computing $f = \mathcal{L}^{-1}[F]$. Starting from the Problem 1, in this paper we suggest a procedure to solve in a stable way the LTI from multi-exponential data. Data-driven approaches are very frequent in several fields (see e.g. [15, 16, 18]) and it is common to adapt the classic fitting models of which the analytical properties and the order of accuracy are known, to the characteristics of the problem under examination, taking information from the data. The LTI is a framework at the base of numerous applications, e.g. Nuclear Magnetic Resonance (NMR) for the study of porous media [9]. Therefore, the literature is wide and diversified so as the applications of the LT in science and engineering problems [12]. In some works the numerical inversion of the LT is based on collocation formulas of the LT *inverse* function [4], or on its development in series of orthogonal functions, particularly polynomials [3, 11, 13] or singular functions [1]. Moreover, the ill-posed nature of the problem and the existence of general software packages for LTI (e.g. [6, 7]), that are *not* specific for applications, lead to use general purpose methods and related mathematical software. Very often multi-exponential data are analysed by a Non-Negative Least Squares (NNLS) procedure e.g. implemented in the software package CONTIN [14]. From the above considerations, comes the idea of a fitting model that enjoys properties similar to those of a LT function making the application of some LTI methods meaningful. Our model is an exponential-polynomial smoothing spline that follows the nature of the data and enjoys some of the main LT properties. The procedure that we propose is made of two phases as synthesized below:

Procedure 1. LTI of multi-exponential data

1: **input:** multi-exponential data (s_i, F_i), $i = 1, \ldots, n$
2: Data-driven definition of the spline space parameters
     definition of the GB-splines via the definition of Bernstein-like bases
     derivation of the smoothing exponential-polynomial spline fitting the input data
3: LT inversion
4: **output:** Laplace transform inversion

The numerical experiment on LTI here presented confirms the correctness of our ideas, even though more accurate numerical experiments on LTI are under investigation. Open numerical issues require: (*i*) a detailed analysis of the stability, connected to the knots distribution; (*ii*) the boundary basis definition, affecting the conditioning of the involved linear systems. These issues will be the object of following investigations.

In Sect. 2 we report a description of the spline model used in our approach and its derivation both from the known models in literature and from the properties we impose to characterize it; in Sect. 3 we present a test that proves the reliability of the model in a LTI procedure applied to exponential data. Last section refers conclusions.

2   Smoothing Spline Approach

It is well known that a wide literature is dedicated to the splines and that the polynomial splines are generalized in several different ways. A generalization of polynomial splines is given by Chebyshev splines and, even further, by *generalized splines* (see [17], for example) whose local pieces belong to different spaces with connecting conditions specified at the knots. GB-splines (i.e. basis splines for the generalized spline spaces) are characterized by minimal support with respect to "degree", smoothness, and domain partition. They are largely used in computer graphics, curve and surface reconstructions but also in image and signal processing.

In this paper we present a special instance of generalized spline model dictated by the multi-exponential decaying of the data also outside the knots and we present a test of LTI based on this model. The results confirm that our spline is a continuous model for data with exponential decay, that allows with success using LTI methods to solve discrete data inversion problems.

In a previous paper [5] a cubic complete smoothing polynomial spline with an exponential-polynomial decay *only* outside the knots and global C^1-regularity was considered. More recently (see [2]) we have defined a natural smoothing exponential-polynomial spline enjoying C^2 regularity in a wider interval including the knots. Here, setting a functional space reflecting the (exponential) nature of the data, an exponential-polynomial smoothing spline (special instance of *L*-splines) is defined minimizing a functional balancing fidelity to the data and regularity request, according to the following:

Definition 2. *Given a partition of the interval* $[a, b]$, $\Delta := \{a < x_1 < \ldots x_N < b\}$ *and a vector* $(y_1, \cdots, y_N) \in \mathbb{R}^N$, *a natural smoothing L-spline, related to a fixed differential operator* $\mathcal{L}_n$ *of order n, on* Δ, *is the solution of the problem*

$$\min_{u \in H^n[a,b]} \left\{ \sum_{i=1}^{N} (w_i[u(x_i) - y_i]^2 + \lambda \int_a^b (\mathcal{L}_n u(x))^2 \, dx \right\}, \qquad (1)$$

with $(w_1, \ldots, w_N)$ *non zero weights,* $\lambda \in \mathbb{R}$ *a regularization parameter and* $H^n[a, b]$ *a suitable Hilbert space.*

The first issue when dealing with a natural smoothing L-spline is given by the selection of the differential operator while the second one is the construction of the GB-splines for the corresponding spline space in order to derive a stable solution of (1) expressed as

$$s(x) = \sum_{j=1}^{N} c_j \varphi_j(x). \tag{2}$$

With the aim to solve the Problem 1, in our model we assume like knots the $x_i \in \Delta$, $i = 1, \ldots, N$, localized at the abscissae of the experimental data. We work with generalized splines with segments in the spaces of exponential-polynomial functions,

$$\mathbb{E}_2 := \mathrm{span}\{e^{-\alpha x},\ x\,e^{-\alpha x}\} \text{ and } \mathbb{E}_4 := \mathrm{span}\{e^{\alpha x},\ x\,e^{\alpha x},\ e^{-\alpha x},\ x\,e^{-\alpha x}\},\ \alpha \in \mathbb{R}^+,$$

and construct the associated GB-splines using Bernstein-like basis. The corresponding spline model is a **S**moothing **G**eneralized **B**-**S**pline on **B**ernestein basis, SGBSB for short. A short algorithmic description of our model construction follows (see [2] for details).

Procedure 2. SGBSB

1: Definition of the differential operator and the related null-spaces $\mathbb{E}_2$ of $\mathcal{L}_2$ and $\mathbb{E}_4$ of $\mathcal{L}_2^* \mathcal{L}_2$,

2: Definition of the GB-spline basis functions, $\{\varphi_\ell(x)\}_{\ell=1,\ldots,N}$, in (2):

(a) the support of each φ_ℓ is compact;
(b) $\varphi_\ell \in C^2[a, b]$;
(c) $\varphi_\ell \in \mathbb{E}_4$ for $x \in (x_i, x_{i+1})$, $i = 1, \ldots, N-1$;
(d) $\varphi_\ell \in \mathbb{E}_2$, for $x \in (a, x_1)$ or $x \in (x_N, b)$;
(e) φ_ℓ, $j = 1, \ldots, N$, are positive and bell shaped functions.

3: Representation of each GB-spline in Bernstein-like basis:
   3.1 : definition of *regular basis functions*:

$$\varphi_\ell(x)|_{[x_j, x_{j+1}]} = \sum_{i=0}^{3} \gamma_{\ell,j,i} \tilde{B}_i(x - x_j), \quad j = \ell-2, \cdots, \ell+1, \quad \ell = 3, \cdots, N-2$$

with $\tilde{B}_i$ Bernstein-like basis functions of $\mathbb{E}_4$, $i = 0, ..., 3$, and C^2 regularity at the internal points $x_{\ell-1}$, x_ℓ, $x_{\ell+1}$;
   3.2 : definition of *boundary left and right basis functions*:
   like the regular ones but with pieces in $\mathbb{E}_2$, outside the knots interval, in $(a, x_1]$ and in $[x_n, b)$ respectively.

Table 1. LTI results. The columns refer, from left to right: evaluation points (t), computed values by applying GS to the real function F (f_F) and to the SGBSB s (f_s); pointwise errros between f_F and f_s $(|f_F - f_s|)$, at the t points.

t	f_F	f_s	$|f_F - f_s|$
2.0000e+00	3.9920e−01	3.9925e−01	4.6485e−05
2.5000e+00	2.5880e−01	2.5881e−01	5.7951e−06
3.0000e+00	1.6189e−01	1.6187e−01	2.1254e−05
3.5000e+00	9.9504e−02	9.9504e−02	1.0816e−07
4.0000e+00	6.0095e−02	6.0115e−02	2.0280e−05
4.5000e+00	3.5306e−02	3.5294e−02	1.2250e−05
5.0000e+00	1.9735e−02	1.9776e−02	4.0631e−05
5.5000e+00	9.9866e−03	1.0065e−02	7.8389e−05
6.0000e+00	3.9346e−03	3.9392e−03	4.5941e−06
6.5000e+00	2.4117e−04	2.4450e−04	3.3348e−06
7.0000e+00	−1.9429e−03	−1.9464e−03	3.5076e−06
7.5000e+00	−3.1609e−03	−3.1904e−03	2.9417e−05
8.0000e+00	−3.7633e−03	−3.8100e−03	4.6736e−05
8.5000e+00	−3.9771e−03	−3.9692e−03	7.9306e−06
9.0000e+00	−3.9505e−03	−3.4085e−03	5.4200e−04
9.5000e+00	−3.7799e−03	−2.4630e−03	1.3168e−03
1.0000e+01	−3.5281e−03	−1.4930e−03	2.0351e−03

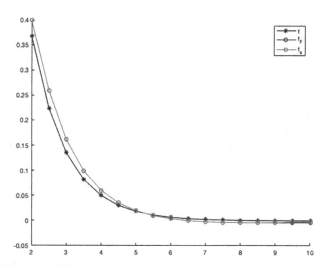

Fig. 1. Curves of the computed LTI, by applying the GS algorithm to F (f_F) and to SGBSB (f_s). The results are comparable and overlapping and they follow the behaviour of the real inverse function f described by the red curve ('*-'). The abscissae are referred in the first column of the Table 1. (Color figure online)

3   Numerical Results

In this section we present some results about a test of LTI. Let be given $N = 60$ samples $(x_i, y_i)_{i=1}^{N}$, uniformly distributed in $[0.05, 3]$. The test was carried out with MATLAB R2018a software on a Intel(R) Core(TM) i5, 1.8 GHz processor. We assume that $y_i = F(x_i)$ with F the following LT:

$$F(x) = e^{-x}/(1 + x),$$

whose LT inverse function is:

$$f(t) = e^{-(t-1)}u(t-1), \ t > 1, \ u(t) \text{ step function.} \tag{3}$$

The described SGBSB, s, requires the definition of some model parameters, that is a, b, α (see [2] for details). In this test we set $a = -1$, $b = 3.5$, and we compute $\alpha = 1.6$ through a nonlinear least-squares regression of the data. The Gaver-Stehfest (GS) algorithm for LTI [8,19] is used to compute an approximation of f at the points

$$t_j \in [2, 10], \ t_1 = 2, \ t_{j+1} = t_j + 0.5, \ j = 1, ..., 16.$$

The Matlab code gavsteh.m is available at the Mathworks File Exchange; the algorithm for the SGBSB is implemented in a proprietary Matlab code, available from the authors.

Numerical results are reported in Table 1 whose columns, from left to right, contain:

1. t: the evaluation points, t_j, $j = 1, \ldots, 17$
2. f_F: the computed values by applying GS to the real function F;
3. f_s: the computed values by applying GS to the SGBSB;
4. $|f_F - f_s|$: the computed pointwise erros between f_F and f_s, at the t points.

The corresponding curves of the computed values, together with the graph of f in (3), are in Fig. 1.

We remark that the GS algorithm requires, as input parameters, the evaluation point t, and the function to be inverted; then, it dynamically evaluates the input function at suitable points, depending, among the other method parameters, on t. In the test here presented, we evaluated also the approximation error between F and the spline model, inverted in its place, and we observe that the maximum absolute approximation error on the LT:

$$\|F - s\|_{\infty} = 9.9464 \times 10^{-4}$$

is at most of the same order of the maximum error on the computed solution, as confirmed by the values referred in the last column of Table 1.

4 Conclusion

In this work, we present the results of a LTI method, when applied to a natural smoothing exponential-polynomial spline modelling exponential decay data. This scenario is found in many applications. A detailed analysis of the parameters and the sensitivity of the model are under investigation and will be object of future studies.

Acknowledgements. The authors are members of the INdAM Research group GNCS and of the Research ITalian network on Approximation (RITA).

References

1. Bertero, M., Brianzi, P., Pike, E.R.: On the recovery and resolution of exponential relaxation rates from experimental data: Laplace transform inversion in weighted spaces. Inverse Probl. **1**, 1–15 (1985)
2. Campagna, R., Conti, C., Cuomo, S.: Smoothing exponential-polynomial splines for multiexponential decay data. Dolomites Res. Notes Approx. **12**, 86–100 (2019)
3. Campagna, R., D'Amore, L., Murli, A.: An efficient algorithm for regularization of Laplace transform inversion in real case. J. Comput. Appl. Math. **210**(1), 84–98 (2007)
4. Cuomo, S., D'Amore, L., Murli, A., Rizzardi, M.: Computation of the inverse Laplace transform based on a collocation method which uses only real values. J. Comput. Appl. Math. **198**(1), 98–115 (2007)
5. D'Amore, L., Campagna, R., Galletti, A., Marcellino, L., Murli, A.: A smoothing spline that approximates Laplace transform functions only known on measurements on the real axis. Inverse Probl. **28**(2), 025007 (2012)
6. D'Amore, L., Campagna, R., Mele, V., Murli, A.: ReLaTIve. An Ansi C90 software package for the real Laplace transform inversion. Numer. Algorithms **63**(1), 187–211 (2013)
7. D'Amore, L., Campagna, R., Mele, V., Murli, A.: Algorithm 946: ReLIADiff-A C++ software package for real Laplace transform inversion based on algorithmic differentiation. ACM Trans. Math. Softw. **40**(4), 31:1–31:20 (2014)
8. D'Amore, L., Mele, V., Campagna, R.: Quality assurance of Gaver's formula for multi-precision Laplace transform inversion in real case. Inverse Probl. Sci. Eng. **26**(4), 553–580 (2018)
9. Galvosas, P., Callaghan, P.T.: Multi-dimensional inverse Laplace spectroscopy in the NMR of porous media. C.R. Phys. **11**(2), 172–180 (2010). Multiscale NMR and relaxation
10. Henrici, P.: Applied and Computational Complex Analysis, Volume 1: Power Series Integration Conformal Mapping Location of Zero (1988)
11. Miller, M.K., Guy, W.T.: Numerical inversion of the Laplace transform by use of Jacobi polynomials. SIAM J. Numer. Anal. **3**(4), 624–635 (1966)
12. Naeeni, M.R., Campagna, R., Eskandari-Ghadi, M., Ardalan, A.A.: Performance comparison of numerical inversion methods for Laplace and Hankel integral transforms in engineering problems. Appl. Math. Comput. **250**, 759–775 (2015)
13. Papoulis, A.: A new method of inversion of the Laplace transform. Q. Appl. Math. **14**(4), 405–414 (1957)

14. Provencher, S.W.: Contin: a general purpose constrained regularization program for inverting noisy linear algebraic and integral equations. Comput. Phys. Commun. **27**(3), 229–242 (1982)
15. Romano, A., Campagna, R., Masi, P., Cuomo, S., Toraldo, G.: Data-driven approaches to predict states in a food technology case study. In: 2018 IEEE 4th International Forum on Research and Technology for Society and Industry (RTSI), pp. 1–5 (2018)
16. Romano, A., Cavella, S., Toraldo, G., Masi, P.: 2D structural imaging study of bubble evolution during leavening. Food Res. Int. **50**(1), 324–329 (2013)
17. Schumaker, L.: Spline Functions: Basic Theory, 3rd edn. Cambridge Mathematical Library, Cambridge University Press, Cambridge (2007)
18. Severino, G., D'Urso, G., Scarfato, M., Toraldo, G.: The IoT as a tool to combine the scheduling of the irrigation with the geostatistics of the soils. Future Gener. Comput. Syst. **82**, 268–273 (2018)
19. Stehfest, H.: Algorithm 368: numerical inversion of Laplace transforms. Commun. ACM **13**(1), 47–49 (1970)

An Adaptive Refinement Scheme
for Radial Basis Function Collocation

Roberto Cavoretto$^{(\boxtimes)}$ and Alessandra De Rossi

Department of Mathematics "Giuseppe Peano", University of Torino,
Via Carlo Alberto 10, 10123 Turin, Italy
{roberto.cavoretto,alessandra.derossi}@unito.it

Abstract. In this paper we present an adaptive refinement algorithm for solving elliptic partial differential equations via a radial basis function (RBF) collocation method. The adaptive scheme is based on the use of an error indicator, which is characterized by the comparison of two RBF collocation solutions evaluated on a coarser set and a finer one. This estimate allows us to detect the domain parts that need to be refined by adding points in the selected areas. Numerical results support our study and point out the effectiveness of our algorithm.

Keywords: Meshfree methods · Adaptive algorithms · Refinement techniques · Elliptic PDEs

1 Introduction

In this paper we present a new adaptive refinement scheme for solving elliptic partial differential equations (PDEs). Our adaptive algorithm is applied to a non-symmetric radial basis function (RBF) collocation method, which was originally proposed by Kansa [5]. This approach has engendered a large number of works, mainly by scientists from several different areas of science and engineering (see e.g. [1–4,7] and references therein). Basically, the adaptive scheme we propose is based on the use of an error indicator characterized by the comparison of two approximate RBF collocation solutions, which are evaluated on a coarser set and a finer one. This estimate allows us to identify the domain parts that need to be refined by adding points in the selected areas. In our numerical experiments we show the efficacy of our refinement algorithm, which is tested by modeling some Poisson-type problems.

The paper is organized as follows. In Sect. 2 we review some basic information on Kansa's collocation method, which is applied to elliptic PDEs. Section 3 describes the adaptive refinement algorithm. In Sect. 4 we show some numerical results carried out to illustrate the performance of the adaptive scheme. Section 5 contains conclusions.

© Springer Nature Switzerland AG 2020
Y. D. Sergeyev and D. E. Kvasov (Eds.): NUMTA 2019, LNCS 11973, pp. 19–26, 2020.
https://doi.org/10.1007/978-3-030-39081-5_3

2   Nonsymmetric RBF Collocation

Given a domain $\Omega \subset \mathbb{R}^d$, we consider a (time independent) elliptic PDE along with its boundary conditions

$$
\begin{aligned}
\mathcal{L}u(\boldsymbol{x}) &= f(\boldsymbol{x}), & \boldsymbol{x} \in \Omega, \\
\mathcal{B}u(\boldsymbol{x}) &= g(\boldsymbol{x}), & \boldsymbol{x} \in \partial\Omega,
\end{aligned}
\tag{1}
$$

where $\mathcal{L}$ is a linear elliptic partial differential operator and $\mathcal{B}$ is a linear boundary operator.

For Kansa's collocation method we choose to represent the approximate solution $\hat{u}$ by a RBF expansion analogous to that used in the field of RBF interpolation [3], i.e. $\hat{u}$ is expressed as a linear combination of basis functions

$$
\hat{u}(\boldsymbol{x}) = \sum_{j=1}^{N} c_j \phi_\varepsilon(||\boldsymbol{x} - \boldsymbol{z}_j||_2),
\tag{2}
$$

where c_j is an unknown real coefficient, $|| \cdot ||_2$ denotes the Euclidean norm, and $\phi_\varepsilon : \mathbb{R}_{\geq 0} \to \mathbb{R}$ is some RBF depending on a *shape parameter* $\varepsilon > 0$ such that

$$
\phi_\varepsilon(||\boldsymbol{x} - \boldsymbol{z}||_2) = \phi(\varepsilon||\boldsymbol{x} - \boldsymbol{z}||_2), \qquad \forall \boldsymbol{x}, \boldsymbol{z} \in \Omega.
$$

In Table 1 we list some examples of popular globally supported RBFs, which are commonly used for solving PDEs (see [3] for details).

Table 1. Some examples of popular RBFs.

RBF	$\phi_\varepsilon(r)$
Gaussian (GA)	$e^{-\varepsilon^2 r^2}$
Inverse MultiQuadric (IMQ)	$(1 + \varepsilon^2 r^2)^{-1/2}$
MultiQuadric (MQ)	$(1 + \varepsilon^2 r^2)^{1/2}$

In (2) we can distinguish between the set $X = \{\boldsymbol{x}_1, \ldots, \boldsymbol{x}_N\}$ of *collocation points* and the set $Z = \{\boldsymbol{z}_1, \ldots, \boldsymbol{z}_N\}$ of *centers*. Additionally, for the sake of convenience we split the set X into a subset X_I of interior points and a subset X_B of boundary points, so that $X = X_I \cup X_B$.

Matching the PDE and the boundary conditions in (1) at the collocation points X, we obtain a linear system of equations

$$
\Phi \boldsymbol{c} = \boldsymbol{v},
$$

where $\boldsymbol{c} = (c_1, \ldots, c_N)^T$ is the vector of coefficients, $\boldsymbol{v} = (v_1, \ldots, v_N)^T$ is the vector of entries

$$
v_i = \begin{cases} f(\boldsymbol{x}_i), & \boldsymbol{x}_i \in X_I, \\ g(\boldsymbol{x}_i), & \boldsymbol{x}_i \in X_B. \end{cases}
$$

and $\Phi \in \mathbb{R}^{N \times N}$ is the collocation matrix

$$\Phi = \begin{bmatrix} \Phi_{\mathcal{L}} \\ \Phi_{\mathcal{B}} \end{bmatrix}. \tag{3}$$

The two blocks in (3) are defined as

$$(\Phi_{\mathcal{L}})_{ij} = \mathcal{L}\phi_\varepsilon(||\boldsymbol{x}_i - \boldsymbol{z}_j||_2), \, \boldsymbol{x}_i \in X_I, \; \boldsymbol{z}_j \in Z,$$
$$(\Phi_{\mathcal{B}})_{ij} = \mathcal{B}\phi_\varepsilon(||\boldsymbol{x}_i - \boldsymbol{z}_j||_2), \, \boldsymbol{x}_i \in X_B, \; \boldsymbol{z}_j \in Z,$$

Since the collocation matrix (3) may be singular for certain configurations of the centers $\boldsymbol{z}_j$, it follows that the nonsymmetric collocation method cannot be well-posed for arbitrary center locations. However, it is possible to find sufficient conditions on the centers so that invertibility of Kansa's matrix is ensured. For a more detailed analysis of Kansa's collocation method and some variations thereof derived from applications, see e.g. [3,6] and references therein.

3   Adaptive Refinement Algorithm

In this section we present the adaptive algorithm proposed to solve time independent PDE problems by Kansa's approach.

Step 1. We define two sets, $X_{N_1^{(0)}}$ and $X_{N_2^{(0)}}$, of collocation points and two sets, $Z_{N_1^{(0)}}$ and $Z_{N_2^{(0)}}$ of centers. Each couple of sets has size $N_1^{(0)}$ and $N_2^{(0)}$, respectively, with $N_1^{(0)} < N_2^{(0)}$ and the symbol $^{(0)}$ identifying the initial iteration. We then split the related sets as follows:

- $X_{N_1^{(0)}} = X_{I,N_1^{(0)}} \cup X_{B,N_1^{(0)}}$ and $X_{N_2^{(0)}} = X_{I,N_2^{(0)}} \cup X_{B,N_2^{(0)}}$ are sets of interior and boundary collocation points, respectively;
- $Z_{N_1^{(0)}} = Z_{I,N_1^{(0)}} \cup Z_{B,N_1^{(0)}}^A$ and $Z_{N_2^{(0)}} = Z_{I,N_2^{(0)}} \cup Z_{B,N_2^{(0)}}^A$ are sets of interior and additional boundary centers, respectively.

Here we assume that $X_{I,N_i^{(0)}} = Z_{I,N_i^{(0)}}$, with $i = 1, 2$, while the set $Z_{B,N_i^{(0)}}^A$ of centers is taken outside the domain Ω as suggested in [3]. However, we note that it is also possible to consider only a set of data as collocation points and centers.

Step 2. For $k = 0, 1, \ldots$, we iteratively find two collocation solutions of the form (2), called $\hat{u}_{N_1^{(k)}}$ and $\hat{u}_{N_2^{(k)}}$, which are respectively computed on $N_1^{(k)}$ and $N_2^{(k)}$ collocation points and centers.

Step 3. We compare the two approximate RBF solutions by evaluating error on the (coarser) set containing $N_1^{(k)}$ points, i.e.

$$|\hat{u}_{N_2^{(k)}}(\boldsymbol{x}_i) - \hat{u}_{N_1^{(k)}}(\boldsymbol{x}_i)|, \qquad \boldsymbol{x}_i \in X_{N_1^{(k)}}.$$

Observe that here we assume that the solution computed on $N_2^{(k)}$ discretization points gives more accurate results than the ones obtained with only $N_1^{(k)}$ points.

Step 4. After fixing a tolerance *tol*, we determine all points $\boldsymbol{x}_i \in X_{N_1(k)}$ such that

$$|\hat{u}_{N_2(k)}(\boldsymbol{x}_i) - \hat{u}_{N_1(k)}(\boldsymbol{x}_i)| > tol. \tag{4}$$

Step 5. In order to refine the distribution of discretization points, we compute the *separation distance*

$$q_{X_{N_1(k)}} = \frac{1}{2} \min_{i \neq j} ||\boldsymbol{x}_i - \boldsymbol{x}_j||_2, \qquad \boldsymbol{x}_i \in X_{N_1(k)}. \tag{5}$$

Step 6. For $k = 0, 1, \ldots$ we update the two sets $X_{N_1(k+1)}$ and $X_{N_2(k+1)}$ of collocation points (and accordingly the corresponding sets $Z_{N_1(k+1)}$ and $Z_{N_2(k+1)}$ of centers) as follows. For each point $\boldsymbol{x}_i \in X_{N_1(k)}$, such that the condition (4) is satisfied, we add to $\boldsymbol{x}_i$:

- four points (the blue circles depicted in the left frame of Fig. 1), thus creating the set $X_{N_1(k+1)}$;
- eight points (the red squares shown in the right frame of Fig. 1), thus generating the set $X_{N_2(k+1)}$.

In both cases the new points are given by properly either adding or subtracting the value of (5) to the components of $\boldsymbol{x}_i$. Furthermore, we remark that in the illustrative example of Fig. 1 the point $\boldsymbol{x}_i$ is marked by a black cross, while the new sets are such that $X_{N_1(k)} \subset X_{N_2(k)}$, for $k = 1, 2, \ldots$.

Step 7. The iterative process stops when having no points anymore which fulfill the condition (4), giving the set $X_{N_2(k^*)}$ back. Note that k^* is here used to denote the last algorithm iteration.

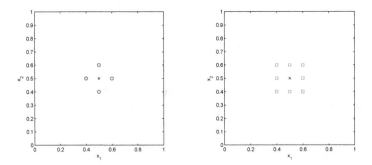

Fig. 1. Illustrative example of refinement for sets $X_{N_1(k)}$ (left) and $X_{N_2(k)}$ (right) in the adaptive algorithm. (Color figure online)

4   Numerical Results

In this section we summarize the results derived from application of our adaptive refinement algorithm, which is implemented in MATLAB environment. All the results are carried out on a laptop with an Intel(R) Core(TM) i7-6500U CPU 2.50 GHz processor and 8 GB RAM.

In the following we restrict our attention on solving some elliptic PDE problems via the nonsymmetric RBF collocation method. In particular, in (1) we consider a few Poisson-type problems, taking the Laplace operator $\mathcal{L} = -\Delta$ and assuming Dirichlet boundary conditions. Hence, the PDE problem in (1) can be defined as follows:

$$
\begin{aligned}
-\Delta u(\boldsymbol{x}) &= f(\boldsymbol{x}), & \boldsymbol{x} \in \Omega, \\
u(\boldsymbol{x}) &= g(\boldsymbol{x}), & \boldsymbol{x} \in \partial\Omega.
\end{aligned}
\tag{6}
$$

Then, we focus on two test problems of the form (6) defined on the domain $\Omega = [0,1]^2$. The exact solutions of such Poisson problems are

$$
\text{P1}: u_1(x_1, x_2) = \sin(x_1 + 2x_2^2) - \sin(2x_1^2 + (x_2 - 0.5)^2),
$$
$$
\text{P2}: u_2(x_1, x_2) = \frac{1}{2} x_2 \left[\cos(4x_1^2 + x_2^2 - 1) \right]^4 + \frac{1}{4} x_1.
$$

A graphical representation of these analytic solutions is shown in Fig. 2.

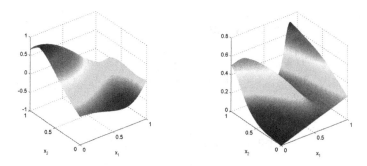

Fig. 2. Graphs of exact solutions u_1 (left) and u_2 (right) of Poisson problems.

In our numerical tests we analyze the performance of the adaptive refinement strategy applied to Kansa's collocation method by using globally supported RBFs such as MQ, IMQ and GA (see Table 1). We remark that the use of compactly supported RBFs is also possible and effective but our tests showed that the most accurate results were obtained with quite large supports. So the use of compactly supported functions does not provide particular benefits w.r.t. globally supported RBFs. For this reason and for the sake of brevity, we do not consider this case in the present paper.

The two starting sets defined in Sect. 3 consist of $N_1{}^{(0)} = 289$ and $N_2{}^{(0)} = 1089$ grid collocation points, while the tolerance in (4) is given by $tol = 10^{-4}$.

In particular, in order to measure the quality of our results, we compute the Maximum Absolute Error (MAE), i.e.,

$$\text{MAE} = \max_{1 \leq i \leq N_{eval}} |u(\boldsymbol{y}_i) - \hat{u}(\boldsymbol{y}_i)|.$$

which is evaluated on a grid of $N_{eval} = 40 \times 40$ evaluation points. Moreover, in regards to the efficiency of the adaptive scheme, we report the CPU times computed in seconds.

In Tables 2, 3 and 4 we present the results obtained, also indicating the final number N_{fin} of collocation points required to achieve the fixed tolerance. Further, as an example, for brevity in only one case for each test problem, we report the "refined grids" after applying iteratively the adaptive algorithm. More precisely, in Fig. 3 we graphically represent the final distribution of points obtained after the last algorithm iteration by applying: the MQ-RBF with $\varepsilon = 4$ for the test problem P1 (left), and the IMQ-RBF with $\varepsilon = 3$ for the test problem P2 (right).

Table 2. MQ, $\varepsilon = 4$, $tol = 10^{-4}$.

Test problem	N_{fin}	MAE	CPU time
P1	856	8.28×10^{-5}	1.2
P2	798	2.66×10^{-4}	1.7

Table 3. IMQ, $\varepsilon = 3$, $tol = 10^{-4}$.

Test problem	N_{fin}	MAE	CPU time
P1	1001	3.10×10^{-5}	1.7
P2	808	1.10×10^{-4}	1.8

Table 4. GA, $\varepsilon = 9$, $tol = 10^{-4}$.

Test problem	N_{fin}	MAE	CPU time
P1	898	3.49×10^{-4}	2.1
P2	843	2.08×10^{-4}	1.9

Analyzing the numerical results, we can observe as the adaptive algorithm allows us to increase the number of points in the regions where the solution is not accurate enough. From the tables we note as MQ and IMQ give more accurate results than GA. In fact, though the number of points required to satisfy the fixed tolerance is quite similar for all used RBFs, we can remark a greater instability of GA that needs a larger value of ε to work effectively. Finally, in terms of computational efficiency the algorithm converges in few seconds in each of the tests carried out.

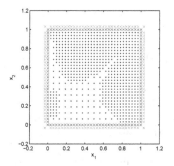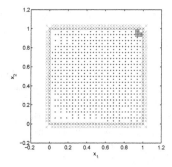

Fig. 3. Final distribution of points obtained after applying the refinement process with MQ, $\varepsilon = 4$, for problem P1 (left) and with IMQ, $\varepsilon = 3$, for problem P2.

5 Conclusions

In this work we presented an adaptive refinement algorithm to solve time independent elliptic PDEs. This refinement strategy is tested on a nonsymmetric RBF collocation scheme, known as Kansa's method. More precisely, here we proposed an adaptive approach based on a refinement technique, which consisted in comparing two collocation solutions computed on a coarser set of collocation points and a finer one. This process allowed us to detect the domain areas where it is necessary to adaptively add points, thus enhancing accuracy of the method. Numerical results supported this study by showing the algorithm performance on some Poisson-type problems.

As future work we are interested in investigating and possibly extending our adaptive schemes to hyperbolic and parabolic PDE problems. Moreover, we are currently working on the optimal selection of the RBF shape parameter in collocation schemes. However, this is out of the scopes of the present paper and it will be dealt with in forthcoming works.

Acknowledgments. The authors sincerely thank the two anonymous referees for carefully reviewing this paper and for their valuable comments and suggestions. They acknowledge support from the Department of Mathematics "Giuseppe Peano" of the University of Torino via Project 2018 "Algebra, geometry and numerical analysis" and Project 2019 "Mathematics for applications". Moreover, this work was partially supported by INdAM – GNCS Project 2019 "Kernel-based approximation, multiresolution and subdivision methods and related applications". This research has been accomplished within RITA (Research ITalian network on Approximation).

References

1. Cavoretto, R., De Rossi, A.: Adaptive meshless refinement schemes for RBF-PUM collocation. Appl. Math. Lett. **90**, 131–138 (2019)
2. Chen, W., Fu, Z.-J., Chen, C.S.: Recent Advances on Radial Basis Function Collocation Methods. Springer Briefs in Applied Science and Technology. Springer, Berlin (2014). https://doi.org/10.1007/978-3-642-39572-7

3. Fasshauer, G.E.: Meshfree Approximation Methods with Matlab. Interdisciplinary Mathematical Sciences, vol. 6. World Scientific Publishing Co., Singapore (2007)
4. Libre, N.A., Emdadi, A., Kansa, E.J., Shekarchi, M., Rahimian, M.: A fast adaptive wavelet scheme in RBF collocation for nearly singular potential PDEs. CMES Comput. Model. Eng. Sci. **38**, 263–284 (2008)
5. Kansa, E.J.: Multiquadrics - a scattered data approximation scheme with applications to computational fluid-dynamics - II solutions to parabolic, hyperbolic and elliptic partial differential equations. Comput. Math. Appl. **19**, 147–161 (1990)
6. Schaback, R.: Convergence of unsymmetric kernel-based meshless collocation methods. SIAM J. Numer. Anal. **45**, 333–351 (2007)
7. Sarra, S.A., Kansa, E.J.: Multiquadric radial basis function approximation methods for the numerical solution of partial differential equations. In: Advances in Computational Mechanics, vol. 2. Tech Science Press, Encino (2009)

A 3D Efficient Procedure for Shepard Interpolants on Tetrahedra

Roberto Cavoretto[1] , Alessandra De Rossi[1(✉)] ,
Francesco Dell'Accio[2] , and Filomena Di Tommaso[2]

[1] Department of Mathematics "Giuseppe Peano", University of Turin,
Via Carlo Alberto 10, 10123 Turin, Italy
{roberto.cavoretto,alessandra.derossi}@unito.it
[2] Department of Mathematics and Computer Science, University of Calabria,
via P. Bucci, Cubo 30A, 87036 Rende (CS), Italy
francesco.dellaccio@unical.it, ditommaso@mat.unical.it

Abstract. The need of scattered data interpolation methods in the multivariate framework and, in particular, in the trivariate case, motivates the generalization of the fast algorithm for triangular Shepard method. A block-based partitioning structure procedure was already applied to make the method very fast in the bivariate setting. Here the searching algorithm is extended, it allows to partition the domain and nodes in cubic blocks and to find the nearest neighbor points that need to be used in the tetrahedral Shepard interpolation.

Keywords: Scattered data interpolation · Tetrahedral Shepard operator · Fast algorithms · Approximation algorithms

1 Introduction

Given a set of values of a function f at certain scattered nodes $X_n = \{x_1, \ldots, x_n\}$ in a compact convex domain $\Omega \subset \mathbb{R}^2$, the triangular Shepard method [8] can be applied efficiently to interpolate the target function $f : \Omega \to \mathbb{R}$. In [5] we proposed a triangular Shepard method which combines triangle-based basis functions with linear combinations of the values $f(x_i)$ at the vertices of the triangles. Moreover, the triangulation can be found in an efficient way by reducing the number of triangles. The triangulation considered is called *compact triangulation* and it allows the triangles to overlap or being disjoint. These triangulations are determined by minimizing the bound of the error of the linear interpolant on the vertices of the triangle, chosen in a set of nearby nodes. For these triangulations a block-based partitioning structure procedure was presented in [3] to make the method very fast, since the vertices of the triangles must be chosen in a set of nearby nodes.

In recent years an increasing attention to the multivariate framework was given. For this reason we propose in this paper a generalization to the 3D setting. More precisely, we propose a fast searching procedure to apply to the tetrahedral

© Springer Nature Switzerland AG 2020
Y. D. Sergeyev and D. E. Kvasov (Eds.): NUMTA 2019, LNCS 11973, pp. 27–34, 2020.
https://doi.org/10.1007/978-3-030-39081-5_4

Shepard interpolation. It allows to partitioning the 3D domain and nodes in cubic blocks and to find the nearest neighbor points to compute the Shepard interpolant on tetrahedra. Similar algorithms were also analized in [2] in the context of trivariate partition of unity methods combined with the use of local radial kernels.

The paper is organized as follows. In Sect. 2 the tetrahedral Shepard method for trivariate interpolation is recalled. In Sect. 3 we give a pseudo-code of the complete interpolation algorithm, presenting the procedures used to identify and search the nearest neighbor points in the 3D interpolation scheme. In Sect. 4 we show some numerical experiments obtained to illustrate the performance of our tetrahedral Shepard algorithm. Finally, Sect. 5 contains conclusions and future work.

2   Tetrahedral Shepard Interpolant

Let be $X_n = \{x_1, \ldots, x_n\}$ a set of data points or nodes of $\mathbb{R}^3$ with an associated set of function data $F_n = \{f_1, \ldots, f_n\}$ and $H = \{h_1, \ldots, h_m\}$ a set of tetrahedra with vertices in X_n. Let us denote by $W_j = \{x_{j_1}, x_{j_2}, x_{j_3}, x_{j_4}\}$ the set of vertices of h_j, $j = 1, \ldots, m$. Moreover, we assume that the set $\{W_j\}_{j=1,\ldots,m}$ constitutes a cover of X_n, that is

$$\bigcup_{j=1}^{m} W_j = X_n.$$

We can associate to each tetrahedra h_j the set of barycentric coordinates of a point $x \in \mathbb{R}^3$, that is

$$\mu_{j,j_1}(x) = \frac{W(x, x_{j_2}, x_{j_3}, x_{j_4})}{W(x_{j_1}, x_{j_2}, x_{j_3}, x_{j_4})}, \quad \mu_{j,j_2}(x) = \frac{W(x_{j_1}, x, x_{j_3}, x_{j_4})}{W(x_{j_1}, x_{j_2}, x_{j_3}, x_{j_4})},$$

$$\mu_{j,j_3}(x) = \frac{W(x_{j_1}, x_{j_2}, x, x_{j_4})}{W(x_{j_1}, x_{j_2}, x_{j_3}, x_{j_4})}, \quad \mu_{j,j_4}(x) = \frac{W(x_{j_1}, x_{j_2}, x_{j_3}, x)}{W(x_{j_1}, x_{j_2}, x_{j_3}, x_{j_4})},$$

where $W(x, y, v, z)$ denotes $\frac{1}{6}$ the signed volume of the tetrahedra h_j. The linear polynomial $\lambda_j(x)$ which interpolates the data at the vertices of the tetrahedra h_j can be expressed in terms of barycentric coordinates in the following form

$$\lambda_j(x) = \sum_{k=1}^{4} \mu_{j,j_k}(x) f_{j_k}, \quad j = 1, \ldots, m. \tag{1}$$

The tetrahedral basis functions are a normalization of the product of the inverse distances from the vertices of the tetrahedra h_j

$$\beta_{\nu,j}(x) = \frac{\prod_{\ell=1}^{4} ||x - x_{j_\ell}||^{-\nu}}{\sum_{k=1}^{m} \prod_{\ell=1}^{4} ||x - x_{k_\ell}||^{-\nu}}, \quad j = 1, \ldots, m, \quad \nu > 0, \tag{2}$$

where $|| \cdot ||$ is the Euclidean norm. The tetrahedral Shepard method is defined by

$$\mathcal{T}_\nu [f] (\boldsymbol{x}) = \sum_{j=1}^{m} \beta_{\nu,j} (\boldsymbol{x}) \lambda_j (\boldsymbol{x}). \tag{3}$$

Tetrahedral basis functions form a partition of unity, as the triangular Shepard ones, and allow the interpolation of functional and derivative values. In fact, the following results hold, the proofs can be easily obtained in analogy with [5, Proposition 2.1].

Proposition 1. *The tetrahedral basis function $\beta_{\nu,j}(\boldsymbol{x})$ and its gradient (that exists for $\nu > 1$) vanish at all nodes $\boldsymbol{x}_i \in X_n$ that are not a vertex of the corresponding tetrahedron h_j. That is,*

$$\beta_{\nu,j}(\boldsymbol{x}_i) = 0, \tag{4}$$

$$\nabla \beta_{\nu,j}(\boldsymbol{x}_i) = 0, \quad \nu > 1, \tag{5}$$

for any $j = 1, \ldots, m$ and $i \notin \{j_1, j_2, j_3, j_4\}$. Moreover, they form a partition of unity, that is

$$\sum_{j=1}^{m} \beta_{\nu,j}(\boldsymbol{x}) = 1 \tag{6}$$

and consequently, for each $i = 1, \ldots, n$,

$$\sum_{j \in J_i} \beta_{\nu,j}(\boldsymbol{x}_i) = 1, \tag{7}$$

$$\sum_{j \in J_i} \nabla \beta_{\nu,j}(\boldsymbol{x}_i) = 0, \quad \nu > 1, \tag{8}$$

where $J_i = \left\{ k \in \{1, \ldots, m\} : i \in \{k_1, k_2, k_3, k_4\} \right\}$ is the set of tetrahedra which have $\boldsymbol{x}_i$ as a vertex.

These properties imply that the operator $\mathcal{T}_\nu$ satisfies the following ones, see [4] for details.

Proposition 2. *The operator $\mathcal{T}_\nu$ is an interpolation operator, that is,*

$$\mathcal{T}_\nu[f](\boldsymbol{x}_i) = f_i, \qquad i = 1, \ldots, n,$$

and reproduces polynomials up to the degree 1.

The procedure to select the compact 3D-triangulation (by tetrahedra) of the node set X_n strongly affects the results of the analysis of the convergence of the operator $\mathcal{T}_\nu [f] (\boldsymbol{x})$.

In order to determine the approximation order of the tetrahedral operator, we denote by $\Omega \subset \mathbb{R}^3$ a compact convex domain containing X_n and by $C^{1,1}(\Omega)$ the class of differentiable functions $f : \Omega \to \mathbb{R}$ whose partial derivative of order 1 are Lipschitz-continuous, equipped with the seminorm

$$\|f\|_{1,1} = \sup \left\{ \frac{\|D^\mu f(\boldsymbol{u}) - D^\mu f(\boldsymbol{v})\|}{\|\boldsymbol{u} - \boldsymbol{v}\|} : \boldsymbol{u}, \boldsymbol{v} \in \Omega, \boldsymbol{u} \neq \boldsymbol{v}, \|\mu\| = 1 \right\}. \tag{9}$$

We also denote by $e_{k,\ell} = x_{j_k} - x_{j_\ell}$, with $k, \ell = 1, 2, 3, 4$, the edge vectors of the tetrahedron h_j. Then, the following result holds (for the proof see [4]).

Proposition 3. *Let $f \in C^{1,1}(\Omega)$ and $h_j \in H$ a tetrahedron of vertices x_{j_1}, x_{j_2}, x_{j_3}, x_{j_4}. Then, for all $x \in \Omega$ we have*

$$|f(x) - \lambda_j(x)| \leq \|f\|_{1,1} \left(3\|x - x_{j_1}\|_2^2 + \frac{27}{2}C_j k_j \|x - x_{j_1}\|_2\right), \tag{10}$$

where $k_j = \max_{k,\ell=1,2,3,4} \|e_{k,\ell}\|$ and C_j is given by the ratio between the maximum edge and the volume, and then is a constant which depends only on the shape of the tetrahedron h_j. The error bound is valid for any vertex.

3   Trivariate Shepard Interpolation Algorithm on Tetrahedra

In this section we present the interpolation algorithm, which performs the tetrahedral Shepard method (3) using the block-based partitioning structure and the associated searching procedure. Here we consider $\Omega = [0, 1]^3$.

INPUTS: n, number of data; $X_n = \{x_1, \ldots, x_n\}$, set of data points; $F_n = \{f_1, \ldots, f_n\}$, set of data values; n_e, number of evaluation points; n_w, localizing parameter.

OUTPUTS: $E_{n_e} = \{\mathcal{T}_\nu[f](z_1), \ldots, \mathcal{T}_\nu[f](z_{n_e})\}$, set of approximated values.

Step 1: Generate a set $Z_{n_e} = \{z_1, \ldots, z_{n_e}\} \subseteq \Omega$ of evaluation points.

Step 2: For each point x_i, $i = 1, \ldots, n$, construct a neighborhood of radius

$$\delta = \frac{\sqrt{3}}{d}, \quad \text{with} \quad d = \left\lfloor \left(\frac{n}{8}\right)^{1/3} \right\rfloor.$$

where the value of d is suitably chosen extending the definition contained in [2]. This phase performs the localization.

Step 3: Compute the number b of blocks (along one side of the unit cube Ω) defined by

$$b = \left\lceil \frac{1}{\delta} \right\rceil.$$

In this way we get the side of each cubic block is equal to the neighborhood radius. This choice enables us to examine in the searching procedure only a small number of blocks, so to reduce the computational cost as compared to the most advanced searching techniques, as for instance the kd-trees [10]. The benefit is proved by the fact that this searching process is carried out in constant time, i.e. $O(1)$. Further, in this partitioning phase we number the cube-shaped blocks from 1 to b^3.

Step 4: Build the partitioning structure on the domain Ω and split the set X_n of interpolation nodes in b^3 cubic blocks. Here we are able to obtain a fast searching procedure to detect the interpolation points nearest to each of nodes.
Step 5: For each neighborhood or point (i.e., the neighborhood centre), solve the containing query and the range search problems to detect all nodes X_{n_k}, $k = 1, \ldots, b^3$, belonging to the k-th block and its twenty-six neighboring blocks (or less in case the block lies on the boundary). This is performed by repeatedly using a *quicksort* routine.
Step 6: For each data point $\boldsymbol{x}_i \in X_n$, fix its n_w nearest neighbors $\mathcal{N}(\boldsymbol{x}_i) \subset X_n$. Among the

$$\frac{n_w \left(n_w - 1\right) \left(n_w - 2\right)}{6}$$

tetrahedra with a vertex in $\boldsymbol{x}_i$, name it $\boldsymbol{x}_{j_1}$ and other three vertices in $\mathcal{N}(\boldsymbol{x}_i)$, choose the one which locally reduces the bound for the error of the local linear interpolant

$$3 \left\| \boldsymbol{x} - \boldsymbol{x}_{j_1} \right\|_2^2 + \frac{27}{2} k_j \frac{k_j^3}{W \left(\boldsymbol{x}_{j_1}, \boldsymbol{x}_{j_2}, \boldsymbol{x}_{j_3}, \boldsymbol{x}_{j_4} \right)} \left\| \boldsymbol{x} - \boldsymbol{x}_{j_1} \right\|_2 .$$

Step 7: Compute the local basis function $\beta_{\nu,j}(\boldsymbol{z})$, $j = 1, \ldots, m$, at each evaluation point $\boldsymbol{z} \in Z_{n_e}$.
Step 8: Compute the linear interpolants $\lambda_j(\boldsymbol{z})$, $j = 1, \ldots, m$, at each evaluation point $\boldsymbol{z} \in Z_{n_e}$.
Step 9: Apply the tetrahedral Shepard method (3) and evaluate the trivariate interpolant at the evaluation points $\boldsymbol{z} \in Z_{n_e}$.

4   Numerical Results

We present here accuracy and efficiency results of the trivariate interpolation algorithm proposed. The algorithm was implemented in MATLAB. All the numerical experiments have been carried out on a laptop with an Intel(R) Core i7 6500U CPU 2.50 GHz processor and 8.00 GB RAM.

In the following we analize the results obtained about several tests carried out. We solved very large interpolation problems by means of the tetrahedral Shepard method (3). To do this we considered two different distributions of irregularly distributed (or scattered) nodes contained in the unit cube $\Omega = [0, 1]^3 \subset \mathbb{R}^3$, and taking a number n of interpolation nodes that varies from 2 500 to 20 000. More precisely, as interpolation nodes we focus on a few sets of uniformly random Halton points generated through the MATLAB program `haltonseq.m` [6], and pseudo-random points obtained by using the `rand` MATLAB command. In addition, the interpolation errors are computed on a grid consisting of $n_e = 21 \times 21 \times 21$ evaluation points, while as localizing parameter we fix the value $n_w = 13$ and $\nu = 2$.

In the various experiments we discuss the performance of our interpolation algorithm assuming the data values are given by the following two trivariate test functions:

$$f_1(x_1, x_2, x_3) = \cos(6x_3)(1.25 + \cos(5.4x_2))/(6 + 6(3x_1 - 1)^2),$$
$$f_2(x_1, x_2, x_3) = \exp(-81/16((x_1 - 0.5)^2 + (x_2 - 0.5)^2 + (x_3 - 0.5)^2))/3.$$

These functions are usually used to test and validate new approximation methods and algorithms (see e.g. [9]).

As a measure of the accuracy of our results, we compute the Maximum Absolute Error (MAE) and the Root Mean Square Error (RMSE), whose formulas are respectively given by

$$\text{MAE} = ||f - \mathcal{T}_\nu[f]||_\infty = \max_{1 \le i \le n_e} |f(z_i) - \mathcal{T}_\nu[f](z_i)|$$

and

$$\text{RMSE} = \frac{1}{\sqrt{n_e}}||f - \mathcal{T}_\nu[f]||_2 = \sqrt{\frac{1}{n_e}\sum_{i=1}^{n_e}|f(z_i) - \mathcal{T}_\nu[f](z_i)|^2},$$

where $z_i \in Z_{n_e}$ is an evaluation point belonging to the domain Ω.

In Tables 1 and 2 we report MAEs and RMSEs that decrease when the number n of interpolation points increases. Comparing then the errors obtained by using the two data distributions, we can note that a (slightly) better accuracy is achieved whenever we employ Halton nodes. This fact is basically due to greater level of regularity of Halton points than pseudo-random MATLAB nodes. Analyzing the error behavior with the test functions f_1 and f_2, we get similar results in terms of accuracy of the interpolation scheme.

Table 1. MAE and RMSE computed on Halton points.

n	f_1		f_2	
	MAE	RMSE	MAE	RMSE
2 500	4.29E−2	4.63E−3	1.37E−2	1.99E−3
5 000	3.75E−2	3.04E−3	1.03E−2	1.14E−3
10 000	2.39E−2	2.05E−3	5.96E−3	7.33E−4
20 000	1.72E−2	1.33E−3	3.41E−3	4.50E−4

Then we compare the performance of the optimized searching procedure based on the partitioning of nodes in cubic blocks with a standard implementation of the algorithm where one computes all the distances between the interpolation nodes. With regard to the efficiency of the 3D Shepard interpolation algorithm the CPU times computed in seconds are around 27 and 55 for the sets of 10000 and 20000 nodes, respectively. Using a standard implementation the seconds increase to about 41 and 318 for the two sets. From this study we highlight a remarkable enhancement in terms of computational efficiency when the new partitioning and searching techniques are applied.

Table 2. MAE and RMSE computed on pseudo-random MATLAB points.

n	f_1		f_2	
	MAE	RMSE	MAE	RMSE
2 500	6.56E−2	5.49E−3	2.28E−2	2.49E−3
5 000	3.89E−2	3.89E−3	1.62E−2	1.63E−3
10 000	4.00E−2	2.71E−3	8.86E−3	1.03E−3
20 000	1.77E−2	1.65E−3	7.60E−3	6.38E−4

5   Conclusions and Future Work

In this paper we presented a new trivariate algorithm to efficiently interpolate scattered data nodes using the tetrahedral Shepard method. Since this interpolation scheme needs to find suitable tetrahedra associated with the nodes, we proposed a fast searching procedure based on the partitioning of domain in cube blocks. Such a technique turned out to be computationally more efficient than a standard one. Numerical experiments showed good performance of our procedure, which enabled us to quickly deal with a large number of nodes.

Another possible extension is given by a spherical triangular or tetrahedral Shepard method, which can be applied on the sphere $\mathbb{S}^2$ or other manifolds (see e.g. [1,11]).

Acknowledgments. The authors acknowledge support from the Department of Mathematics "Giuseppe Peano" of the University of Torino via Project 2019 "Mathematics for applications". Moreover, this work was partially supported by INdAM – GNCS Project 2019 "Kernel-based approximation, multiresolution and subdivision methods and related applications". This research has been accomplished within RITA (Research ITalian network on Approximation).

References

1. Allasia, G., Cavoretto, R., De Rossi, A.: Hermite-Birkhoff interpolation on scattered data on the sphere and other manifolds. Appl. Math. Comput. **318**, 35–50 (2018)
2. Cavoretto, R., De Rossi, A., Perracchione, E.: Efficient computation of partition of unity interpolants through a block-based searching technique. Comput. Math. Appl. **71**, 2568–2584 (2016)
3. Cavoretto, R., De Rossi, A., Dell'Accio, F., Di Tommaso, F.: Fast computation of triangular Shepard interpolants. J. Comput. Appl. Math. **354**, 457–470 (2019)
4. Cavoretto R., De Rossi A., Dell'Accio F., Di Tommaso F.: An efficient trivariate algorithm for tetrahedral Shepard interpolation (2019, submitted)
5. Dell'Accio, F., Di Tommaso, F., Hormann, K.: On the approximation order of triangular Shepard interpolation. IMA J. Numer. Anal. **36**, 359–379 (2016)
6. Fasshauer, G.E.: Meshfree Approximation Methods with Matlab. World Scientific Publishing Co., Singapore (2007)

7. Fasshauer, G.E., McCourt, M.J.: Kernel-Based Approximation Methods using Matlab. Interdisciplinary Mathematical Sciences, vol. 19, World Scientific Publishing Co., Singapore (2015)

8. Little, F.F.: Convex combination surfaces. In: Barnhill R.E., Boehm, W. (eds.) Surfaces in Computer Aided Geometric Design, Amsterdam, North-Holland, pp. 99–108 (1983)

9. Renka, R.J.: Multivariate interpolation of large sets of scattered data. ACM Trans. Math. Softw. **14**, 139–148 (1988)

10. Wendland, H.: Scattered Data Approximation. Cambridge Monographs on Applied and Computational Mathematics, vol. 17. Cambridge University Press, Cambridge (2005)

11. Zhang, M., Liang, X.-Z.: On a Hermite interpolation on the sphere. Appl. Numer. Math. **61**, 666–674 (2011)

Interpolation by Bivariate Quadratic Polynomials and Applications to the Scattered Data Interpolation Problem

Francesco Dell'Accio$^{(\boxtimes)}$ and Filomena Di Tommaso

Department of Mathematics and Computer Science, University of Calabria,
Rende, Italy
francesco.dellaccio@unical.it, ditommaso@mat.unical.it

Abstract. As specified by Little [7], the triangular Shepard method can be generalized to higher dimensions and to set of more than three points. In line with this idea, the hexagonal Shepard method has been recently introduced by combining six-points basis functions with quadratic Lagrange polynomials interpolating on these points and the error of approximation has been carried out by adapting, to the case of six points, the technique developed in [4]. As for the triangular Shepard method, the use of appropriate set of six-points is crucial both for the accuracy and the computational cost of the hexagonal Shepard method. In this paper we discuss about some algorithm to find useful six-tuple of points in a fast manner without the use of any triangulation of the nodes.

Keywords: Multinode Shepard methods · Rate of convergence · Approximation order

1 Introduction

Let $X = \{x_1, ..., x_n\}$ be a set of n distinct points of a compact convex domain $\Omega \subset \mathbb{R}^2$ with associated function evaluation data $f_i = f(x_i), i = 1, \ldots, n$. We assume that points in X are scattered, that is they do not obey any structure or order between their relative locations. The problem of reconstruction of an unknown function from these kind of data is well known and well studied in approximation theory and several methods both which require a mesh (here we mention that ones based on multivariate splines, finite elements, box splines) or meshless methods, mainly based on radial basis functions, have been developed with this goal and are successfully applied in different contexts. In this paper, however, we focus on some methods which are variations of the classic Shepard method (for a survey see [5]) and, in particular, to the triangular Shepard method. Our goal is to extend the triangular Shepard method to set of more than three points.

© Springer Nature Switzerland AG 2020
Y. D. Sergeyev and D. E. Kvasov (Eds.): NUMTA 2019, LNCS 11973, pp. 35–46, 2020.
https://doi.org/10.1007/978-3-030-39081-5_5

2   Triangular Shepard Method

Let $T = \{t_1, t_2, \ldots, t_m\}$ be a triangulation of X, where $t_j = [x_{j_1}, x_{j_2}, x_{j_3}]$ denotes the triangle with vertices $x_{j_1}, x_{j_2}, x_{j_3} \in X$. The *triangular Shepard interpolant* is defined by

$$K_\mu[f](x) = \sum_{j=1}^m B_{\mu,j}(x) L_j[f](x), \qquad x \in \Omega$$

where

$$L_j[f](x) = \lambda_{j,j_1}(x)f(x_{j_1}) + \lambda_{j,j_2}(x)f(x_{j_2}) + \lambda_{j,j_3}(x)f(x_{j_3})$$

is the linear polynomial based on the vertices of t_j in barycentric coordinates and the *triangular Shepard basis function* is given by

$$B_{\mu,j}(x) = \frac{\displaystyle\prod_{k=1}^3 |x - x_{j_k}|^{-\mu}}{\displaystyle\sum_{k=1}^m \prod_{l=1}^3 |x - x_{k_l}|^{-\mu}}, \qquad j = 1, \ldots, m, \quad \mu > 0.$$

The operator $K_\mu[f]$ interpolates $f(x_i)$ for each $i = 1, \ldots, n$ and reproduces linear polynomials. Little noticed that it surpasses Shepard's method greatly in aesthetic behaviour but did not give indications neither on the choice of the triangulation nor on the approximation order of the method. In a recent paper [4] we have tackled these two problems and noticed that in order to reach an adequate order of approximation, the triangulation must satisfy particular conditions. For instance, the triangles can form a Delaunay triangulation of X or a so called *compact triangulation*, that is they may overlap or being disjoint. In the second case T is composed by a significantly smaller number of triangles with respect to the Delaunay triangulation (see Figs. 1 and 2) and the accuracy of approximation of the two cases are comparable [4]. The routine to detect a compact triangulation can be organized in a fast algorithm using the localizing searching technique [2] and the computational cost to implement the triangular Shepard scheme is $O(n \log n)$ [1].

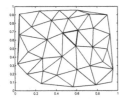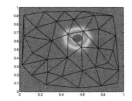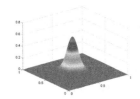

Fig. 1. Triangular Shepard basis function $B_{\mu,i}(x)$ with respect to the triangle in bold for the Delaunay triangulation of X

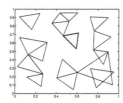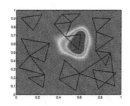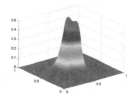

Fig. 2. Triangular Shepard basis function $B_{\mu,i}(x)$ with respect to the triangle in bold for the a compact triangulation of X

The goal of the paper is to discuss on the extension of the triangular Shepard method to a method based on six-point configurations of nodes in Ω. The interest relies in the better accuracy of approximation, with respect to the triangular Shepard method, provided that the local configurations of nodes are identified in a proper manner. At a first glance, however, it is not clear at all which are the configurations of six nodes that allow the solution of the Lagrange interpolation problem in a set of scattered data (without the computation of the Vandermonde determinant). The possible generalization of the triangular Shepard method to set of more than three points has already been announced by Little [7] without any suggestion on how to realize it.

3   Hexagonal Shepard Method

Let $S = \{s_1, s_2, \ldots, s_m\}$ be a cover of X by means of six-point subsets of X: each $s_j = \{x_{j_k}\}_{k=1,\ldots,6}$ is a set of pairwise distinct nodes $x_{j_1}, x_{j_2}, \ldots, x_{j_6} \in X$ and

$$\bigcup_{j=1}^{m} \{j_1, j_2, j_3, j_4, j_5, j_6\} = \{1, 2, \ldots, n\}.$$

With $S = \{s_1, s_2, \ldots, s_m\}$ we denote also the set of six-tuples $s_j = (x_{j_k})_{k=1,\ldots,6}$ that we identify with the hexagon (possibly with self-intersections) bounded by the finite chain of straight line segments $[x_{j_k}, x_{j_{k+1}}]$, $k = 1, \ldots, 6$, $x_{j_7} = x_{j_1}$, respectively. It will be clear from the context if we are dealing with subsets or six-tuples, depending on the need of the order of the nodes in the subsets. We assume that *the Lagrange interpolation problem associated to each s_j is poised in the space Π_2^2* of bivariate polynomials of degree less than or equal to 2, that is the Lagrange interpolation polynomial exists and it is unique. For each $\mu > 0$ the *hexagonal Shepard operator* is defined by

$$H_\mu [f](x) = \sum_{j=1}^{m} E_{\mu,j}(x) L_j[f](x), \qquad x \in \Omega$$

where $L_j[f](x)$ is the quadratic Lagrange interpolation polynomial on the six-tuple s_j, $j = 1, \ldots, m$ and the *hexagonal Shepard basis functions* with respect to the set S are given by

$$E_{\mu,j}(x) = \frac{\prod_{k=1}^{6} |x - x_{j_k}|^{-\mu}}{\sum_{k=1}^{m} \prod_{l=1}^{6} |x - x_{k_l}|^{-\mu}}, \qquad j = 1, \ldots, m, \quad \mu > 0.$$

The six-points Shepard basis functions satisfy the following properties:

1. $E_{\mu,j}(x) \geq 0$,
2. $\sum_{j=1}^{m} E_{\mu,j}(x) = 1$,
3. $E_{\mu,j}(x_i) = 0, \forall j = 1, \ldots, m, \ \forall i : x_i \notin s_j$,

As a direct consequence of properties 2 and 3 we have

$$\forall x_i \in s_j, \quad \sum_{j=1}^{m} E_{\mu,j}(x_i) = 1.$$

It follows that

(i) $H_\mu[f](x)$ interpolates function evaluations at each node x_i;
(ii) $H_\mu[f](x)$ reproduces all polynomials up to degree 2

It is of interest to see the graphic behaviour of the basis functions: for a six-tuple which has a point (in green) rounded by the others 5 (in red), the hexagonal basis function has the shape displayed Fig. 3. Analogously, for a six-tuple which has a point near to the boundary (in green) and the others 5 (in red) on its right side, the hexagonal basis function has the shape displayed Fig. 4.

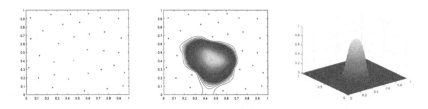

Fig. 3. Hexagonal basis function for a six-tuple which has a point (in green) rounded by the others 5 (in red) (Color figure online)

3.1 Quadratic Lagrange Interpolation and Detection of Appropriate Six-Tuple Set

The first problem to solve is that of the explicit representation of the quadratic polynomial interpolant on the points of a six-tuple $s_j = \{x_{j_i}\}_{i=1,\ldots,6} \in S$.

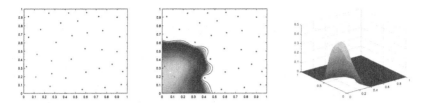

Fig. 4. Hexagonal basis function for a six-tuple which has a point near to the boundary (in green) and the others 5 (in red) on its right side (Color figure online)

A useful choice is the Lagrange form

$$L_j[f](x) = \sum_{i=1}^{6} \lambda_{j,j_i}(x) f(x_{j_i}), \quad j = 1, \ldots, m$$

which requires the computation of the Lagrange basis functions $\lambda_{j,j_i}(x)$, $i = 1, \ldots, 6$ through the Kronecker's delta property

$$\lambda_{j,j_i}(x_{j_k}) = \begin{cases} 1, & i = k \\ 0, & \text{otherwise.} \end{cases}$$

We then focus on the quadratic polynomial

$$\lambda_{j,j_1}(x) = \frac{\ell_1(x)}{\ell_1(x_{j_1})}$$

which vanishes at all points in s_j with the exception of x_{j_1}: its analytical expression can be written such that the remaining polynomials $\lambda_{j,j_i}(x)$, $i = 2, \ldots, 6$ are obtained by means of the permutation $i \rightarrow i \,(\mathrm{mod}\, 6) + 1$ of the indices $1, \ldots, 6$. We denote by

$$A(x, y, z) = \begin{vmatrix} 1 & x_1 & x_2 \\ 1 & y_1 & y_2 \\ 1 & z_1 & z_2 \end{vmatrix}$$

twice the signed area of triangle of vertices $x, y, z \in \mathbb{R}^2$. Since $\ell_1(x)$ is different from zero on x_{j_1} and vanishes on $x_{j_2}, \ldots, x_{j_6}$, its analytical expression can be obtained by linear combination

$$\ell_1(x) = \alpha Q_0(x) + \beta Q_1(x), \tag{1}$$

of the quadratic polynomials $Q_0(x) = A(x, x_{j_2}, x_{j_3}) A(x, x_{j_5}, x_{j_6})$ and $Q_1(x) = A(x, x_{j_2}, x_{j_6}) A(x, x_{j_3}, x_{j_5})$ both vanishing in $x_{j_2}, x_{j_3}, x_{j_5}, x_{j_6}$ (Figs. 5 and 6). The coefficients α and β which assure the vanishing of $\ell_1(x)$ in x_{j_4} are straightforward:

$$\alpha = A(x_{j_4}, x_{j_3}, x_{j_5}) A(x_{j_4}, x_{j_6}, x_{j_2}) \text{ and } \beta = A(x_{j_4}, x_{j_2}, x_{j_3}) A(x_{j_4}, x_{j_5}, x_{j_6})$$

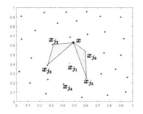
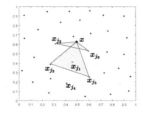

Fig. 5. Triangles involved in the definition of polynomials $Q_0(x)$ and $Q_1(x)$

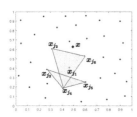
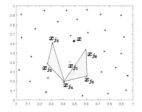

Fig. 6. Triangles involved in the definition of constants α and β

For reasons of stability and accuracy, the crucial point in the definition of the hexagonal Shepard method is the choice of the set of six-tuples S. We can take into account an algorithm proposed by J. Dalik in the paper [3] on 2008. In this paper he focuses on quadratic interpolation in vertices of unstructured triangulations of a bounded closed domain $\Omega \subset \mathbb{R}^2$ and gives conditions which guarantee the existence, the uniqueness and the optimal-order of local interpolation polynomials. More precisely, he proposes an algorithm to identify useful six-tuples of points which are vertices of a triangulation T of Ω satisfying specific properties: (i) the triangles of T must have no obtuse angles; (ii) the triangles of T must have area greater than, or equal to, a fixed constant times h^2, where h is the *meshsize* of T, i.e. the longest length of the sides of the triangles in T. The core of the algorithm is the following procedure which allows to select a set of 5 counterclockwise nodes around each interior vertex x_{j_1} of T in a set of ν nearby nodes as follows. Let us consider the ν nodes sharing with x_{j_1} a common side of the triangulation ordered counterclockwise.

- If $\nu = 5$ we associate to x_{j_1} the five nodes around it;
- if $\nu > 5$ we select the 5 nodes around x_{j_1} as follows: at each step k ($1 \leq k \leq \nu - 5$) we eliminate the node (in red in the Fig. 7) such that

$$\alpha_i + \alpha_{i+1} = min\{\alpha_j + \alpha_{j+1} : j = 1, \ldots, \nu - k\}$$

- if $\nu < 5$, according to the orientation, we add $5 - \nu$ nodes (in red in the Fig. 8) which share an edge with one of the ν triangles around x_{j_1}.

Nevertheless, to overcome the necessity of a preliminary triangulation, it is possible to use fast algorithms to find six-tuple of points around a node. As an example, if $x_j \in X$ and $N_j = \{x_{j_1}, \ldots, x_{j_\nu}\} \subset X$ is the sequence of $\nu \geq 5$

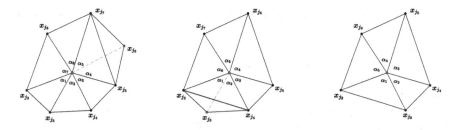

Fig. 7. Procedure to select a set of 5 counterclockwise nodes around x_{j_1} if $\nu > 5$

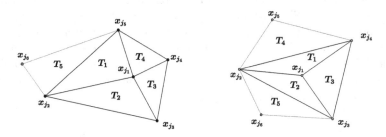

Fig. 8. Procedure to add $5 - n$ nodes which share an edge with one of the n triangles around x_{j_1} if $\nu < 5$ (Color figure online)

nodes nearest to x_j, enumerated with respect to the increasing distances from x_j, we can always re-enumerate the sequence in order that the nodes in N_j are counterclockwise around x_j. A simple algorithm can be organized as follows:

Algorithm 1. Re-enumerating neighbours of x_j counterclockwise around it

---

INPUT x_j, N_j
for all $l = 1 : \nu$ **do**
  $N_u(l, :) = [(x_{j_l} - x_j)/\|x_{j_l} - x_j\|, l]$
end for
sort $N_u(N_u(:, 2) \geq 0)$ in decreasing order with respect to the first column
sort $N_u(N_u(:, 2) < 0)$ in increasing order with respect to the first column
concatenate the two vectors $[N_j(N_u(:, 3)); N_j(N_u(:, 3))]$

In order to select the set of 5 counterclockwise nodes around each interior vertex x_j we can adapt the Dalik's procedure for maximizing the angles $x_{j_k}\widehat{x_j x}_{j_{k+1}}$, $k = 1, \ldots, \nu$, $x_{j_{\nu+1}} = x_{j_1}$ or use some other technique which allows to reduce the bound of the remainder term in the Lagrange interpolation on s_j (see Theorem 1).

In Table 1 we compare the approximation accuracies of $\mathcal{H}_4$ with those of the triangular Shepard operator K_2 using 1000 Halton interpolation nodes on $[0, 1]^2$. The six-tuple set S is computed by means of the Dalik's algorithm applied to the Delaunay triangulation of the nodes. The numerical experiments are realized by considering the set of test functions generally used in this field, defined in [8] (see Fig. 9).

The numerical results show that the accuracy of approximation achieved by $\mathcal{H}_4$ based on the Dalik's algorithm is comparable but not better than those of K_2. It is worth noting that it is not always possible to extract a strongly regular triangulation from X; moreover, by applying the Dalik's algorithm to the Delaunay triangulation of X, we observe the presence of six-tuples of points whose hexagons are not relatively small (see Fig. 10) and of a large number of hexagons.

In order to improve the accuracy of approximation of $\mathcal{H}_4$, in line with the case of the triangular Shepard method, we need to (i) avoid six-tuples s_j which contain nodes not relatively near; (ii) be able to reduce the number of hexagons of the cover S. Then it is necessary to find a procedure to compare two or more six-tuples of points sharing a common vertex. We reach our goal by minimizing the error bound for $L_j[f](x)$. In fact, the Lagrange interpolation polynomials at the points $\{\{x_{j_k}\}_{k=1}^6\}_{j=1}^m$ is given by

$$L_j[f](x) = \sum_{i=1}^6 \lambda_{j,j_i}(x)f(x_{j_i}), \quad j=1,\ldots,m$$

so that by considering the second order Taylor expansion of $f(x_{j_i})$, $i=2,\ldots,6$ centered at x_{j_1} with integral remainder, we get

$$L_j[f](x) = T_2[f, x_{j_1}](x) + \delta_j(x)$$

where

$$T_2[f, x_{j_1}](x) = \lambda_{j,j_1}(x)f(x_{j_1}) + \sum_{i=2}^6 \lambda_{j,j_i}(x)(f(x_{j_1}) + (x_{j_i} - x_{j_1}) \cdot \nabla f(x_{j_1})$$
$$+ (x_{j_i} - x_{j_1})Hf(x_{j_1})(x_{j_i} - x_{j_1})^T)$$

and

$$\delta_j(x) = \sum_{i=2}^6 \lambda_{j,j_i}(x) \frac{|x_{j_i} - x_{j_1}|^3}{2} \int_0^1 \frac{\partial^3 f(x_{j_1} + t(x_{j_i} - x_{j_1}))}{\partial \nu_{j_i}^3}(1-t)^2 \, dt.$$

Therefore

$$|f(x) - L_j[f](x)| \le |f(x) - T_2[f, x_{j_1}](x)| + |\delta_j(x)|.$$

We can bound the remainder term $\delta_j(x)$ [6].

Theorem 1. *For each $x \in \Omega$ we have*

$$|\delta_j(x)| \le \frac{1}{6}\|f\|_{2,1} K_j h_j |x - x_{j_{max}}|^2$$

where $|x - x_{j_{max}}| = \max_{i=2,\ldots,6} |x - x_{j_i}|$ and K_j is the sum of constants which control the shape of the triangles involved in the definition of the Lagrange basis functions.

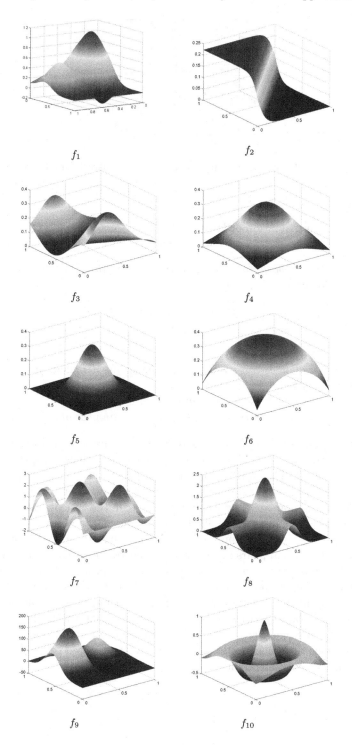

f_1

f_2

f_3

f_4

f_5

f_6

f_7

f_8

f_9

f_{10}

Fig. 9. Test functions

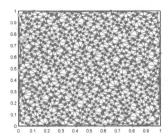

Fig. 10. Six-tuples of points whose hexagons are not relatively small (in red) (Color figure online)

The cubic approximation order of $H_\mu[f]$ is guaranteed by the following theorem [6].

Theorem 2. *Let Ω be a compact convex domain which contains X. If $f \in C^{2,1}(\Omega)$ then for each $\mu > 5/6$ we have*

$$||f - H_\mu[f]|| \leq KM ||f||_{2,1} h^3,$$

where

- $h > 0$ *is as smaller as the nodes and the hexagons distributions are uniform and the sizes of the hexagons are small;*
- $K > 0$ *is a as smaller as the constants K_j are*
- $M > 0$ *is as smaller as the maximum of the local number of hexagons is small.*

After the application of the Dalik's algorithm, we use the previous bounds to associate to each node x_i, which is vertex of two or more hexagons, the one with the smallest bound. This procedure certainly generates a cover $\{s_j\}_{j=1}^\nu$ of X by means of six-point subsets and the number of hexagons is considerably lower with respect to the number of nodes. For example, in the case of the 1000 Halton points, we observe the presence of 355 hexagons versus the 976 hexagons used to define the operator $\mathcal{H}_4$ based only on the Dalik's algorithm. As a consequence the constant M which appears in the previous Theorem is smaller since we avoid clustered hexagons. The accuracy of approximation produced by this operator are reported in the second column of Table 1. Finally, the results in the third column are obtained by applying the fast Algorithm 1 for re-enumerating 9 neighbours of x_j counterclockwise around it, by applying the Dalik's procedure to select a set of 5 counterclockwise nodes around x_j and by using previous bounds to associate to each node x_j, which is vertex of two or more hexagons, the one with the smallest bound.

Table 1. Numerical comparison among the approximation accuracies of the operator $\mathcal{H}_4$ on the six-tuples set generated by the Dalik's algorithm, the operator $\mathcal{H}_4$ on the six-tuples set generated by the refined Dalik's algorithm, the operator $\mathcal{H}_4$ on the six-tuples set generated by the fast algorithm and the operator K_2 on the compact triangulation of X

		$\mathcal{H}_4$ Dalik	$\mathcal{H}_4$ refined Dalik	$\mathcal{H}_4$ fast refined	K_2
f_1	e_{max}	8.93e−02	3.19e−03	7.86e−03	2.97e−02
	e_{mean}	2.53e−04	2.27e−04	2.73e−04	1.22e−03
	e_{MS}	2.00e−03	3.85e−04	5.13e−04	2.21e−03
f_2	e_{max}	1.45e−02	1.59e−02	4.43e−03	6.04e−03
	e_{mean}	9.18e−05	1.15e−04	1.09e−04	3.40e−04
	e_{MS}	4.47e−04	4.45e−04	2.89e−04	7.26e−04
f_3	e_{max}	4.42e−03	2.83e−04	4.66e−04	3.87e−03
	e_{mean}	2.09e−05	1.86e−05	2.14e−05	2.43e−04
	e_{MS}	1.09e−04	2.80e−05	3.28e−05	4.05e−04
f_4	e_{max}	7.86e−03	8.48e−04	7.49e−04	9.05e−03
	e_{mean}	2.10e−05	8.10e−06	8.42e−06	3.02e−04
	e_{MS}	2.24e−04	2.86e−05	2.36e−05	5.35e−04
f_5	e_{max}	1.81e−03	1.24e−03	2.57e−03	1.34e−02
	e_{mean}	4.17e−05	5.92e−05	6.58e−05	3.90e−04
	e_{MS}	9.06e−05	1.11e−04	1.34e−04	7.98e−04
f_6	e_{max}	4.93e−03	7.19e−04	6.41e−04	8.66e−03
	e_{mean}	3.46e−05	3.07e−05	3.29e−05	2.94e−04
	e_{MS}	1.81e−04	5.58e−05	5.64e−05	5.24e−04
f_7	e_{max}	1.19e+00	1.13e−01	1.11e−01	2.50e−01
	e_{mean}	5.27e−03	3.86e−03	4.44e−03	1.89e−02
	e_{MS}	3.13e−02	6.27e−03	7.53e−03	2.92e−02
f_8	e_{max}	2.96e−01	2.38e−02	3.36e−02	1.35e−01
	e_{mean}	1.07e−03	9.39e−04	1.10e−03	4.32e−03
	e_{MS}	6.82e−03	1.81e−03	2.12e−03	8.25e−03
f_9	e_{max}	1.20e+01	2.06e+00	1.48e+00	8.18e+00
	e_{mean}	4.94e−02	6.28e−02	7.20e−02	3.08e−01
	e_{MS}	2.18e−01	1.28e−01	1.46e−01	6.61e−01
f_{10}	e_{max}	1.72e−01	3.35e−02	6.88e−02	1.54e−01
	e_{mean}	8.49e−04	9.42e−04	1.09e−03	4.26e−03
	e_{MS}	4.15e−03	2.14e−03	2.65e−03	8.07e−03

4   Conclusion and Prospectives of Research

Using a procedure to select five nodes around a sixth one which guarantees the existence and uniqueness of the Lagrange interpolation and a criterion to compare the bounds of Lagrange interpolants on two or more six-tuples of points sharing a common vertex, we have generalized the triangular Shepard method to the hexagonal Shepard method, that is a fast method based on six points with cubic approximation order and which compares well with the triangular Shepard method in numerical accuracies. Further researches will be devoted to the generalization of the hexagonal Shepard method to the case of scattered data of $\mathbb{R}^3$.

Acknowledgments. This work was partially supported by the INdAM-GNCS 2019 research project "Kernel-based approximation, multiresolution and subdivision methods and related applications". This research has been accomplished within RITA (Research ITalian network on Approximation).

References

1. Cavoretto, R., De Rossi, A., Dell'Accio, F., Di Tommaso, F.: Fast computation of triangular Shepard interpolants. J. Comput. Appl. Math. **354**, 457–470 (2019)
2. Cavoretto, R., De Rossi, A., Perracchione, E.: Efficient computation of partition of unity interpolants through a block-based searching technique. Comput. Math. Appl. **71**(12), 2568–2584 (2016)
3. Dalík, J.: Optimal-order quadratic interpolation in vertices of unstructured triangulations. Appl. Math. **53**(6), 547–560 (2008)
4. Dell'Accio, F., Di Tommaso, F., Hormann, K.: On the approximation order of triangular Shepard interpolation. IMA J. Numer. Anal. **36**, 359–379 (2016)
5. Dell'Accio, F., Di Tommaso, F.: Scattered data interpolation by Shepard's like methods: classical results and recent advances. Dolomites Res. Notes Approx. **9**, 32–44 (2016)
6. Dell'Accio, F., Di Tommaso, F.: On the hexagonal Shepard method. Appl. Numer. Math. **150**, 51–64 (2020). https://doi.org/10.1016/j.apnum.2019.09.005. ISSN 0168-9274
7. Little, F.: Convex combination surfaces. In: Barnhill, R.E., Boehm, W. (eds.) Surfaces in Computer Aided Geometric Design, North-Holland, vol. 1479, pp. 99–108 (1983)
8. Renka, R.J.: Algorithm 790: CSHEP2D: cubic shepard method for bivariate interpolation of scattered data. ACM Trans. Math. Softw. **25**(1), 70–73 (1999)

Comparison of Shepard's Like Methods with Different Basis Functions

Francesco Dell'Accio[1] (ID), Filomena Di Tommaso[1][(✉)] (ID),
and Domenico Gonnelli[2]

[1] Department of Mathematics and Computer Science, University of Calabria,
Via P. Bucci, Cubo 30A, Rende, Italy
{francesco.dellaccio,filomena.ditommaso}@unical.it
[2] NTT DATA Italia, Contrada Cutura, Rende, Italy
Domenico.Gonnelli@nttdata.com

Abstract. The problem of reconstruction of an unknown function from a finite number of given scattered data is well known and well studied in approximation theory. The methods developed with this goal are several and are successfully applied in different contexts. Due to the need of fast and accurate approximation methods, in this paper we numerically compare some variation of the Shepard method obtained by considering different basis functions.

Keywords: Shepard methods · Interpolation · Quasi-interpolation

1 Introduction

Let be X a convex domain of $\mathbb{R}^2$, $X_n = \{x_1, x_2, \ldots, x_n\} \subset X$ a set of nodes and $F_n = \{f_1, \ldots, f_n\} \subset \mathbb{R}$ a set of associated function values. The problem of reconstruction of a continuous function from the data (X_n, F_n) is well known and well studied in approximation theory. When the points of X_n bear no regular structure at all, we talk about scattered data approximation problem [17]. Several methods have been developed with this goal and are successfully applied in different contexts. Some of these methods require a mesh, some others are meshless, mainly based on radial basis functions (see for example [11,12]).

Recently, it has been pointed out the need of fast approximation methods which overcome the high computational cost and the slowness of interpolation schemes which entail the solution of large linear systems or the use of elaborated mathematical procedures to find the values of parameters needed for setting those schemes [15]. The Shepard method [16] and some of its variations [4,15] belong to this class of methods.

The Shepard scheme consists in the construction of the function

$$x \to S_{\phi,n}[f](x) = \frac{\displaystyle\sum_{j=1}^{n} \phi(x - x_j)f_j}{\displaystyle\sum_{j=1}^{n} \phi(x - x_j)} \tag{1}$$

Y. D. Sergeyev and D. E. Kvasov (Eds.): NUMTA 2019, LNCS 11973, pp. 47–55, 2020.
https://doi.org/10.1007/978-3-030-39081-5_6

to approximate a target function f in X whose values $f_j = f(x_j)$ at the nodes are known. In the original scheme [16] ϕ is the λ-power ($\lambda > 0$) of the inverse distance function $x \to |x|^{-1}$ in the Euclidean space $\mathbb{R}^2$. Since ϕ has a singularity in $x = 0$, the Shepard approximant (1) interpolates the values of the target function at the node x_j, $j = 1, \ldots, n$, but the reproduction quality of the interpolation operator $S_{\phi,n}[\cdot]$ is limited to constant functions. Moreover its approximation accuracy is poor and the presence of flat spots in the neighborhood of all nodes, if $\lambda \geq 1$, or cusps at each x_j, if $\lambda < 1$, is not visually pleasing. In the years several variations of the Shepard method have been proposed to improve its approximation accuracy. They are based on the substitution of the function values f_j with the values at x of interpolation polynomials at x_j and on the use of local support basis functions. These schemes are called combined Shepard methods (for more details see [10] and the references therein) and give the possibility to interpolate additional derivative data if these are provided [1–3,6,7,9]. More recent methods focus both on readily implementation and efficiency of the approximation. The scaled Shepard methods [15] are based on the general scheme (1) but, instead of $|x|^{-1}$, they make use of basis functions $K(x)$ satisfying

- $K : \mathbb{R}^2 \to \mathbb{R}_+$ is continuous on $\mathbb{R}^2$;
- $\min\limits_{|x| \leq 1} K(x) > 0$;
- for a fixed $\alpha > 0$, there holds

$$K(x) \leq \kappa \left(1 + |x|^2\right)^{-\alpha}, \quad x \in \mathbb{R}^2, \tag{2}$$

where κ is a constant independent on x.

For the optimality of the approximation order, they take into account the fill distance [17]

$$h_{X_n} = \sup_{x \in X} \inf_{y \in X_n} |x - y|$$

through a dilation factor β_n, by setting $\phi(x) = K(\beta_n x)$. The continuity of the basis function at $x = 0$ causes the lost of the interpolation feature and the resulting approximant is *quasi-interpolant*, since it preserves the reproduction property of constant functions.

Another improvement of the Shepard method (1) is the triangular Shepard operator [13], obtained by combining triangle-based basis functions with local linear interpolants on the vertices of a triangulation of X_n. The triangular Shepard is an interpolation operator which reproduces polynomials up to the degree 1.

Being motivated by the need of fast and accurate approximation methods, as specified above, in this paper we provide an explicit numerical comparison among the accuracies and the CPU times (in seconds) of the Shepard, scaled Shepard and triangular Shepard approximants. In Sect. 2, after recalling the definitions of the parameters necessary for rightly setting the scaled Shepard method, we make a comparison among Shepard and scaled Shepard approximants on a set of 10 test functions commonly used in this field [14]. In Sect. 3 we compare the triangular Shepard method with its scaled versions. All tests have been carried out on a laptop with an Intel(R) Core i7 5500U CPU 2.40 GHz processor and 8.00 GB RAM. Finally, in Sect. 4 we formulate conclusions about our investigations.

2   Scaled Shepard Approximants

As pointed out in [15], the scaled Shepard approximants perform better if the points of X_n are *well-separated* and X_n is *quasi-uniformly distributed* in X. These conditions are fulfilled as soon as constants c and C, independent on n, exist such that

$$\inf_{\substack{x,y \in X_n \\ x \neq y}} |x - y| \geq cn^{-1/2}$$

and

$$h_{X_n} \leq Cn^{-1/2}.$$

Under these assumption, the dilation factor $\beta_n = C^{-1}\sqrt{n} \geq 1$ can be computed and the condition (2) is equivalent to

$$K(\beta_n x) \leq \kappa \min\left(1, (\beta_n |x|)^{-2\alpha}\right), \quad x \in \mathbb{R}^2. \tag{3}$$

The scaled Shepard approximant is

$$x \rightarrow S_{K,n}[f](x) = \frac{\displaystyle\sum_{j=1}^{n} K(\beta_n(x - x_j))f(x_j)}{\displaystyle\sum_{j=1}^{n} K(\beta_n(x - x_j))}. \tag{4}$$

The following numerical experiments confirm the better accuracy of approximation achieved by the quasi-interpolant operator (4) with respect to those of the original Shepard operator (1) according to the theoretical results proven in [15]. The experiments are realized by considering the following functions

$$
\begin{aligned}
K_1(x) &= e^{-a^2|x|^2} &&\text{Gaussian function,} \\
K_2(x) &= e^{-a|x|} &&C^0\text{-Matérn function,} \\
K_3(x) &= \left(1 + |x|^2\right)^{-a} &&\text{Generalized inverse multiquadric function}
\end{aligned} \tag{5}
$$

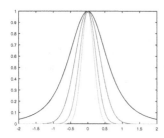 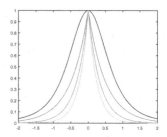

Fig. 1. Behaviour of function $K_1(x)$ (left) and $K_2(x)$ (right) for $a = 2, 3, 4$ (in blue, red and yellow) and of function $(1 + |x|^2)^{-2}$ (in black). (Color figure online)

which are continuous and positive in $\mathbb{R}^2$ with positive minimum in the closed ball $|x| \leq 1$ for each $a > 0$. In the case of the Gaussian function $K_1(x)$, by setting $|x| = r$, condition (2) becomes

$$e^{-a^2 r^2} \leq \kappa(1 + r^2)^{-\alpha}, r > 0$$

and for $r = 1$ we get $e^{-a^2} < \kappa 2^{-\alpha}$, which is satisfied for each $a^2 \geq \alpha \geq 1$ and $\kappa = 1$. In Fig. 1 (left) we display the behaviour of the function $K_1(x)$ for the values of parameter $a = 2, 3, 4$ (in blue, red and yellow) and of the function $(1 + |x|^2)^{-2}$ (in black). Analogously, in the case of the C^0-Matérn function $K_2(x)$, condition (2) becomes

$$e^{-ar} \leq \kappa(1 + r^2)^{-\alpha}, r > 0$$

and for $r = 1$ we get $e^{-a} < \kappa 2^{-\alpha}$ which is satisfied for each $a \geq \alpha \geq 1$ and $\kappa = 1$. In Fig. 1 (right) we display the behaviour of function $K_2(x)$ for $a = 2, 3, 4$ (in blue, red and yellow) and of function $(1 + |x|^2)^{-2}$ (in black). Finally, in the case of the generalized inverse multiquadric function $K_3(x)$, condition (2) is trivially satisfied for each $a \geq \alpha$ and $\kappa = 1$. In Table 1 we compare the approximation accuracies of the original Shepard scheme $S_{|\cdot|^{-2}, n}[f](x)$ with those of the quasi-interpolant schemes $S_{K_j, n}[f](x)$, $j = 1, 2, 3$ by setting $a = 50$. The numerical experiments are performed by considering the 10 test functions defined in [14]. In Table 1 we report the absolute values of the maximum error e_{max}, the average error e_{mean}, and the mean square error e_{MS} computed by using the $n = 1000$ Halton interpolation nodes [18] in the unit square $[0, 1] \times [0, 1]$, for which we report the value $\beta_n \approx 2.6087$. The errors are computed at the $n_e = 101 \times 101$ points of a regular grid of the square $[0, 1] \times [0, 1]$. The obtained numerical results show that quasi-interpolant operators $S_{K_j, n}$, $j = 1, 2, 3$ perform better than the original Shepard method $S_{|\cdot|^{-2}, n}$ but similar to the operator $S_{|\cdot|^{-6}, n}$. It is then meaningful to compare above operators in terms of CPU times and we report the results, computed in seconds, in Table 2.

Table 1. Comparison among the original Shepard approximant $S_{|\cdot|^{-2},n}$, $S_{|\cdot|^{-6},n}$ and the quasi-interpolant Shepard schemes $S_{K_i,n}, i = 1, 2, 3$ with parameter $a = 50$.

		$S_{|\cdot|^{-2},n}$	$S_{|\cdot|^{-6},n}$	$S_{K_1,n}$	$S_{K_2,n}$	$S_{K_3,n}$
f_1	e_{max}	$2.26e-1$	$7.03e-2$	$8.48e-2$	$6.81e-2$	$7.75e-2$
	e_{mean}	$3.71e-2$	$7.42e-3$	$9.13e-3$	$5.13e-3$	$8.74e-3$
	e_{MS}	$5.02e-2$	$1.13e-2$	$1.42e-2$	$8.16e-3$	$1.33e-2$
f_2	e_{max}	$4.66e-2$	$3.36e-2$	$3.87e-2$	$3.25e-2$	$3.13e-2$
	e_{mean}	$1.56e-2$	$1.74e-3$	$2.12e-3$	$1.26e-3$	$2.82e-3$
	e_{MS}	$1.76e-2$	$3.89e-3$	$4.86e-3$	$2.96e-3$	$4.90e-3$
f_3	e_{max}	$9.57e-2$	$2.16e-2$	$2.45e-2$	$2.13e-2$	$2.77e-2$
	e_{mean}	$1.48e-2$	$2.94e-3$	$3.63e-3$	$1.98e-3$	$2.69e-3$
	e_{MS}	$1.98e-2$	$4.00e-3$	$5.03e-3$	$2.83e-3$	$4.15e-3$
f_4	e_{max}	$6.67e-2$	$1.45e-2$	$1.65e-2$	$1.42e-2$	$1.70e-2$
	e_{mean}	$1.59e-2$	$3.19e-3$	$3.91e-3$	$2.20e-3$	$3.07e-3$
	e_{MS}	$1.97e-2$	$3.90e-3$	$4.89e-3$	$2.82e-3$	$4.29e-3$
f_5	e_{max}	$1.08e-1$	$2.22e-2$	$3.40e-2$	$1.96e-2$	$1.86e-2$
	e_{mean}	$1.46e-2$	$2.42e-3$	$3.03e-3$	$1.62e-3$	$2.41e-3$
	e_{MS}	$2.12e-2$	$4.23e-3$	$5.43e-3$	$2.90e-3$	$3.99e-3$
f_6	e_{max}	$1.52e-1$	$3.74e-2$	$3.58e-2$	$3.78e-2$	$5.86e-2$
	e_{mean}	$1.37e-2$	$3.29e-3$	$3.96e-3$	$2.39e-3$	$4.07e-3$
	e_{MS}	$1.91e-2$	$4.58e-3$	$5.45e-3$	$3.74e-3$	$6.86e-3$
f_7	e_{max}	$1.41e+0$	$6.93e-1$	$6.38e-1$	$6.84e-1$	$9.48e-1$
	e_{mean}	$2.94e-1$	$7.99e-2$	$9.65e-2$	$5.79e-2$	$1.24e-1$
	e_{MS}	$3.80e-1$	$1.03e-1$	$1.27e-1$	$7.81e-2$	$1.57e-1$
f_8	e_{max}	$8.43e-1$	$1.86e-1$	$2.62e-1$	$1.52e-1$	$2.10e-1$
	e_{mean}	$1.10e-1$	$2.15e-2$	$2.66e-2$	$1.48e-2$	$2.81e-2$
	e_{MS}	$1.50e-1$	$3.35e-2$	$4.25e-2$	$2.35e-2$	$4.02e-2$
f_9	e_{max}	$5.27e+1$	$1.69e+1$	$2.39e+1$	$1.48e+1$	$1.31e+1$
	e_{mean}	$7.65e+0$	$1.53e+0$	$1.88e+0$	$1.06e+0$	$1.97e+0$
	e_{MS}	$1.10e+1$	$2.60e+0$	$3.28e+0$	$1.86e+0$	$3.18e+0$
f_{10}	e_{max}	$5.82e-1$	$1.45e-1$	$2.14e-1$	$1.16e-1$	$3.14e-1$
	e_{mean}	$6.78e-2$	$1.59e-2$	$1.93e-2$	$1.14e-2$	$2.68e-2$
	e_{MS}	$9.18e-2$	$2.38e-2$	$3.01e-2$	$1.68e-2$	$3.64e-2$

3 Scaled Triangular Shepard Approximants

The triangular Shepard scheme [13] requires the definition of a triangulation $T = \{t_1, \ldots, t_m\}$ of the node set X_n and, in line with previous notation, it consists in the construction of the function

Table 2. CPU times in seconds computed on the 1000 Halton interpolation points in the case of function f_1.

Operator	CPU time
$S_{|\cdot|^2,n}$	0.2928
$S_{|\cdot|^{-6},n}$	1.0640
$S_{K_1,n}$	0.7624
$S_{K_2,n}$	0.8485
$S_{K_3,n}$	1.5148

$$x \to T_{\phi,m}[f](x) = \frac{\sum_{j=1}^{m} \prod_{\ell=1}^{3} \phi(x - x_{j_\ell}) L_j[f](x)}{\sum_{k=1}^{m} \prod_{\ell=1}^{3} \phi(x - x_{k_\ell})} \tag{6}$$

where ϕ is the λ-power ($\lambda > 0$) of the inverse distance function $x \to |x|^{-1}$ in the Euclidean space $\mathbb{R}^2$ and $L_j[f](x)$ is the linear interpolation polynomial at the vertices $\{x_{j_1}, x_{j_2}, x_{j_3}\}$ of the triangle $t_j \in T$. The triangular Shepard scheme reaches better accuracy of approximation with respect to the original Shepard one both for a Delaunay triangulation or a compact triangulation of the node set X_n. A compact triangulation of X_n, in particular, consists of a set of triangles which may overlap or being disjoint. These triangles are determined in order to reduce the error bound of the local linear interpolant and its number is about $1/3$ the number of triangles in the Delaunay triangulation [8]. The routine to detect a compact triangulation can be organized in a fast algorithm using the localizing searching technique developed in [5] and the computational cost to implement the scheme (6) is $O(n \log n)$ [4]. In line with Sect. 2, we can consider the quasi-interpolants

$$x \to T_{K,m}[f](x) = \frac{\sum_{j=1}^{m} \prod_{\ell=1}^{3} K(\beta_n(x - x_{j_\ell})) L_j[f](x)}{\sum_{k=1}^{m} \prod_{\ell=1}^{3} K(\beta_n(x - x_{k_\ell}))} \tag{7}$$

which reproduce polynomials up to the degree 1 and numerically compare, in accuracy of approximation and CPU time, these operators with the triangular Shepard one. We use the same data and functions of previous experiments and report the obtained numerical results in Tables 3 and 4.

Table 3. Comparison among the triangular Shepard approximant $T_{|\cdot|^{-2},m}$ and the quasi-interpolant triangular Shepard schemes $T_{K_i,n}, i = 1, 2, 3$ with parameter $\beta = 50$.

		$T_{|\cdot|^{-2},m}$	$T_{K_1,m}$	$T_{K_2,m}$	$T_{K_3,m}$
f_1	e_{max}	$1.62e-2$	$1.81e-2$	$1.46e-2$	$1.90e-2$
	e_{mean}	$1.18e-3$	$1.63e-3$	$1.36e-3$	$1.57e-3$
	e_{MS}	$1.99e-3$	$2.55e-3$	$2.11e-3$	$2.53e-3$
f_2	e_{max}	$6.78e-3$	$9.20e-3$	$8.01e-3$	$6.49e-3$
	e_{mean}	$3.48e-4$	$4.82e-4$	$4.00e-4$	$4.47e-4$
	e_{MS}	$7.42e-4$	$1.01e-3$	$8.41e-4$	$8.85e-4$
f_3	e_{max}	$5.14e-3$	$5.92e-3$	$3.77e-3$	$4.79e-3$
	e_{mean}	$2.93e-4$	$3.99e-4$	$3.34e-4$	$3.71e-4$
	e_{MS}	$4.95e-4$	$5.94e-4$	$4.95e-4$	$5.87e-4$
f_4	e_{max}	$3.22e-3$	$2.96e-3$	$2.54e-3$	$2.93e-3$
	e_{mean}	$2.38e-4$	$3.19e-4$	$2.67e-4$	$3.12e-4$
	e_{MS}	$3.78e-4$	$4.50e-4$	$3.78e-4$	$4.57e-4$
f_5	e_{max}	$1.12e-2$	$9.99e-3$	$9.53e-3$	$1.08e-2$
	e_{mean}	$3.91e-4$	$5.59e-4$	$4.61e-4$	$4.98e-4$
	e_{MS}	$7.91e-4$	$1.01e-3$	$8.27e-4$	$9.72e-4$
f_6	e_{max}	$8.37e-3$	$5.95e-3$	$6.91e-3$	$7.10e-3$
	e_{mean}	$2.92e-4$	$3.19e-4$	$2.74e-4$	$3.95e-4$
	e_{MS}	$4.81e-4$	$4.45e-4$	$3.96e-4$	$5.35e-4$
f_7	e_{max}	$2.42e-1$	$3.05e-1$	$2.49e-1$	$2.33e-1$
	e_{mean}	$1.86e-2$	$2.58e-2$	$2.17e-2$	$2.38e-2$
	e_{MS}	$2.75e-2$	$3.65e-2$	$3.05e-2$	$3.29e-2$
f_8	e_{max}	$1.16e-1$	$1.15e-1$	$1.08e-1$	$1.17e-1$
	e_{mean}	$4.31e-3$	$6.07e-3$	$5.06e-3$	$5.61e-3$
	e_{MS}	$8.04e-3$	$1.07e-2$	$8.85e-3$	$9.82e-3$
f_9	e_{max}	$5.73e+0$	$8.07e+0$	$5.43e+0$	$5.88e+0$
	e_{mean}	$2.91e-1$	$4.27e-1$	$3.55e-1$	$3.71e-1$
	e_{MS}	$5.65e-1$	$7.82e-1$	$6.42e-1$	$6.97e-1$
f_{10}	e_{max}	$1.36e-1$	$1.71e-1$	$1.35e-1$	$1.56e-1$
	e_{mean}	$4.20e-3$	$5.81e-3$	$4.91e-3$	$5.44e-3$
	e_{MS}	$7.66e-3$	$9.64e-3$	$8.30e-3$	$9.78e-3$

Table 4. CPU times in seconds computed on the 1000 Halton interpolation points in the case of function f_1.

Operator	CPU time
$T_{|\cdot|^2,n}$	0.5610
$T_{K_1,n}$	1.6901
$T_{K_2,n}$	1.6012
$T_{K_3,n}$	2.1759

4   Conclusion

In this paper we numerically compared classic Shepard method with some of its variations obtained by considering different basis functions. The classic Shepard method with exponent $\lambda = 2$ is the fastest but compares worse with respect to all others. For $\lambda = 6$ its approximation accuracy is comparable with that one of the quasi-interpolants recently introduced in [15] but the CPU time increases notably. At the end of the day, the triangular Shepard method is a good compromise between computational efficiency and accuracy of approximation, as demonstrated by comparing all numerical tables.

Acknowledgments. This work was partially supported by the INdAM-GNCS 2019 research project "Kernel-based approximation, multiresolution and subdivision methods and related applications". This research has been accomplished within RITA (Research ITalian network on Approximation).

References

1. Caira, R., Dell'Accio, F., Di Tommaso, F.: On the bivariate Shepard-Lidstone operators. J. Comput. Appl. Math. **236**(7), 1691–1707 (2012)
2. Cătinaş, T.: The combined Shepard-Lidstone bivariate operator. In: Mache, D.H., Szabados, J., de Bruin, M.G. (eds.) Trends and Applications in Constructive Approximation, pp. 77–89. Birkhäuser Basel, Basel (2005)
3. Cătinaş, T.: The bivariate Shepard operator of Bernoulli type. Calcolo **44**(4), 189–202 (2007)
4. Cavoretto, R., De Rossi, A., Dell'Accio, F., Di Tommaso, F.: Fast computation of triangular Shepard interpolants. J. Comput. Appl. Math. **354**, 457–470 (2019)
5. Cavoretto, R., De Rossi, A., Perracchione, E.: Efficient computation of partition of unity interpolants through a block-based searching technique. Comput. Math. Appl. **71**(12), 2568–2584 (2016)
6. Costabile, F.A., Dell'Accio, F., Di Tommaso, F.: Enhancing the approximation order of local Shepard operators by Hermite polynomials. Comput. Math. Appl. **64**(11), 3641–3655 (2012)
7. Costabile, F.A., Dell'Accio, F., Di Tommaso, F.: Complementary Lidstone interpolation on scattered data sets. Numer. Algorithms **64**(1), 157–180 (2013)
8. Dell'Accio, F., Di Tommaso, F., Hormann, K.: On the approximation order of triangular Shepard interpolation. IMA J. Numer. Anal. **36**, 359–379 (2016)

9. Dell'Accio, F., Di Tommaso, F.: Complete Hermite-Birkhoff interpolation on scattered data by combined Shepard operators. J. Comput. Appl. Math. **300**, 192–206 (2016)
10. Dell'Accio, F., Di Tommaso, F.: Scattered data interpolation by Shepard's like methods: classical results and recent advances. Dolomites Res. Notes Approx. **9**, 32–44 (2016)
11. Fasshauer, G.E.: Meshfree Approximation Methods with Matlab. World Scientific Publishing Co. Inc., Singapore (2007)
12. Franke, R.: Scattered data interpolation: tests of some methods. Math. Comput. **38**(157), 181–200 (1982)
13. Little, F.: Convex combination surfaces. In: Barnhill, R.E., Boehm, W. (eds.) Surfaces in Computer Aided Geometric Design, North-Holland, vol. 1479, pp. 99–108 (1983)
14. Renka, R.J.: Algorithm 790: CSHEP2D: cubic Shepard method for bivariate interpolation of scattered data. ACM Trans. Math. Softw. **25**(1), 70–73 (1999)
15. Senger, S., Sun, X., Wu, Z.: Optimal order Jackson type inequality for scaled Shepard approximation. J. Approx. Theor. **227**, 37–50 (2018)
16. Shepard, D.: A two-dimensional interpolation function for irregularly-spaced data. In: Proceedings of the 1968 23rd ACM National Conference, ACM 1968, pp. 517–524. ACM, New York (1968)
17. Wendland, H.: Scattered Data Approximation, vol. 17. Cambridge University Press, Cambridge (2004)
18. Wong, T.T., Luk, W.S., Heng, P.A.: Sampling with Hammersley and Halton points. J. Graph. Tools **2**(2), 9–24 (1997)

A New Remez-Type Algorithm for Best Polynomial Approximation

Nadaniela Egidi[ID], Lorella Fatone[✉][ID], and Luciano Misici[ID]

Dipartimento di Matematica, Università degli Studi di Camerino,
Via Madonna delle Carceri 9, 62032 Camerino, MC, Italy
{nadaniela.egidi,lorella.fatone,luciano.misici}@unicam.it

Abstract. The best approximation problem is a classical topic of the approximation theory and the Remez algorithm is one of the most famous methods for computing minimax polynomial approximations. We present a slight modification of the (second) Remez algorithm where a new approach to update the trial reference is considered. In particular at each step, given the local extrema of the error function of the trial polynomial, the proposed algorithm replaces all the points of the trial reference considering some "ad hoc" oscillating local extrema and the global extremum (with its adjacent) of the error function. Moreover at each step the new trial reference is chosen trying to preserve a sort of equidistribution of the nodes at the ends of the approximation interval. Experimentally we have that this method is particularly appropriate when the number of the local extrema of the error function is very large. Several numerical experiments are performed to assess the real performance of the proposed method in the approximation of continuous and Lipschitz continuous functions. In particular, we compare the performance of the proposed method for the computation of the best approximant with the algorithm proposed in [17] where an update of the Remez ideas for best polynomial approximation in the context of the chebfun software system is studied.

Keywords: Best polynomial approximation · Remez algorithm · Interpolation

1 Introduction

A classical problem of the approximation theory, going back to Chebyshev himself, is to look for a polynomial among those of a fixed degree that minimizes the deviation in the supremum norm from a given continuous function on a given interval. It is known that this polynomial exists and is unique, and is known as the best, uniform, Chebyshev or minimax approximation to the function.

In 1934, Evgeny Yakovlevich Remez published in a series of three papers the algorithm, that now bears his name, for the computation of the best polynomial approximation [20–22]. With the help of three female students at the University of Kiev, Remez obtained the coefficients and the errors of the best approximations to $|x|$ by polynomials of degree 5, 7, 9, and 11 accurate to about 4 decimal

© Springer Nature Switzerland AG 2020
Y. D. Sergeyev and D. E. Kvasov (Eds.): NUMTA 2019, LNCS 11973, pp. 56–69, 2020.
https://doi.org/10.1007/978-3-030-39081-5_7

places. In that period both theoretical properties and computational aspects of best polynomial approximation were deeply investigated and the Remez algorithm can be considered as one of the first nontrivial algorithms for a nonlinear computational problem that was solved before the invention of computers.

With the advent of computers, in the mid-20th century, researches on the computation of best approximations increased (some publications report examples of best approximations with degrees of a couple of dozens) but, after this period, the interest diminished probably due to the lack of minimax approximation applications aside from digital signal processing and approximation of special functions.

In recent years best polynomial approximation became a mandatory topic in books of approximation theory but not always as useful as one might image in practical applications. In fact, in practice, other types of polynomial interpolation (for example Chebyshev interpolants) are often as good or even better minimax polynomial approximations (see, e.g. [1,15]). Nevertheless the best approximation is a fundamental topic of approximation theory, going back more than 150 years, and we believe that the study and development of robust algorithms to compute this kind of approximation is an interesting line of research. Moreover best approximation is a challenging problem and some new additional ideas sometimes could indicate directions for its improvement.

In this paper we present a new version of the Remez algorithm that increases its robustness (see [10]). Moreover this new version is reliable and able to compute high accuracy best polynomial approximations with degrees in the hundreds or thousands to different functions. In particular we propose a slight modification of the Remez algorithm presented by the authors in [17] in the context of the chebfun software system (see [2,17] and, for more details, https://www.chebfun. org).

The paper is organized as follows. In Sect. 2 we review the classical Remez algorithm specifying how the computation of a trial reference and a trial polynomial is performed. In particular the Remez algorithm proposed in the chebfun software system is presented. Section 3 shows a new strategy to adjust the trial reference from the error of the trial polynomial. Finally in Sect. 4 we compare the performance of the chebfun Remez algorithm and the slight variation proposed here.

2   Classical Remez Algorithm

Let $\mathbb{P}_n$ be the set of polynomials of degree less than or equal to $n \in \mathbb{N}$ and having real coefficients. Let f be a continuous function defined on a finite closed interval $I = [a,b]$, $a, b \in \mathbb{R}$, $a < b$, that is $f \in \mathcal{C}(I)$, and let $\|\cdot\|_\infty$ be the supremum norm of $\cdot$ on I, that is

$$\|f\|_\infty = \max_{x \in I} |f(x)|. \tag{1}$$

The classical problem of best polynomial approximation can be stated as follow: given $f \in \mathcal{C}(I)$ find $q_n^* \in \mathbb{P}_n$ such that:

$$\|f - q_n^*\|_\infty \leq \|f - p_n\|_\infty, \quad \forall p_n \in \mathbb{P}_n. \tag{2}$$

It is well known that in $\mathcal{C}(I)$ there exists a unique best uniform approximation polynomial $q_n^* \in \mathbb{P}_n$ of f (see, for example, [1, 6, 14–16, 19] and the references therein).

Since the operator that assigns to each continuous function its best polynomial approximation q_n^*, although continuous, is nonlinear (see, e.g., [14]), iterative methods have been developed to compute q_n^*. The Remez algorithm is one of these methods (see [20–22]). Given an initial grid, the Remez algorithm consists of an iterative procedure that, modifying at every step the grid of trial interpolants, converges quadratically to q_n^*, under some suitable assumptions on f.

The Remez algorithm is essentially based on the following two theorems.

Theorem 1. *(Chebyshev Equioscillation Theorem) Let $f \in \mathcal{C}(I)$. A polynomial $q_n \in \mathbb{P}_n$ is the best approximation to f (that is $q_n = q_n^*$) if and only if there exists a set of $n + 2$ distinct points, $a \leq x_0 < x_1 < \cdots < x_{n+1} \leq b$ in I such that*

$$f(x_i) - q_n(x_i) = \lambda \sigma_i \|f - q_n^*\|_\infty, \qquad i = 0, 1, \ldots, n + 1, \tag{3}$$

where $\sigma_i = (-1)^i$ and $\lambda = 1$ or $\lambda = -1$ is a fixed constant.

For the proof of this fundamental theorem (in original or generalized form) see, for example, [4–6, 8, 12, 13, 16, 18].

The above theorem concerns the property of equioscillation of $f - q_n^*$. Next theorem provides a lower estimate of the best approximation error and it is very useful in numerical methods for finding the polynomial of best uniform approximation.

Theorem 2. *(De La Vallée Poussin) Let $f \in \mathcal{C}(I)$. Suppose we have a polynomial $q_n \in \mathbb{P}_n$ which satisfies*

$$f(y_i) - q_n(y_i) = (-1)^i e_i, \qquad i = 0, 1, \ldots, n + 1, \tag{4}$$

where $a \leq y_0 < y_1 < \cdots < y_{n+1} \leq b$, and all e_i, $i = 0, 1, \ldots, n + 1$, are nonzero and of the same sign. Then, for every $p_n \in \mathbb{P}_n$, we have:

$$\min_i |f(y_i) - q_n(y_i)| \leq \max_i |f(y_i) - p_n(y_i)|, \tag{5}$$

and, in particular:

$$\min_i |f(y_i) - q_n(y_i)| \leq \|f - q_n^*\|_\infty \leq \|f - q_n\|_\infty. \tag{6}$$

For the proof see [7].

We define a set of $n + 2$ ordered points $\underline{x} = (x_0, x_1, \ldots, x_{n+1})$ in I that satisfies (3) a reference equioscillation set or simply a reference.

Estimates of the best uniform approximation error in terms of the smoothness of the function f can be found in [1] (see also [15]).

Let

$$r_n(x) = f(x) - q_n(x), \quad x \in I = [a, b],$$

be the remainder (i.e. the error function) when we approximate f with $q_n \in \mathbb{P}_n$. If $r_n(a) \neq 0$ and $r_n(b) \neq 0$ and r_n oscillates exactly $n + 2$ times then r_n is called standard remainder, otherwise it is called not standard remainder. We note that when r_n is standard then it has exactly $n+2$ local minima or maxima. Chebyshev equioscillation theorem ensures that the residual $r_n^* = f - q_n^*$ is standard.

From Theorems 1 and 2 we have that a polynomial $q_n \in \mathbb{P}_n$ whose error function oscillates $n + 2$ times (i.e. (4) holds), is a "near-best" approximation of f, in the sense that

$$\|f - q_n\|_\infty \leq C \|f - q_n^*\|_\infty, \quad C = \frac{\|f - q_n\|_\infty}{\min_i |f(y_i) - q_n(y_i)|} \geq 1. \quad (7)$$

On the basis of the above considerations it is possible to deduce methods that often give a good estimate of the minimax approximation. Among all the near-minimax approximations we mention the least-squares approximation, the interpolation at the Chebyshev nodes and, perhaps the most used, forced-oscillation best approximation. For more details see, e.g., [1] and the references therein.

The Remez algorithm is one of the most famous methods for the computation of minimax polynomial approximations [20–22]. This algorithm constructs a sequence of trial references $\underline{x}^{(k)} = (x_0^{(k)}, x_1^{(k)}, \ldots, x_{n+1}^{(k)}) \in \mathbb{R}^{n+2}$, with $a \leq x_0^{(k)} < x_1^{(k)} < \cdots < x_{n+1}^{(k)} \leq b$, and a sequence of trial polynomials $q_n^{(k)}(x) \in \mathbb{P}_n$, $k = 0, 1, \ldots$, that satisfy the alternation condition (4) in such a way that $C^{(k)}$, given by (7) when we replace q_n with $q_n^{(k)}$, approaches to 1 when $k \to +\infty$.

Remez Algorithm

Let $\{\phi_i, \ i = 0, 1, \ldots, n\}$ be a basis of $\mathbb{P}_n$, and let

$$\underline{x}^{(0)} = (x_0^{(0)}, x_1^{(0)}, \ldots, x_{n+1}^{(0)}) \in \mathbb{R}^{n+2}, \quad (8)$$

such that $a \leq x_0^{(0)} < x_1^{(0)} < \cdots < x_{n+1}^{(0)} \leq b$. Given $\underline{x}^{(k)}$ at each kth step, $k = 0, 1, \ldots$, perform the following steps.

1. *Computation of the trial polynomial.*

Construct $q_n^{(k)}(x) = \sum_{i=0}^{n} a_{n,i}^{(k)} \phi_i(x) \in \mathbb{P}_n$ such that

$$f\left(x_j^{(k)}\right) - q_n^{(k)}\left(x_j^{(k)}\right) = (-1)^j E_n^{(k)}, \quad j = 0, 1, \ldots, n + 1, \quad (9)$$

where $E_n^{(k)}$ is called the levelled error (it may be positive or negative).

The polynomial $q_n^{(k)}(x)$ is obtained from the solution of linear system (9) that has $n + 2$ equations and $n + 2$ unknowns: the levelled error $E_n^{(k)}$ plus $a_{n,i}^{(k)}$, $i = 0, 1, \ldots, n$. Note that $\left| E_n^{(k)} \right| \leq \| f - q_n^* \|_\infty$.

2. If $\left| E_n^{(k)} \right| = \| f - q_n^* \|_\infty$ from the Theorem 1 we have that $q_n^* = q_n^{(k)}$ and the algorithm stops.

3. *Adjustment of the trial reference from the error of the trial polynomial.*

If $0 < \left| E_n^{(k)} \right| < \| f - q_n^* \|_\infty$, the goal is to construct a new trial reference $\underline{x}^{(k+1)}$ such that the residual

$$r_n^{(k)}(x) = f(x) - q_n^{(k)}(x), \tag{10}$$

oscillates at $x_j^{(k+1)}$, $j = 0, 1, \ldots, n$, that is

$$f\left(x_j^{(k+1)}\right) - q_n^{(k)}\left(x_j^{(k+1)}\right) = (-1)^j E_{n,j}^{(k+1)}, \qquad j = 0, 1, \ldots, n+1, \tag{11}$$

with $E_{n,j}^{(k+1)}$, $j = 0, 1, \ldots, n + 1$, having all the same sign and such that $\left| E_{n,j}^{(k+1)} \right| \geq \left| E_n^{(k)} \right|$, $j = 0, 1, \ldots, n + 1$.

This last step will be described in more details in the next section both for the classical Remez algorithm and for the proposed slight modification.

We note that theorem of De La Vallée Poussin guarantees the existence of this new trial reference. In fact, to be sure of increasing the levelled error, the replacement of the old trial reference with the new one must satisfies:

$$|E_n^{(k)}| \leq \min_j |E_{n,j}^{(k+1)}| \leq \| f - q_n^* \|_\infty \leq \left\| f - q_n^{(k+1)} \right\|_\infty. \tag{12}$$

This implies that the new polynomial $q_n^{(k+1)}$ equioscillates with a levelled error greater in modulus than the previous one, i.e. $\left| E_n^{(k+1)} \right| > \left| E_n^{(k)} \right|$. The monotonic increase of the modulus of the levelled error is the key observation in order to show that the algorithm converges to q_n^*, i.e. $q_n^{(k)}$ converges uniformly to q_n^* when $k \to +\infty$ (for more details see, for example, [19]).

Remez proposed two strategies to perform Step 3. of the previous algorithm, that is, he suggested two approaches to construct the new trial reference $\underline{x}^{(k+1)}$ knowing $\underline{x}^{(k)}$ and $q_n^{(k)}$. In the first Remez algorithm, $\underline{x}^{(k+1)}$ is constructed by moving one of the points of $\underline{x}^{(k)}$ to the abscissa of the global extremum of $r_n^{(k)}$ while keeping the sign alternation. In the second Remez algorithm $\underline{x}^{(k+1)}$ is constructed by replacing all the points of the trial reference with alternating abscissa of the extrema of $r_n^{(k)}$ including the global extremum. The linear convergence of the second Remez algorithm can be proved for each continuous function (see [18]). Furthermore it is possible to prove that if f is twice differentiable, the error decays with a quadratic rate at every $n+2$ steps in the case of the first Remez algorithm [19] and at every step in the case of the second Remez algorithm [23].

In the next section the second Remez algorithm is explained in detail and a slight modification of the second Remez algorithm is proposed, where a new approach to update the trial reference is considered. We will see that this new strategy is particularly effective when the trial polynomial has a very large number of extrema.

Obviously some stopping criteria for the Remez algorithm must be considered (see Step 2.). Among the others we mention the control of the difference between the absolute value of the levelled error and the maximum error of the trial polynomial (see [2]), the monitoring of the relative error between the trial polynomial and the function (especially when the approximation is used to compute a given function on a computer that uses floating point arithmetic) and, obviously, the restriction on the maximum number of iterations.

It is worthwhile to note that the choice of basis $\{\phi_i,\ i=0,1,\ldots,n\}$ of $\mathbb{P}_n$ in Step 1. of the Remez algorithm is a critical point in the construction of best approximants. In fact the basis used to represent polynomials determines the numerical properties of the linear system (9) used for the computation of the trial polynomials (see [9]). For example, the monomial basis, is a bad choice, in fact the condition number of the resulting Vandermonde matrix, generally, grows exponentially [11]. Usually the Chebyshev polynomial basis is used, but also this choice can give an ill-conditioned system for arbitrary sets of points.

To overcome this difficulty instead of solving at each Remez iteration k, $k=0,1,\ldots$, the linear system (9) we can use the following strategy. Let us construct two standard interpolating polynomials $s_n^{(k)}$, $t_n^{(k)}$ of degree n, i.e. $s_n^{(k)}$, $t_n^{(k)} \in \mathbb{P}_n$, such that:

$$s_n^{(k)}\left(x_j^{(k)}\right) = f\left(x_j^{(k)}\right), \quad t_n^{(k)}\left(x_j^{(k)}\right) = (-1)^j, \qquad j=0,1,\ldots,n. \qquad (13)$$

That is, $s_n^{(k)}$ interpolates f and $t_n^{(k)}$ interpolates the ordinates $(-1)^j$, $j=0,1,\ldots,n$, at the first $n+1$ nodes of the trial reference. Let $E_n^{(k)}$ be given by the following formula

$$s_n^{(k)}\left(x_{n+1}^{(k)}\right) - t_n^{(k)}\left(x_{n+1}^{(k)}\right) E_n^{(k)} = f\left(x_{n+1}^{(k)}\right) - (-1)^{n+1} E_n^{(k)}, \qquad (14)$$

then it is easy to prove that $E_n^{(k)}$ is the levelled error at step k and the corresponding trial polynomial is given by the following linear combination

$$q_n^{(k)}(x) = s_n^{(k)}(x) - t_n^{(k)}(x) E_n^{(k)}, \qquad (15)$$

that is also a polynomial of degree n, i.e. $q_n^{(k)} \in \mathbb{P}_n$, satisfying Eq. (9).

Note that the interpolants $s_n^{(k)}$, $t_n^{(k)} \in \mathbb{P}_n$, can be constructed with any classical interpolation method. In [17] the authors proposed to use the Lagrange basis and evaluate the trial polynomial with an ad hoc barycentric formula. This barycentric Lagrange formulation is enough stable and effective for the evaluation of high-degree polynomials. For more details about barycentric interpolation formulas, see, for example, [3].

Going into details in [17] the authors present an update of the second Remez algorithm in the context of the chebfun software system. This algorithm carries out numerical computing with functions rather than numbers. The chebfun system is a free/open-source software system written in MATLAB for numerical computation with functions of one real variable. It was introduced in its original form in [2]. The command remez, inside chebfun, allows a practical computation of a minimax approximation by using the second Remez algorithm. In particular, command remez, at each iteration step, computes the levelled error $E_n^{(k)}$ of the Remez algorithm thanks to an explicit formula resulting from the barycentric Lagrange representation formula used in the manipulation of trial polynomials (see Step 1. of the Remez algorithm) and uses chebfun global rootfinding to compute all the local extrema of $r_n^{(k)}$ from which the new trial reference is constructed (see Step 3. of the Remez algorithm).

Here and in [10] we also adopt the barycentric Lagrange representation formula proposed in [17] for the construction of trial polynomials and we also use chebfun global rootfinding to compute all the local extrema of $r_n^{(k)}$ from which the new trial reference is constructed. In particular, in the next section we present two algorithms for the construction of the new trial reference from the local extrema of $r_n^{(k)}$, that is, the one implemented in the command remez in chebfun software system and the new version proposed in this paper.

3   From a Trial Polynomial to a Trial Reference: A New Approach

This section provides more insights on Step 3. of the second Remez algorithm outlined in the previous section. In particular at each Remez iteration, after having computed a trial polynomial, a new trial reference must be constructed. Thus, let us suppose that at each Remez iteration k, $k = 0, 1, \ldots$, given a trial reference $\underline{x}^{(k)}$, there is a polynomial $q_n^{(k)} \in \mathbb{P}_n$ defining the error function $r_n^{(k)}$ in (10). Moreover let us suppose to correctly locate the oscillating extrema of $r_n^{(k)}$.

In the classical second Remez algorithm the new trial reference is constructed as follows.

Algorithm 1: the second Remez algorithm to compute a trial reference.

Let m be the number of the local maxima or minima of $r_n^{(k)}$ defined in (10) and let $\underline{z}^{(k)} \in \mathbb{R}^m$ be the vector of their abscissas sorted in ascending order.

(i) If $m = n+2$, that is if $r_n^{(k)}$ has exactly $n+2$ local extrema, then $\underline{x}^{(k+1)} = \underline{z}^{(k)}$.

(ii) If $m > n + 2$, let $\underline{r}^{k,n} \in \mathbb{R}^m$ be the vector defined as $r_i^{k,n} = r_n^{(k)}\left(z_i^{(k)}\right)$, $i = 1, 2, \ldots m$. The new trial reference $\underline{x}^{(k+1)}$ is obtained refining $\underline{z}^{(k)}$ as follows.

(iia) Delete from $\underline{z}^{(k)}$ the components where $\left|r_i^{k,n}\right| < \left|E_n^{(k)}\right|$. Then, for each resulting set of consecutive points with the same sign, consider only those where $\underline{r}^{k,n}$ attains the largest value and delete the others. Let $\overline{m} \geq n+2$ be the number of the remaining extrema abscissas and $\underline{y}^{(k)} \in \mathbb{R}^{\overline{m}}$ be the corresponding vector obtained from $\underline{z}^{(k)}$ after this refining. Note that by construction $\underline{y}^{(k)}$ contains the abscissas of oscillating extrema.

(iib) Given $\underline{y}^{(k)} \in \mathbb{R}^{\overline{m}}$ obtained in (iia) choose $n+2$ consecutive points among the components of $\underline{y}^{(k)} \in \mathbb{R}^{\overline{m}}$ including the global extremum $\overline{z}$ of $\underline{r}_n^{(k)}$, defined as

$$|r_n^{(k)}(\overline{z})| = ||\underline{r}^{k,n}||_\infty, \tag{16}$$

by using the following rules:
- if $\overline{m} = n+2$ set $\underline{x}^{(k+1)} = \underline{y}^{(k)}$,
- if $\overline{m} > n+2$ and on the left of $\overline{z}$ there are less than $n+2$ points, then $\underline{x}^{(k+1)}$ is equal to the first $n+2$ components of $\underline{y}^{(k)}$,
- if $\overline{m} > n+2$ and on the left of $\overline{z}$ there are more than $n+2$ points, then $\underline{x}^{(k+1)}$ is equal to the first $n+1$ components of $\underline{y}^{(k)}$ on the left of $\overline{z}$ and $\overline{z}$ itself.

In this paper we propose to modify Step (iib) of Algorithm 1 as follows.
Algorithm 2: a new approach to compute a trial reference.

(iib)' Choose $n+2$ consecutive points among the components of $\underline{y}^{(k)} \in \mathbb{R}^{\overline{m}}$ that has been obtained after refining (iia) including the global extremum $\overline{z}$ of $\underline{r}_n^{(k)}$ defined in (16), by using the following rules.
- If $\overline{m} = n+2$ set $\underline{x}^{(k+1)} = \underline{y}^{(k)}$.
- If $\overline{m} = n+3$ and $\overline{z}$ is among the first $\lfloor (n+2)/2 \rfloor$ components of $\underline{y}^{(k)}$, the new trial reference $\underline{x}^{(k+1)}$ is equal to the first $n+2$ components of $\underline{y}^{(k)}$, otherwise $\underline{x}^{(k+1)}$ is equal to the last $n+2$ components of $\underline{y}^{(k)}$, by indicating with $\lfloor \cdot \rfloor$ the integer part of $\cdot$.
- If $\overline{m} > n+3$ and $\overline{m}$ and $n+2$ are even numbers than consider the following rules.
  - If $\overline{z}$ belongs to the first $(n+2)/2$ components of $\underline{y}^{(k)}$ or to the last $(n+2)/2$ components of $\underline{y}^{(k)}$, then $\underline{x}^{(k+1)}$ is the vector given by the first $(n+2)/2$ components of $\underline{y}^{(k)}$ followed by the last $(n+2)/2$ components of $\underline{y}^{(k)}$.
  - If $\overline{z}$ belongs to the components of $\underline{y}^{(k)}$ between the position $(n+2)/2+1$ and the position $\overline{m}-(n+2)/2$, then $\underline{x}^{(k+1)}$ is the vector having increasing components formed by the global extremum $\overline{z}$, the component of $\underline{y}^{(k)}$ just on the left of $\overline{z}$, plus n components chosen between the other components of $\underline{y}^{(k)}$ so that both the equidistribution of the nodes at the ends of $\underline{y}^{(k)}$ and the alternation of signs of the error function are maintained.
- Similar rules apply when $\overline{m} > n+3$ and $\overline{m}$, $n+2$ are odd or when $\overline{m} > n+3$ and $\overline{m}$ is even and $n+2$ is odd (or vice versa) (see [10]). In this paper we only note that special attention must be paid when $n+2$ is odd and $\overline{m}$ is even (or vice versa), since in these cases keeping the sign alternation of the error function is not a trivial matter. For more details see [10].

4   Numerical Results

Let us denote by $\mathcal{R}$ and $\mathcal{R}_{new}$ the Remez algorithm within the chebfun software system obtained by using, respectively, the Algorithm 1 and the Algorithm 2 of Sect. 3 to compute the trial reference. Note that algorithm $\mathcal{R}$ is implemented in the command **remez** in chebfun software system (see [2, 17]), while $\mathcal{R}_{new}$ is the slight modification of the Remez algorithm proposed in this paper and further examined in [10].

In this section we compare the performance of the two algorithms $\mathcal{R}$ and $\mathcal{R}_{new}$ through the computation of some best approximants. To do this comparison we consider some of the functions used in [17] besides other functions obtained from the previous ones by adding an oscillating noise.

In particular we use $\mathcal{R}$ and $\mathcal{R}_{new}$ to compute minimax polynomial approximation in the domain $[a, b] = [-1, 1]$. For the sake of brevity in this paper we report only numerical results obtained when $n = 30$, but very similar results have been achieved for different other choices of n (see [10]).

In addition to a maximum number of iteration K_{max}, we use the following stopping criterion for the two Remez algorithms: given a tolerance *tol*, the algorithm stops at the first iteration k where

$$||r_n^{(k)}||_\infty - |E_n^{(k)}| \leq tol. \tag{17}$$

Note that the quantity on the left of (17) is always positive. In the numerical experiments we use $K_{max} = 100$ and $tol = 10^{-7}$.

Both the algorithms $\mathcal{R}$ and $\mathcal{R}_{new}$ start from the following trial reference:

$$\underline{x}^{(0)} = (x_0^{(0)}, x_1^{(0)}, \ldots, x_{n+1}^{(0)}) \in \mathbb{R}^{n+2}, \tag{18}$$

defined as

$$x_j^{(0)} = \cos\left(\frac{j\pi}{n+1}\right), \quad j = 0, 1, \ldots, n+1, \tag{19}$$

and so they construct the same $q_n^{(0)}$. That is, the two Remez algorithms $\mathcal{R}$ and $\mathcal{R}_{new}$ construct the same initial trial polynomial $q_n^{(0)}$.

We observe that $x_j^{(0)}$, $j = 0, 1, \ldots, n+1$, in (19) are the $n+2$ maxima of $|T_{n+1}(x)|$, where T_{n+1} is the Chebyshev polynomial of degree $n+1$. It is worthwhile to note that this trial reference is considered an excellent choice for a near-minimax approximation, see [1].

Tables 1, 2 show the numerical results obtained looking for best polynomial approximations by polynomials of degree $n = 30$ to some Hölder continuous functions. In these tables, for each function f, we report the following quantities: $\left\|f - q_n^{(0)}\right\|_\infty$, i.e. the infinite norm of the error obtained approximating f with the initial trial polynomial $q_n^{(0)}$, $\|f - q_n^*\|_\infty$, that is the approximated minimax error obtained applying Remez algorithms $\mathcal{R}$ and $\mathcal{R}_{new}$, and finally K, i.e. the number of iterations necessary to obtain the corresponding minimax error. When the algorithm is unstable we simply type *not stable* in the table, avoiding to report the minimax error.

Note that the functions listed in Table 2 are obtained from those reported in Table 1 by adding an oscillating noise of the form $\epsilon \sin(Ax)$, where ϵ represents the magnitude of the error.

As we can see from Table 1, without the noise, the two Remez algorithms $\mathcal{R}$ and $\mathcal{R}_{\text{new}}$ have comparable performances. In fact, they converge in the same iterations number K, reaching the same approximated minimax error. Similar results are obtained also for different choices of n.

Table 1. Best polynomial approximation q_n^* of degree $n = 30$ to f obtained with the two algorithms $\mathcal{R}$ and $\mathcal{R}_{\text{new}}$. The variable K denotes the number of iterations necessary to obtain the corresponding minimax error.

$f(x)$	$\left\| f - q_n^{(0)} \right\|_\infty$	$\mathcal{R}$ $\|f - q_n^*\|_\infty$	K	$\mathcal{R}_{\text{new}}$ $\|f - q_n^*\|_\infty$	K
$\|x\|$	0.0322581	0.0093325	5	0.0093325	5
$\sqrt{x+1}$	0.0134517	0.0066025	4	0.0066025	4
$\sqrt{\|x - 0.1\|}$	0.1908033	0.0636224	6	0.0636224	6
$1 - \sin(5\|x - 0.5\|)$	0.1230295	0.0406541	7	0.0406541	7

Table 2. Best polynomial approximation q_n^* of degree $n = 30$ to f obtained with the two algorithms $\mathcal{R}$ and $\mathcal{R}_{\text{new}}$. When the algorithm is unstable we write *not stable*, while when it converges we report the number of iterations K necessary to obtain the corresponding minimax error.

$f(x)$	$\left\| f - q_n^{(0)} \right\|_\infty$	$\mathcal{R}$ $\|f - q_n^*\|_\infty$	$\mathcal{R}_{\text{new}}$ $\|f - q_n^*\|_\infty$	K
$\|x\| + \epsilon \sin(Ax)$	0.1072307	*not stable*	0.0541669	29
$\sqrt{x+1} + \epsilon \sin(Ax)$	0.1071668	*not stable*	0.0500000	32
$\sqrt{\|x - 0.1\|} + \epsilon \sin(Ax)$	0.1916529	*not stable*	0.1008867	25
$1 - \sin(5\|x - 0.5\|) + \epsilon \sin(Ax)$	0.1239266	*not stable*	0.0659363	25

Problems arise when the functions exhibit a great number of oscillations and n is sufficiently large. To simulate this situation we add to the functions listed in Table 1 a very oscillating noise. The results obtained considering a particular noise of the form $\epsilon \sin(Ax)$ with $\epsilon = 0.05$, $A = 100$ are shown in Table 2, but very similar findings are achieved when other types of oscillating errors and/or other choices of n are taken into account (see further details in [10]). From Table 2 we have that, unlike $\mathcal{R}_{\text{new}}$, the classical Remez algorithm is unstable, in fact

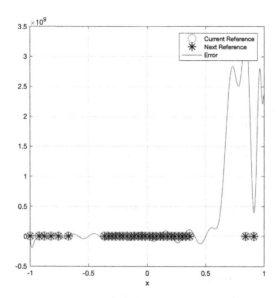

Fig. 1. Error obtained at the last $K_{max} = 100$ iteration of the Remez algorithm $\mathcal{R}$ when computing the best polynomial approximation of degree $n = 30$ to $f(x) = 1 - \sin(5|x - 0.5|) + 0.05\sin(100x)$, $x \in [-1, 1]$.

the algorithm $\mathcal{R}$ doesn't converge when applied to this kind of functions. The instability is mainly due to the fact that the noise added to the functions has a great number of oscillations and, as a consequence, the number of the local extrema of the error function is very large. In such a case, an ad hoc choice of the trial reference (as the one proposed here) is essential for the stability of the Remez algorithm.

To have an idea of what happens in these cases, we plot in Figs. 1, 2 the error $r_n^{(k)}$ defined in (10) obtained applying the Remez algorithms $\mathcal{R}$ and $\mathcal{R}_{new}$ to compute the minimax approximation to function $f(x) = 1 - \sin(5|x - 0.5|) + 0.05\sin(100x)$, $x \in [-1, 1]$ when $n = 30$.

In this case $\mathcal{R}_{new}$ converges after $K = 25$ iterations (see Table 2) and the graph of the corresponding error function at the 25th iteration is shown in Fig. 2. As can be seen from Fig. 2, since the new algorithm $\mathcal{R}_{new}$ converges, the error curve equioscillates in $n + 2 = 32$ points.

Instead the classical Remez algorithm $\mathcal{R}$ is unstable and it doesn't converge after $K_{max} = 100$ iterations (see Table 2). In Fig. 1 we can see the instability of the algorithm $\mathcal{R}$ reflected on the magnitude of the error obtained at the last iteration fixed in the computation.

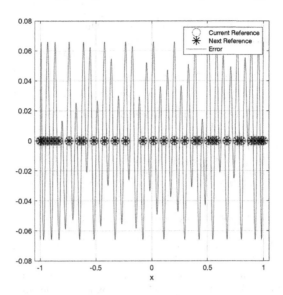

Fig. 2. Error of best polynomial approximation of degree $n = 30$ to $f(x) = 1 - \sin(5|x - 0.5|) + 0.05 \sin(100x)$, $x \in [-1, 1]$, obtained by using the Remez algorithm $\mathcal{R}_{\text{new}}$.

5 Conclusions

In this paper a slight modification of the (second) Remez algorithm, for the computation of the best approximation to a function f, has been presented, developed and implemented in the Remez algorithm $\mathcal{R}_{\text{new}}$. In this algorithm, a new approach to update the trial reference has been considered. In particular at each step of the algorithm, given the local extrema of the error function, the new trial reference is chosen in such a way to contain the global extremum (with its adjacent) of the error function together with some other ad hoc oscillating local extrema that must be equidistributed in the approximation interval.

This slight modification has been compared with the classical (second) Remez algorithm $\mathcal{R}$ implemented in the chebfun system (a free/open-source software system written in MATLAB).

Numerical results show that the two algorithms $\mathcal{R}$ and $\mathcal{R}_{\text{new}}$ perform similarly well on classical examples used to test algorithms for minimax approximation. However, when the degree of the approximant polynomial n is large and the function f has many oscillations, the classical algorithm $\mathcal{R}$ is not stable whereas the new slight modification $\mathcal{R}_{\text{new}}$ easily computes the searched minimax approximant.

In conclusion when the error curve for a best polynomial approximation by polynomials of degree n has many more oscillating points than the $n + 2$ points of equioscillations of the minimax approximant, then the new Remez algorithm $\mathcal{R}_{new}$ often outperforms the Remez algorithm $\mathcal{R}$.

References

1. Atkinson, K.E.: An Introduction to Numerical Analysis, 2nd edn. Wiley, New York (1989)
2. Battles, Z., Trefethen, L.N.: An extension of MATLAB to continuous functions and operators. SIAM J. Sci. Comput. **25**(5), 1743–1770 (2004)
3. Berrut, J.P., Trefethen, L.N.: Barycentric Lagrange interpolation. SIAM Rev. **46**, 501–517 (2004)
4. Blichfeldt, H.F.: Note on the functions of the form $f(x) \equiv \varphi(x) + a_1 x^{n-1} + \cdots + a_n$. Trans. Am. Math. Soc. **2**, 100–102 (1901)
5. Chebyshev, P.L.: Théorie des mécanismes connus sous le nom de parallélogrammes, Mémoires de l'Académie impériale des sciences de St. Pétersbourg, vol. 7, pp. 539–564 (1854)
6. Davis, P.: Interpolation and Approximation. Ginn (Blaisdell), Boston (1963)
7. De La Vallée Poussin, Ch.J.: Leçons sur l'approximation des fonctions d'une variable réelle. Gautier-Villars, Paris (1919)
8. Devore, R.A., Lorentz, G.G.: Constructive Approximation. Grundlehren der mathematischen Wissenschaften, vol. 303. Springer, Berlin (1993)
9. Dunham, C.B.: Choice of basis for Chebyshev approximation. ACM Trans. Math. Softw. **8**, 21–25 (1982)
10. Egidi, N., Fatone, L., Misici, L.: A stable Remez algorithm for minimax approximation (in preparation)
11. Higham, N.J.: Accuracy and Stability of Numerical Algorithms, 2nd edn. SIAM, Philadelphia (2002)
12. Kirchberger, P.: Über Tschebysheff'sche Annäherungsmethoden, dissertation, Göttingen (1902)
13. Korneichuk, N.: Exact Constants in Approximation Theory, Encyclopedia of Mathematics and its Applications, vol. 38. Cambridge University Press, Cambridge (1991)
14. Lorentz, G.G.: Approximation of Functions. Holt, Rinehart and Winston, New York (1966)
15. Mastroianni, G., Milovanovic, G.V.: Interpolation Processes. Basic Theory and Applications. Springer Monographs in Mathematics. Springer, Berlin (2008). https://doi.org/10.1007/978-3-540-68349-0
16. Meinardus, G.: Approximation of Functions: Theory and Numerical Methods. Springer, New York (1967). https://doi.org/10.1007/978-3-642-85643-3. (transl. L. Schumaker)
17. Pachón, R., Trefethen, L.N.: Barycentric-Remez algorithms for best polynomial approximation in the chebfun system. BIT Numer. Math. **49**(4), 721–741 (2009)
18. Petrushev, P.P., Popov, V.A.: Rational Approximation of Real Function, Encyclopedia of Mathematics and its Applications, vol. 28. Cambridge University Press, Cambridge (1987)
19. Powell, M.J.D.: Approximation Theory and Methods. Cambridge University Press, Cambridge (1981)
20. Remez, E.Y.: Sur la détermination des polynomes d'approximation de degré donné. Commun. Kharkov Math. Soc. **10**, 41–63 (1934)

21. Remez, E.Y.: Sur le calcul effectif des polynomes d'approximation de Tchebychef. Comptes rendus de l'Académie des Sciences **199**, 337–340 (1934)

22. Remez, E.Y.: Sur un procédé convergent d'approximations successives pour déterminer les polynomes d'approximation. Comptes rendus de l'Académie des Sciences **198**, 2063–2065 (1934)

23. Veidinger, L.: On the numerical determination of the best approximation in the Chebyshev sense. Numer. Math. **2**, 99–105 (1960)

An SVE Approach for the Numerical Solution of Ordinary Differential Equations

Nadaniela Egidi$^{(\boxtimes)}$ and Pierluigi Maponi

Scuola di Scienze e Tecnologie, Università di Camerino, 62032 Camerino, MC, Italy
{nadaniela.egidi,pierluigi.maponi}@unicam.it

Abstract. The derivative operator is reformulated as a Volterra integral operator of the first kind. So, the singular value expansion (SVE) of the kernel of such integral operator can be used to obtain new numerical methods to solve differential equations. We present such ideas in the solution of initial value problems for ordinary differential equations of first order. In particular, we develop an iterative scheme where global error in the solution of this problem is gradually reduced at each step. The global error is approximated by using the system of the singular functions in the aforementioned SVE.

Some experiments are used to show the performances of the proposed numerical method.

Keywords: Approximation · Ordinary differential equation · Numerical differentiation · Singular value expansion · Volterra integral equation

1 Introduction

Differential equations are one of the most important mathematical tools used in many scientific fields such as physics, chemistry, biology, economics. So, each advancement in the numerical solution of differential models has great influence on applied sciences. In particular, numerical methods for solving initial value problems for ordinary differential equations of first order have a central role in numerical analysis. The most popular numerical methods for solving such problems can be organised in two classes: multistep methods [1–3], and Runge-Kutta methods [5–7]. Both these classes share the well known Euler's method.

The approximation of derivatives is an important tool in the solution of ordinary differential equations (ODEs) and allows the definition of algebraic equations for the corresponding numerical solution [9]. In this paper, we propose an iterative method to solve ODEs where the global error in the solution is gradually reduced at each step, and the initial guess is obtained by the Euler's method. This iterative method is based on the singular value expansion of a particular Volterra integral operator that gives a reformulation of the derivative

© Springer Nature Switzerland AG 2020
Y. D. Sergeyev and D. E. Kvasov (Eds.): NUMTA 2019, LNCS 11973, pp. 70–85, 2020.
https://doi.org/10.1007/978-3-030-39081-5_8

operator. Hence, such a SVE can be a useful tool for the approximation of the derivative operator and for the construction of new numerical methods to solve differential equations.

In Sect. 2 we recall the formulation of the derivative operator as a Volterra integral operator, and the SVE of the corresponding kernel. In Sect. 3 we present the iterative method to solve initial value problems for ODEs of first order. In Sect. 4 we present the numerical results obtained on some classical ODEs. In Sect. 5 we give conclusions and future developments.

2 The Derivative Operator as a Volterra Integral Operator

In this section we give the reformulation of the derivative operator as a Volterra integral operator, and we recall the SVE of the obtained integral operator.

2.1 The Volterra Integral Operator

Let $\nu \geq 1$ be a given integer number, and let $g : [0,1] \to \mathbb{R}$ be a continuously differentiable function up to order ν. Let $g^{(j)}$ be the jth derivative of g. We suppose that $g^{(j)}(0)$, $j = 0, 1, \ldots, \nu - 1$, are known or already calculated with lower values of ν and let

$$h(t) = g(t) - \sum_{j=0}^{\nu-1} \frac{g^{(j)}(0)}{j!} t^j, \quad t \in [0, 1]. \tag{1}$$

We note that $h^{(j)}(0) = 0$, $j = 0, 1, \ldots, \nu - 1$ and $h^{(\nu)}(t) = g^{(\nu)}(t)$, $t \in [0, 1]$. Hence, from standard arguments on Maclaurin formula, we have that the derivation problem for h (and so for g) can be formulated as the following Volterra integral equation of the first kind:

$$h(t) = \int_0^t \frac{(t-s)^{\nu-1}}{(\nu-1)!} h^{(\nu)}(s) ds, \quad t \in [0, 1]. \tag{2}$$

Therefore $v = h^{(\nu)}$ is the unique solution (see [11] for details) of the following integral equation

$$\int_0^1 K(t, s) v(s) ds = h(t), \quad t \in [0, 1], \tag{3}$$

where h is a known function, and the integral kernel $K : [0, 1] \times [0, 1] \to \mathbb{R}$ is defined as follows:

$$K(t, s) = \begin{cases} \frac{(t-s)^{\nu-1}}{(\nu-1)!}, & 0 \leq s < t \leq 1, \\ 0, & 0 \leq t \leq s \leq 1. \end{cases} \tag{4}$$

Let $\mathcal{K}$ be the integral operator having kernel K defined by (4), then Eq. (3) can be rewritten as

$$\mathcal{K}v = h. \tag{5}$$

From standard arguments on integral operators with square integrable kernels, we have that there exists a Singular Value Expansion (SVE) of the kernel (4), that is

$$K(t, s) = \sum_{l=1}^{\infty} \mu_l u_l(t) v_l(s), \quad t, s \in [0, 1], \tag{6}$$

where the non-null values $\mu_1 \geq \mu_2 \geq \ldots$ are the singular values of K, and for $l = 1, 2, \ldots$, the functions u_l and v_l are respectively the left-singular function and right-singular function associated with μ_l, see [4,10] for details.

Let $\mathcal{K}^*$ be the adjoint integral operator of $\mathcal{K}$, then its kernel is

$$K^*(s, t) = K(t, s), \quad t, s \in [0, 1], \tag{7}$$

and, from (6) we easily obtain

$$\mathcal{K}v_l = \mu_l u_l, \quad \mathcal{K}^* u_l = \mu_l v_l, \quad l = 1, 2, \ldots, \tag{8}$$

moreover

$$v_l = \mu_l \frac{d^\nu}{dt^\nu} u_l, \qquad u_l = (-1)^\nu \mu_l \frac{d^\nu}{dt^\nu} v_l. \tag{9}$$

and from the orthonormality properties of the singular functions, we obtain that the solution of (5) is

$$h^{(\nu)}(t) = \sum_{l=1}^{\infty} \frac{\langle h, u_l \rangle}{\mu_l} v_l(t), \tag{10}$$

where $\langle \cdot, \cdot \rangle$ denotes the inner product on the space of real square integrable functions on $[0, 1]$.

2.2 The Singular Value Expansion of the Integral Operator

In this section we describe the fundamental formulas for the computation of the SVE of kernel (4), these formulas have been obtained in [8], where the reader can find the necessary details.

Let $\rho_2 : \mathbb{Z} \to \{0, 1\}$ be the function such that, for $k \in \mathbb{Z}$, $\rho_2(k)$ gives the reminder after the division of k by 2, and let $\rho_2^-(k) = 1 - \rho_2(k)$. For $k, q, j \in \mathbb{Z}$, $\gamma \in \mathbb{R}$, we define

$$\theta_q = \frac{2q + \rho_2(\nu)}{2\nu}\pi = \begin{cases} \frac{q\pi}{\nu}, & \text{if } \nu \text{ is even,} \\ \frac{(2q+1)\pi}{2\nu}, & \text{if } \nu \text{ is odd,} \end{cases}$$

$$c_q = \cos\theta_q, \qquad s_q = \sin\theta_q, \qquad z_q = e^{\iota\theta_q} = c_q + \iota s_q$$

$$c_{k,q} = \cos((k+1)\theta_q), \qquad s_{k,q} = \sin((k+1)\theta_q),$$

$$c_q^{(\gamma)} = \cos(\gamma s_q), \qquad s_q^{(\gamma)} = \sin(\gamma s_q),$$

$$c_{k,q}^{(\gamma)} = \cos((k+1)\theta_q - \gamma s_q) = c_{k,q}c_q^{(\gamma)} + s_{k,q}s_q^{(\gamma)}, \tag{11}$$

$$s_{k,q}^{(\gamma)} = \sin((k+1)\theta_q - \gamma s_q) = s_{k,q}c_q^{(\gamma)} - c_{k,q}s_q^{(\gamma)}, \tag{12}$$

$$\alpha_q^{(\gamma)} = (-1)^q e^{\gamma c_q},$$

$$\eta = \frac{\nu - \rho_2(\nu)}{2},$$

$$\underline{c}_{\cdot,j}^{(\gamma)} = \left(c_{0,j}^{(\gamma)}, c_{1,j}^{(\gamma)}, \dots, c_{\nu-1,j}^{(\gamma)}\right)^T \in \mathbb{R}^\nu,$$

$$\underline{s}_{\cdot,j}^{(\gamma)} = \left(s_{0,j}^{(\gamma)}, s_{1,j}^{(\gamma)}, \dots, s_{\nu-1,j}^{(\gamma)}\right)^T \in \mathbb{R}^\nu,$$

$$\underline{c}_{\cdot,j} = \underline{c}_{\cdot,j}^{(0)} \in \mathbb{R}^\nu, \quad \underline{s}_{\cdot,j} = \underline{s}_{\cdot,j}^{(0)} \in \mathbb{R}^\nu.$$

We have the following results.

Theorem 1. *Let $\mu_l > 0$ be a singular value of the integral operator $\mathcal{K}$ defined by its kernel (4), and let $\gamma_l = 1/\sqrt[\nu]{\mu_l}$. Then, the singular functions corresponding to μ_l are*

$$u_l(t) = \sum_{p=0}^{\nu-\rho_2(\nu)} e^{\gamma_l c_p t}\left(C_p^{(u)}\cos(\gamma_l s_p t) + S_p^{(u)}\sin(\gamma_l s_p t)\right), \quad t \in [0,1], \tag{13}$$

$$v_l(t) = \sum_{p=0}^{\nu-\rho_2(\nu)} e^{\gamma_l c_p t}\left(C_p^{(v)}\cos(\gamma_l s_p t) + S_p^{(v)}\sin(\gamma_l s_p t)\right), \quad t \in [0,1], \tag{14}$$

where coefficients $C_p^{(\cdot)}, S_p^{(\cdot)} \in \mathbb{R}$, $p = 0, 1, \dots, \nu - \rho_2(\nu)$, are defined by the following relations:

– *if ν is odd*

$$C_p^{(u)} = (-1)^{p+1}S_p^{(v)}, \quad S_p^{(u)} = (-1)^p C_p^{(v)}, \tag{15}$$

$$\sum_{p=0}^{\nu-1}\left(C_p^{(v)}c_{k,p} - S_p^{(v)}s_{k,p}\right) = 0, \quad k = 0, 1, \dots, \nu - 1, \tag{16}$$

$$\sum_{p=0}^{\nu-1}\alpha_p^{(\gamma_l)}\left(S_p^{(v)}c_{k,p}^{(\gamma_l)} + C_p^{(v)}s_{k,p}^{(\gamma_l)}\right) = 0, \quad k = 0, 1, \dots, \nu - 1; \tag{17}$$

– *if ν is even*

$$S_0^{(v)} = S_\nu^{(v)} = S_0^{(u)} = S_\nu^{(v)} = 0, \tag{18}$$

$$C_p^{(u)} = (-1)^p C_p^{(v)}, \ S_p^{(u)} = (-1)^p S_p^{(v)}, \tag{19}$$

$$\sum_{p=0}^{\nu} \left(C_p^{(v)} c_{k,p} - S_p^{(v)} s_{k,p} \right) = 0, \ k = 0, 1, \ldots, \nu - 1, \tag{20}$$

$$\sum_{p=0}^{\nu} \alpha_p^{(\gamma_l)} \left(C_p^{(v)} c_{k,p}^{(\gamma_l)} - S_p^{(v)} s_{k,p}^{(\gamma_l)} \right) = 0, \ k = 0, 1, \ldots, \nu - 1. \tag{21}$$

Proof. See [8] for the proof of this Theorem. □

Theorem 2. *When μ_l, $l = 1, 2, \ldots$, is a singular value of integral operator $\mathcal{K}$ defined by its kernel (4), then $\gamma_l = 1/\sqrt[\xi]{\mu_l}$ is a zero of the function*

$$h_\nu : \mathbb{R}^+ \to \mathbb{R}, \qquad h_\nu(\gamma) = \det\left(M(\gamma)\right)$$

where $M(\gamma)$ is the coefficients matrix of linear system (15)–(17) or (18)–(21) when we substitute γ_l with γ.
    When $\nu = 1$ we have

$$h_1(\gamma) = -\cos(\gamma), \tag{22}$$

when $\nu = 2$ we have

$$h_2(\gamma) = -4\left(1 - \cos(\gamma)\cosh(\gamma)\right), \tag{23}$$

and when $\nu \geq 3$ we have that the function h_ν satisfies the following asymptotic relation:

$$h_\nu(\gamma) = (-1)^{\eta+1} d^2 2^{2\eta-1} (2\rho_2(\nu) - \rho_2^-(\nu)) \cos(\gamma) e^{\gamma\xi} + g_\nu(\gamma), \tag{24}$$

where $g_\nu(\gamma) = \mathcal{O}(e^{\gamma\xi_0})$ when $\gamma \to +\infty$, $d > 0$ is given by

$$d = \prod_{\eta+1 \leq q \leq \nu-1} \left(\rho_2(\nu) + 2\rho_2^-(\nu)(1 + c_q)\right) s_q(-c_q) \cdot$$

$$\prod_{\eta+1 \leq p < q \leq \nu-1} \left(|z_p - \bar{z}_q|^2 |z_p - z_q|^2\right), \tag{25}$$

and

$$\xi = 2\sum_{i=0}^{\eta-1} c_i - \rho_2^-(\nu)c_0, \tag{26}$$

$$\xi_0 = c_{\eta-1} + 2\sum_{i=0}^{\eta-2} c_i - \rho_2^-(\nu)c_0. \tag{27}$$

Proof. See [8] for the proof of this Theorem.   □

We note that, given γ_l a zero of the function h_ν, the singular functions associated with μ_l respect to the base functions (13)–(14) are given by a non trivial solution of the corresponding linear system (15)–(17) or (18)–(21) with coefficients matrix $M(\gamma_l)$. So that Theorems (1) and (2) give the main tools for the computation of the SVE of operator K, in fact they respectively provides the formula for the computation of the singular functions and singular values.

So, this computation of the SVE can be used to approximate the derivative of a given function but also to solve a differential equation as shown in the next section, where, for the sake of simplicity, we consider only the case $\nu = 1$. In particular in this case we have that $h_1 : \mathbb{R}^+ \to \mathbb{R}$ is given in (22) and for $l = 1, 2, \dots$ and $t \in [0, 1]$ the SVE is

$$\gamma_l = \frac{\pi}{2} + (l - 1)\pi, \quad \mu_l = \frac{1}{\gamma_l}, \quad u_l(t) = \sqrt{2}\sin(\gamma_l t), \quad v_l(t) = \sqrt{2}\cos(\gamma_l t), \quad (28)$$

we note that $u_l(0) = v_l(1) = 0$, $v_l = \mu_l u_l'$, $u_l = -\mu_l v_l'$.

3   Initial Value Problem for First Order Ordinary Differential Equation

We consider an initial value problem for first order ordinary differential equation

$$\begin{cases} y'(x) = f(x, y(x)), \ x \in (0, 1) \\ y(0) = 0, \end{cases} \quad (29)$$

where $f : [0, 1] \times \mathbb{R} \to \mathbb{R}$ is a known continuous function, and the solution $y : [0, 1] \to \mathbb{R}$ is a continuously differentiable function.

Let $\tilde{y}$ be an approximation to y solution of (29) such that $\tilde{y}(0) = 0$, let

$$e(x) = \tilde{y}(x) - y(x) \quad (30)$$
$$r(x) = \tilde{y}'(x) - f(x, \tilde{y}(x)) \quad (31)$$

then

$$e'(x) = \tilde{y}'(x) - y'(x) = r(x) + f(x, \tilde{y}(x)) - f(x, y(x)) \quad (32)$$
$$\approx r(x) + f_y'(x, \tilde{y}(x))e(x) \quad (33)$$

3.1   Numerical Solution of ODEs

Let

$$e(x) = \sum_{l=1}^{\infty} e_l u_l(x), \quad 0 \le x \le 1, \quad (34)$$

then from formula (10)

$$e'(x) = \sum_{l=1}^{\infty} e_l \gamma_l v_l(x), \qquad 0 \leq x < 1. \tag{35}$$

We note that this last relation cannot be valid for $x = 1$, since in general $e'(1) \neq 0$ but $v_l(1) = 0$ for all l (see formula (28)). On the other hand, this relation can provide a good approximation of e' when this function is small.

The coefficients e_l, $l = 1, 2, \ldots$ of (28) are computed as the minimizer of the following problem

$$\min_e \left\| e'(x) - f_y'(x, \tilde{y}(x))e(x) - r(x) \right\|_2^2, \tag{36}$$

where $\|\cdot\|_2$ denotes the 2-norm. In particular, from the first-order conditions for the minimum, for $l = 1, 2, \ldots$, these coefficients must satisfy

$$\int_0^1 \left(e'(x) - f_y'(x, \tilde{y}(x))e(x) - r(x) \right) \left(\gamma_l v_l(x) - f_y'(x, \tilde{y}(x))u_l(x) \right) dx = 0,$$

that is

$$\sum_{m=1}^{\infty} e_m \int_0^1 \left(\gamma_l \gamma_m v_l v_m - f_y'(x, \tilde{y}(x)) \left(\gamma_m v_m u_l + \gamma_l v_l u_m \right) + \right.$$

$$\left. \left(f_y'(x, \tilde{y}(x)) \right)^2 u_m u_l \right) dx = \int_0^1 \left(\gamma_l v_l - f_y'(x, \tilde{y}(x))u_l \right) r(x) dx \tag{37}$$

We propose the following method, based on a truncation of formulas (34) and (37).

Given y_k, $k = 0, 1, \ldots$, an approximation of the solution y of (29) such that $y_k(0) = 0$, we define

$$r_k(x) = y_k'(x) - f(x, y_k(x))$$

Given $L > 0$, let $\underline{e}_k = (e_{k,1}, e_{k,2}, \ldots, e_{k,L})^T \in \mathbb{R}^L$ be the solution of

$$M\underline{e}_k = \underline{b} \tag{38}$$

where $\underline{b} = (b_1, b_2, \ldots, b_L)^T$ is given by

$$b_l = \int_0^1 \left(\gamma_l v_l(x) - f_y'(x, y_k(x))u_l(x) \right) r_k(x) dx, \qquad l = 1, 2, \ldots, L, \tag{39}$$

and $M \in \mathbb{R}^{L,L}$ has entries for $l, m = 1, 2, \ldots, L$ given by

$$M_{l,m} = \int_0^1 \left(\gamma_l \gamma_m v_l(x)v_m(x) - f_y'(x, y_k(x)) \left(\gamma_m v_m(x)u_l(x) \right. \right.$$

$$\left. \left. + \gamma_l v_l(x)u_m(x) \right) + \left(f_y'(x, y_k(x)) \right)^2 u_m(x)u_l(x) \right) dx, \tag{40}$$

where we can observe that M is a symmetric matrix.

We note that the matrix M seems to be almost positive definite, in the sense that if we neglect the first few rows, the other rows are diagonally dominant with positive diagonal entries. This fact suggest that, under suitable conditions, M is non singular and that system (38) can be solved efficiently by a proper modification of the Cholesky method.

Given

$$e_k(x) = \sum_{l=1}^{L} e_{k,l} u_l(x), \tag{41}$$

we note that by construction $e_k(x) \simeq y_k(x) - y(x)$ so that we define

$$y_{k+1}(x) = y_k(x) - e_k(x). \tag{42}$$

The above mentioned important properties of matrix M and the convergence analysis of $\{y_k\}_k$ deserve a future study of the proposed method that will allow also the evaluation of the applicability for this new method.

3.2   Implementation Details

Let $N > 0$, $h = 1/N$, and $x_j = jh$, $j = 0, 1, \ldots, N$ or $j = 1/2, 3/2, \ldots, N - 1/2$, we note that $x_0 = 0$ and $x_N = 1$. Let $y_k(x)$ be an approximation of $y(x)$ then the integrals in (39) and (40) are computed by using the midpoint rule with quadrature nodes $x_{j+1/2} \in (0, 1)$, $j = 0, 1, \ldots N - 1$, that is $x_{1/2} < x_{3/2} < \cdots < x_{(2N-1)/2}$, note that $0 = x_0 < x_{1/2} < x_1 < x_{3/2} < \cdots < x_{(2N-1)/2} < x_N = 1$.

Let

$$\tilde{\underline{Y}}_k = (Y_{k,0}, Y_{k,1}, \ldots, Y_{k,N})^T \in \mathbb{R}^{N+1},$$

$$\underline{Y}_k = (Y_{k,1/2}, Y_{k,3/2}, \ldots, Y_{k,(2N-1)/2})^T \in \mathbb{R}^N,$$

where $Y_{k,j} \simeq y_k(x_j)$.

The initial guess $y^{(0)}(x)$ of the recursive procedure is obtained by the Euler method, in particular we construct:

$$Y_{0,0} = 0, \qquad Y_{0,i} = Y_{0,i-1} + hf(x_{i-1}, Y_{0,i-1}), \qquad i = 1, \ldots N,$$

$$Y_{0,i+1/2} = \frac{Y_{0,i} + Y_{0,i+1}}{2}, \qquad i = 0, 1, \ldots N - 1,$$

$$y_0'(x_{i+1/2}) = f(x_i, Y_{0,i}), \qquad i = 0, 1, \ldots N - 1.$$

We note that, at each step k, we need $y_k'(x_{i+1/2})$, $i = 0, 1, \ldots N - 1$, because these values are necessary to compute the vector $\underline{b}$ by (39) and (31). Given $\underline{Y}_k$ and $y_k'(x_{i+1/2})$, $i = 0, 1, \ldots N - 1$, we find $\underline{e}_k \in \mathbb{R}^L$ as solution of system (38) and we construct $\underline{Y}_{k+1}$ from (41) and (42). Moreover we construct $y_{k+1}'(x_{i+1/2})$, $i = 0, 1, \ldots N - 1$, by differentiating (41) and (42), we note that differentiating (41) we obtain a truncation of (35). Also the first N entries of $\tilde{\underline{Y}}_k$ are obtained by using (41) and (42), instead $\tilde{Y}_{k+1,N} = Y_{k+1,N-1/2} + (h/2)f(x_{k+1,N-1/2}, Y_{k+1,N-1/2})$, because in this case we cannot use (35).

4   Numerical Results

We present the results of some numerical experiments to evaluate the performance of the proposed method.

In these experiments, we have considered the implementation described in the previous section. The system (38) has been solved by using two different strategies: the Gaussian elimination method and Gauss-Seidel method. In the Gauss-Seidel method a fixed number P of iterations has been performed. In particular, at each iteration $k = 0, 1, \ldots$ of the proposed method, and each iteration $p = 0, 1, \ldots, P$ of the Gauss-Seidel method, the vector $\underline{e}_k^{(p)}$ has been constructed, where the initial approximation $\underline{e}_k^{(0)}$ is chosen equal to the null vector for $k = 0$, instead for $k \geq 1$ we have $\underline{e}_k^{(0)} = \underline{e}_{k-1}^{(P)}$, finally the solution of system (38) is $\underline{e}_k = \underline{e}_k^{(P)}$.

Let $\underline{\tilde{Y}} \in \mathbb{R}^{N+1}$ be an approximation of the exact solution of problem (29) at nodes x_i, $i = 0, 1, \ldots, N$, i.e. $\underline{y} = (y(x_0), y(x_1), \ldots, y(x_N))^T \in \mathbb{R}^{N+1}$.

We consider the following performance indices given by the infinite norm and the quadratic mean of the error, that is:

$$E_\infty(\underline{\tilde{Y}}) = \left\| \underline{\tilde{Y}} - \underline{y} \right\|_\infty, \qquad E_2(\underline{\tilde{Y}}) = \frac{\left\| \underline{\tilde{Y}} - \underline{y} \right\|_2}{\sqrt{N+1}}. \tag{43}$$

In particular we use the following notation:

$$E_\infty^E = E_\infty(\underline{\tilde{Y}}_0), \qquad E_2^E = E_2(\underline{\tilde{Y}}_0), \tag{44}$$

are the errors for the initial approximation $\underline{\tilde{Y}}_0$ obtained with the Euler method;

$$E_\infty^{G,K} = E_\infty(\underline{\tilde{Y}}_K^G), \qquad E_2^{G,K} = E_2(\underline{\tilde{Y}}_K^G), \tag{45}$$

are the errors for the approximation $\underline{\tilde{Y}}_K^G$ obtained with the proposed algorithm after K iterations and solving the linear systems (38) by the Gaussian elimination method;

$$E_\infty^{GS,K} = E_\infty(\underline{\tilde{Y}}_K^{GS}), \qquad E_2^{GS,K} = E_2(\underline{\tilde{Y}}_K^{GS}), \tag{46}$$

are the errors for the approximation $\underline{\tilde{Y}}_K^{GS}$ obtained with the proposed algorithm after K iterations and solving the linear systems (38) by the Gauss-Seidel method, with $P = 3$;

To test the proposed method we use the following examples.

Example 1. *The following initial-value problem*

$$\begin{cases} y'(x) = y + 1, & x \in (0, 1), \\ y(0) = 0, \end{cases} \tag{47}$$

has solution $y(x) = e^x - 1$.

Example 2. *The following initial-value problem for Bernoulli equation*

$$\begin{cases} y'(x) = (y+1)\left(A - \frac{\pi}{2}\tan\left(\frac{\pi}{4}x\right)\right), & x \in (0,1), \\ y(0) = 0, \end{cases} \tag{48}$$

where $A \in \mathbb{R}$, *has solution*

$$y(x) = e^{Ax}\cos^2\left(\frac{\pi}{4}x\right) - 1.$$

Example 3. *The following initial-value problem for the Riccati equation*

$$\begin{cases} y'(x) = f_2(x)y^2 + f_1(x)y + f_0(x), & x \in (0,1), \\ y(0) = 0, \end{cases} \tag{49}$$

where for $A > 1$, *and*

$$f_2(x) = 2x, \tag{50}$$
$$f_1(x) = 2x(1 - 2A(x-1)^2), \tag{51}$$
$$f_0(x) = 2A(x-1) - 2A^2x(x-1)^4 - 2Ax(x-1)(1 - 2A(x-1)^2), \tag{52}$$

has solution

$$y(x) = A\left((x-1)^2 - \frac{1}{A + (1-A)e^{-x^2}}\right).$$

Table 1. The errors for Example 1 having solution $y(x)$, when the approximation is computed by: the Euler's method (E^E); the proposed iterative method with K iterations and Gaussian elimination method $(E^{G,K})$; the proposed iterative method with K iterations and Gauss-Seidel method with $P = 3$ $(E^{GS,K})$. The notation $x(y)$ means $x \cdot 10^y$.

$y(x) = e^x - 1$						
N	L	E_∞^E	$E_\infty^{G,1}$	$E_\infty^{GS,1}$	$E_\infty^{G,3}$	$E_\infty^{GS,3}$
25	5	5.24(−2)	1.45(−2)	1.61(−2)	1.45(−2)	1.44(−2)
50	10	2.67(−2)	3.98(−3)	5.33(−3)	3.98(−3)	3.86(−3)
N	L	E_2^E	$E_2^{G,1}$	$E_2^{GS,1}$	$E_2^{G,3}$	$E_2^{GS,3}$
25	5	2.51(−2)	6.25(−3)	7.16(−3)	6.25(−3)	6.20(−3)
50	10	1.26(−2)	1.67(−3)	2.46(−3)	1.67(−3)	1.61(−3)

In the numerical experiments we choose $L = 5, 10$ and $N = 25, 50$, the results are reported in Tables 1, 2 and 3 and Figs. 1, 2 and 3. From these results, we can observe that a few iterations of the proposed method is able to substantially decrease the error in the initial approximation $\underset{\sim}{Y}_0$ computed with the Euler's method. So, the good results obtained with the simple implementation described

Table 2. The errors for Example 2, with $A = 0.5$, having solution $y(x)$, when the approximation is computed by: the Euler's method (E^E); the proposed iterative method with K iterations and Gaussian elimination method ($E^{G,K}$); the proposed iterative method with K iterations and Gauss-Seidel method with $P = 3$ ($E^{GS,K}$). The notation $x(y)$ means $x \cdot 10^y$.

$y(x) = e^{x/2} \cos^2\left(\frac{\pi}{4}x\right) - 1$						
N	L	E_∞^E	$E_\infty^{G,1}$	$E_\infty^{GS,1}$	$E_\infty^{G,3}$	$E_\infty^{GS,3}$
25	5	2.25(−2)	8.14(−5)	7.41(−5)	8.14(−5)	8.14(−5)
50	10	1.11(−2)	2.84(−5)	2.43(−5)	2.84(−5)	2.84(−5)
N	L	E_2^E	$E_2^{G,1}$	$E_2^{GS,1}$	$E_2^{G,3}$	$E_2^{GS,3}$
25	5	1.48(−2)	4.24(−5)	3.76(−5)	4.24(−5)	4.24(−5)
50	10	7.35(−3)	1.37(−5)	1.08(−5)	1.37(−5)	1.37(−5)

Table 3. The errors for Example 3, with $A = 10$, having solution $y(x)$, when the approximation is computed by: the Euler's method (E^E); the proposed iterative method with K iterations and Gaussian elimination method ($E^{G,K}$); the proposed iterative method with K iterations and Gauss-Seidel method with $P = 3$ ($E^{GS,K}$). The notation $x(y)$ means $x \cdot 10^y$.

$y(x) = 10\left((x-1)^2 - \frac{1}{10 - 9e^{-x^2}}\right)$						
N	L	E_∞^E	$E_\infty^{G,1}$	$E_\infty^{GS,1}$	$E_\infty^{G,3}$	$E_\infty^{GS,3}$
25	5	3.57(−1)	4.82(−2)	4.85(−2)	4.65(−2)	4.65(−2)
50	10	1.71(−1)	1.35(−3)	1.39(−3)	1.27(−3)	1.27(−3)
N	L	E_2^E	$E_2^{G,1}$	$E_2^{GS,1}$	$E_2^{G,3}$	$E_2^{GS,3}$
25	5	1.48(−1)	1.96(−2)	1.96(−2)	1.93(−2)	1.93(−2)
50	10	7.10(−2)	4.65(−4)	5.47(−4)	6.12(−4)	6.12(−4)

in Sect. 3.2 deserve further study to evaluate the effective potential of the proposed method and its eventual application in the solution of partial differential equations.

There is no much difference between performing the iterative method with one or three iterations, independently from the method used to solve the linear systems (38), so that in Figs. 1, 2 and 3 we report the results only for one iteration of the method $K = 1$.

Moreover, the results obtained by solving the linear systems (38) with the Gaussian elimination method or with the Gauss-Seidel method with $P = 3$ are very similar, as we can see from Tables 1, 2 and 3 and in Figs. 1, 2 and 3.

So a good strategy is to implement the proposed iterative method with a single iteration $K = 1$, by solving systems (38) with $P = 3$ iterations of the Gauss-Seidel method.

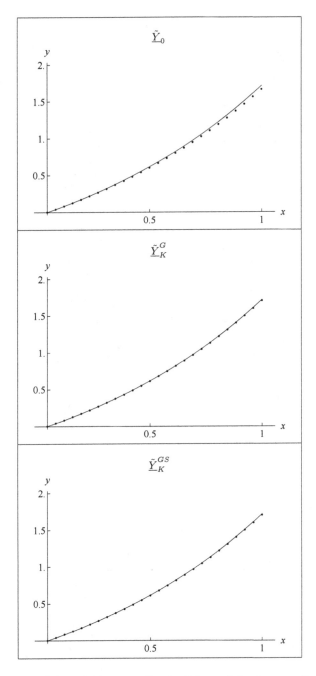

Fig. 1. The graphs of the solution $y = e^x - 1$ of Example 1 when $x \in [0, 1]$ (the lines), of its initial approximation $\tilde{\underline{Y}}_0$ (the dots at the top), and of its approximations, $\tilde{\underline{Y}}_K^G$ and $\tilde{\underline{Y}}_K^{GS}$ (the dots at the middle and at the bottom, respectively), obtained with the proposed method for $N = 25$, $L = 5$, $K = 1$, the corresponding errors are given in Table 1.

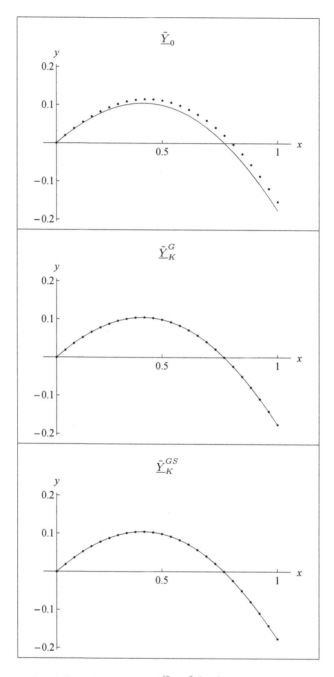

Fig. 2. The graphs of the solution $y = e^{x/2} \cos^2 \left(\frac{\pi}{4}x \right) - 1$ of Example 2 when $A = 0.5$ and $x \in [0, 1]$ (the lines), of its initial approximation $\underline{\tilde{Y}}_0$ (the dots at the top), and of its approximations, $\underline{\tilde{Y}}_K^G$ and $\underline{\tilde{Y}}_K^{GS}$ (the dots at the middle and at the bottom, respectively), obtained with the proposed method for $N = 25$, $L = 5$, $K = 1$, the corresponding errors are given in Table 2.

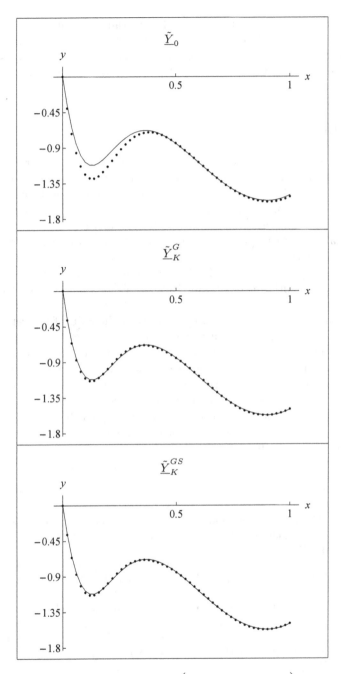

Fig. 3. The graphs of the solution $y(x) = 10\left((x-1)^2 - \frac{1}{10-9e^{-x^2}}\right)$ of Example 3 when $A = 10$ and $x \in [0,1]$ (the lines), of its initial approximation $\tilde{\underline{Y}}_0$ (the dots at the top), and of its approximations, $\tilde{\underline{Y}}_K^G$ and $\tilde{\underline{Y}}_K^{GS}$ (the dots at the middle and at the bottom, respectively), obtained with the proposed method for $N = 50$, $L = 10$, $K = 1$, the corresponding errors are given in Table 3.

We note that, for the first two considered examples, where the ODEs are linear, the errors do not change by performing one or more iterations of the proposed method, when the Gaussian elimination method is used to solve the linear systems (38). In the third example, the errors rapidly reach their limit values.

5 Conclusions

The derivative operator is reformulated as a Volterra integral operator, and its SVE is used to approximate the derivative operator and to obtain a new iterative method for the solution of ODEs. This iterative method is constructed in order to reduce, at each step, the global error of the approximating solution. As initial approximation is used the one obtained by Euler's method.

Numerical experiments show that the proposed iterative method gives satisfactory results at the first iteration, with a small number of singular functions and a small number of iterations in the Gauss-Seidel solution of the linear systems. In fact, the initial error obtained by the Euler's method is decreased of one order of magnitude or more.

These numerical results confirm that by using the SVE of the reformulated derivative operator we can construct new numerical methods to solve differential equations. In particular, this initial study has shown interesting results for the proposed method, even if a number of questions must be addressed, such as the refinement of the approximation techniques defined in Sect. 3.2 to improve the efficiency of the proposed method, the application of similar ideas to partial differential equations, the study of convergence and applicability of the proposed method, finding a proper modification of the Cholesky method to direct solve the envolved linear systems.

References

1. Dahlquist, G.: Convergence and stability in the numerical integration of ordinary differential equations. Math. Scand. **4**, 33–53 (1956). https://doi.org/10.7146/math.scand.a-10454
2. Henrici, P.: Discrete Variable Methods in ODEs. Wiley, New York (1962)
3. Lambert, J.D.: Computational Methods in Ordinary Differential Equations. Wiley, New York (1973)
4. Jörgens, K.: Linear Integral Operators. Pitman Advanced Pub. Program, Boston (1982)
5. Atkinson, K.E.: An Introduction To Numerical Analysis. Wiley, New York (1978)
6. Butcher, J.C.: Numerical Methods for Ordinary Differential Equations. Wiley, New York (2003)
7. Lapidus, L., Seinfeld, J.: Numerical Solution of Ordinary Differential Equations. Academic Press, New York (1971)
8. Egidi, N., Maponi, P.: The singular value expansion of the Volterra integral equation associated to a numerical differentiation problem. J. Math. Anal. Appl. **460**, 656–681 (2018)

9. Fernández, D.C.D.R., Hicken, J.E., Zingg, D.W.: Review of summation-by-parts operators with simultaneous approximation terms for the numerical solution of partial differential equations. Comput. Fluids **95**, 171–196 (2014)
10. Kress, R.: Linear Integral Equations. Springer, Berlin (1989). https://doi.org/10.1007/978-3-642-97146-4
11. Zhang, H., Zhang, Q.: Sparse discretization matrices for Volterra integral operators with applications to numerical differentiation. J. Integr. Equ. Appl. **23**(1), 137–156 (2011)

Uniform Weighted Approximation by Multivariate Filtered Polynomials

Donatella Occorsio[1,2(✉)] and Woula Themistoclakis[2]

[1] Department of Mathematics, Computer Science and Economics, University of Basilicata, v.le dell'Ateneo Lucano 10, 85100 Potenza, Italy
`donatella.occorsio@unibas.it`
[2] C.N.R. National Research Council of Italy, I.A.C. Istituto per le Applicazioni del Calcolo "Mauro Picone", via P. Castellino, 111, 80131 Napoli, Italy
`woula.themistoclakis@cnr.it`

Abstract. The paper concerns the weighted uniform approximation of a real function on the $d-$cube $[-1,1]^d$, with $d > 1$, by means of some multivariate filtered polynomials. These polynomials have been deduced, via tensor product, from certain de la Vallée Poussin type means on $[-1,1]$, which generalize classical delayed arithmetic means of Fourier partial sums. They are based on arbitrary sequences of filter coefficients, not necessarily connected with a smooth filter function. Moreover, in the continuous case, they involve Jacobi–Fourier coefficients of the function, while in the discrete approximation, they use function's values at a grid of Jacobi zeros. In both the cases we state simple sufficient conditions on the filter coefficients and the underlying Jacobi weights, in order to get near–best approximation polynomials, having uniformly bounded Lebesgue constants in suitable spaces of locally continuous functions equipped with weighted uniform norm. The results can be useful in the construction of projection methods for solving Fredholm integral equations, whose solutions present singularities on the boundary. Some numerical experiments on the behavior of the Lebesgue constants and some trials on the attenuation of the Gibbs phenomenon are also shown.

Keywords: Weighted polynomial approximation · de la Vallée Poussin means · Filtered approximation · Lebesgue constants · Projection methods for singular integral equations · Gibbs phenomenon

1 Introduction

Many problems arising in the applied science can be modeled by integral equations on $D = [-1,1]^d$ and in many cases, since the solution is unbounded on ∂D, they are uniquely solvable in some Banach space of locally continuous functions, equipped with weighted uniform norm (see, e.g. [15,16,18] and the references therein). In such cases, the classical projection methods fail in view of the unboundedness of the corresponding Lebesgue constants (LC) associated with the applied projection (typically, Lagrange and Fourier projections). In one dimension [17] a solution has been offered by certain quasi–polynomial projections, which generalize trigonometric de la Vallée Poussin means (briefly VP means) and fall into the class of *filtered polynomials* (see e.g. [5,10,13,22–24]). They have been successfully applied in [17] to construct a numerical method which is uniformly convergent and stable in $[-1,1]$. Here we provide the basic theorems for the extension to higher dimensions.

We introduce and study multivariate filtered polynomial sequences deduced, via tensor product, from the univariate generalized VP means recently studied in [24]. We state simple sufficient conditions in order that these polynomial sequences converge (in a given weighted uniform norm) to the function they aim to approximate with uniformly bounded related LC. This remarkable behavior is especially desirable in proving the convergence and the stability of some numerical methods for solving several functional equations (see e.g. [4,6,7,9,19,25]).

Moreover, from the general approximation view–point, our results offer many different possibilities of factorizing the function in the product of a weight function to compensate the singular factor, and a smooth factor, which can be approximated very well by polynomials.

The numerical experiments focus on the bidimensional case $D = [-1,1]^2$ where several optimal systems of nodes (e.g. Padua points, Xu points) have been already constructed. Nevertheless, most of them cannot be used in the case of functions unbounded at ∂D, since some nodes lie on the boundary of D (see e.g. [1–3,26]). Comparisons with these kinds of polynomial approximation can be found in the forthcoming paper by the authors [20], being them outside the scope of this paper. Indeed, our focus mainly deals with the uniform weighted approximation of functions which can be also unbounded at the border of the square.

The outline of the paper is the following. Section 2 introduces some notation and it deals with classical Fourier and Lagrange polynomial projections. Section 3 is devoted to filtered approximation in the continuous and the discrete case. Finally, Sect. 4 contains some numerical experiments.

2   Basic Definitions and Properties

In order to approximate a function **f** on D, we consider multivariate polynomials in the variable $\mathbf{x} := (x_1, \ldots, x_d) \in D$, which are of degree at most m_k w.r.t. the variable x_k, for any $k = 1, \ldots, d$ (briefly $\forall k \in \mathcal{N}_1^d$). Setting $\mathbf{m} := (m_1, \ldots, m_d) \in \mathbb{N}^d$, this polynomial space will be denoted by $\mathbf{P_m}$.

Denoting by $v^{\gamma,\delta}$ the Jacobi weight

$$v^{\gamma,\delta}(z) := (1-z)^\gamma(1+z)^\delta, \qquad z \in (-1,1), \quad \gamma, \delta > -1,$$

throughout the paper, for any $\gamma_k, \delta_k \geq 0$, $k \in \mathcal{N}_1^d$, we set

$$\mathbf{u}(\mathbf{x}) = v^{\gamma_1,\delta_1}(x_1)v^{\gamma_2,\delta_2}(x_2)\cdots v^{\gamma_d,\delta_d}(x_d) := \prod_{k=1}^d u_k(x_k), \tag{1}$$

and we consider functions $\mathbf{f}$ that can be also unbounded on ∂D, but they are such that

$$\|\mathbf{f}\mathbf{u}\|_\infty = \max_{\mathbf{x} \in D} |\mathbf{f}(\mathbf{x})|\mathbf{u}(\mathbf{x}) < \infty.$$

We denote by $L_\mathbf{u}^\infty := \{\mathbf{f} : \|\mathbf{f}\mathbf{u}\|_\infty < \infty\}$ the Banach space equipped with the previous norm.

Moreover, we denote by $E_\mathbf{m}(\mathbf{f})_\mathbf{u}$ the error of best weighted polynomial approximation by means of polynomials in $\mathbf{P_m}$, i.e.

$$E_\mathbf{m}(\mathbf{f})_\mathbf{u} = \inf_{\mathbf{P} \in \mathbf{P_m}} \|(\mathbf{f} - \mathbf{P})\mathbf{u}\|_\infty.$$

The extension of the Weierstrass theorem to D ensures that

$$\lim_{\mathbf{m} \to \infty} E_\mathbf{m}(\mathbf{f})_\mathbf{u} = 0, \qquad \forall \mathbf{f} \in C_\mathbf{u},$$

where $C_\mathbf{u} \subset L_\mathbf{u}^\infty$ denotes the subspace of all functions that are continuous in the interior of D and tend to zero as $\mathbf{x}$ tends to the edges of D (i.e. as $x_k \to \pm 1$, in correspondence of some $\gamma_k, \delta_k > 0$, $k \in \mathcal{N}_1^d$).

Along all the paper the constant $\mathcal{C}$ will be used several times, having different meaning in different formulas. Moreover we write $\mathcal{C} \neq \mathcal{C}(a, b, \ldots)$ in order to say that $\mathcal{C}$ is a positive constant independent of the parameters $a, b, \ldots$

Finally, for $\mathbf{r} = (r_1, \ldots, r_d)$ and $\mathbf{s} = (s_1, \ldots, s_d)$ in $\mathbb{N}^d$, we set

$$\mathbb{N}[\mathbf{r}, \mathbf{s}] := \{\mathbf{i} = (i_1, \ldots, i_d) \in \mathbb{N}^d : r_k \le i_k \le s_k, \; k \in \mathcal{N}_1^d\},$$

where for $\mathbf{r} = \mathbf{0}, \mathbf{1}$ we agree that $\mathbf{0} = (0, \ldots, 0)$ and $\mathbf{1} = (1, \ldots, 1)$.

Along the paper, at the occurrence, other notations will be introduced. For the convenience of the reader a list of notations is given at the end of the paper.

2.1   Optimal Polynomial Projectors

Setting

$$w_k := v^{\alpha_k, \beta_k}, \qquad \alpha_k, \beta_k > -1, \qquad k = 1, \ldots, d,$$

we denote by $\{p_m(w_k)\}_{m=0}^\infty$ the corresponding sequence of the univariate orthonormal Jacobi polynomials with positive leading coefficients.

For any $\mathbf{x} = (x_1, \ldots, x_d)$, we set

$$\mathbf{w}(\mathbf{x}) := \prod_{k=1}^d w_k(x_k) = \prod_{k=1}^d (1 - x_k)^{\alpha_k} (1 + x_k)^{\beta_k},$$

and for any $\mathbf{m} = (m_1, \ldots, m_d) \in \mathbb{N}^d$, we consider in $\mathbf{P_m}$ the following polynomial basis

$$\mathbf{P_m}(\mathbf{w}, \mathbf{x}) := \prod_{k=1}^d p_{m_k}(w_k, x_k).$$

Then the d–dimensional Fourier polynomial $\mathbf{S_m}(\mathbf{w}, \mathbf{f}, \mathbf{x})$ of $\mathbf{f}$ w.r.t. $\mathbf{w}$ is given by

$$\mathbf{S_m}(\mathbf{w}, \mathbf{f}, \mathbf{x}) = \sum_{\mathbf{i} \in \mathbb{N}[0,\mathbf{m}]} c_\mathbf{i}(\mathbf{w}, \mathbf{f}) \mathbf{p_i}(\mathbf{w}, \mathbf{x}),$$

$$c_\mathbf{i}(\mathbf{w}, \mathbf{f}) = \int_D \mathbf{f}(\mathbf{y}) \mathbf{p_i}(\mathbf{w}, \mathbf{y}) \mathbf{w}(\mathbf{y}) d\mathbf{y}. \tag{2}$$

This polynomial can be deduced via tensor product from one dimensional Jacobi–Fourier projections.

By using [14, Th. 4.34, p. 276], the following result can be easily proved

Theorem 1. *Let $\mathbf{w}$ and $\mathbf{u}$ satisfy the following bounds*

$$\begin{cases} \dfrac{\alpha_k}{2} + \dfrac{1}{4} < \gamma_k \leq \dfrac{\alpha_k}{2} + \dfrac{3}{4} \quad and \ \ 0 \leq \gamma_k < \alpha_k + 1, \\[2mm] \dfrac{\beta_k}{2} + \dfrac{1}{4} < \delta_k \leq \dfrac{\beta_k}{2} + \dfrac{3}{4} \quad and \ \ 0 \leq \delta_k < \beta_k + 1, \end{cases} \quad \forall k \in \mathcal{N}_1^d. \tag{3}$$

Then for any $\mathbf{f} \in L_\mathbf{u}^\infty$ and $\mathbf{m} \in \mathbb{N}^d$ it follows that

$$\|\mathbf{S_m}(\mathbf{w}, \mathbf{f})\mathbf{u}\|_\infty \leq \mathcal{C} \|\mathbf{f}\mathbf{u}\|_\infty \log \mathbf{m}, \qquad \mathcal{C} \neq \mathcal{C}(\mathbf{f}, \mathbf{m}), \tag{4}$$

where $\log \mathbf{m} := \prod_{k=1}^d \log m_k$.

This result and the invariance property

$$\mathbf{S_m}(\mathbf{w}, \mathbf{f}) \equiv \mathbf{f}, \qquad \forall \mathbf{f} \in \mathbf{P_m}, \tag{5}$$

yield the following

Corollary 1. *Under the assumptions of Theorem 1, we have*

$$E_\mathbf{m}(\mathbf{f})_\mathbf{u} \leq \|[f - \mathbf{S_m}(\mathbf{w}, \mathbf{f})]\mathbf{u}\|_\infty \leq \mathcal{C} E_\mathbf{m}(\mathbf{f})_\mathbf{u} \log \mathbf{m}, \qquad \mathcal{C} \neq \mathcal{C}(\mathbf{f}, \mathbf{m}). \tag{6}$$

Denoted by $\{x_{m_k,\ell}(w_k)\}_{\ell=1}^{m_k}$ the zeros of the univariate polynomial $p_{m_k}(w_k)$ and by $\{\lambda_{m_k,\ell}(w_k)\}_{\ell=1}^{m_k}$ the associated Christoffel numbers, for any $\mathbf{i} = (i_1, \ldots, i_d) \in \mathbb{N}[1, \mathbf{m}]$ we set

$$\mathbf{x_i^{(m)}}(\mathbf{w}) = (x_{m_1,i_1}(w_1), \ldots, x_{m_k,i_k}(w_k), \ldots, x_{m_d,i_d}(w_d)),$$

$$\Lambda_\mathbf{i}^{(\mathbf{m})}(\mathbf{w}) = \prod_{k=1}^d \lambda_{m_k,i_k}(w_k).$$

Then the Gauss-Jacobi cubature formula of order $\mathbf{m}$ obtained via tensor product of univariate Gauss-Jacobi rules, is given by

$$\int_D \mathbf{f}(\mathbf{x}) \mathbf{w}(\mathbf{x}) d\mathbf{x} = \sum_{\mathbf{i} \in \mathbb{N}[1,\mathbf{m}]} \Lambda_\mathbf{i}^{(\mathbf{m})}(\mathbf{w}) \mathbf{f}(\mathbf{x_i^{(m)}}(\mathbf{w})) + R_\mathbf{m}(\mathbf{w}, \mathbf{f}), \tag{7}$$

where the remainder term $R_m(\mathbf{w}, \mathbf{f})$ vanishes when $\mathbf{f} \in \mathbf{P}_{2m-1}$.

By discretizing the Fourier coefficients of $\mathbf{S_m}(\mathbf{w}, \mathbf{f})$ in (2) by the previous rule (7) of order $(\mathbf{m} + \mathbf{1})$, we get the following polynomial

$$\mathbf{L_m}(\mathbf{w}, \mathbf{f}, \mathbf{x}) = \sum_{\mathbf{i} \in \mathbb{N}[1,\mathbf{m}]} \Lambda_{\mathbf{i}}^{(\mathbf{m}+1)}(\mathbf{w}) K_m(\mathbf{w}, \mathbf{x}, \mathbf{x}_{\mathbf{i}}^{(\mathbf{m}+1)}(\mathbf{w})) \mathbf{f}(\mathbf{x}_{\mathbf{i}}^{(\mathbf{m}+1)}(\mathbf{w})), \quad (8)$$

where

$$K_m(\mathbf{w}, \mathbf{x}, \mathbf{y}) := \sum_{\mathbf{r} \in \mathbb{N}[0,\mathbf{m}]} \mathbf{p_r}(\mathbf{w}, \mathbf{y}) \mathbf{p_r}(\mathbf{w}, \mathbf{x}).$$

By definition it follows that $\mathbf{L_m}(\mathbf{w}, \mathbf{f}) \in \mathbf{P_m}$ and

$$\mathbf{L_m}(\mathbf{w}, \mathbf{f}) \equiv \mathbf{f}, \qquad \forall \mathbf{f} \in \mathbf{P_m}.$$

Moreover, it is easy to check that the following interpolation property holds

$$\mathbf{L_m}(\mathbf{w}, \mathbf{f}, \mathbf{x}_{\mathbf{i}}^{(\mathbf{m}+1)}(\mathbf{w})) \equiv \mathbf{f}(\mathbf{x}_{\mathbf{i}}^{(\mathbf{m}+1)}(\mathbf{w})), \qquad \forall \mathbf{i} \in \mathbb{N}[1, (\mathbf{m}+1)],$$

i.e. $\mathbf{L_m}(\mathbf{w}, \mathbf{f})$ is the multivariate Lagrange polynomial in $\mathbf{P_m}$ interpolating $\mathbf{f}$ at the zeros of $\mathbf{p_{m+1}}(\mathbf{w})$. The following approximation theorem generalizes [18, Prop. 2.1] obtained in the case $d = 2$.

Theorem 2. *Let* $\mathbf{w}$ *and* $\mathbf{u}$ *be such that*

$$\begin{cases} \max\left(0, \dfrac{\alpha_k}{2} + \dfrac{1}{4}\right) \le \gamma_k \le \dfrac{\alpha_k}{2} + \dfrac{5}{4}, \\ \max\left(0, \dfrac{\beta_k}{2} + \dfrac{1}{4}\right) \le \delta_k \le \dfrac{\beta_k}{2} + \dfrac{5}{4}, \end{cases} \qquad k \in \mathcal{N}_1^d. \qquad (9)$$

Then, for any $\mathbf{f} \in L_{\mathbf{u}}^{\infty}$ *and* $\mathbf{m} \in \mathbb{N}^d$, *we have*

$$\|\mathbf{L_m}(\mathbf{w}, \mathbf{f})\mathbf{u}\|_\infty \le \mathcal{C}\|\mathbf{f}\mathbf{u}\|_\infty \log \mathbf{m}, \qquad (10)$$

and

$$\|[\mathbf{f} - \mathbf{L_m}(\mathbf{w}, \mathbf{f})]\mathbf{u}\|_\infty \le \mathcal{C} E_{\mathbf{m}}(\mathbf{f})_{\mathbf{u}} \log \mathbf{m}, \qquad (11)$$

where in both the estimates $\mathcal{C} \ne \mathcal{C}(\mathbf{f}, \mathbf{m})$.

3   Main Results

Setting $\mathbf{N} = (N_1, \ldots, N_d) \in \mathbb{N}^d$ and $\mathbf{M} = (M_1, \ldots, M_d) \in \mathbb{N}^d$, with $N_k < M_k$ for all $k \in \mathcal{N}_1^d$, let us consider d uniformly bounded sequences of filter coefficients $\{h_\ell^{N_k, M_k}\}_{l=1,2,\ldots}$, such that

$$h_\ell^{N_k, M_k} = \begin{cases} 1, & if \ \ell \in [0, N_k] \\ 0, & if \ \ell \notin [0, M_k], \end{cases} \qquad l = 1, 2, \ldots, \qquad \forall k \in \mathcal{N}_1^d. \qquad (12)$$

As pointed out in [24] these coefficients can be either samples of filter functions having different smoothness (see e.g. [12,21] and the references therein) or they can be connected to no filter function. This last one is the case of some generalized de la Vallée Poussin means, firstly introduced in [11] by using the convolution structure of orthogonal polynomials.

In what follows we assume $\mathbf{N} \sim \mathbf{M}$ which means

$$N_k < M_k \leq \mathcal{C}N_k, \qquad \forall k \in \mathcal{N}_1^d, \qquad \mathcal{C} \neq \mathcal{C}(\mathbf{N}, \mathbf{M}).$$

Moreover, we set

$$\mathbf{h_i^{N,M}} := \prod_{k=1}^{d} h_{i_k}^{N_k, M_k}, \qquad \mathbf{i} = (i_1, \ldots, i_d) \in \mathbb{N}^d.$$

3.1 Continuous Filtered Approximation

By means of the previous filter coefficients, we define the following filtered Fourier sum (or generalized de la Vallée Poussin mean)

$$\mathcal{V}_{\mathbf{N}}^{\mathbf{M}}(\mathbf{w}, \mathbf{f}, \mathbf{x}) = \sum_{\mathbf{i} \in \mathbb{N}[0,M]} \mathbf{h_i^{N,M}} \mathbf{c_i}(\mathbf{w}, \mathbf{f}) \mathbf{p_i}(\mathbf{w}, \mathbf{x}). \tag{13}$$

Note that this polynomial is a weighted delayed mean of the previous Fourier sums. Indeed, the following summation by part formula

$$\sum_{j=N}^{M} a_j b_j = a_M s_M + \sum_{j=N}^{M-1} s_j(a_j - a_{j+1}), \qquad s_j := \sum_{r=N}^{j} b_r, \tag{14}$$

and the assumptions in (12), yield

$$\mathcal{V}_{\mathbf{N}}^{\mathbf{M}}(\mathbf{w}, \mathbf{f}, \mathbf{x}) = \sum_{\mathbf{r} \in \mathbb{N}[N,M]} \mathbf{d_r^{N,M}} \mathbf{S_r}(\mathbf{w}, \mathbf{f}, \mathbf{x}), \tag{15}$$

where we set

$$\mathbf{d_r^{N,M}} = \prod_{k=1}^{d} (h_{r_k}^{N_k, M_k} - h_{r_k+1}^{N_k, M_k}), \qquad \mathbf{r} = (r_1, \ldots, r_d).$$

We observe that $\mathcal{V}_{\mathbf{N}}^{\mathbf{M}}(\mathbf{w}, \mathbf{f}) \in \mathbf{P_M}$. Moreover, by (12) we easily get

$$\mathcal{V}_{\mathbf{N}}^{\mathbf{M}}(\mathbf{w}, \mathbf{f}) = \mathbf{S_N}(\mathbf{w}, \mathbf{f}) = \mathbf{f}, \qquad \forall \mathbf{f} \in \mathbf{P_N}. \tag{16}$$

The following theorem supplies conditions under which the operator $\mathcal{V}_{\mathbf{N}}^{\mathbf{M}}(\mathbf{w}) : \mathbf{f} \in L_{\mathbf{u}}^{\infty} \to \mathcal{V}_{\mathbf{N}}^{\mathbf{M}}(\mathbf{w}, \mathbf{f}) \in \mathbf{P_M} \subset L_{\mathbf{u}}^{\infty}$ is uniformly bounded w.r.t. $\mathbf{N} \sim \mathbf{M}$. It can be derived from the univariate case in [24, Theorem 3.1].

Theorem 3. *Assume that the Jacobi weights* **u** *and* **w** *satisfy*

$$\left| \gamma_k - \delta_k - \frac{\alpha_k - \beta_k}{2} \right| < 1, \qquad \forall k \in \mathcal{N}_1^d \tag{17}$$

and

$$\begin{cases} \dfrac{\alpha_k}{2} - \dfrac{1}{4} < \gamma_k \leq \dfrac{\alpha_k}{2} + \dfrac{5}{4} \quad and \ \ 0 \leq \gamma_k < \alpha_k + 1, \\[2mm] \dfrac{\beta_k}{2} - \dfrac{1}{4} < \delta_k \leq \dfrac{\beta_k}{2} + \dfrac{5}{4} \quad and \ \ 0 \leq \delta_k < \beta_k + 1, \end{cases} \qquad \forall k \in \mathcal{N}_1^d. \tag{18}$$

Moreover, assume that $\mathbf{N} \sim \mathbf{M}$ *and that besides (12), the filter coefficients defining* $\mathcal{V}_{\mathbf{N}}^{\mathbf{M}}(\mathbf{w}, \mathbf{f})$ *satisfy*

$$\sum_{\ell=N_k}^{M_k} \left| \Delta^2 h_\ell^{N_k, M_k} \right| \leq \frac{\mathcal{C}}{N_k}, \quad \mathcal{C} \neq \mathcal{C}(N_k, M_k), \qquad k \in \mathcal{N}_1^d, \tag{19}$$

where, as usual, $\Delta^2 h_\ell^{N_k, M_k} = h_{\ell+2}^{N_k, M_k} - 2h_{\ell+1}^{N_k, M_k} + h_\ell^{N_k, M_k}, \ \forall k \in \mathcal{N}_1^d.$
Then for any $\mathbf{f} \in L_{\mathbf{u}}^\infty$, *we have*

$$\|\mathcal{V}_{\mathbf{N}}^{\mathbf{M}}(\mathbf{w}, \mathbf{f})\mathbf{u}\|_\infty \leq \mathcal{C}\|\mathbf{f}\mathbf{u}\|_\infty, \quad \mathcal{C} \neq \mathcal{C}(\mathbf{N}, \mathbf{M}, \mathbf{f}). \tag{20}$$

By the previous theorem, in view of the invariance property (16), the following corollary comes down

Corollary 2. *Under the assumptions of Theorem 3, for all* $\mathbf{f} \in L_{\mathbf{u}}^\infty$, *it is*

$$E_{\mathbf{M}}(\mathbf{f})_{\mathbf{u}} \leq \|[f - \mathcal{V}_{\mathbf{N}}^{\mathbf{M}}(\mathbf{w}, \mathbf{f})]\mathbf{u}\|_\infty \leq \mathcal{C}E_{\mathbf{N}}(\mathbf{f})_{\mathbf{u}}, \quad \mathcal{C} \neq \mathcal{C}(\mathbf{f}, \mathbf{N}, \mathbf{M}). \tag{21}$$

3.2   Discrete Filtered Approximation

Now we introduce the following discrete version of (13)

$$\mathbf{V}_{\mathbf{N}}^{\mathbf{M}}(\mathbf{w}, \mathbf{f}, \mathbf{x}) = \sum_{i \in \mathbb{N}[0, M]} h_i^{\mathbf{N}, \mathbf{M}} c_i^{(m)}(\mathbf{w}, \mathbf{f}) p_i(\mathbf{w}, \mathbf{x}) \tag{22}$$

$$c_i^{(m)}(\mathbf{w}, \mathbf{f}) = \sum_{r \in \mathbb{N}[1, m]} \lambda_r^{(m)}(\mathbf{w}) \mathbf{f}(\mathbf{x}_r^{(m)}) p_i(\mathbf{w}, \mathbf{x}_r^{(m)}),$$

obtained by approximating the coefficients $c_i(\mathbf{w}, \mathbf{f})$ in (13) by the $\mathbf{m} - th$ Gauss-Jacobi rule in (7), with

$$\mathbf{m} = \left\lceil \frac{\mathbf{M} + \mathbf{N} + 1}{2} \right\rceil.$$

This choice of **m** assures that

$$\mathbf{V}_{\mathbf{N}}^{\mathbf{M}}(\mathbf{w}, \mathbf{f}, \mathbf{x}) = \mathcal{V}_{\mathbf{N}}^{\mathbf{M}}(\mathbf{w}, \mathbf{f}) = \mathbf{f}, \qquad \forall \mathbf{f} \in \mathbf{P}_{\mathbf{N}}.$$

Moreover, from [24, Theorem 4.1], the following near–best approximation result can be deduced.

Theorem 4. *Assume that* **u** *and* **w** *satisfy (17) and*

$$
\begin{cases}
\dfrac{\alpha_k}{2} - \dfrac{1}{4} < \gamma_k \le \dfrac{\alpha_k}{2} + \dfrac{5}{4} & \text{and } \gamma_k \ge 0, \\[2mm]
\dfrac{\beta_k}{2} - \dfrac{1}{4} < \delta_k \le \dfrac{\beta_k}{2} + \dfrac{5}{4} & \text{and } \delta_k \ge 0,
\end{cases}
\qquad \forall k \in \mathcal{N}_1^d. \qquad (23)
$$

Moreover, let be **N** ~ **M**, *and assume that the filter coefficients defining* $V_N^M(\mathbf{w}, \mathbf{f})$ *satisfy (12) and (19).*
Then for any $\mathbf{f} \in L_\mathbf{u}^\infty$, *we have*

$$
\|V_N^M(\mathbf{w}, \mathbf{f})\mathbf{u}\|_\infty \le \mathcal{C}\|\mathbf{fu}\|_\infty, \quad \mathcal{C} \ne \mathcal{C}(\mathbf{N}, \mathbf{M}, \mathbf{f}) \qquad (24)
$$

and

$$
E_\mathbf{M}(\mathbf{f})_\mathbf{u} \le \|[\mathbf{f} - V_N^M(\mathbf{w}, \mathbf{f})]\mathbf{u}\|_\infty \le \mathcal{C}E_\mathbf{N}(\mathbf{f})_\mathbf{u}, \quad \mathcal{C} \ne \mathcal{C}(\mathbf{M}, \mathbf{N}, \mathbf{f}). \qquad (25)
$$

In conclusion, under the assumptions of the previous theorems, we have that for any $\mathbf{f} \in C_\mathbf{u}$, both the polynomials $\mathcal{V}_N^M(\mathbf{w}, \mathbf{f})$ and $V_N^M(\mathbf{w}, \mathbf{f})$ are near–best approximation polynomials converging to **f** as **N** → ∞ and **M** ~ **N**, with the same order of the best polynomial approximation of **f**.

4 Numerical Experiments

Now we propose some tests exploiting the behaviors of the discrete filtered polynomials previously introduced, in the case $D = [-1, 1]^2$.

We point out that in all the experiments we focus on the number $m_1 \cdot m_2$ of nodes determined by $\mathbf{m} = (m_1, m_2)$. For simplicity, we fix $m_1 = m_2 = m$ and, for any given $0 < \theta < 1$, we take

$$
\mathbf{N} = (1 - \theta)\mathbf{m}, \qquad \text{and} \qquad \mathbf{M} = (1 + \theta)\mathbf{m},
$$

so that the assumption **N** ~ **M** is satisfied. Moreover, unless stated otherwise, we set $\theta = 0.6$.

As filter coefficients, we take

$$
h_\ell^{N_k, M_k} = h\left(1 + \frac{\ell - N_k}{M_k - N_k + 1}\right), \qquad \ell \in \mathbb{N}, \qquad k \in \{1, 2\},
$$

where the filter function $h(x)$ vanishes outside of $[0, 2]$, it is equal 1 on $[0, 1]$ and for $x \in (1, 2]$ it is given by

$$
h(x) = \begin{cases}
2 - x, & \text{classic de la Vallée Poussin filter (VP filter)}, \\[2mm]
\dfrac{\sin(\pi(x-1))}{\pi(x-1)}, & \text{Lanczos filter}, \\[2mm]
\dfrac{1 + \cos(\pi(x-1))}{2}, & \text{raised-cosine filter}.
\end{cases} \qquad (26)
$$

Note that all the previous filter choices satisfy the hypotheses (12) and (19), required in our main results.

Moreover, all the computations have been performed in double-machine precision ($eps \sim 2.220446049250313e - 16$) by MatLab version R2018a.

In the first test we want to compare the approximation provided by the discrete filtered polynomial $\mathbf{V}^{N,M}(\mathbf{w}, \mathbf{f})$ corresponding to the classical VP filter, with that one of the Lagrange polynomial $\mathbf{L}_{m-1}(\mathbf{w}, \mathbf{f})$ based on the same nodes. To this aim, for increasing m, we compute the respective Lebesgue constant (LC), namely the following operator norms

$$\|\mathbf{L}_{m-1}(\mathbf{w})\|_{L_u^\infty} := \sup_{\mathbf{f} \neq 0} \frac{\|\mathbf{L}_{m-1}(\mathbf{w}, \mathbf{f})\mathbf{u}\|_\infty}{\|\mathbf{f}\mathbf{u}\|_\infty},$$

$$\|\mathbf{V}^{N,M}(\mathbf{w})\|_{L_u^\infty} := \sup_{\mathbf{f} \neq 0} \frac{\|\mathbf{V}^{N,M}(\mathbf{w}, \mathbf{f})\mathbf{u}\|_\infty}{\|\mathbf{f}\mathbf{u}\|_\infty},$$

according to the formulas

$$\|\mathbf{L}_{m-1}(\mathbf{w})\|_{L_u^\infty} = \sup_{\mathbf{x} \in D} \sum_{i \in \mathbb{N}[1,m]} \Lambda_i^{(m)} \left| K_m(\mathbf{w}, \mathbf{x}, \mathbf{x}_i^{(m)}(\mathbf{w})) \right| \frac{\mathbf{u}(\mathbf{x})}{\mathbf{u}(\mathbf{x}_i^{(m)}(\mathbf{w}))},$$

$$\|\mathbf{V}_N^M(\mathbf{w})\|_{L_u^\infty} = \sup_{\mathbf{x} \in D} \sum_{i \in \mathbb{N}[1,m]} \Lambda_i^{(m)} \left| v^{N,M}(\mathbf{w}, \mathbf{x}, \mathbf{x}_i^{(m)}(\mathbf{w})) \right| \frac{\mathbf{u}(\mathbf{x})}{\mathbf{u}(\mathbf{x}_i^{(m)}(\mathbf{w}))},$$

where

$$K_m(\mathbf{w}, \mathbf{x}, \mathbf{y}) := \sum_{r \in \mathbb{N}[0,m]} p_r(\mathbf{w}, \mathbf{x}) p_r(\mathbf{w}, \mathbf{y}),$$

$$v^{N,M}(\mathbf{w}, \mathbf{x}, \mathbf{y}) := \sum_{r \in \mathbb{N}[0,M]} h_r^{N,M} p_r(\mathbf{w}, \mathbf{x}) p_r(\mathbf{w}, \mathbf{y}).$$

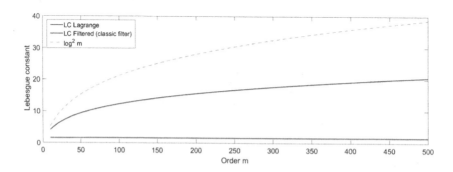

Fig. 1. LC of Lagrange and filtered operators for $\alpha_k = \beta_k = -0.7, \gamma_k = \delta_k = 0$, $k = 1, 2$.

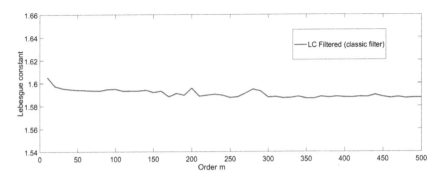

Fig. 2. LC of filtered operator for $\alpha_k = \beta_k = -0.7$, $\gamma_k = \delta_k = 0$, $k = 1, 2$.

The results are displayed in Fig. 1 for the unweighted case $\mathbf{u}(\mathbf{x}) = 1$ and, for the sake of clarity, the plot of LC associated with filtered approximation is repeated in Fig. 2.

The second experiment concerns the weighted approximation, namely the case $\mathbf{u}(\mathbf{x}) \neq 1$. In order to test the goodness of the convergence ranges in the assumptions (17) and (23), we computed, for increasing m, the LC of the filtered operator associated with the classical VP filter for two different choices of $\mathbf{w}$. Figure 3 shows the resulting behaviors in a case where the previous assumptions hold (left plot), and in another case where they do not (right plot).

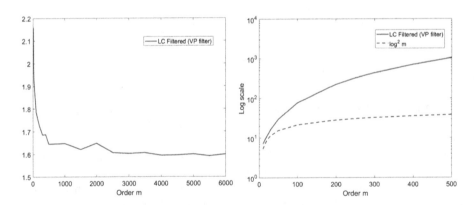

Fig. 3. LC for parameters $\alpha_k = \beta_k = 0.5$, $\gamma_k = \delta_k = 1$ ($k = 1, 2$) on the left hand side, and parameters (out of ranges) $\alpha_k = \beta_k = 0.6$, $\gamma_k = \delta_k = 2$ ($k = 1, 2$) on the right–hand side

In the third experiment we are going to fix the couple of weights $u_k = 1$, $w_k = v^{-\frac{1}{2},-\frac{1}{2}}$, $k = 1, 2$, and we let to vary the filter coefficients or the parameter θ. More precisely, in Fig. 4 we show the behavior of $\|\mathbf{V}^{\mathbf{N,M}}(\mathbf{w})\|_{L_u^\infty}$ for the different filter functions in (26), while Fig. 5 displays the LC corresponding to the classic VP filter and θ varying in $\{0.1 : 0.1 : 0.9\}$.

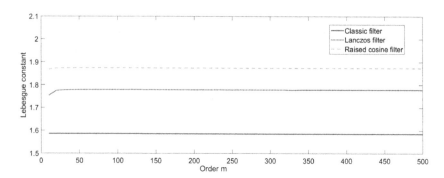

Fig. 4. LC for different filters and $\alpha_k = \beta_k = -0.5$, $\gamma_k = \delta_k = 0$, $k \in \mathcal{N}_1^2$.

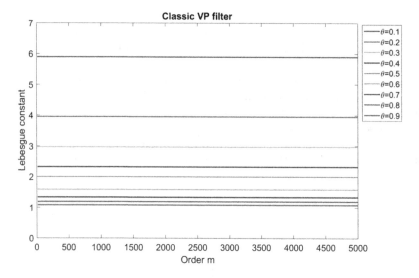

Fig. 5. LC for several theta and $\alpha_k = \beta_k = -0.5$, $\gamma_k = \delta_k = 0$, $k = 1, 2$.

Finally, the case of classical VP filter and variable θ is also shown in Fig. 6 for a different choice of the weight $\mathbf{w}$ defining the filtered operator $\mathbf{V}^{N,M}(\mathbf{w})$.

The last experiment deals with the Gibbs phenomenon. We consider the approximation of the following two bounded variation functions

$$\mathbf{f}_1(\mathbf{x}) = \text{sign}(x_1) + \text{sign}(x_2), \qquad \mathbf{x} = (x_1, x_2) \in [-1, 1]^2,$$

and

$$\mathbf{f}_2(\mathbf{x}) = \begin{cases} 1, \text{ if } & x_1^2 + x_2^2 \le 0.6^2 \\ 0, otherwise \end{cases} \qquad \mathbf{x} = (x_1, x_2) \in [-1, 1]^2,$$

the last function being also considered in [8].

In these cases it is well–known that Lagrange polynomials present overshoots and oscillations close to the singularities, which are preserved also in the regular

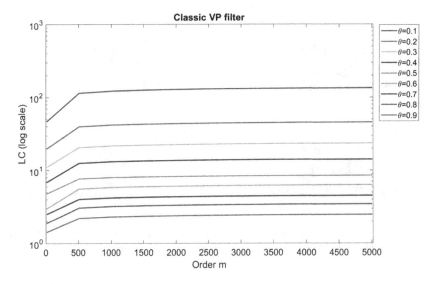

Fig. 6. LC for several theta and $\alpha_k = \beta_k = 0.2$, $\gamma_k = \delta_k = 0$, $k = 1, 2$.

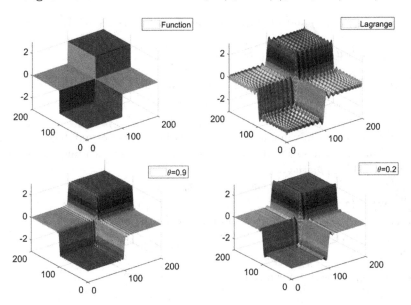

Fig. 7. Lagrange and filtered (VP filter) polynomials of the function $\mathbf{f}_1$ (plotted at the top) for $m = 50$ (2500 nodes) and parameters $\alpha_k = \beta_k = \gamma_k = \delta_k = 0$ $(k = 1, 2)$.

part of the function. Figures 7 and 8 show that this phenomenon can be strongly reduced if we take the discrete filtered (VP filter) polynomial based on the same nodes set and a suitable value of the parameter $\theta \in (0, 1)$.

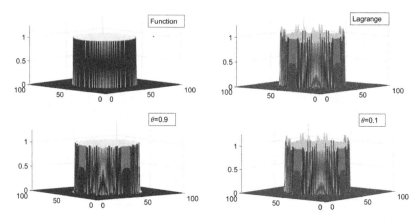

Fig. 8. Lagrange and filtered (VP filter) polynomial of the function $\mathbf{f}_2$ (plotted at the top) for $m = 300$ and parameters $\alpha_k = \beta_k = -0.5$, $\gamma_k = \delta_k = 0$ ($k = 1, 2$).

List of Notations

- $\mathbf{x} = (x_1, x_2, \ldots, x_d)$ and similarly for other bold letters ($\mathbf{N}, \mathbf{M}, \mathbf{i}$ etc.) denoting vectors
- $D = [-1, 1]^d$, $d > 1$ ($d = 2$ in Sect. 4)
- $\mathcal{N}_1^d = \{1, \ldots, d\}$
- $\mathbb{N}[\mathbf{r}, \mathbf{s}] = \{\mathbf{i} = (i_1, \ldots, i_d) \in \mathbb{N}^d \ : \ i_k \in [r_k, s_k], \ \forall k \in \mathcal{N}_1^d\}$
- $\mathbf{u}(\mathbf{x}) = v^{\gamma_1, \delta_1}(x_1) \cdot v^{\gamma_2, \delta_2}(x_2) \cdots v^{\gamma_d, \delta_d}(x_d) = \prod_{k=1}^d u_k(x_k)$
- $\mathbf{w}(\mathbf{x}) = v^{\alpha_1, \beta_1}(x_1) \cdot v^{\alpha_2, \beta_2} \cdots v^{\alpha_d, \beta_d}(x_d) = \prod_{k=1}^d w_k(x_k)$
- $p_m(w, t)$ denotes the value at t of the univariate orthonormal Jacobi polynomial of degree m associated with w and having positive leading coefficient
- $p_{\mathbf{i}}(\mathbf{w}, \mathbf{x}) = \prod_{k=1}^d p_{i_k}(w_k, x_k)$
- $x_{m,\ell}(w)$ ($\ell = 1, \ldots, m$) are the zeros of $p_m(w, t)$
- $\mathbf{x}_{\mathbf{i}}^{(m)}(\mathbf{w}) = (x_{m_1, i_1}(w_1), x_{m_2, i_2}(w_2), \ldots, x_{m_d, i_d}(w_d))$,
- $\lambda_{m,\ell}(w)$ ($\ell = 1, \ldots, m$) are the univariate Christoffel numbers of order m associated with w, i.e. $\lambda_{m,\ell}(w) = [\sum_{j=0}^{m-1} p_j^2(w, x_{m,\ell}(w))]^{-1}$
- $\Lambda_{\mathbf{i}}^{(m)}(\mathbf{w}) = \prod_{k=1}^d \lambda_{m_k, i_k}(w_k)$
- $\log \mathbf{m} = \prod_{k=1}^d \log m_k$, for any $\mathbf{m} = (m_1, \ldots, m_d) \in \mathbb{N}^d$
- $\mathbf{h}_{\mathbf{i}}^{\mathbf{N}, \mathbf{M}} = \prod_{k=1}^d h_{i_k}^{N_k, M_k}$

Acknowledgments. The authors are grateful to the anonymous referees for carefully reviewing this paper and for their valuable comments and suggestions.

This research has been accomplished within Rete ITaliana di Approssimazione (RITA) and partially supported by the GNCS-INdAM funds 2019, project "Discretizzazione di misure, approssimazione di operatori integrali ed applicazioni".

References

1. Bos, L., Caliari, M., De Marchi, S., Vianello, M.: A numerical study of the Xu polynomial interpolation formula in two variables. Computing **76**, 311–324 (2006)
2. Bos, L., Caliari, M., De Marchi, S., Vianello, M., Xu, Y.: Bivariate Lagrange interpolation at the Padua points: the generating curve approach. J. Approx. Theory **143**, 15–25 (2006)
3. Caliari, M., De Marchi, S., Vianello, M.: Bivariate Lagrange interpolation at the Padua points: computational aspects. J. Comput. Appl. Math. **221**, 284–292 (2008)
4. Capobianco, M.R., Criscuolo, G., Junghanns, P., Luther, U.: Uniform convergence of the collocation method for Prandtl's integro-differential equation. ANZIAM J. **42**(1), 151–168 (2000)
5. Capobianco, M.R., Themistoclakis, W.: Interpolating polynomial wavelets on $[-1, 1]$. Adv. Comput. Math. **23**(4), 353–374 (2005)
6. De Bonis, M.C., Occorsio, D.: Quadrature methods for integro-differential equations of Prandtl's type in weighted uniform norms. AMC (to appear)
7. De Bonis, M.C., Occorsio, D.: On the simultaneous approximation of a Hilbert transform and its derivatives on the real semiaxis. Appl. Numer. Math. **114**, 132–153 (2017)
8. De Marchi, S., Erb, W., Marchetti, F.: Spectral filtering for the reduction of the gibbs phenomenon for polynomial approximation methods on Lissajous curves with applications in MPI. Dolom. Res. Notes Approx. **10**, 128–137 (2017)
9. Fermo, L., Russo, M.G., Serafini, G.: Numerical methods for Cauchy bisingular integral equations of the first kind on the square. J. Sci. Comput. **79**(1), 103–127 (2019)
10. Filbir, F., Mhaskar, H.N., Prestin, J.: On a filter for exponentially localized kernels based on Jacobi polynomials. J. Approx. Theory **160**, 256–280 (2009)
11. Filbir, F., Themistoclakis, W.: On the construction of de la Vallée Poussin means for orthogonal polynomials using convolution structures. J. Comput. Anal. Appl. **6**, 297–312 (2004)
12. Gottlieb, D., Shu, C.-W.: On the Gibbs phenomenon and its resolution. SIAM Rev. **39**(4), 644–668 (1997)
13. Mhaskar, H.N.: Localized summability kernels for Jacobi expansions. In: Rassias, T.M., Gupta, V. (eds.) Mathematical Analysis, Approximation Theory and Their Applications. SOIA, vol. 111, pp. 417–434. Springer, Cham (2016). https://doi.org/10.1007/978-3-319-31281-1_18
14. Mastroianni, G., Milovanovic, G.: Interpolation Processes. Basic Theory and Applications. Springer Monographs in Mathematics. Springer, Heidelberg (2008). https://doi.org/10.1007/978-3-540-68349-0
15. Mastroianni G., Milovanović, G., Occorsio, D.: A Nyström method for two variables Fredholm integral equations on triangles. Appl. Math. Comput. **219**, 7653–7662 (2013)
16. Mastroianni, G., Russo, M.G., Themistoclakis, W.: The boundedness of the Cauchy singular integral operator in weighted Besov type spaces with uniform norms. Integr. Eqn. Oper. Theory **42**, 57–89 (2002)
17. Mastroianni, G., Themistoclakis, W.: A numerical method for the generalized airfoil equation based on the de la Vallée Poussin interpolation. J. Comput. Appl. Math. **180**, 71–105 (2005)
18. Occorsio, D., Russo, M.G.: Numerical methods for Fredholm integral equations on the square. Appl. Math. Comput. **218**(5), 2318–2333 (2011)

19. Occorsio, D., Russo, M.G.: Nyström methods for Fredholm integral equations using equispaced points. Filomat **28**(1), 49–63 (2014)
20. Occorsio, D., Themistoclakis, W.: Uniform weighted approximation on the square by polynomial interpolation at Chebyshev nodes. Submitted to a Special Iusse of NUMTA 2019
21. Sloan, I.H., Womersley, R.S.: Filtered hyperinterpolation: a constructive polynomial approximation on the sphere. Int. J. Geomath. **3**, 95–117 (2012)
22. Themistoclakis, W.: Weighted L^1 approximation on $[-1, 1]$ via discrete de la Vallée Poussin means. Math. Comput. Simul. **147**, 279–292 (2018)
23. Themistoclakis, W.: Uniform approximation on $[-1, 1]$ via discrete de la Vallée Poussin means. Numer. Algorithms **60**, 593–612 (2012)
24. Themistoclakis, W., Van Barel, M.: Generalized de la Vallée Poussin approximations on $[-1, 1]$. Numer. Algorithms **75**, 1–31 (2017)
25. Themistoclakis, W., Vecchio, A.: On the solution of a class of nonlinear systems governed by an M -matrix. Discrete Dyn. Nat. Soc. **2012**, 12 (2012)
26. Xu, Y.: Lagrange interpolation on Chebyshev: points of two variables. J. Approx. Theory **87**(2), 220–238 (1996)

Computational Methods for Data Analysis

A Travelling Wave Solution for Nonlinear Colloid Facilitated Mass Transport in Porous Media

Salvatore Cuomo[1]($^\boxtimes$)(iD), Fabio Giampaolo[1](iD), and Gerardo Severino[2](iD)

[1] Department of Mathematics, University of Naples - Federico II, Naples, Italy
salvatore.cuomo@unina.it, fab.giampaolo@gmail.com
[2] Department of Agricultural Sciences, University of Naples - Federico II,
Naples, Italy
severino@unina.it

Abstract. Colloid facilitated solute transport through porous media is investigated. Sorption on the matrix is modelled by the linear equilibrium isotherm whereas sorption on colloidal sites is regulated by nonlinearly equilibrium *vs* nonequilibrium. A travelling wave-type solution is obtained to describe the evolution in both the liquid and colloidal concentration.

Keywords: Nonlinear transport · Colloids · Travelling wave

1   Introduction

Aquifers' contamination and adverse effects on the environment have become a matter of considerable concern. Hence, it would be desirable having a predicting model to asses the effects of such a contamination risk [13]. A large number of mathematical models have been developed in the past [11,12,15]. Current models often emphasize either physical or chemical aspects of transport [2,8]. However, a general approach relies upon multicomponent transport modelling, and it has been only recently formulated in [6]. In this case, both adsorption and complexation in solution were taken into account along with precipitation/dissolution phenomena. In particular, being concerned with complex chemistry, such a model comes with non linear transport equations. The main interest related to the solution for such a system of PDEs is that it allows one to analyze the sensitivity of mass transport to the variation of different parameters. Unfortunately, generally analytical (closed form) solutions are not achievable, and concurrently one has to resort with numerical approximation. However, such a stand point may arise two serious issues. One is about the discretization in order to suppress instability and numerical dispersion [9]. The other is that in some situations (typically at the very beginning of the transport process) very steep concentration(s) gradients may develop. To accurately monitor these gradients, very fine discretizations may be necessary: an undesirable

© Springer Nature Switzerland AG 2020
Y. D. Sergeyev and D. E. Kvasov (Eds.): NUMTA 2019, LNCS 11973, pp. 103–108, 2020.
https://doi.org/10.1007/978-3-030-39081-5_10

situation in view of computational times [4]. Alternatively, one can introduce a few approximations which enable one to gain analytical solution. In particular, in the present paper a travelling wave solution is derived for the case of colloid facilitaed mass transport with nonlinear sorption/desorption.

2   The Transport Model

We consider steady flow with constant velocity U in a one dimensional porous medium that carries simultaneously solute and colloidal particles. In such a system the mass flux vectors for the colloids and solute are

$$J_c(z,t) = nC_c(z,t)U_c(z,t), \quad J(z,t) = nC(z,t)U(z,t) - \overline{D}\frac{\partial C(z,t)}{\partial z} \quad (1)$$

respectively, where C_c and C are the concentrations (per unit volume of fluid) of colloidal and solute particles, $\overline{D}$ is the pore scale dispersion and n the porosity. Generally, advection colloidal velocity U_c is larger than U, due to "exclusion phenomena". We model this 'exclusion' process by assuming that U_c depends linearly upon U through a constant coefficient $R_e \geq 1$, i.e. $U_c = R_e U$. Colloidal generation/removal is quantified by the mass balance equation for colloidal particles. However, in the context of the present paper we neglect both generation and removal, such that one can assume that the concentration C_c of colloidal particles is uniform. The applicability of such an assumption is throughly discussed in [5].

Let S denote the solute concentration on colloids, defined per unit colloidal concentration; the actual solute concentration C_c^s on colloidal particles is SC_c. Furthermore, we denote with N the solute concentration (mass of sorbed solute per unit bulk volume) sorbed on the porous matrix. Thus, the total solute concentration writes as $C_t = n(C + C_c^s) + N$, and concurrently the (solute) mass balance equation is

$$\frac{\partial}{\partial t}[n(C + C_c^s) + N] + \frac{\partial}{\partial z}(J_c + J) = 0. \quad (2)$$

Substitution of (1) into (2) leads to the reaction diffusion equation

$$\frac{\partial C}{\partial t} + \frac{\partial C_c^s}{\partial t} + U\frac{\partial}{\partial z}(R_e C_c^s + C) = -\frac{1}{n}\frac{\partial N}{\partial t} + D\frac{\partial^2 C}{\partial z^2} \quad \left(D = \frac{\overline{D}}{n}\right) \quad (3)$$

where, for simplicity, we have regarded n and $\overline{D}$ as constant. Generally, both C_c^s and N depend upon C in a very complex fashion, however we shall assume that sorption on the porous medium is governed only by the linear equilibrium model, i.e. $N = nK_d C$ (K_d is the linear partitioning coefficient between the fluid and sorbed phase), and we mainly focus on the effect of mass exchange between the fluid and colloidal solute concentration. For this reason, we consider a quite general dependence of C_c^s upon C accounting for a nonlinear mass exchange, i.e.

$$\frac{\partial}{\partial t} C_c^s = L\left[\varphi\left(C\right) - C_c^s\right], \tag{4}$$

being L a given rate transfer coefficient. We assume that solute is continuously injected at $z = 0$:

$$C\left(0, t\right) = C_0, \tag{5}$$

whit zero initial C-concentration. The nonlinear reaction function φ describes the equilibrium between the two phases, and its most used expressions are those of Langmuir and Freudlich [8]. When kinetics is fast enough, i.e. $L \gg 1$, the left hand side of (4) may be neglected, up to a transitional boundary layer [10], therefore leading to a non linear equilibrium sorption process.

3   Travelling Wave Solution

Overall, the system of Eqs. (3)–(4) can not be solved analytically. To obtain a simple solution (which nevertheless keeps the main physical insights), we consider a travelling wave type solution. Specifically, we assume that a travelling wave solution, that generally occurs after a large time. This allows one to assume that C at the inlet boundary is approximately equal to the feed concentration C_0, so that the initial/boundary conditions may be approximate as follows

$$C(z, t) \simeq \begin{cases} C_0 & \text{for } z = -\infty \\ 0 & \text{for } z = +\infty, \end{cases} \tag{6}$$

for any $t \geq 0$. We introduce the moving coordinate system $\eta = z - \alpha t$, where α represents the constant speed of the travelling wave (that will be determined later on). We assume that each concentration, both in the liquid and in the colloidal phase, moves with the same velocity α, which authorizes to write $C(z, t) = C(\eta)$ and $C_c^s(z, t) = C_c^s(\eta)$. Of course this approximation works better and better as the time increases. With these assumptions, the system (3)–(4) is reduced to an ODE-problem

$$\begin{cases} -\alpha \dfrac{\mathrm{d}}{\mathrm{d}\eta}\left(RC + C_c^s\right) + U\dfrac{\mathrm{d}}{\mathrm{d}\eta}\left(C + C_c^s\right) = 0 \\[2ex] -\alpha \dfrac{\mathrm{d}}{\mathrm{d}\eta} C_c^s = L\left[\varphi\left(C\right) - C_c^s\right] \end{cases} \tag{7}$$

with boundary conditions given by

$$C\left(-\infty\right) = C_0, \qquad C\left(+\infty\right) = 0 \tag{8}$$
$$C_c^s\left(-\infty\right) = \varphi_0, \qquad C_c^s\left(+\infty\right) = \varphi\left(0\right) = 0, \tag{9}$$

where $\varphi_0 \equiv \varphi\left(C_0\right)$. We now identify the wave velocity α by integrating the first of (7) over η

$$- \alpha \left(RC + C_c^s\right) + U \left(C + C_c^s\right) = A = \text{const.} \tag{10}$$

From boundary conditions (8)–(9), one has

$$A = 0 \quad \Rightarrow \quad \alpha = U \left(\frac{C_0 + \varphi_0}{RC_0 + \varphi_0}\right). \tag{11}$$

that inserted in (7) yields, after some algebra, the following boundary value problem for C:

$$\begin{cases} \dfrac{\mathrm{d}}{\mathrm{d}\eta}C = \tau_0 \left[\dfrac{\varphi_0}{C_0}C - \varphi\left(C\right)\right] \\ C\left(-\infty\right) = C_0, \quad C\left(+\infty\right) = 0, \end{cases} \tag{12}$$

with the constant τ_0 defined as:

$$\tau_0 = \frac{L}{\alpha} = \frac{L}{U}\left(\frac{\varphi_0 + RC_0}{\varphi_0 + C_0}\right). \tag{13}$$

Solution of (12) is possible if $\varphi \equiv \varphi\left(x\right)$ is a "convex" isotherm (i.e. $\mathrm{d}^2\varphi/\mathrm{d}x^2 < 0$). This is always the case for Freundlich/Langmuir type φ-function. Moreover, the presence of the reaction rate L gives smooth concentration profiles that become steeper and steeper as L increases, in particular for $L \to \infty$ solution of (12) asymptotically behaves as the one pertaining to the (nonlinear) equilibrium [8]. For the application to the quantification of the pollution risk, it is important to quantify the extent of the concentration front S_f. This latter exists if and only if

$$\int_0^\varepsilon \frac{\mathrm{d}x}{\varphi\left(x\right)} < \infty \qquad \text{for any } \varepsilon > 0. \tag{14}$$

The differential equation in (12) is expressed in integral form upon integration with respect to an arbitrary reference $C\left(\eta_r\right) = C_r$. The final result is

$$\tau_0\left(\eta_r - \eta\right) = \int_C^{C_r} \frac{\mathrm{d}x}{\dfrac{\varphi_0}{C_0}x - \varphi\left(x\right)} = G\left(C, C_r\right), \tag{15}$$

being the shape of the G-function depending upon the structure of the reaction function φ. Since the solution (15) is given as function of an arbitrary reference value $C_r \equiv C\left(\eta_r\right)$, it is important to identify the position η_r pertaining to such a concentration value. The adopted approach relies on the method of moments [1, 3, 7]. In particular, we focus on the zero-order moments [L], i.e.

$$\eta_X = \frac{1}{C_0}\int_{-\infty}^{+\infty} \mathrm{d}\eta\, C\left(\eta\right), \qquad \eta_Y = \frac{1}{\varphi_0}\int_{-\infty}^{+\infty} \mathrm{d}\eta\, \varphi\left(\eta\right). \tag{16}$$

Before proceeding further, we wish noting here that when the number (say n) of unknown parameters is greater than 1, one has to sort with moments of order $0, 1, \ldots, n-1$. The moment η_X can be regarded as the distance along which C has increased from 0 to C_0 (at a fixed time t). A similar physical insight can be attached to the moment η_Y pertaining to the sorbed concentration on colloidal sites. While integrals (16) are not bounded, their difference it is. Indeed, integration from $\eta = -\infty$ to $\eta = +\infty$ of the ODE in (12) by virtue of (16) leads to $\eta_Y - \eta_X = \frac{C_0}{\tau_0 \, \varphi_0}$. We can now focus on how to use moments (16) to identify the reference point η_r. Thus, we integrate (15) from 0 to C_0, and division by C_0 results in $\eta_r - \eta_X = \frac{\widetilde{G}}{\tau_0 \, C_0}$, with $\widetilde{G} = \int_0^{C_0} \mathrm{d}x \, G(x, C_r)$. Once $\widetilde{G}$ has been computed, the reference position is uniquely fixed.

4   Concluding Remarks

A travelling wave solution for colloid facilitated mass transport through porous media has been obtained. A linear, reversible kinetic relation has been assumed to account for mass transfer from/toward colloidal particles. This leads to a simple BV-problem, that can be solved by means of a standard finite difference numerical method. The present study is also the fundamental prerequisite to investigate the dispersion mechanisms of pullatants under under (more complex) flow configurations along the lines of [14]. Some of them are already part of ongoing research projects.

References

1. Aris, R.: On the dispersion of linear kinematic waves. Proc. Roy. Soc. Lond. Ser. A Math. Phys. Sci. **245**(1241), 268–277 (1958)
2. Bellin, A., Severino, G., Fiori, A.: On the local concentration probability density function of solutes reacting upon mixing. Water Resour. Res. **47**(1), W01514 (2011). https://doi.org/10.1029/2010WR009696
3. Bolt, G.: Movement of solutes in soil: principles of adsorption/exchange chromatography. In: Developments in Soil Science, vol. 5, pp. 285–348 (1979)
4. Campagna, R., Cuomo, S., Giannino, F., Severino, G., Toraldo, G.: A semi-automatic numerical algorithm for turing patterns formation in a reaction-diffusion model. IEEE Access **6**, 4720–4724 (2018)
5. Cvetkovic, V.: Colloid-facilitated tracer transport by steady random ground-water flow. Phys. Fluids **12**(9), 2279–2294 (2000)
6. Severino, G., Campagna, R., Tartakovsky, D.: An analytical model for carrier-facilitated solute transport in weakly heterogeneous porous media. Appl. Math. Model. **44**, 261–273 (2017). https://doi.org/10.1016/j.apm.2016.10.064
7. Severino, G., Comegna, A., Coppola, A., Sommella, A., Santini, A.: Stochastic analysis of a field-scale unsaturated transport experiment. Adv. Water Resour. **33**(10), 1188–1198 (2010). https://doi.org/10.1016/j.advwatres.2010.09.004
8. Severino, G., Dagan, G., van Duijn, C.: A note on transport of a pulse of nonlinearly reactive solute in a heterogeneous formation. Comput. Geosci. **4**(3), 275–286 (2000). https://doi.org/10.1023/A:1011568118126
9. Severino, G., Leveque, S., Toraldo, G.: Uncertainty quantification of unsteady source flows in heterogeneous porous media. J. Fluid. Mech. **870**, 5–26 (2019). https://doi.org/10.1017/jfm.2019.203

10. Severino, G., Monetti, V., Santini, A., Toraldo, G.: Unsaturated transport with linear kinetic sorption under unsteady vertical flow. Transp. Porous Media **63**(1), 147–174 (2006). https://doi.org/10.1007/s11242-005-4424-0
11. Severino, G., Tartakovsky, D., Srinivasan, G., Viswanathan, H.: Lagrangian models of reactive transport in heterogeneous porous media with uncertain properties. Proc. Roy. Soc. A Math. Phys. Eng. Sci. **468**(2140), 1154–1174 (2012)
12. Severino, G., Cvetkovic, V., Coppola, A.: On the velocity covariance for steady flows in heterogeneous porous formations and its application to contaminants transport. Comput. Geosci. **9**(4), 155–177 (2005). https://doi.org/10.1007/s10596-005-9005-3
13. Severino, G., Cvetkovic, V., Coppola, A.: Spatial moments for colloid-enhanced radionuclide transport in heterogeneous aquifers. Adv. Water Resour. **30**(1), 101–112 (2007). https://doi.org/10.1016/j.advwatres.2006.03.001
14. Severino, G., De Bartolo, S., Toraldo, G., Srinivasan, G., Viswanathan, H.: Travel time approach to kinetically sorbing solute by diverging radial flows through heterogeneous porous formations. Water Resour. Res. **48**(12) (2012). https://doi.org/10.1029/2012WR012608
15. Severino, G., Santini, A., Monetti, V.M.: Modelling water flow water flow and solute transport in heterogeneous unsaturated porous media. In: Pardalos, P.M., Papajorgji, P.J. (eds.) Advances in Modeling Agricultural Systems, vol. 25, pp. 361–383. Springer, Boston (2009). https://doi.org/10.1007/978-0-387-75181-8_17

Performance Analysis of a Multicore Implementation for Solving a Two-Dimensional Inverse Anomalous Diffusion Problem

Pasquale De Luca[1]([✉])[iD], Ardelio Galletti[2][iD], Giulio Giunta[2][iD],
Livia Marcellino[2][iD], and Marzie Raei[3]

[1] Department of Computer Science, University of Salerno, Fisciano, Italy
p.deluca16@studenti.unisa.it
[2] Department of Science and Technology, University of Naples Parthenope,
Naples, Italy
{ardelio.galletti,giulio.giunta,livia.marcellino}@uniparthenope.it
[3] Department of Mathematics, Malek Ashtar University of Technology, Isfahan, Iran
marzie.raei@gmail.com

Abstract. In this work we deal with the solution of a two-dimensional inverse time fractional diffusion equation, involving a Caputo fractional derivative in his expression. Since we deal with a huge practical problem with a large domain, by starting from an accurate meshless localized collocation method using RBFs, here we propose a fast algorithm, implemented in a multicore architecture, which exploits suitable parallel computational kernels. More in detail, we firstly developed, a C code based on the numerical library LAPACK to perform the basic linear algebra operations and to solve linear systems, then, due to the high computational complexity and the large size of the problem, we propose a parallel algorithm specifically designed for multicore architectures and based on the Pthreads library. Performance analysis will show accuracy and reliability of our parallel implementation.

Keywords: Fractional models · Multicore architecture · Parallel algorithms

1 Introduction

In recent decades, fractional calculus has become highly attractive due to wide applications in science and engineering. Indeed, fractional models are beneficial and powerful mathematical tools to describe the inherent properties of processes in mechanics, chemistry, physics, and other sciences. Meshless methods represent a good technique for solving these models in high-dimensional and complicated computational domains. In particular, in the current work we deal with the

© Springer Nature Switzerland AG 2020
Y. D. Sergeyev and D. E. Kvasov (Eds.): NUMTA 2019, LNCS 11973, pp. 109–121, 2020.
https://doi.org/10.1007/978-3-030-39081-5_11

solution of a two-dimensional inverse time fractional diffusion equation [1,2], defined as follows:

$$_{0}^{c}D_{t}^{\alpha}v(\mathbf{x},t) = \kappa\Delta v(\mathbf{x},t) + f(\mathbf{x},t), \qquad \mathbf{x} = (x,y) \in \Omega \subseteq R^{2}, \quad t \in]0,T], \quad (1)$$

with the following initial and Dirichlet boundary conditions:

$$\begin{aligned}
v(\mathbf{x},0) &= \varphi(\mathbf{x}), & \mathbf{x} &\in \Omega, \\
v(\mathbf{x},t) &= \psi_{1}(\mathbf{x},t), & \mathbf{x} &\in \Gamma_{1}, \quad t \in]0,T], \\
v(\mathbf{x},t) &= \psi_{2}(\mathbf{x})\rho(t), & \mathbf{x} &\in \Gamma_{2}, \quad t \in]0,T],
\end{aligned} \qquad (2)$$

and the non-local boundary condition

$$\iint\limits_{\Omega} v(\mathbf{x},t)d\mathbf{x} = h(t), \qquad t \in]0,T], \qquad (3)$$

where $v(\mathbf{x},t)$ and $\rho(t)$ are unknown functions and $_{0}^{c}D_{t}^{\alpha} = \frac{\partial^{\alpha}}{\partial t^{\alpha}}$ denotes the Caputo fractional derivative [12,13] of order $\alpha \in]0,1]$. It is obvious and well-known that in such problems, collocation methods based on global radial basis functions lead to ill-conditioned coefficient matrices. Also, if we deal with a huge practical problem with a large domain, the computational cost and ill-conditioning will grow dramatically. Therefore, the use of fast algorithms and parallel computational kernels become unavoidable and necessary. So, in this work, by starting from an accurate meshless localized collocation method for approximating the solution of (1) based on the local radial point interpolation (LRPI) technique, a parallel procedure exploiting the multicore environment capabilities is proposed. The parallel algorithm is based on a *functional* decomposition approach in order to perform an asynchronous kind of parallelism, by solving in parallel different tasks of the overall work, in order to obtain a meaningful gain in terms of performance.

The rest of the paper is organized as follows: in Sect. 2 the numerical procedure to discretize the inverse time fractional diffusion equation is shown; Sect. 3 deals with the description of both sequential and parallel implementation details of the algorithm; in Sect. 4 we provide tests and experiments that prove accuracy and efficiency, in terms of performance, of the parallel implementation; finally, in Sect. 5, we draw conclusions.

2   Numerical Background: The Time Discretization and Meshless Localized Collocation

In this section, the numerical approach to discretize the problem (1) is summarized. Following [1,2], we firstly consider an implicit time stepping procedure discretize the fractional model (1) in time direction, then we make use of a meshless localized collocation method to evaluate the unknown functions in some collocation points.

2.1   The Time Discretization Approximation

In order to discretize the fractional model (1) in time direction an implicit time stepping procedure is employed. More in particular, by choosing a time step $\tau > 0$ and setting $t^n = n\tau$, for $n = 0, \ldots, T/\tau$ (assume that T/τ is an integer), and by substituting $t = t^{n+1}$ in the Eq. (1), the following relation is obtained:

$$\begin{smallmatrix}c\\0\end{smallmatrix}D_t^\alpha v(\mathbf{x}, t^{n+1}) = \kappa \Delta v(\mathbf{x}, t^{n+1}) + f(\mathbf{x}, t^{n+1}), \quad (\mathbf{x}, t^{n+1}) \in \Omega \times (0, T], \quad (4)$$

Therefore we make use of the following second-order time discretization for the Caputo derivative of $v(\mathbf{x}, t)$ at point $t = t^{n+1}$ [3,4]:

$$\left[{}_0^c D_t^\alpha v(\mathbf{x}, t) \right]_{t=t^{n+1}} = \sum_{j=0}^{n+1} \frac{\omega^\alpha(j)}{\tau^\alpha} v(\mathbf{x}, t^{n+1-j}) - \frac{t^{-\alpha}}{\Gamma(1-\alpha)} v(\mathbf{x}, 0) + O(\tau^2), \quad (5)$$

where

$$\omega^\alpha(j) = \begin{cases} \dfrac{\alpha+2}{2} p_0^\alpha, & j = 0, \\ \dfrac{\alpha+2}{2} p_j^\alpha - \dfrac{\alpha}{2} p_{j-1}^\alpha, & j > 0, \end{cases} \quad \text{and} \quad p_j^\alpha = \begin{cases} 1, & j = 0, \\ \left(1 - \dfrac{\alpha+1}{j}\right) p_{j-1}^\alpha, & j \geq 1. \end{cases}$$

By substituting the Eqs. (5) in (4) the following relation is obtained:

$$\frac{\omega^\alpha(0)}{\tau^\alpha} v^{n+1} - \kappa \Delta v^{n+1} = -\sum_{j=1}^{n} \frac{\omega^\alpha(j)}{\tau^\alpha} v^{n+1-j} + \frac{t^{-\alpha}}{\Gamma(1-\alpha)} v^0 + f^{n+1}, \quad (6)$$

with $f^{n+1} = f(\mathbf{x}, t^{n+1})$ and $v^{n+1-j} = v(\mathbf{x}, t^{n+1-j})$ $(j = 0, \ldots, n+1)$.

2.2   The Meshless Localized Collocation Method

In last decades, meshless methods have been employed in many different fields of science and engineering [8–11]. This is because these methods can compute numerical solutions without using any given mesh of the problem domain. In the meshless localized collocation method, the global domain Ω is partitioned into local sub-domains Ω_i $(i = 1, \ldots, N)$ corresponding to every point. These sub-domains ordinarily are circles or squares and cover the entire global domain Ω. Then the radial point interpolation shape functions, ϕ_i, are constructed locally over each Ω_i by combining radial basis functions and the monomial basis function [5] corresponding to each local field point $\mathbf{x}_i$. In the current work, one of the most popular RBFs, i.e., the generalized multiquadric radial basis function (GMQ-RBF) is used as follows:

$$\phi(r) = (r^2 + c^2)^q, \quad (7)$$

where c is the shape parameter. We highlight that other RBF basis function could be also used (for example Gaussian or Wendland RBFs) without changing the description structure below. However, as shown in the experimental section, GMQ-RBFs guarantee good results in terms of accuracy. The local radial point interpolation shape function generates the $N \times N$ sparse matrix Φ. Therefore v can be approximated by

$$v(\mathbf{x}) = \sum_{i=1}^{N} \phi_i(\mathbf{x}) v_i \tag{8}$$

where $\phi_i(\mathbf{x}) = \phi(\|\mathbf{x} - \mathbf{x}_i\|_2)$ (the norm $\|\mathbf{x} - \mathbf{x}_i\|_2$ denotes the Euclidean distance between $\mathbf{x}$ and field point $\mathbf{x}_i$). Substituting approximation formula (8) in Eqs. (6), (2) and (3) yields:

$$\sum_{i=1}^{N} \left[\frac{\omega^\alpha(0)}{\tau^\alpha} \phi_i(\mathbf{x}_j) - \kappa \left[\frac{\partial^2 \phi_i}{\partial x^2} + \frac{\partial^2 \phi_i}{\partial y^2} \right] (\mathbf{x}_j) \right] v_i^{n+1} = -\sum_{j=1}^{n} \frac{\omega^\alpha(j)}{\tau^\alpha} \sum_{i=1}^{N} \phi_i(\mathbf{x}_j) v_i^{n+1-j}$$

$$+ \frac{t^{-\alpha}}{\Gamma(1-\alpha)} \sum_{i=1}^{N} \phi_i(\mathbf{x}_j) v^0 + f^{n+1}, \qquad j = 1, \ldots, N_\Omega \tag{9}$$

$$\sum_{i=1}^{N} \phi_i(\mathbf{x}_j) v_i^{n+1} = \psi_1^{n+1}(\mathbf{x}_j), \qquad j = N_\Omega + 1, \ldots, N_\Omega + N_{\Gamma_1}, \tag{10}$$

$$\sum_{i=1}^{N} \phi_i(\mathbf{x}_j) v_i^{n+1} = \psi_2^{n+1}(\mathbf{x}_j) \rho^{n+1}, \quad j = N_\Omega + N_{\Gamma_1} + 1, \ldots, N_\Omega + N_{\Gamma_1} + N_{\Gamma_2}, \tag{11}$$

$$\sum_{i=1}^{N} \left(\int_\Omega \phi_i(\mathbf{x}_j) d\Omega \right) v_i^{n+1} = h^{n+1}. \tag{12}$$

The collocation equations (9) are referred to the N_Ω interior points in Ω, while the N_{Γ_1} equations (10) and the N_{Γ_2} equations (11) (involving also the unknown $\rho^{n+1} = \rho(t^{n+1})$) arise from the initial and Dirichlet boundary conditions. Finally, a further equation is obtained by applying 2D Gaussian-Legendre quadrature rules of order 15. Therefore, the time discretization approximation and the local collocation strategy construct a linear system of $N + 1$ linear equations with $N + 1$ unknown coefficients ($N = N_\Omega + N_{\Gamma_1} + N_{\Gamma_2}$). The unknown coefficients

$$\mathbf{v}^{(n+1)} = (v_1^{n+1}, \ldots, v_N^{n+1}, \rho^{n+1})$$

are obtained by solving the sparse linear system:

$$\mathcal{A} \mathbf{v}^{(n+1)} = \mathcal{B}^{(n+1)}, \tag{13}$$

where $\mathcal{A}$ is a $(N+1) \times (N+1)$ coefficient matrix and $\mathcal{B}^{(n+1)}$ is the $(N+1)$ vector. In this regard it should be noted that, unlike $\mathcal{B}^{(n+1)}$, the coefficient matrix $\mathcal{A}$ does not change its entries along the time steps. Moreover, due to the local approach, where only nearby points of each field point $\mathbf{x}_i$ are considered, each related equation of the linear systems involves few unknown values and, consequently the coefficient matrix $\mathcal{A}$ is sparse.

Previous discussion allows us to introduce the following scheme, Algorithm 1, which summarizes the main steps needed to solve the numerical problem.

Algorithm 1. Pseudo-code for problem (1)

Input: $\kappa, \alpha, T, \tau, \varphi, \Psi_1, \Psi_2, h,$

$\quad\{\mathbf{x}_i\}_{i=1}^{N_\Omega},$ $\qquad\qquad$ % interior points

$\quad\{\mathbf{x}_{i+N_\Omega}\}_{i=1}^{N_{\Gamma_1}}$ $\qquad$ % Γ_1 boundary points

$\quad\{\mathbf{x}_{i+N_\Omega+N_{\Gamma_1}}\}_{i=1}^{N_{\Gamma_2}}$ $\quad$ % Γ_2 boundary points

Output: $\left\{\{v_i^{n+1}\}_{i=1}^N\right\}_{n=0}^{T/\tau-1},$

$\quad\{\rho^{n+1}\}_{n=0}^{T/\tau-1}$

1: build **A** $\qquad$ % by following (9,10,11,12)
2: for $n = 0, 1, \ldots, T/\tau - 1$ $\quad$ % loop on time slices
3: $\quad$ build $\mathbf{B}^{(n+1)}$ $\qquad$ % by following (9,10,11,12)
4: $\quad$ compute $\mathbf{v}^{(n+1)}$: $\qquad$ % solution of $\mathcal{A}\mathbf{v}^{(n+1)} = \mathcal{B}^{(n+1)}$ in (13)
5: endfor

3 Sequential and Parallel Implementation

To perform an effective implementation of Algorithm 1, we firstly developed a sequential version of the code, then a multicore parallel version. More precisely, in the following we present a detailed description of the sequential algorithm implementation:

(ds1) generate input points, called CenterPoints, in a 2D regular grid;
(ds2) partition CenterPoints in interior points center_I and boundary points center_b1 and center_b2;
(ds3) find, for each fixed interior point $(i = 1, \ldots, \text{size}(\text{center\_I}))$ its local neighbors and, by evaluating the Laplacian of the local RBF interpolating function, we build the i-th row of A_total (i.e. **A**). We highlight that this step requires to solve multiple linear systems of small size (number of neighbors) for each point in center_I);
(ds4) build next size(center_b1) + size(center_b2) rows of A_total (by using (10) and (11));
(ds5) build last row of A_total by evaluating the integral in (3) by means of 2D Gaussian-Legendre quadrature rules of order 15;
(ds6) generate a discrete 1D time interval tt= $[0 : \tau : T]$, with step τ;
(ds7) for each time in tt, build the right-hand side vector B (i.e. $\mathbf{B}^{(n+1)}$) and solve the sparse linear system A_total · sol = B, where sol is the computed value of $\mathbf{v}^{(n+1)}$;
(ds8) finally, the condition number of A_total is also estimated.

Previous steps are carried over the following Algorithm 2, where: the multiple linear systems, presented in (ds3), are built at lines 9, 10, 11 and then solved by using the routine `dgesv` of the LAPACK library based on the LU factorization method, while to solve the sparse linear system, at line 16, a specific routine of the CSPARSE library is employed [15], i.e. the `cs_lusol` routine, typical for linear systems characterized by sparse coefficient matrices. Finally, to evaluate the condition number of the sparse matrix at line 21, we used the `condition_simple1` routine of the CONDITION library [14].

Algorithm 2. Sequential algorithm

1: STEP 0: *input phase*

2: `generate CenterPoints`
3: `find Center_I`   % *interior points*
4: `find Center_b1`   % *boundary points*
5: `find Center_b2`   % *boundary points*

6: STEP 1: *construction of the coefficient matrix*

7: `for each point of CenterPoints`
8:     `find its local neighbors`
9:     `solve multiple linear systems`   % *one for each point of* `center_I`
10:    `solve multiple linear systems`   % *one for each point of* `center_b1`
11:    `solve multiple linear systems`   % *one for each point of* `center_b2`
12: `endfor`

13: STEP 2: *loop on time*

14: **for** $n = 0; n < T/\tau; step = 1$ **do**
15:    `build B`   % (by using results of lines 8,9,10,11)
16:    `solve A_total · sol = B`
17:    `set sol_M[n+1]:=sol`
18: **end for**

19: STEP 3: *output phase and condition number evaluation*

20: `reshape matrix sol_M`
21: `compute cond`   % *condition number of matrix* `A_total`

Despite the good accuracy achieved in our experiments, we observed that executing the software on a latest-generation CPU requires very large execution times. In order to improve the performance, a parallel approach is introduced. To be specific, we have chosen to parallelize main kernels in Algorithm 2 by using the well-known powerful of multicore processors, that are widely used across many application domains, including general-purpose, embedded, network and digital signal processing [6,7].

Our parallel algorithm is based on a *functional decomposition* combined with a classical *domain decomposition* approach. In other words, the pool of threads works exploiting an asynchronous parallelism by solving different tasks of the overall work in a parallel way. In the following, the parallel strategy implemented is described in detail.

The input phase in STEP 0 uses a parallel definition of the local neighbors by considering the classical domain decomposition approach of the CenterPoint set and using the information related to the id_thread, i.e. the index of each thread. In this way, we can define a local size n_loc, by dividing the global size of the data structure by the thread number n_thread. Hence, we deduce the start and the final positions of each local structure as:

$$\text{start = id\_thread} \times \text{n\_loc}, \text{end = (id\_thread + 1)} \times \text{n\_loc}.$$

The local block, for each subset of point which each thread deals with, are found and collected by a suitable synchronization barrier. In fact, since this phase uses several system calls to copy the local chunk into the global neighbors, which are stored in the shared memory, we manage this *critical region* by protecting the shared data to a race condition, using *semaphores* (see libraries in [17,18]).

The STEP 1, is related to the construction of the coefficient matrix A_total. In order to parallelize this task, we firstly use the domain decomposition approach in which each sub-set of the CenterPoint finds its local neighbors, as in the STEP 0; then, for the first time an asynchronous approach to menage the work of threads is used. Every threads works in parallel to solve the multiple linear systems shown at lines 9, 10 and 11 of Algorithm 2, by using the dgesv routine of the LAPACK library. After this task, still a critical region occurs. Then, to ensure the correct data writing a similar synchronization mechanism to the one used for the STEP 0, has been implemented. To complete this phase a check at the barrier happens, and then a semaphore to unlock STEP 1 is activated.

The STEPS 2 and 3, which provide the sparse linear systems solution and the condition number computation are executed in a similar way to what happens

---

Algorithm 3. parallel sparse linear systems resolution and condition number evaluation - asynchronous approach

---

1: shared variables: NG, A_total
2: **one thread execution**
3: compute the condition number of the A_total matrix in serial way into cond
4:
5: **other threads execution**
6: **for** $n = 0; n < T/\tau; step = 1$ **do**
7:     build B
8:     solve A_total · sol = B
9:     set sol_M[n+1]:=sol
10: **end for**
11: **check barrier for all threads**

---

in STEP 1. This phase is shown in Algorithm 3 and it is the main computational kernel of the parallel algorithm.

Again, a combination of domain decomposition and functional decomposition approach is used to perform in parallel these final STEPS. In fact, while a single thread compute the condition number of the sparse matrix, the other threads build in parallel the B vectors, for each time step, breaking down the work by means of a domain decomposition scheme. Moreover, they solve the sparse linear system, at each time, by using the multicore parallel version of the cs_lusol routine of CSPARSE library.

To be specific, for the second time we exploit the power of such environment by using the possibility to operate in an asynchronous way. Well-suited semaphores help the execution in order to manage the critical regions and to protect the routines which are not *thread-safe*, i.e. those routines that cannot be called from multiple threads without unwanted interaction between them. This ensures, always, that the parallel software has the correct data consistency.

4   Implementation Details and Numerical Tests

Our parallel algorithm is developed and tested on two CPU Intel Xeon with 6 cores, E5-2609v3, 1.9 Ghz, 32 GB of RAM memory, 4 channels 51Gb/s memory bandwidth. We use the Pthreads library [16], for UNIX systems, as specific interface for the IEEE POSIX 1003.1c standard: a set of C language programming types and procedures for managing the synchronization and concurrency in a multicore environment.

4.1   Accuracy

Here we are interested in measuring the error due to the numerical approximation introduced by using both time discretization and the local collocation approach. Following results are obtained by using the parallel algorithm with 4 threads. It is also useful to point out that the accuracy does not depend on the number of threads. As measures of the approximation error we use both the relative root mean square ϵ_v and maximum absolute error ε_ρ, defined as:

$$\epsilon_v = \max_{n=1,\ldots,T/\tau} \sqrt{\sum_{i=1}^{N} \left(\frac{\tilde{v}(\mathbf{x}_i, t^n) - v(\mathbf{x}_i, t^n)}{v(\mathbf{x}_i, t^n)} \right)^2}, \qquad \epsilon_\rho = \max_{n=1,\ldots,T/\tau} |\tilde{\rho}(t^n) - \rho(t^n)|,$$

where $\tilde{v}$ and $\tilde{\rho}$ denote the computed values of the true solutions v and ρ, respectively. All results in this section are referred to the following case study:

- as set of points we use a uniform two-dimensional grid in $\Omega = [0,1]^2$ so that $N = (1 + h^{-1})^2$, being h the space discretization step along both x and y directions;
- the boundaries are $\Gamma_2 = \{(x,y)|x = 1, 0 < y < 1\}$ and $\Gamma_1 = \partial\Omega \setminus \Gamma_2$;

– the source function is:

$$f(\mathbf{x}, t) = 2 \left(\frac{t^{2-\alpha}}{\Gamma(3-\alpha)} - t^2 \right) \exp(x+y),$$

while the initial and Dirichlet boundary conditions are set by taking:

$$\varphi(\mathbf{x}) = 0, \qquad \psi_1(\mathbf{x}, t) = t^2 \exp(x+y) \quad \text{and} \quad \psi_2(\mathbf{x}) = \exp(x+y)$$

– the true solutions of the problem are:

$$v(\mathbf{x}, t) = t^2 \exp(x+y) \quad \text{and} \quad \rho(t) = t^2.$$

Table 1 shows the error behaviour in terms of time step size (τ) by letting the number of points $N = 400$ and two different values of the fractional order α. This table illustrates the convergence and the accuracy of the proposed method while decreasing τ. In Table 2 the error estimates and the condition number of the coefficient matrix are reported by increasing the number of points N and taking $\tau = 0.01$ for two different values of the fractional order α. The results show that the accuracy of the method is improved by increasing N and the condition number indicates that the coefficient matrix $\mathbf{A}$ has an acceptable condition number. Figure 1 demonstrates the approximate solution and the point-wise absolute error for $v(\mathbf{x}, t)$ by letting $N = 625$, $\tau = 0.01$ and $\alpha = 0.75$. Also, local sub-domain for a center point and sparsity pattern of $\mathbf{A}$ are shown in Fig. 2.

Table 1. Behaviour of the condition number μ and of the approximation error for $\alpha = 0.25$, $\alpha = 0.85$, for $N = 400$ and for several values of τ.

α	τ	ϵ_v	ϵ_ρ	α	τ	ϵ_v	ϵ_ρ
0.25	$\frac{1}{10}$	$9.45 \cdot 10^{-5}$	$9.96 \cdot 10^{-5}$	**0.85**	$\frac{1}{10}$	$9.29 \cdot 10^{-5}$	$9.79 \cdot 10^{-5}$
	$\frac{1}{20}$	$5.11 \cdot 10^{-5}$	$5.39 \cdot 10^{-5}$		$\frac{1}{20}$	$4.92 \cdot 10^{-5}$	$5.19 \cdot 10^{-5}$
	$\frac{1}{40}$	$3.82 \cdot 10^{-5}$	$4.02 \cdot 10^{-5}$		$\frac{1}{40}$	$3.74 \cdot 10^{-5}$	$3.91 \cdot 10^{-5}$
	$\frac{1}{80}$	$3.61 \cdot 10^{-5}$	$3.66 \cdot 10^{-5}$		$\frac{1}{80}$	$3.55 \cdot 10^{-5}$	$3.58 \cdot 10^{-5}$
	$\frac{1}{160}$	$3.56 \cdot 10^{-5}$	$3.56 \cdot 10^{-5}$		$\frac{1}{160}$	$3.51 \cdot 10^{-5}$	$3.49 \cdot 10^{-5}$
	$\frac{1}{320}$	$3.54 \cdot 10^{-5}$	$3.54 \cdot 10^{-5}$		$\frac{1}{320}$	$3.49 \cdot 10^{-5}$	$3.47 \cdot 10^{-5}$

4.2   Performance Analysis

This section deals with the efficiency analysis of our parallel algorithm in terms of execution times, in order to highlight the gain achieved.

In Table 3 several executions of our software are showed, by varying the input size and the threads number t. We can see a significant gain with respect to the serial version, already starting from the run with 2 cores.

Table 2. Behaviour of the condition number μ and of the approximation error for $\alpha = 0.35$, $\alpha = 0.75$, for $\tau = 0.01$ and for several values of N.

α	N	ϵ_v	ϵ_ρ	μ	α	N	ϵ_v	ϵ_ρ	μ
0.35	36	$7.07 \cdot 10^{-4}$	$8.64 \cdot 10^{-4}$	$1.98 \cdot 10^{4}$	**0.75**	36	$7.06 \cdot 10^{-4}$	$8.63 \cdot 10^{-4}$	$1.88 \cdot 10^{4}$
	81	$2.16 \cdot 10^{-4}$	$2.44 \cdot 10^{-4}$	$8.32 \cdot 10^{4}$		81	$2.15 \cdot 10^{-4}$	$2.44 \cdot 10^{-4}$	$8.15 \cdot 10^{4}$
	144	$6.56 \cdot 10^{-5}$	$7.18 \cdot 10^{-5}$	$3.50 \cdot 10^{5}$		144	$6.58 \cdot 10^{-5}$	$7.21 \cdot 10^{-5}$	$3.22 \cdot 10^{5}$
	196	$2.35 \cdot 10^{-5}$	$2.30 \cdot 10^{-5}$	$6.69 \cdot 10^{5}$		196	$2.34 \cdot 10^{-5}$	$2.33 \cdot 10^{-5}$	$6.15 \cdot 10^{5}$
	289	$2.91 \cdot 10^{-5}$	$1.57 \cdot 10^{-5}$	$1.49 \cdot 10^{6}$		289	$2.88 \cdot 10^{-5}$	$1.53 \cdot 10^{-5}$	$1.38 \cdot 10^{6}$
	400	$3.59 \cdot 10^{-5}$	$3.63 \cdot 10^{-5}$	$2.93 \cdot 10^{6}$		400	$3.56 \cdot 10^{-5}$	$3.58 \cdot 10^{-5}$	$2.86 \cdot 10^{6}$
	576	$4.96 \cdot 10^{-5}$	$5.18 \cdot 10^{-5}$	$6.23 \cdot 10^{6}$		576	$4.92 \cdot 10^{-5}$	$5.14 \cdot 10^{-5}$	$6.05 \cdot 10^{6}$

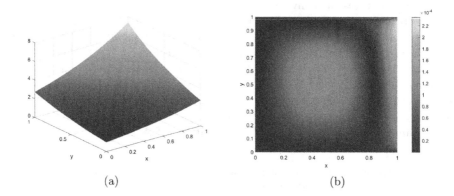

(a)   (b)

Fig. 1. (a) Exact solution $v(\mathbf{x}, t)$; (b) absolute error ($N = 625$, $\tau = 0.01$, $\alpha = 0.75$)

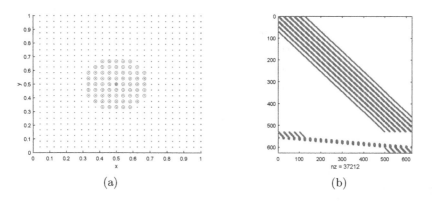

(a)   (b)

Fig. 2. (a) Local sub-domain for a given center point; (b) sparsity pattern of $\mathbf{A}$

Table 3. Execution times in seconds (s) achieved by varying both number of threads t (t = 1, 2, 4, 6, 9, 12) and size of the problem N.

N	Serial time (s)	Parallel time (s)					
		1	2	4	6	9	12
8.1×10^3	1.2×10^4	30.02	15.02	10.97	9.03	8.94	7.98
1.0×10^4	3.19×10^4	39.98	20.01	14.97	12.08	11.99	11.01
1.69×10^4	1.66×10^5	97.06	56.92	41.95	35.05	33.99	29.01
1.98×10^4	2.46×10^5	114.98	85.02	56.97	45.64	40.81	38.80
2.25×10^4	3.9×10^5	195.96	116.93	83.94	77.05	74.96	68.01

We observe that, the execution time of the parallel version is improved especially for the use of an "ad hoc" memory allocation and a suitable scheduling policy. More precisely, the gain is so large, by using two cores or more, for the massive use of shared-memory combined with a suitable employment of the local stack memory level of each thread. The strong performance, compared to the sequential version, is confirmed also by increasing the number of cores until six, while, for a larger number of cores, the performance degrades. This can also be observed through Tables 4 and 5, which show the speed-up and the efficiency, respectively. This is due to the hardware characteristics of our supercomputer (two CPUs, with 6 cores) and, precisely, to the fact that while dual CPU systems setups pack many core counts and outshine single processor servers by a large margin, our tests show a marginal performance increase over single CPU configurations, caused by the fact that the CPUs worked on the same data, at the same time, in the shared memory. In other words, the synchronization, which needs the access to the shared memory by the CPUs, slows down the performance and decreases the earnings expectations inevitably.

Table 4. Speed-up achieved by varying both number of threads t (t = 2, 4, 6, 9, 12) and size of the problem N.

N	Parallel Speed-up				
	2	4	6	9	12
8.1×10^3	1.99	2.73	3.32	3.35	3.76
1.0×10^4	1.99	2.67	3.30	3.34	3.63
1.69×10^4	1.70	2.31	2.76	2.93	3.34
1.98×10^4	1.35	2.01	2.51	2.81	2.96
2.25×10^4	1.67	2.33	2.54	2.61	2.88

Table 5. Efficiency achieved by varying both number of threads t (t = 2, 4, 6, 9, 12) and size of the problem N.

N	Parallel efficiency				
	2	4	6	9	12
8.1×10^3	0.99	0.68	0.55	0.37	0.31
1.0×10^4	0.99	0.66	0.55	0.37	0.30
1.69×10^4	0.85	0.57	0.43	0.32	0.27
1.98×10^4	0.67	0.50	0.41	0.31	0.24
2.25×10^4	0.83	0.58	0.42	0.29	0.24

5   Conclusion and Future Work

In this paper, we proposed a parallel code in order to solve a two-dimensional inverse time fractional diffusion equation. The algorithm implements a numerical procedure which is based on the discretization of the Caputo fractional derivative and on the use of a meshless localized collocation method exploiting the radial basis functions properties. Because the high computational complexity and the large size of the problem, the parallel algorithm is designed for multicore architectures and, as seen in the experiments, this code has proved to be very efficient and accurate.

References

1. Yan, L., Yang, F.: The method of approximate particular solutions for the time-fractional diffusion equation with a non-local boundary condition. Comput. Math. Appl. **70**, 254–264 (2015)
2. Abbasbandy, S., Ghehsareh, H.R., Alhuthali, M.S., Alsulami, H.H.: Comparison of meshless local weak and strong forms based on particular solutions for a non-classical 2-D diffusion model. Eng. Anal. Boundary Elem. **39**, 121–128 (2014)
3. Tian, W.Y., Zhou, H., Deng, W.H.: A class of second order difference approximations for solving space fractional diffusion equations. Math. Comput. **84**, 1703–1727 (2015)
4. Kilbas, A., Srivastava, M.H., Trujillo, J.J.: Theory and Application of Fractional Differential Equations, North Holland Mathematics Studies, vol. 204 (2006)
5. Liu, Q., Gu, Y.T., Zhuang, P., Liu, F., Nie, Y.F.: An implicit RBF meshless approach for time fractional diffusion equations. Comput. Mech. **48**, 1–12 (2011)
6. Cuomo, S., De Michele, P., Galletti, A., Marcellino, L.: A GPU parallel implementation of the local principal component analysis overcomplete method for DW image denoising. In: Proceedings - IEEE Symposium on Computers and Communications, vol. 2016, pp. 26–31, August 2016
7. D'Amore, L., Arcucci, R., Marcellino, L., Murli, A.: A parallel three-dimensional variational data assimilation scheme. In: AIP Conference Proceedings, vol. 1389, pp. 1829–1831 (2011)

8. Cuomo, S., Galletti, A., Giunta, G., Marcellino, L.: Reconstruction of implicit curves and surfaces via RBF interpolation. Appl. Numer. Math. **116**, 157–171 (2017). https://doi.org/10.1016/j.apnum.2016.10.016

9. Ala, G., Ganci, S., Francomano, E.: Unconditionally stable meshless integration of time domain Maxwell's curl equations. Appl. Math. Comput. **255**, 157–164 (2015)

10. Cuomo, S., Galletti, A., Giunta, G., Starace, A.: Surface reconstruction from scattered point via RBF interpolation on GPU. In: 2013 Federated Conference on Computer Science and Information Systems, FedCSIS 2013, art. no. 6644037, pp. 433–440 (2013)

11. Francomano, E., Paliaga, M., Galletti, A., Marcellino, L.: First experiences on an accurate SPH method on GPUs. In: Proceedings - 13th International Conference on Signal-Image Technology and Internet-Based Systems, SITIS 2017, 2018-January, pp. 445–449 (2018). https://doi.org/10.1109/SITIS.2017.79

12. Ala, G., Di Paola, M., Francomano, E., Pinnola, F.: Electrical analogous in viscoelasticity. Commun. Nonlinear Sci. Numer. Simul. **19**(7), 2513–2527 (2014)

13. Podlubny, I.: Fractional Differential Equation. Academic Press, San Diego (1999)

14. https://people.sc.fsu.edu/jburkardt/c_src/condition/condition.html

15. http://faculty.cse.tamu.edu/davis/suitesparse.html

16. https://computing.llnl.gov/tutorials/pthreads/

17. https://pubs.opengroup.org/onlinepubs/7908799/xsh/semaphore.h.html

18. https://pubs.opengroup.org/onlinepubs/009695399/functions/pthread_mutex_lock.html

Adaptive RBF Interpolation for Estimating Missing Values in Geographical Data

Kaifeng Gao[1], Gang Mei[1(✉)], Salvatore Cuomo[2], Francesco Piccialli[2], and Nengxiong Xu[1]

[1] China University of Geosciences (Beijing), Beijing, China
[2] University of Naples Federico II, Naples, Italy
gang.mei@cugb.edu.cn

Abstract. The quality of datasets is a critical issue in big data mining. More interesting things could be found for datasets with higher quality. The existence of missing values in geographical data would worsen the quality of big datasets. To improve the data quality, the missing values are generally needed to be estimated using various machine learning algorithms or mathematical methods such as approximations and interpolations. In this paper, we propose an adaptive Radial Basis Function (RBF) interpolation algorithm for estimating missing values in geographical data. In the proposed method, the samples with known values are considered as the data points, while the samples with missing values are considered as the interpolated points. For each interpolated point, first, a local set of data points are adaptively determined. Then, the missing value of the interpolated point is imputed via interpolating using the RBF interpolation based on the local set of data points. Moreover, the shape factors of the RBF are also adaptively determined by considering the distribution of the local set of data points. To evaluate the performance of the proposed method, we compare our method with the commonly used k-Nearest Neighbor (kNN) interpolation and Adaptive Inverse Distance Weighted (AIDW) interpolation, and conduct three groups of benchmark experiments. Experimental results indicate that the proposed method outperforms the kNN interpolation and AIDW interpolation in terms of accuracy, but worse than the kNN interpolation and AIDW interpolation in terms of efficiency.

Keywords: Data mining · Data quality · Data imputation · RBF interpolation · kNN

1 Introduction

Datasets are the key elements in big data mining, and the quality of datasets has an important impact on the results of big data analysis. For a higher quality dataset, some hidden rules can often be mined from it, and through these rules we

© Springer Nature Switzerland AG 2020
Y. D. Sergeyev and D. E. Kvasov (Eds.): NUMTA 2019, LNCS 11973, pp. 122–130, 2020.
https://doi.org/10.1007/978-3-030-39081-5_12

can find some interesting things. At present, big data mining technology is widely used in various fields, such as geographic analysis [10,16], financial analysis, smart city and biotechnology. It usually needs a better dataset to support the research, but in fact there is always noise data or missing value data in the datasets [9,14,17]. In order to improve data quality, various machine learning algorithms [7,15] are often required to estimate the missing value.

RBF approximation techniques combined with machine learning algorithms such as neural networks can be used to optimize numerical algorithms. Besides, RBF interpolation algorithm is a popular method for estimating missing values [4–6]. In large-scale computing, the cost can be minimized by using adaptive scheduling method [1]. RBF is a distance-based function, which is meshless and dimensionless, thus it is inherently suitable for processing multidimensional scattered data. Many scholars have done a lot of work on RBF research. Skala [13] used CSRBF to analyze big datasets, Cuomo et al. [2,3] studied the reconstruction of implicit curves and surfaces by RBF interpolation. Kedward et al. [8] used multiscale RBF interpolation to study mesh deformation. In RBF, the shape factor is an important factor affecting the accuracy of interpolation. Some empirical formulas for optimum shape factor have been proposed by scholars.

In this paper, our objective is to estimate missing values in geographical data. We proposed an adaptive RBF interpolation algorithm, which adaptively determines the shape factor by the density of the local dataset. To evaluate the performance of adaptive RBF interpolation algorithm in estimating missing values, we used three datasets for verification experiments, and compared the accuracy and efficiency of adaptive RBF interpolation with that of kNN interpolation and AIDW.

The rest of the paper is organized as follows. Section 2 mainly introduces the implementation process of the adaptive RBF interpolation algorithm, and briefly introduces the method to evaluate the performance of adaptive RBF interpolation. Section 3 introduces the experimental materials, and presents the estimated results of missing values, then discusses the experimental results. Section 4 draws some conclusions.

2   Methods

In this paper, our objective is to develop an adaptive RBF interpolation algorithm to estimate missing values in geospatial data, and compare the results with that of kNN and AIDW. In this section, we firstly introduce the specific implementation process of the adaptive RBF interpolation algorithm, then briefly introduces the method to evaluate the performance of adaptive RBF interpolation.

The basic ideas behind the RBF interpolation are as follows. Constructing a high-dimensional function $f(x), x \in R^n \pounds$, suppose there is a set of discrete points $x_i \in R^n, i = 1, 2, \cdots N_\$$ with associated data values $f(x_i) \in R, i = 1, 2, \cdots N$. Thus, the function $f(x)$ can be expressed as a linear combination of

RBF in the form (Eq. (1)):

$$f(x) = \sum_{j=1}^{N} a_j \phi \left(\|x - x_j\|_2 \right) \tag{1}$$

where N is the number of interpolation points, $\{a_j\}$ is the undetermined coefficient, the function ϕ is a type of RBF.

The kernel function selected in this paper is Multi-Quadric RBF(MQ-RBF), which is formed as (Eq. (2)):

$$\phi(r) = \sqrt{(r^2 + c^2)} \tag{2}$$

where r is the distance between the interpolated point and the data point, c is the shape factor. Submit the data points (x_i, y_i) into Eq. (1), then the interpolation conditions become (Eq. (3)):

$$y_i = f(x_i) = \sum_{j=1}^{N} a_j \phi \left(\|x_i - x_j\|_2 \right), i = 1, 2, \cdots N \tag{3}$$

When using the RBF interpolation algorithm in a big dataset, it is not practical to calculate an interpolated point with all data points. Obviously, the closer the data point is to the interpolated point, the greater the influence on the interpolation result and the data point far from the interpolated point to a certain distance, its impact on the interpolated point is almost negligible. Therefore, we calculate the distances from an interpolated point to all data points, and select 20 points with the smallest distances as a local dataset for the interpolated point.

In Eq. (2), the value of the shape factor c in MQ-RBF has a significant influence on the calculation result of interpolation. We consult the method proposed by Lu and Wang [11,12], adaptively determining the value c of the interpolated points by the density of the local dataset. The expected density D_{exp} is calculated by the function (Eq. (4)):

$$D_{exp} = \frac{N_{dp}}{(X_{max} - X_{min})(Y_{max} - Y_{min})} \tag{4}$$

where N_{dp} is the number of data points in the dataset, X_{max} is the maximum value of x_i for the data points in the dataset, X_{min} is the minimum value of x_i in dataset, Y_{max} is the maximum value of y_i in dataset, Y_{min} is the minimum value of y_i in dataset.

And the local density D_{loc} is calculated by (Eq. (5)):

$$D_{loc} = \frac{N_{loc}}{(x_{max} - x_{min})(y_{max} - y_{min})} \tag{5}$$

where N_{loc} is the number of data points in the local dataset, in this paper, we set N_{loc} as 20. x_{max} is the maximum value of x_i for the data points in local dataset, x_{min} is the minimum value of x_i in local dataset, y_{max} is the maximum value of y_i in local dataset, y_{min} is the minimum value of y_i in local dataset.

With both the local density and the expected density, the local density statistic D can be expressed as (Eq. (6)):

$$D\left(s_0\right) = \frac{D_{loc}}{D_{exp}} \tag{6}$$

where s_0 is the location of an interpolated point. Then normalize the $D\left(s_0\right)$ measure to μ_D by a fuzzy membership function (Eq. (7)):

$$\mu_D = \begin{cases} 0 & D\left(s_0\right) \leq 0 \\ 0.5 - 0.5\cos\left[\frac{\pi}{2}D\left(s_0\right)\right] & 0 \leq D\left(s_0\right) \leq 2 \\ 1 & D\left(s_0\right) \geq 2 \end{cases} \tag{7}$$

Finally, determine the shape factor c by a triangular membership function. The calculation process of triangular membership functions with different adaptive shape factors is shown in Fig. 1. For example, if μ_D is 0.45, according to the proportional relationship, the value of shape factor c will be $(0.25 \times c_2 + 0.75 \times c_3)$; see Eq. (8).

$$c = \begin{cases} c_1 & 0.0 \leq \mu_D \leq 0.1 \\ c_1\left[1 - 5\left(\mu_D - 0.1\right)\right] + 5c_2\left(\mu_D - 0.1\right) & 0.1 \leq \mu_D \leq 0.3 \\ 5c_3\left(\mu_D - 0.3\right) + c_2\left[1 - 5\left(\mu_D - 0.3\right)\right] & 0.3 \leq \mu_D \leq 0.5 \\ c_3\left[1 - 5\left(\mu_D - 0.5\right)\right] + 5c_4\left(\mu_D - 0.5\right) & 0.5 \leq \mu_D \leq 0.7 \\ 5c_5\left(\mu_D - 0.7\right) + c_4\left[1 - 5\left(\mu_D - 0.7\right)\right] & 0.7 \leq \mu_D \leq 0.9 \\ c_5 & 0.9 \leq \mu_D \leq 1.0 \end{cases} \tag{8}$$

where c_1, c_2, c_3, c_4, c_5 are five levels of shape factor.

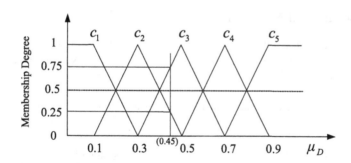

Fig. 1. Triangular membership function for different degrees of the adaptive shape factor [11]

After determining the shape factor c, the next steps are the same as the general RBF calculation method. The specific process of the adaptive RBF interpolation algorithm is illustrated in Fig. 2.

In order to evaluate the computational accuracy of the adaptive RBF interpolation algorithm, we use the metric, Root Mean Square Error (RMSE) to measure the accuracy. The RMSE evaluates the error accuracy by comparing

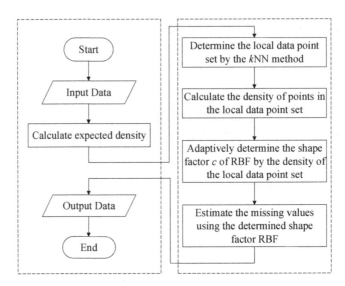

Fig. 2. The flow chart of the adaptive RBF interpolation algorithm

the deviation between the estimated value and the true value. In addition, we record each calculation time as a basis for evaluating the efficiency of interpolation calculation. Then, we compare the accuracy and efficiency of adaptive RBF estimator with the results of kNN and AIDW estimators.

3   Results And Discussion

3.1   Experimental Materials

To evaluate the performance of the presented adaptive RBF interpolation algorithm, we use three datasets to test it. The details of the experimental environment are listed in Table 1.

Table 1. Basic information of experimental environment

Specification	Details
OS	Windows 7. Professional
CPU	Intel (R) i5-4210U
CPU Frequency	1.70 GHz
CPU RAM	8 GB
CPU Core	4

In our experiments, we use three datasets from three cities' Digital Elevation Model (DEM) images; see Fig. 3. The range of three DEM images are the same.

Figure 3(a) is the DEM image of Beijing working area. The landform of this area is mountainous in a small part of Northwest and plains in other areas. Figure 3(b) is the DEM image of Chongqing city. There are several mountains in the northeast-southwest direction, and the southeast area is a mountainous area. Figure 3(c) is the DEM image of Longyan city, which is hilly and high in the east and low in the west.

We randomly select 10% observed samples from each dataset as the samples with missing values, and the rest as the samples with known values. It should be noted that the samples with missing values have really elevation values in fact, but for testing, we assume the elevations are missing. Basic information of the datasets is listed in Table 2.

Table 2. Experimental data

Dataset	Number of known values	Number of missing values	Illustration
Beijing	1,111,369	123,592	Fig. 3(a)
Chongqing	1,074,379	97,525	Fig. 3(b)
Longyan	1,040,670	119,050	Fig. 3(c)

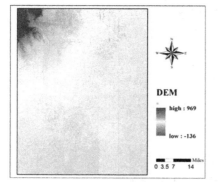

(a) The DEM map of Beijing City, China

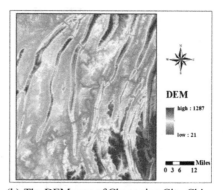

(b) The DEM map of Chongqing City, China

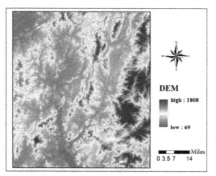

(c) The DEM map of Longyan City, China

Fig. 3. The DEM maps of three cities for the experimental tests

3.2   Experimental Results and Discussion

We compare the accuracy and efficiency of adaptive RBF estimator with that of
kNN and AIDW estimators.

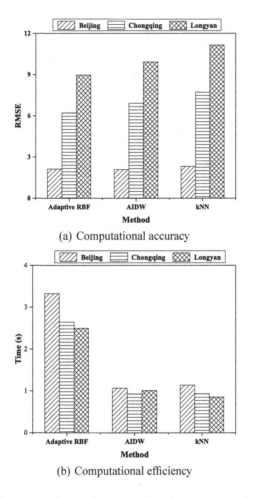

(a) Computational accuracy

(b) Computational efficiency

Fig. 4. The comparisons of computational accuracy and efficiency

In the Fig. 4, we find that the accuracy of the adaptive RBF estimator is
the best performing, and the kNN estimator with the lowest accuracy. With
the number of known data points in the datasets decreases, the accuracy of
three estimators decreases significantly. Moreover, the computational efficiency
of adaptive RBF estimator is worse than that of kNN estimator and AIDW estimator. Among the three methods, kNN has the best computational efficiency.
With the increase of data quantity, the disadvantage of the computational efficiency of kNN estimator becomes more and more obvious.

The data points selected from DEM images are evenly distributed, and the shape factor c of the adaptive RBF interpolation algorithm is adapted according to the density of the points in the local dataset. Therefore, when the missing data is estimated in a dataset with a more uniform data point, the advantages of the adaptive RBF interpolation algorithm may not be realized. We need to do further research in datasets with uneven datasets.

4   Conclusions

In this paper, we have proposed an adaptive RBF interpolation algorithm for estimating missing values in geographical data and evaluated its computational accuracy and efficiency. We have conducted three groups of experiments. The results show that the accuracy of the adaptive RBF interpolation performs better than kNN interpolation and AIDW in regularly distributed datasets. Therefore, we consider that the proposed adaptive method can effectively improve the accuracy of estimating missing data in geographical data by adaptively determining the shape factor.

Acknowledgments. This research was jointly supported by the National Natural Science Foundation of China (11602235), and the Fundamental Research Funds for China Central Universities (2652018091, 2652018107, and 2652018109). The authors would like to thank the editor and the reviewers for their contribution.

References

1. Barone, G.B., et al.: An approach to forecast queue time in adaptive scheduling: how to mediate system efficiency and users satisfaction. Int. J. Parallel Program. **45**(5), 1–30 (2016)
2. Cuomo, S., Galletti, A., Giunta, G., Marcellino, L.: Reconstruction of implicit curves and surfaces via rbf interpolation. Appl. Numer. Math. **116**, 60–63 (2016)
3. Cuomo, S., Gallettiy, A., Giuntay, G., Staracey, A.: Surface reconstruction from scattered point via RBF interpolation on GPU (2013)
4. Ding, Z., Mei, G., Cuomo, S., Tian, H., Xu, N.: Accelerating multi-dimensional interpolation using moving least-squares on the GPU. Concurr. Comput. **30**(24), e4904 (2018)
5. Ding, Z., Gang, M., Cuomo, S., Li, Y., Xu, N.: Comparison of estimating missing values in IoT time series data using different interpolation algorithms. Int. J. Parallel Programm. 1–15 (2018)
6. Ding, Z., Gang, M., Cuomo, S., Xu, N., Hong, T.: Performance evaluation of gpu-accelerated spatial interpolation using radial basis functions for building explicit surfaces. Int. J. Parallel Programm. **157**, 1–29 (2017)
7. Gao, D., Liu, Y., Meng, J., Jia, Y., Fan, C.: Estimating significant wave height from SAR imagery based on an SVM regression model. Acta Oceanol. Sinica **37**(3), 103–110 (2018)
8. Kedward, L., Allen, C.B., Rendall, T.C.S.: Efficient and exact mesh deformation using multiscale rbf interpolation. J. Computat. Phys. **345**, 732–751 (2017)

9. Keogh, R.H., Seaman, S.R., Bartlett, J.W., Wood, A.M.: Multiple imputation of missing data in nested case-control and case-cohort studies. Biometrics **74**(4), 1438–1449 (2018)
10. Liang, Z., Na, Z., Wei, H., Feng, Z., Qiao, Q., Luo, M.: From big data to big analysis: a perspective of geographical conditions monitoring. Int. J. Image Data Fusion **9**(3), 194–208 (2018)
11. Lu, G.Y., Wong, D.W.: An adaptive inverse-distance weighting spatial interpolation technique. Comput. Geosci. **34**(9), 1044–1055 (2008)
12. Mei, G., Xu, N., Xu, L.: Improving GPU-accelerated adaptive IDW interpolation algorithm using fast KNN search. Springerplus **5**(1), 1389 (2016)
13. Skala, V.: RBF interpolation with CSRBF of large data sets. Proc. Comput. Sci. **108**, 2433–2437 (2017)
14. Sovilj, D., Eirola, E., Miche, Y., Bjork, K.M., Rui, N., Akusok, A., Lendasse, A.: Extreme learning machine for missing data using multiple imputations. Neurocomputing **174**(PA), 220–231 (2016)
15. Tang, T., Chen, S., Meng, Z., Wei, H., Luo, J.: Very large-scale data classification based on k-means clustering and multi-kernel SVM. Soft Comput. **1**, 1–9 (2018)
16. Thakuriah, P.V., Tilahun, N.Y., Zellner, M.: Big data and urban informatics: innovations and challenges to urban planning and knowledge discovery. In: Thakuriah, P.V., Tilahun, N., Zellner, M. (eds.) Seeing Cities Through Big Data. SG, pp. 11–45. Springer, Cham (2017). https://doi.org/10.1007/978-3-319-40902-3_2
17. Tomita, H., Fujisawa, H., Henmi, M.: A bias-corrected estimator in multiple imputation for missing data. Stat. Med. **47**(1), 1–16 (2018)

Stochastic Mechanisms of Information Flow in Phosphate Economy of *Escherichia coli*

Ozan Kahramanoğulları[1,3]([envelope]) [ID], Cansu Uluşeker[2] [ID],
and Martin M. Hancyzc[3,4] [ID]

[1] Department of Mathematics,
University of Trento, Trento, Italy
ozan.kah@gmail.com
[2] Department of Chemistry, Bioscience and Environmental Engineering,
University of Stavanger, Stavanger, Norway
[3] Department of Cellular, Computational and Integrative Biology,
University of Trento, Trento, Italy
[4] Chemical and Biological Engineering, University of New Mexico, Albuquerque, USA

Abstract. In previous work, we have presented a computational model and experimental results that quantify the dynamic mechanisms of auto-regulation in *E. coli* in response to varying external phosphate levels. In a cycle of deterministic ODE simulations and experimental verification, our model predicts and explores phenotypes with various modifications at the genetic level that can optimise inorganic phosphate intake. Here, we extend our analysis with extensive stochastic simulations at a single-cell level so that noise due to small numbers of certain molecules, e.g., genetic material, can be better observed. For the simulations, we resort to a conservative extension of Gillespie's stochastic simulation algorithm that can be used to quantify the information flow in the biochemical system. Besides the common time series analysis, we present a dynamic visualisation of the time evolution of the model mechanisms in the form of a video, which is of independent interest. We argue that our stochastic analysis of information flow provides insights for designing more stable synthetic applications that are not affected by noise.

Keywords: Synthetic biology · *E. coli* · Modelling · Stochasticity · Noise

1 Introduction

The rapidly growing field of synthetic biology, at the crossroads of molecular biology, genetics and quantitative sciences, aims at developing living technologies by re-engineering the makeup of organisms. The applications in this field are designed by channeling the quantitative understanding of the molecular processes to a methodological workflow that can be compared to the use of mechanics in civil engineering. The aim, in these applications, is to modify the organisms

© Springer Nature Switzerland AG 2020
Y. D. Sergeyev and D. E. Kvasov (Eds.): NUMTA 2019, LNCS 11973, pp. 131–145, 2020.
https://doi.org/10.1007/978-3-030-39081-5_13

to enhance and benefit from their natural capacity for certain tasks. For example, in enhanced biological phosphorus removal (EBPR) micro-organisms such as *E. coli* are used to profit from their inherent regulatory mechanisms that efficiently respond to phosphate starvation. Achieving a quantitative understanding of the molecular mechanisms involved in such processes from signal transduction to gene regulation has implications in biotechnology applications.

In previous work [1], we have presented a computational model of the dynamic mechanisms in *E. coli* phosphate economy. Our model, based on a chemical reaction representation, explores the biochemical auto-regulation machinery that relays the information on extracellular inorganic phosphate (P_i) concentration to the genetic components. The ordinary differential equation simulations with our model quantify the dynamic response to varying external P_i levels of *E. coli* with which it optimises the expression of the proteins that are involved in the P_i intake. The analysis of the simulations with our model and their experimental verification showed that our model captures the variations in phenotype resulting from modifications on the genetic components. This allowed us to explore a spectrum of synthetic applications that respond to various external P_i concentrations with varying levels of gene expression.

Besides the deterministic processes that are faithfully captured by ordinary differential equations, many aspects of gene expression employ stochastic processes [2]. In particular, the randomness in transcription and translation due to small molecule numbers of genetic material can result in significant fluctuations in mRNA and protein numbers in individual cells. This, in return, can lead to cell-to-cell variations in phenotype with consequences for function [3].

For the case of synthetic applications that involve modifications in the genetic makeup of the cells, the stochasticity in the biochemical processes introduces yet another parameter that needs to be monitored and can even be exploited beyond the development of more reliable synthetic devices [4]. In this regard, the stochastic effects in gene expression have been the topic of extensive research. In particular, the fluctuations that depend on gene network structure and the biochemical affinity of the interacting biochemical components have been investigated both experimentally and theoretically, e.g., [5,6].

Here we extend our analysis in [1] with extensive stochastic simulations with Gillespie's SSA algorithm [6,7]. Our simulations, at a single-cell level allow us to monitor the noise due to small numbers of molecules. This way, we quantify the effect of the model parameters corresponding to various synthetic promoter designs on signal robustness in conditions of different regimes of external P_i concentrations. For the simulations, we resort to a conservative extension of SSA that can be used to quantify the information flow [8,9]. Our analysis reveals the distribution of the system resources and the resulting information flow in terms of species fluxes between system components in response to external P_i signal at different time intervals. Based on this, we provide a quantification of the noise in the system due to stochastic processes in different conditions. We argue that our analysis provides insights that can guide the design of synthetic applications, where the effect of stochasticity can be predicted and controlled.

We present a visualisation of the dynamic evolution of the system fluxes in the form of a video, which is of independent interest. In particular, we show that such visualisations can provide insights in the analysis of stochastic simulations with chemical reaction networks beyond the time series they produce.

2 Phosphate Economy of *E. coli*

The regulatory mechanisms in *E. coli* that control the inorganic phosphate (P_i) uptake involve the interplay between two complementary mechanisms. When the external P_i concentration is above the millimolar range, P_i is transported into the cell mainly by the low-affinity P_i transporter (Pit) system, which is constitutively expressed and dependent on the proton motive force [10]. However, when the external P_i concentration falls below the 0.2 mM range, the high-affinity Phosphate specific transport (Pst) system is induced. This triggers the expression of an operon that includes an ABC transporter, which actively transports P_i by ATP-consumption. The Pst system involves a positive feedback loop, and it induces its own expression via a two-component system (TCS) consisting of the histidine kinase PhoR and the transcription factor PhoB. Both Pit and Pst are highly specific for P_i.

P_i intake by Pst system is a negative process, whereby a high external P_i concentration turns the system off; the activation is the default state. The current evidence suggests that the TCS mechanism is turned off by the PhoU protein that monitors the ABC transporter activity. In mechanistic terms, when there is sufficient P_i flux, PhoU stabilises the PhoR and this prevents the TCS mechanism from relaying the signal to the transcription factor PhoB. Contrarily, when the external P_i concentration is limited, PhoU does not inhibit the TCS. As a result of the decrease in the external P_i concentration, the concentration of PhoR molecules that are not inhibited by PhoU increases. Thus, the auto-cross-phosphorylation activity of PhoR dimers provides a proxy for the external P_i concentration signal. This is because the Pst signal is relayed by auto-cross-phosphorylation of PhoR dimers that are not inhibited by PhoU.

The Chemical Reaction Network (CRN) model in [1] is displayed in the Appendix section. The model describes the signal transduction processes downstream of PhoU to the genetic components, and the feedback of the gene expression to the Pst system. Our model makes use of the interaction mechanism between PhoU and PhoR by employing a scalar factor for the PhoR auto-cross-phosphorylation activity: the reactions r01, r02, r03, and r04 model the signal transduction from PhoR, where fc is this factor describing the PhoR activity resulting from the external P_i concentration. The fc = 1.0 models the starvation response to the external P_i concentration of $0\,\mu$M. An increase in the external P_i concentration and the resulting inhibition of PhoR by PhoU is modelled by a decrease in the fc. Thus, fc = 0 models a P_i concentration over 0.2 mM.

Following this process, phosphorylated PhoR activates PhoB by phosphotransferase (r05, r06, r07, r08, r09, r10). Phosphorylated PhoB dimerises to constitute an active transcription factor (r11, r12) and binds the promoter region of PhoA and PhoB genes to activate their transcription (r16, r17, r18, r19). The factors bf and uf in reactions r16, r18, r17, and r19 model the affinity of the active transcription factor to the promoter region. The default value of 1.0 for these factors results in the control model, whereas variations in bf and uf model synthetic promoters that can be stronger or weaker.

The histidine kinase PhoR is a bifunctional enzyme that performs two opposing tasks: on one hand, it activates the PhoB dimers as described above. On the other hand, it dephosphorylates the phosphorylated PhoB (r13, r14, r15). The activated promoters transcribe the mRNA molecules for the expression of PhoA, PhoB, and PhoR (r20, r21, r22, r23, r24), which can be subject to degradation or dilution (r25, r26, r27, r28, r29).

The control model in [1] is parameterised within the biologically feasible range and the parameter values are narrowed down by random restart least-squares optimisation by fitting the model dynamics to experimental data. The deterministic simulation plots in Fig. 1 display the concentration dynamics of the active transcription factor dimers DiPhoBpp, the active promoter pPhoAa, and the protein PhoA, which is the yield of the system. As described above, the external P_i concentration is modelled by the fold change fc applied to the auto-cross-phosphorylation rate of the reactions r01 and r03 as this rate is a function of the ABC transporter activity. These simulations show that our model captures the mean behaviour of the system components in agreement with fluorescence readings in experiments. The plots also show that the active transcription factor DiPhoBpp concentration and the active promoter pPhoAa concentration are as expected functions of the external P_i concentration. More importantly, these signals are not affected from the changes in other protein concentrations.

3 Stochastic Analysis and Quantifying Information Flow

Stochastic simulations with Chemical Reaction Networks (CRNs) are commonly performed by using one of the various versions of Gillespie's stochastic simulation

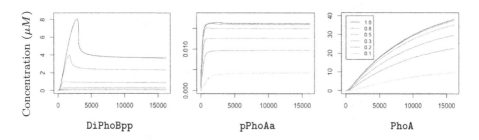

Fig. 1. The time-series plots with ordinary differential equation simulations display the response to the variations in external P_i concentration. The external P_i concentration is given with the fold-change fc. A higher external P_i concentration is modelled with a smaller factor and vice versa. The different fc values are color coded in the legend. (Color figure online)

algorithm (SSA) [7]. Given an initial state as a vector of species quantities, the algorithm constructs a trajectory of the network with respect to the underlying continuous time Markov chain semantics. At each simulation step, the algorithm performs a Monte Carlo procedure to sample from the probability distribution of the possible reaction instances at that state to pick a reaction and its time. The algorithm updates the state and continues in the same way until the end-time is reached. The simulation terminates after logging the trajectory to a file.

Deterministic and stochastic simulations reflect the two facets of the CRNs with respect to the underlying mass action dynamics. Because a stochastic simulation trajectory represents one of the many possible "realisations" of the system, it can capture the fluctuations in species numbers and possible extinctions that may arise due to low species numbers. The deterministic simulations, on the other hand, reflect the mean behaviour of the network, thus they do not capture noise or extinction events. Consequently, the stochastic simulations, at their limit of large numbers, overlap with the deterministic differential equation simulations. The stochastic simulation plots depicted in Fig. 2 exemplify this idea in comparison with the deterministic simulation plots in Fig. 1.

As described above, the SSA generates a stochastic simulation trajectory by sequentially sampling a reaction instance one after another from the distribution of available reactions. The time between two reaction instances is obtained by sampling from an exponential distribution, which is a function of the reaction propensities available at that state. Each reaction instance modifies the system state. The algorithm then continues to pick a reaction instance until it reaches the end-time. The algorithm logs the reaction instances, which provides the common time series representation of the simulations. As a result of this sequential procedure, the timestamps of the reaction instances follow a total order.

However, when we inspect the dependencies of each reaction instance on the available resources at that state, a different point of view arises. This is because each reaction instance consumes reactants as resources that were produced by another reaction at some previous time point. The reaction instance

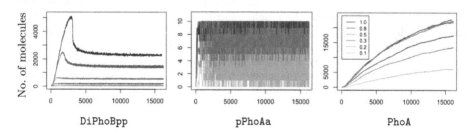

Fig. 2. Stochastic time series with different external P_i concentrations, where the horizontal and vertical axes are the time in seconds and the number of molecules. As in the simulations in Fig. 1, a higher external P_i concentration is given with a smaller factor `fc`, color coded in the legend. The number of promoters, given by 10 plasmids, gives rise to a greater noise in the number of active promoters `pPhoAa` in comparison to those in active transcription factor `DiPhoBpp` and `PhoA`. PhoA quantifies the yield.

produces other resources that become available for consumption at any time later on. Thus, the product of a reaction remains available for consumption, but it is not necessarily consumed by its immediate successor. Conesequently, the production and consumption relationships between the reaction instances in a simulation follow a partial order instead of a total order. The standard SSA loses these relationships, which provide the information on the resource dependencies between the reaction instances. However, these dependencies can be used to quantitatively describe the causality as well as the information flow in the system.

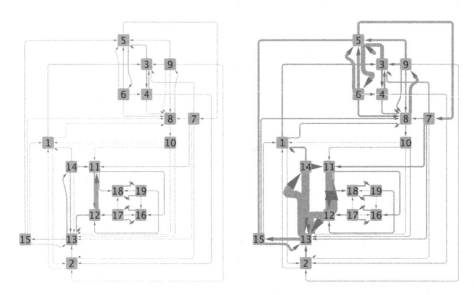

Fig. 3. Flux graphs obtained from a simulation with the CRN in the Appendix section. The graph on the left displays the fluxes in the first 1000 simulated seconds, whereas the second displays the fluxes in the following 1000 simulated seconds. The graphs are rendered with Cytoscape [11] using yFiles. The thickness of the arrows is proportional with the weight of the fluxes. The graphs show that, within the first 1000 s, the fast times-scale dimerisation of the transcription factor PhoB, given by reaction 11, functions as an attractor for the majority of the fluxes. Within the next 1000 s, as the system approaches the steady state, the fluxes to reactions 13, 14 and 15 increase. The concomitant feedback from the genetic circuit to the TCS results in an increased activity in the reactions 03, 04, 05, 06, 07, 08, 09, and 10, performing TCS activation. (Color figure online)

In previous work, we have introduced a conservative extension of Gillespie's stochastic simulation algorithm [7], called fSSA, that keeps track of the resource dependencies between reaction instances [8,9]. In addition to the time series, fSSA logs the dependency information as a partial order by introducing a constant cost to each simulation step. The algorithm introduces two data structures, one for logging the source reaction of the available resources at the current state as a look-up table and another for logging the quantity of the simulated dependencies between reaction instances. The latter structure results in a dag, called *flux graph*, that provides a quantification of the flow of resources between reaction instances for any user-specified time interval. The flux graph is updated at every simulation step by sampling from the table of available resources.

The flux graphs are edge-coloured directed graphs that consist of the flux edges from a time point $t \geq 0$ to another $t' > t$. Each edge of the graph is of the form $p \xrightarrow{x,n} q$, where p and q are nodes representing the reactions of the CRN, x is a network species, and n is the weight. The edge colour x, n on the arrow denotes that between time points t and t', species x flowed from p to q with a weight of n. The weight n denotes the multiplicity of the species x that are logged to have flowed from p to q within the chosen time interval.

The flux graphs in Fig. 3 are obtained from a simulation with the CRN in the Appendix section with an `fc` value of 1.0. The two graphs display the fluxes within the first 1000 simulated seconds and within the following 1000 s.

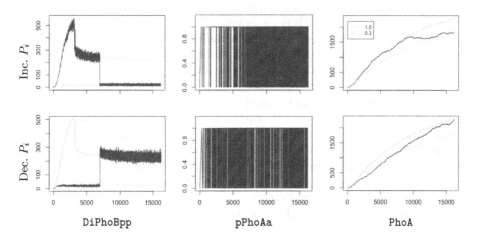

Fig. 4. Stochastic time series of simulations. At 7000 simulated seconds, the phosphorylation fold change `fc` is decreased from 1.0 to 0.3 (top row) or increased from 0.3 to 1.0 (bottom row) in relation to a change in P_i concentration. For comparison, the deterministic trajectories are plotted with dashed lines. The stochastic simulations are scaled down to one tenth of the *E. coli* volume such that there is a single promoter on a plasmid per simulation, and the binding and unbinding effects on the promoter become observable in the plots of the active promoter `pPhoAa`. An increase in the unbinding events results in the fully painted area in the `pPhoAa` plot. A decrease introduces gaps to the painted area. The right-most column displays the adjustment of the system's yield, given by `PhoA`, in response to the change in external P_i levels.

4   Coherence in Response to Phosphate Concentration

In *E. coli*, the TCS mechanism relays the information on external P_i concentration to the genetic level. Thus, the activity level of the transcription factor DiPhoBpp provides an internal representation of the external P_i concentration. In agreement with experimental observations, the simulations with the differential equations in [1] show that the mean behaviour of our model of the TCS mechanism in response to external P_i concentration remains robust to perturbations in many of the system parameters. As shown in the simulations in Figs. 1 and 2, the system maintains a certain steady state in accordance with the external P_i levels also with the feedback provided by the expression of the TCS components that are regulated by this transcription factor: the increased activity of the transcription factor results in the expression of the transcription factor itself as well as the histidine kinase PhoR. Still, the steady state level of PhoB dimers remains in equilibrium as a function of only the input signal. This phenomenon is a consequence of the bi-functional role of PhoR, which participates in both phosphorylation and dephosphorylation of its cognate response regulator PhoB [12–14]. This dual role of PhoR is a mechanism that enhances signal robustness.

In experiments, it has been shown that the phosphatase activity in the TCS provides a rapid dephosphorylation mechanism that tunes the system when it becomes subject to changes, and thereby restores it to the original state [15]. To verify this with our model, we have performed simulations, whereby the system is exposed to a change in external P_i after the active transcription factor DiPhoBpp has reached deterministic equilibrium: the model is first instantiated with an auto-cross-phosphorylation value (fc). In two sets of stochastic simulations, the fc value is then decreased or increased at 7000 simulated seconds, corresponding to a sudden change in the external P_i concentration. The stochastic simulation results, depicted in Fig. 4, show that after these perturbations the system tunes its transcription factor levels accordingly. As a result of this, the promoter activity and the yield of the system (PhoA) adjusts to the modified activity.

Figure 5 displays, at 100 s intervals, the system fluxes just before the perturbation, i.e., $\mathcal{F}[6900, 7000]$, just after the perturbation, i.e., $\mathcal{F}[7000, 7100]$, and the equilibrium interval $\mathcal{F}[16100, 16200]$. Here, $\mathcal{F}[t, t']$ denotes the flux graph between time t and t'. The arrows are scaled proportional with the flux weights.

The left-most flux graph in Fig. 5 displays the flow of the system resources within the 100-s intervals from 6900 to 7000 s. This is the time interval immediately before the perturbation in the top row of Fig. 4, where the system is initiated at the starvation condition. In this interval, just before the increase in the external P_i concentration, the transcription factor activity, given by DiPhoBpp, is at steady state. The flux graph shows that the system is performing cycles of phosphorylation and dephosphorylation of PhoB as well as its complexation to form the active dimer DiPhoBpp and the decomplexation of the dimer back to the monomer PhoBp. The balance between these events maintains the steady state levels of the active transcription factor DiPhoBpp.

The middle flux graph in Fig. 5 displays the fluxes within the 100 s intervals from 7000 to 7100 s. This is the time interval immediately after the perturbation in the top row of Fig. 4, where the external P_i concentration suddenly increases. The flux graph indicates a shift of resources to feed flux a cycle between the reactions 2, 4, 6, 8, and 9. As a consequence of this increase, the transcription factor activity shifts to a new lower steady state. This, in return, accommodates the reduction in phosphorylation of PhoB and the consequent reduction of the transcription factor activity.

The right-most flux graph in Fig. 5 displays the fluxes from 16100 to 16200 s, which is the period at the end of the simulation in the top row of Fig. 4. Here, the system has already adapted to the increase in the external P_i concentration and the transcription factor activity, given by DiPhoBpp, has reached a new steady state. Thus, the situation in the flux graph is similar to the one in the left-most flux graph, whereby the activation and inactivation events are balanced. However, we observe a reduction in the fluxes to the dimerisation reaction 11, which explains the reduced transcription factor activity.[1]

5   Stochasticity in Promoter Design

As demonstrated by the simulations above, the TCS transcription factor activity, that is, the concentration of the phosphorylated PhoB dimers, serves as a proxy

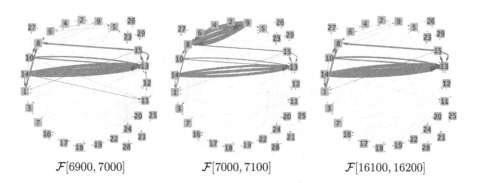

$\mathcal{F}[6900, 7000]$ $\mathcal{F}[7000, 7100]$ $\mathcal{F}[16100, 16200]$

Fig. 5. Stochastic simulation fluxes with a chemical reaction network that models the two-component system response in *E. coli* to a change in external phosphate concentration. The graphs display the fluxes before and after the perturbation and at the end of the simulation. The notation $\mathcal{F}[t, t']$ denotes the flux graph between the time points t and t'. The numbers are the reactions of the CRN in the Appendix section. For visual clarity, flux species are omitted. For a one-minute-long video of the complete simulation fluxes, see: https://youtu.be/PiKRCYyR57k. The network simulates the first 4,5 h after the starvation signal. At 7000 s the phosphate concentration increases and the network responds by lowering `DiPhoBpp` activity as in Fig. 4, top row.

---

[1] For an exposure to the changes in system fluxes throughout the simulation, we refer to the online video of the complete simulation: https://youtu.be/PiKRCYyR57k.

for the external P_i concentration, given by the fc value. The resulting active transcription factor signal activates the promoter and this feeds back as the expression of the TCS components as well as other proteins, e.g., PhoA. This process thus provides the specific adaptation of gene expression dependent on the external P_i response stimuli by providing the appropriate promoter activity. In this setting, the promoter activity, pPhoAa and pPhoBa, is proportional to the affinity of the promoter to the active transcription factor DiPhoBpp as well as its concentration, as described by mass action law.

The binding rate of the active transcription factor to the promoter is determined by the specific nucleotide sequence of the promoter, which also determines how long the promoter remains bound, thus activated, after binding. A mutation in a single nucleotide can result in a drastic modification of the binding and unbinding rates [16–18]. Many applications in synthetic biology are based on exploiting such mechanisms by introducing random mutations to the promoter sequence and, this way, generating libraries of promoters with desired strengths.

In [1], to explore the effect of variations in promoter strength on protein expression, we have performed a class of deterministic simulations. In these simulations, we have measured the PhoA protein yield of the system in conditions of different external P_i concentrations. For each external P_i concentration, we have scanned 100 different promoter designs by varying the promoter binding factors, given by bf in the reactions r16 and r18, and the promoter unbinding rates, given by uf in the reactions r17 and r19, in a spectrum of 10 different values for each. A representative heat-map for these simulations that displays

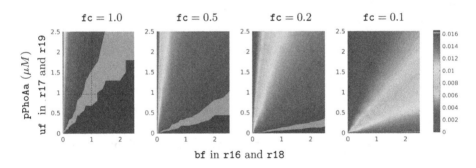

Fig. 6. Heatmaps for the activity of various promoter designs as in [1]. The heatmaps are ordered according to the external P_i concentration given by the fold changes fc applied to the PhoR autphosphorylation reactions. The left most column with 1.0 as the fc value is the starvation condition with $0\mu M$ external P_i. Each heatmap scans 100 simulations by applying 10 different fold change values to the promoter binding rates, given with bf in r16 and r18, as well as 10 different fold change values to the promoter unbinding rates, given with uf in r17 and r18. The heatmaps display the resulting steady state levels of the active promoter pPhoAa in deterministic ordinary differential equation simulations. The intersection of the dashed lines in the left column delivers the experimentally observed regime reported in [1]. The levels of this regime that display the starvation response are highlighted in all the heatmaps.

the mean promoter activity pPhoAa as in [1] is depicted in Fig. 6. These simulations show that in order to obtain the starvation response in the conditions with higher external P_i concentration, promoter binding rates need to be increased and unbinding decreased via the appropriate nucleotide sequence.

Cells with the same genetic makeup can exhibit phenotypic variation in the expression of their different proteins. Some of this variation is attributed to noise that is extrinsic to the protein expression machinery, characterised as the fluctuations in other cellular components. On the other hand, the biochemical process of gene expression can be a source of significant intrinsic noise that results in loss of coherence in the output signal, especially in the context of low molecule numbers [2,19]. The differential equation simulations capture the mean deterministic behaviour that would emerge within a population that employs such mechanisms. However, they do not capture the extent of fluctuations in individuals and the possible variation within the population.

To detect the intrinsic noise in gene expression in the system described by our model, we have run a set of repeated stochastic simulations under three external P_i concentration conditions, modelled as fc values 1.0, 0.3, and 0.1. We have then varied the binding and unbinding rates of the transcription factor and the promoter by applying the factors bf $\in \{0.5, 1.0, 1.5\}$ for the reactions r16, r18 and uf $\in \{0.5, 1.0, 1.5\}$ for the reactions r17, r19. The time series of 27 sets of simulations for 5 repeats each are depicted in Fig. 7.

In accordance with [1], in these simulations we observe that a concomitant increase in binding rates and decrease in unbinding rates provide higher mean levels of active promoter, given with pPhoAa. However, a fine-tuned balance of these rates is required for the system in order not to overshoot the mean gene expression levels in lower external P_i concentration conditions, given with fc values closer to 1.0. From a biological point of view, such an overshoot can have implications on function and introduce a selective pressure.

Stochastic simulations with our model demonstrate an appreciable increase in fluctuations with an increase in unbinding rates (uf) and a moderate decrease in the fluctuations with an increase in binding rates (bf). For a quantification of noise in the system, we have computed the coefficient of variation (CV) for the active promoter pPhoAa and mRNAa. We observed that the CV value for pPhoAa increased with a decrease in fc. However, within all the external P_i concentration regimes, the noise given with CV value for pPhoAa increased together with an increase in the unbinding factor uf. Similarly, an increase in the binding factor bf consistently reduced the noise in promoter activity in all the regimes. The highest fidelity in the promoter activity signal was obtained with bf = 1.5 and uf = 0.5. For the mRNAa signal, however, a significant consistent change in CV value as a result of a change in unbinding factor uf is observable only with fc values 0.3 and 0.1, corresponding to higher P_i concentrations. Similarly, an increase in binding factor bf resulted in a decrease in noise in terms of CV for the mRNAa signal only at higher P_i concentrations.

The CV value provides a quantification of noise for individual species. However, they may not be representative of the noise in the complete machinery. For a more general quantification of noise in the complete system, we have measured the distance between the complete simulation fluxes of the simulations within the same regime. For this, we have computed the mean distance between the normalised flux graphs of the simulations within the same regime. We have then compared these values in relation to variations in the external P_i concentration as well as variations in binding and unbinding rates. We have computed the distance between the two flux graphs $\mathcal{F}$ and $\mathcal{F}'$ as the sum of squared differences

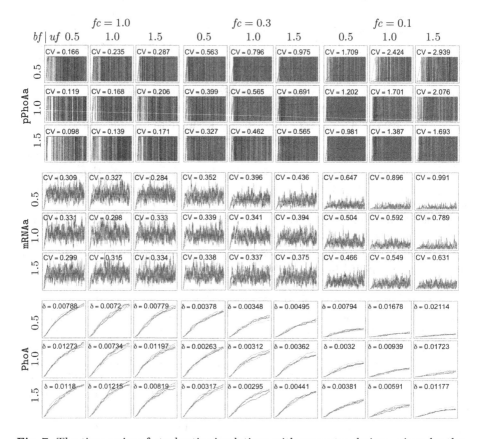

Fig. 7. The time series of stochastic simulations with promoter designs, given by the factors **bf** and **uf**. The top panel displays the promoter activity, i.e., **pPhoAa**, the mid panel displays the levels of PhoA mRNA, i.e., **mRNAa**, and bottom panel displays the yield of the system as PhoA expression, i.e., **PhoA**. The plots are ordered according to the external P_i concentration given by the fold changes **fc** applied to the PhoR aut-phosphorylation rate. The three left most columns with **fc** value of 1.0 is the starvation condition with 0 μM external P_i concentrations, followed by those for increasing levels with **fc** values 0.3 and 0.1, respectively. For each external P_i concentration, the **bf** and **uf** are scanned for values 0.5, 1.0, and 1.5 for the promoter binding rates **r16** and **r18** and the unbinding rates **r17** and **r18**. The instantiation of the factors as **fc** = 1.0, **bf** = 1.0, and **uf** = 1.0 delivers the control regime.

of the edge weights [20]. If an edge does not exist, its weight is assigned a value of 0. The distance function δ, which is a metric [20], is thus defined as follows:

$$\delta(\mathcal{F}_1, \mathcal{F}_2) = \mathsf{sqrt}\left(\sum_{p \xrightarrow{x} q \, \in \, \mathcal{F}_1 \cup \mathcal{F}_2} (w_1 - w_2)^2 \ \text{ for } i = 1, 2, \ w_i = \begin{cases} n_i, \ p \xrightarrow{x, n_i} q \in \mathcal{F}_i \\ 0, \ otherwise \end{cases} \right)$$

The mean distance δ of the simulations are displayed in the lower panel of Fig. 7 together with the time series of PhoA. The distances follow a trend that is similar to CV of mRNAa signal: a consistent change in the mean distance δ between two simulations within the same regime as a result of a change in unbinding factor uf and binding factor bf is observable only with fc values 0.3 and 0.1. Again, the highest fidelity in terms of noise is obtained with bf = 1.5 and uf = 0.5. However, in accordance with the observations in Fig. 6, a more optimal design that prevents an overshoot in promoter activity is obtained with uf > 0.5. This indicates that there is a trade-off between the design of promoters that are capable of exhibiting a starvation response in higher external P_i concentration conditions and the noise in the synthetic system.

6   Discussion

We have presented an analysis of the noise in the biochemical machinery in the phosphate economy of *E. coli* based on stochastic simulations with the SSA algorithm. Besides the common time evolution of the biochemical species quantities, our simulations capture the mechanisms with which information flows between different system components, thereby permitting for a quantification of the fluctuations, not only in the quantities of individual species, but also for the complete system. Because the bacterial two-component system in our model is a central component of many bacterial auto-regulatory processes, our model and its deterministic and stochastic analysis should serve as a template for studying other similar systems involved in the uptake of extracellular molecules. This, in return, should pave the way for the systematic development of a class of synthetic biology technologies that exploit such bacterial processes. In this regard, the visualisation of the time evolution of the system fluxes that we have presented should be relevant for such applications as well as for other models that use a chemical reactions network representation.

Acknowledgements. This work has been partially funded by the European Union's Horizon 2020 research and innovation programme under the grant agreement No 686585 – LIAR, Living Architecture.

Appendix

The CRN in [1] that models the auto-regulation mechanism of *E. coli* in response to varying external phosphate concentrations. The time unit of the reactions is in seconds. The fold-change fc factor in reactions r01 and r03 model the variations in external P_i concentration. The fc = 1.0 value corresponds to the starvation condition and a lower fc value corresponds to a higher external P_i concentration. The binding factor bf in reactions r16, r18 and unbinding factor uf in reactions r17, r19 are scalar factors. They represent the affinity of the active transcription factor to the promoter region. In the control model, the default values of bf = 1.0 and uf = 1.0 are used.

```
reactions
r01 : DiPhoR                    ->     DiPhoRp              , 25.3658*fc;
r02 : DiPhoRp                   ->     DiPhoR               , 8.1165;
r03 : DiPhoRp                   ->     DiPhoRpp             , 25.3658*fc;
r04 : DiPhoRpp                  ->     DiPhoRp              , 8.1165;
r05 :    DiPhoRpp + PhoB        ->     DiPhoRpp_PhoB        , 100;
r06 : DiPhoRpp_PhoB             -> DiPhoRpp + PhoB          , 44.9411;
r07 : DiPhoRpp_PhoB             -> DiPhoRp + PhoBp          , 21.3718;
r08 :    DiPhoRp + PhoB         ->     DiPhoRp_PhoB         , 100;
r09 : DiPhoRp_PhoB              -> DiPhoRp + PhoB           , 94.9411;
r10 : DiPhoRp_PhoB              -> DiPhoR + PhoBp           , 21.3718;
r11 :    PhoBp + PhoBp          ->     DiPhoBpp             , 100;
r12 : DiPhoBpp                  -> PhoBp + PhoBp            , 24.9411;
r13 :    DiPhoR + PhoBp         ->     DiPhoR_PhoBp         , 100;
r14 : DiPhoR_PhoBp              -> DiPhoR + PhoBp           , 34.9411;
r15 : DiPhoR_PhoBp              -> DiPhoR + PhoB            , 12.95;
r16 :    DiPhoBpp + pPhoA       ->     pPhoAa               , 10000*bf;
r17 : pPhoAa                    -> DiPhoBpp + pPhoA         , 1000*uf;
r18 :    DiPhoBpp + pPhoB       ->     pPhoBa               , 10000*bf;
r19 : pPhoBa                    -> DiPhoBpp + pPhoB         , 1000*uf;
r20 : pPhoAa                    -> pPhoAa + mRNAa           , 0.0540;
r21 : mRNAa                     -> mRNAa + PhoA             , 0.0302;
r22 : pPhoBa                    -> pPhoBa + mRNAb           , 0.130;
r23 : mRNAb                     -> mRNAb + PhoB             , 0.036;
r24 : mRNAb                     -> mRNAb + DiPhoR           , 0.0302;
r25 : PhoA                      ->                          , 0.0001;
r26 : PhoB                      ->                          , 0.0001;
r27 : DiPhoR                    ->                          , 0.0001;
r28 : mRNAa                     ->                          , 0.0055;
r29 : mRNAb                     ->                          , 0.0055;

initial state
0.22 DiPhoR;      0.22 PhoB;      0.0166 pPhoA;      0.0166 pPhoB;
```

References

1. Uluşeker, C., Torres-Bacete, J., Garcia, J.L., Hancyzc, M.M., Nogales, J., Kahramanoğulları, O.: Quantifying dynamic mechanisms of auto-regulation in *Escherichia coli* with synthetic promoter in response to varying external phosphate levels. Sci. Rep. **9**, 2076 (2019)
2. Elowitz, M.B., Levine, A.J., Siggia, E.D., Swain, P.S.: Stochastic gene expression in a single cell. Science **297**(5584), 1183–1186 (2002)
3. Raj, A., van Oudenaarden, A.: Nature, nurture, or chance: stochastic gene expression and its consequences. Cell **135**(2), 216–226 (2008)
4. Bandiera, L., Furini, S., Giordano, E.: Phenotypic variability in synthetic biology applications: dealing with noise in microbial gene expression. Front Microbiol. **7**, 479 (2016)
5. Yu, J., Xiao, J., Ren, X., Lao, K., Xie, X.S.: Probing gene expression in live cells, one protein molecule at a time. Science **311**(5767), 1600–1603 (2006)
6. McAdams, H.H., Arkin, A.: Stochastic mechanisms in gene expression. PNAS **94**(3), 814–819 (1997)
7. Gillespie, D.T.: Exact stochastic simulation of coupled chemical reactions. J. Phys. Chem. **81**(25), 2340–2361 (1977)
8. Kahramanoğulları, O., Lynch, J.: Stochastic flux analysis of chemical reaction networks. BMC Syst. Biol. **7**, 133 (2013)
9. Kahramanoğulları, O.: Quantifying information flow in chemical reaction networks. In: Figueiredo, D., Martín-Vide, C., Pratas, D., Vega-Rodríguez, M.A. (eds.) AlCoB 2017. LNCS, vol. 10252, pp. 155–166. Springer, Cham (2017). https://doi.org/10.1007/978-3-319-58163-7_11
10. Harris, R.M., Webb, D.C., Howitt, S.M., Cox, G.B.: Characterization of PitA and PitB from *Escherichia coli*. J Bacteriol. **183**, 5008–5014 (2001)
11. Shannon, P., et al.: Cytoscape: a software environment for integrated models of biomolecular interaction networks. Genome Res. **13**(11), 2498–504 (2003)
12. Miyashiro, T., Goulian, M.: High stimulus unmasks positive feedback in an autoregulated bacterial signaling circuit. PNAS **105**, 17457–17462 (2008)
13. Tiwari, A., et al.: Bistable responses in bacterial genetic networks: designs and dynamical consequences. Math. Biosci. **231**(1), 76–89 (2011)
14. Shinar, G., Milo, R., Matinez, M.R., Alon, U.: Input output robustness in simple bacterial signaling systems. PNAS **104**(50), 19931–19935 (2007)
15. Mukhopadhyay, A., Gao, R., Lynn, D.G.: Integrating input from multiple signals: the VirA/VirG two-component system of Agrobacterium tumefaciens. Chembiochem **5**, 1535–1542 (2004)
16. Hawley, D.K., McClure, W.R.: Compilation and analysis of *Escherichia coli* promoter DNA sequences. Nucleic Acids Res. **11**(8), 2237–2255 (1983)
17. Jensen, P.R., Hammer, K.: The sequence of spacers between the consensus sequences modulates the strength of prokaryotic promoters. Appl. Environ. Microbiol. **64**(1), 82–87 (1998)
18. Jensen, P.R., Hammer, K.: Artificial promoters for metabolic optimization. Biotechnology Bioeng. **58**(2–3), 191–195 (1998)
19. Leveau, J.H., Lindow, S.E.: Predictive and interpretive simulation of green fluorescent protein expression in reporter bacteria. J. Bacteriol **183**(23), 6752–6762 (2001)
20. Kahramanoğulları, O.: On quantitative comparison of chemical reaction network models. In: Proceedings of PERR 2019 3rd Workshop on Program Equivalence and Relational Reasoning, volume 296 of EPTC, pp. 14–27 (2019)

NMR Data Analysis of Water Mobility in Wheat Flour Dough: A Computational Approach

Annalisa Romano[1,2] , Rosanna Campagna[1(✉)] , Paolo Masi[1,2] ,
and Gerardo Toraldo[2,3]

[1] Department of Agricultural Sciences, University of Naples Federico II,
Portici, Naples, Italy
{anromano,rosanna.campagna,paolo.masi}@unina.it
[2] Centre for Food Innovation and Development in the Food Industry (CAISIAL),
University of Naples Federico II, Portici, Naples, Italy
toraldo@unina.it
[3] Department of Mathematics and Applications "R. Caccioppoli",
University of Naples Federico II, Naples, Italy

Abstract. The understanding of the breadmaking process requires to understand the changes in water mobility of dough. The dough ingredients as well as the processing conditions determine the structure of baked products which in turn is responsible for their apparence, texture, taste and stability. The transition from wheat flour to dough is a complex process in which several transformations take place, including those associated with changes in water distribution [13]. The molecular mobility of water in foods can be studied with proton Nuclear Magnetic Resonance (1H NMR). In this study, the measured transverse relaxation times (T2) were considered to investigate the wheat dough development during mixing. The interactions of the flour polymers with water during mixing reduce water mobility and result in different molecular mobilities in dough. The molecular dynamics in heterogeneous systems are very complex. From a mathematical point of view the NMR relaxation decay is generally modelled by the linear superposition of a few exponential functions of the relaxation times. This could be a too rough model and the classical fitting approaches could fail to describe physical reality. A more appealing procedure consists in describing the NMR relaxation decay in integral form by the Laplace transform [2]. In this work a *discrete* Inverse Laplace Transform procedure is considered to obtain the relaxation times distribution of a dataset provided as case study.

Keywords: Bread making · Dough · Laplace transform

1 Introduction

The water molecular mobility in flour dough system is of paramount importance because it has direct influence on the rheological properties of the dough and

© Springer Nature Switzerland AG 2020
Y. D. Sergeyev and D. E. Kvasov (Eds.): NUMTA 2019, LNCS 11973, pp. 146–157, 2020.
https://doi.org/10.1007/978-3-030-39081-5_14

its baking performance. During mixing, the ingredients are transformed into a dough through the formation of gluten, a continuous cohesive viscoelastic protein network in which are dispersed starch granules. At the dough stage, flour particles are hydrated, and the transition from wheat flour to dough is a complex process in which several transformations take place, including those associated with changes in water distribution [13, 21]. Generally these transformations take place during the first minutes of mixing and can be reasonably monitored with a Brabender farinograph that also allows an empirical characterization of the flour on the basis of few dough mixing parameters, including water absorption capacity, dough development time, dough stability time and mixing tolerance index [19, 22]. Although very practical for routine and industrial applications, this approach does not provide any direct information on the physical state of water imbibed by the flour at the molecular and supra-molecular level. The Nuclear Magnetic Resonance (NMR) can provide information about phenomena that involve single molecules or relatively small clusters of molecules in dough and bread. The NMR relaxation times highlight the material composition and are used for quality control. In particular, the molecular mobility of water and biopolymers in food products can be studied with proton nuclear magnetic resonance (1H NMR). Low-resolution (LR) 1H NMR has been mainly used to measure transverse relaxation time T2 in dough and bread (see [1, 3, 12]). When water is bound tightly to the substrate (e.g. flour), it is highly immobilized and shows reduced T2; whereas free water is mobile and has relatively long T2 [14]. Thus, useful information on the strength or degree of water binding can be obtained. In this study, the relaxation time T2, measured by LR 1H NMR, was used to investigate wheat dough development during mixing.

A NMR-based analysis of water mobility in wheat flour dough is carried out and a Laplace Transform Inversion (LTI) recovery test is presented. The LTI functional formulation is indeed useful for the analysis of the NMR relaxation times, resulting in a relaxation time spectra which may provide information on the different water molecules population involved in a process. The drawback of the direct application of LTI numerical methods is the needed of a continuous model describing the data [11]. In order to overcome this issue, the behaviour of the true signal generating the data, is deduced by piecewise functions, fitting sequences of discrete data, also of long duration, that cannot be fitted by a single function, such as a polynomial. Spline functions are preferred, since they do not require equally time intervals and therefore may be used to fit gaps in data files. In literature smoothing spline models reflecting the exponential decay taking into account main Laplace Transform (LT) properties can be found in [4]. This approach allows to overcome the limits of the analysis of discrete data [20], since the problem is converted in a continuous form so that general software packages for LTI (see [9, 10]) can be used. However, in the applications, the most used approach in relaxation studies is a Regularized Inverse Laplace Transform (RILT) algorithm based on weighted least squares solution (see [18]), that allows to solve discrete inverse problems described by integral equations of first species. The integral form of the decaying magnetization signal gives the LT functional relationship between the relaxation signal s and the corresponding distribution function g of the relaxation times. A LTI of s reveals the frequency distribution of relaxation times; in this relaxation time spectrum, the integral of each peak reveals the contribution of each species exhibiting the corresponding specific spin relaxation.

Regarding materials and methods, the sample data were prepared by using: soft wheat flour (Caputo "00", Italy), deionised water, salt and compressed yeast (Mastro Fornaio, Paneangeli®). Dough formulation was: soft wheat flour (60.35%), salt (1.2%), yeast (1.1%), deionised water (37.35%) according to farinographic absorption. In fact, the amount of water (absorption) required to obtain a stability period at 500 Brabender Units (BU) was used. All doughs were prepared in a Brabender Farinograph (O. H. Duisburg, Germany), equipped with a 300 g bowl. Mixing time and temperature were kept constant and equal to $1, 2, 3$ and 13 min and 25 °C, respectively. Proton transverse (T2) relaxation times were measured with a low resolution (20 MHz) 1H NMR spectrometer (the minispec mq-20, Bruker, Milano, Italy) operating at 25 ± 1 °C. 3 g of sample were placed into a 10 mm external diameter NMR tube (filled with sample to about 3 cm height). NMR tubes were immediately sealed with Parafilm® to prevent moisture loss during the NMR experiment. The transverse relaxation times, 1H T2, were determined with a CPMG pulse sequence (Carr, Purcell, Meiboom and Gill) using a recycle delay of 0.4 s and an interpulse spacing of 0.04 ms. The number of data points acquired was 1500 for the CPMG sequence and 20 for the Inversion Recovery. Three NMR tubes were analyzed for each sample of dough. Each average value represents the mean of 3 independent measurements. In Sect. 2 the mathematical modelling and the numerical scheme are presented; Sect. 3 gives the results for a dataset of real samples. In the last section there are some conclusions.

2   Mathematical Modelling

The numerical scheme that we use to analyse the effects of the NMR spectroscopy consists in: (a) a continuous description of the magnetization decay, as a function of the time, like the sum of exponential decay functions; (b) a LTI of the relaxation function, resulting in a relaxation time distribution, characteristic for the samples under investigation. For sake of completeness some definitions are reported.

Definition 1 (Laplace Transform). *Let f be a given integrable function and $s_0 \in \mathbb{C}$ such that*

$$\int_0^\infty e^{-s_0 t} f(t) dt < \infty.$$

Then, for $s \in \mathbb{C} \; : \; Re(s) > Re(s_0)$ we have

$$\int_0^\infty e^{-st} f(t) dt < \infty.$$

Defining

$$\mathcal{C}(f) := \left\{ s \in \mathbb{C}, \ \int_0^{\infty} e^{-st} f(t) dt < \infty \right\}$$

(the so called region of convergence*) the complex valued function F:*

$$F: \ s \in \mathcal{C}(f) \ \rightarrow \ F(s) = \int_0^{\infty} e^{-st} f(t) dt \ \in \mathbb{C}$$

is the Laplace Transform *of f, generally referred to as* $F(s) = \mathcal{L}[f(t)]$.

Definition 2 (Laplace Transform inversion problem). *Given F, compute f such that* $F(s) = \mathcal{L}[f(t)]$.

Definition 3 (Discrete data inversion). *Given*

$$F_i = F(s_i) + \varepsilon_i, \quad i = 1, ..., N$$

with ε_i, $i = 1, ..., N$ *unknown noises and* $a, b \in \mathbb{R}$. *Compute f, or some values of f, where*

$$F(s) = \int_a^b K(s, t) f(t) dt,$$

with K suitable kernel.

The *real* LTI is an ill-posed problem, according to Hadamard definition, since the solution doesn't depend continuously on the data. Due to the ill conditioning of the discrete problem, strong amplifications of the errors occur on the final result, thus regularization techniques are needed. When $K(s, t) = e^{-st}$, the Definition 3 is referred to as *Discrete Laplace transform inversion* (DLTI).

To solve the NMR data analysis by DLTI, first of all the magnetization is modeled as a decreasing function of time, formulated as the sum of exponential terms, depending on the amplitudes g_j and relaxation times, T_j of the components, as follows:

$$s(t) = \sum_{j=1}^{M} g_j e^{-t/T_j} \tag{1}$$

where: t is the experimental time, M is the number of micro-domains having the same spin density, g_j are the amplitudes and the relaxation time T_j of the different components; T_j stands for the transverse relaxation time $(T_2)_j$. Usually, the time constants vary from a few microseconds up to several seconds, depending on the material under investigation. The sum is over the i different water environments. The number of exponential components, M, can range from 1 to a large number, *especially in highly heterogeneous systems*. In the applications, most relaxation curves can be well described with 2 to 3 exponential terms, even if this number strongly affects the solution. In complex multiphase systems, the decay constants have values close enough to each other, so a *continuous distribution* of the different water sites better represents the signal in integral form as

$$s(t) = \int_0^{\infty} K(t, \tau) g(\tau) d\tau.$$

Under these assumptions, LTI methods are mandatory to analyse the NMR relaxation decay. Nevertheless LTI methods require the LT function analytically known almost everywhere in its domain [5–7,17], so a preprocessing of the data, is needed to bridge the discrete data with the continuous model. For our problem we describe the finite dataset of samples like:

$$s_i = \int_a^b K(t_i, \tau) g(\tau) d\tau + \varepsilon_i, \quad a, b \in \mathbb{R}^+, \quad \text{with} \quad s_i = s(t_i), \quad i = 1, \ldots, N,$$
(2)

with ϵ_i noise components. Once assumed the integral form (2), the solution g is computed by solving a *constrained quadratic programming problem* that is defined by regularizing the following problem:

$$\min_{\mathbf{g}} \|\mathbf{s} - A\mathbf{g}\|^2$$
(3)

$$s.t. \quad g(\tau) \geq 0$$
(4)

where $\|\cdot\|$ is the Euclidean norm, $\mathbf{s}$ is the data vector and $\mathbf{g}$ the discrete solution, following the next procedure:

Procedure. DLTI for NMR data analysis

1: Approximate the integrals (2) by numerical integration:

$$s_i = \sum_{m=1}^{N_g} a_m K(t_i, \tau_m) g(\tau_m) + \varepsilon_i, \quad i = 1, \ldots, N$$
(5)

with a_m weights of the quadrature formula, N_g number of the τ_m grid points. The computation of the s_i in (5) requires the solution of the linear system:

$$s_i = \sum_{m=1}^{N_g} A_{i,m} g_m + \varepsilon_i, \quad i = 1, \ldots, N$$
(6)

with $g_m := g(\tau_m)$ and $A_{i,m} = a_m K(t_i, \tau_m)$, $m = 1, \ldots, N_g$, $i = 1, \ldots, N$.

2: Constraints on the solution; in order to reduce the degree of freedoms on the solution, the non-negative inverse function is required, equal to zero at the extremes:

$$g_m \geq 0, \quad m = 1, \ldots, N_g$$
(7)

$$g_1 = 0, \quad g_{N_g} = 0$$
(8)

3: Regularization; moving from (2) to (4) the ill-posedness becomes ill-conditioning, so a regularized solution of (6) is computed by solving the weighted least-square problem

$$\min_{\mathbf{g}} \|M_\epsilon^{-1/2}(\mathbf{s} - A\mathbf{g})\|^2 + \alpha^2 \|Dg\|^2 \text{ subject to (7) and (8)},$$
(9)

Procedure. DLTI for NMR data analysis (cont.)

where M_ϵ is the (positive definite) covariance matrix of the error distribution ε_i and D is the matrix depending on the discretization scheme adopted for the second order derivatives of g. Statistical prior knowledge is needed in order to consider the randomness on the solution due to the noise. The penalty term $\|Dg\|^2$ in (9), is imposed, forcing the smoothness, selecting the solution with minimum number of peaks. The regularization parameter α is critical to the quality of the inversion.

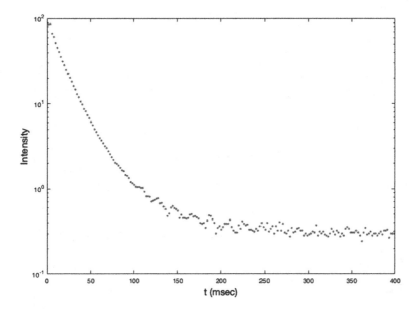

Fig. 1. Transverse relaxation time (T2) for water protons in flour doughs at mixing time 13 min long. $N = 200$ samples.

As concern the choice of α, several methods have been proposed in literature: Discrepancy Principle, Generalized Cross Validation, L-Curve Method. In our numerical tests we choose it empirically, and a more rigorous estimation procedure will be considered in future studies. The numerical solution of the problem (9) by *ad hoc* algorithms for quadratic programming (see [24,25] and references therein), is under investigation.

3   Numerical Experiments

In this section we present the data set and the corresponding inversion results. A typical plot of data for determining T2 for water protons in flour doughs with dough consistency of 500 BU is shown in Fig. 1. The relaxation curve should

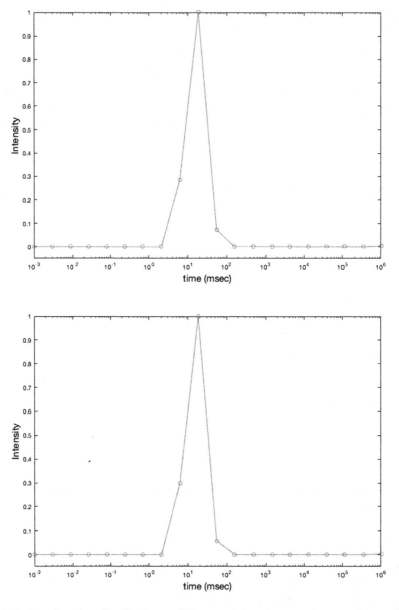

Fig. 2. Relaxation time distribution at different mixing time equal to 1 and 2 min (from up to down).

reveal two components characterized by a short and a long relaxation time: the first component represents the less mobile, or more tightly bound water fraction, while the second one represents the more mobile water fraction [16]. The most mobile water fraction is responsible for the large value of relative humidity

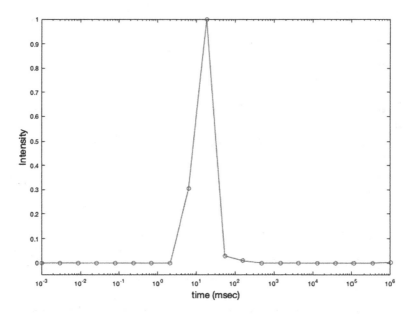

Fig. 3. Relaxation time distribution at different mixing time equal to 3 min.

(RH) and water activity of the dough (above 95%) [8] allowing optimization of storage condition and development of process condition which may result in an extension of shelf life of this product family. The other water fraction behaves like a structure component of the starch and the gluten: this fraction is hard to remove and does not contribute to the overall RH of the system (see [23]). Each of these water fractions is indeed split in several families of molecules that specifically interact with the various components of the dough. Therefore the information on water molecules with large or short relaxation times may be correlated to bread staling processes.

The numerical experiments about the LTI of the mean values of three different acquisitions, each of $N = 200$ samples, after a mixing time of 1, 2, 3 and 13 min long, were carried out with MATLAB R2018a software on a Intel(R) Core(TM) i5, 1.8 GHz processor. The RILT is computed by the Matlab code rilt.m available at the Mathworks File Exchange. We remark that the LTI is an ill-posed problem so only a possible solution is given, satisfying some constraints imposed by the user, according to the data. We fixed the default constraints (7)–(8). The relaxation times distribution $\{g_j\}_j$ is the DLTI of the sampled signal s, calculated by the described procedure. The minimization is computed by the Matlab function fminsearch that uses the Nelder-Mead simplex algorithm as described in [15]. In the Figs. 2 and 3 we describe the computed relaxation time spectrum of the magnetization decay, corresponding to different mixing times, equal to 1, 2 and 3 minutes, respectively. Due to the material dependent variety of the NMR relaxation times the time scale of the relaxation time spectrum has to be optimized for the samples under investigation. The starting relaxation time

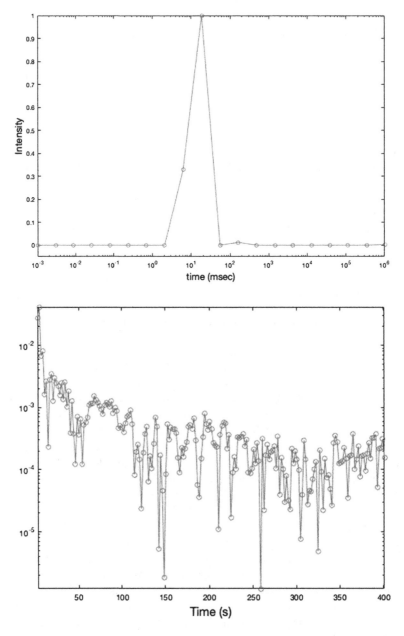

Fig. 4. Relaxation time distribution (top) and corresponding normalized residuals $|s_i - (A\tilde{g})_i|/N$, $i = 1, \ldots, N$, between the measured s_i and the computed magnetization decay values $(A\tilde{g})_i$ (bottom), after a time mixing 13 min long.

should be in the order of the time resolution of the experiment, but larger than zero, the end of the time scale in the order of twice the largest expected relaxation time component. In our experiments the maximum mixing time is 13 min, so the end of the time scale is fixed to 10^6 milliseconds (ms). In combination with the time scale, the resolution can be specified by the number of points, which determines the calculation time of the ILT in a roughly quadratic way. The number of points of the relaxation time spectrum is set to 20. The relaxation spectrum presents some peaks, related to the relaxation time: the larger the relaxation time, the wider the peak. The integral of a peak in the relaxation spectrum corresponds therefore to the contribution of the species exhibiting this particular relaxation behaviour. In Fig. 4 (top) we report the relaxation time distribution after a time mixing 13 min long. The curve proves that one component is clearly resolved in 13 min.

In order to get an estimate of the goodness of the ILT, a plot of the residuals as function of experiment time, between the measured and the calculated magnetization decay corresponding to a mixing time of 13 min, is also displayed in Fig. 4 (bottom).

4   Conclusions

The interactions of the flour polymers with water during mixing reduce water mobility and result in different molecular mobilities in dough. The results showed that 1H NMR is an effective method to have a deeper view of wheat dough development during mixing. A NMR-inversion recovery experiment based on a RILT algorithm furnishes the relaxation time distribution and reveals the dominance of one component responsible in the changes in water mobility of dough. In the LTI algorithm the choice of the regularization parameter α and the more suitable algorithm for the research of the optimal solution, are still open issues. The calibration on the data tailoring the model on the exponential sum defining the relaxation decay function will be object of future studies.

Acknowledgements. This work was partially supported by Gruppo Nazionale per il Calcolo Scientifico - Istituto Nazionale di Alta Matematica (GNCS-INdAM) and by the Centre for Food Innovation and Development in the Food Industry (CAISIAL), University of Naples Federico II, Portici (Naples), Italy. Rosanna Campagna was also supported by the MOD_CELL_DEV Project - Programma di finanziamento della Ricerca di Ateneo, University of Naples Federico II.

References

1. Balestra, F., Laghi, L., Saa, D.T., Gianotti, A., Rocculi, P., Pinnavaia, G.: Physicochemical and metabolomic characterization of kamut®khorasan and durum wheat fermented dough. Food Chem. **187**, 451–459 (2015)
2. Barone, P., Ramponi, A., Sebastiani, G.: On the numerical inversion of the Laplace transform for nuclear magnetic resonance relaxometry. Inverse Prob. **17**(1), 77–94 (2001)

3. Bosmans, G.M., Lagrain, B., Deleu, L.J., Fierens, E., Hills, B.P., Delcour, J.A.: Assignments of proton populations in dough and bread using NMR relaxometry of starch, gluten, and flour model systems. J. Agric. Food Chem. **60**(21), 5461–5470 (2012)
4. Campagna, R., Conti, C., Cuomo, S.: Smoothing exponential-polynomial splines for multiexponential decay data. Dolomites Res. Notes Approx. **12**, 86–100 (2019)
5. Cuomo, S., D'Amore, L., Murli, A.: Error analysis of a collocation method for numerically inverting a laplace transform in case of real samples. J. Comput. Appl. Math. **210**(1), 149–158 (2007). Proceedings of the Numerical Analysis Conference 2005
6. Cuomo, S., D'Amore, L., Murli, A., Rizzardi, M.: Computation of the inverse Laplace Transform based on a collocation method which uses only real values. J. Comput. Appl. Math. **198**(1), 98–115 (2007)
7. Cuomo, S., D'Amore, L., Rizzardi, M., Murli, A.: A modification of weeks' method for numerical inversion of the Laplace transform in the real case based on automatic differentiation. In: Bischof, C.H., Bücker, H.M., Hovland, P., Naumann, U., Utke, J. (eds.) Advances in Automatic Differentiation. LNCSE, pp. 45–54. Springer, Heidelberg (2008). https://doi.org/10.1007/978-3-540-68942-3_5
8. Czuchajowska, Z., Pomeranz, Y.: Differential scanning calorimetry, water activity, and moisture contents in crumb center and near-crust zones of bread during storage. Cereal Chem. **66**, 305–309 (1989)
9. D'Amore, L., Campagna, R., Mele, V., Murli, A.: ReLaTIve. An Ansi C90 software package for the real Laplace transform inversion. Numer. Algorithms **63**(1), 187–211 (2013)
10. D'Amore, L., Campagna, R., Mele, V., Murli, A.: Algorithm 946: ReLIADiff-A C++ software package for real laplace transform inversion based on algorithmic differentiation. ACM Trans. Math. Softw. **40**(4), 31:1–31:20 (2014)
11. D'Amore, L., Mele, V., Campagna, R.: Quality assurance of Gaver's formula for multi-precision Laplace transform inversion in real case. Inverse Probl. Sci. Eng. **26**(4), 553–580 (2018)
12. Doona, C., Baik, M.Y.: Molecular mobility in model dough systems studied by time-domain nuclear magnetic resonance spectroscopy. J. Cereal Sci. **45**(3), 257–262 (2007)
13. Goesaert, H., Slade, L., Levine, H., Delcour, J.A.: Amylases and bread firming: an integrated review. J. Cereal Sci. **50**(3), 345–352 (2009). Special Section: Enzymes in Grain Processing
14. Hills, B.: NMR studies of water mobility in foods. In: Roos, Y., Leslie, R., Lillford, P. (eds.) Water Management in the Design and Distribution of Quality Foods, pp. 107–131. CRC Press, USA (1999)
15. Lagarias, J.C., Reeds, J.A., Wright, M.H., Wright, P.E.: Convergence properties of the Nelder-Mead simplex method in low dimensions. SIAM J. Optim. **9**, 112–147 (1998)
16. Leung, H.K., Magnuson, J.A., Bruinsma, B.L.: Pulsed nuclear magnetic resonance study of water mobility in flour doughs. J. Food Sci. **44**(5), 1408–1411 (1979)
17. Murli, A., Cuomo, S., D'Amore, L., Galletti, A.: Numerical regularization of a real inversion formula based on the Laplace transform's eigenfunction expansion of the inverse function. Inverse Probl. **23**(2), 713–731 (2007)
18. Provencher, S.W.: A constrained regularization method for inverting data represented by linear algebraic or integral equations. Comput. Phys. Commun. **27**(3), 213–227 (1982)

19. Raiola, A., Romano, A., Shanakhat, H., Masi, P., Cavella, S.: Impact of heat treatments on technological performance of re-milled semolina dough and bread. LWT **117**, 108607 (2020)
20. Romano, A., Campagna, R., Masi, P., Cuomo, S., Toraldo, G.: Data-driven approaches to predict states in a food technology case study. In: 2018 IEEE 4th International Forum on Research and Technology for Society and Industry (RTSI), pp. 1–5, September 2018
21. Romano, A., Cavella, S., Toraldo, G., Masi, P.: 2D structural imaging study of bubble evolution during leavening. Food Res. Int. **50**(1), 324–329 (2013)
22. Romano, A., Cesarano, L., Sarghini, F., Masi, P.: The influence of arabinogalactan on wheat dough development during mixing and leavening process. Chem. Eng. Trans. **32**, 1765–1770 (2013)
23. Schiraldi, A., Fessas, D.: Classical and knudsen thermogravimetry to check states and displacements of water in food systems. J. Therm. Anal. Calorim. **71**(1), 225–235 (2003)
24. di Serafino, D., Toraldo, G., Viola, M., Barlow, J.: A two-phase gradient method for quadratic programming problems with a single linear constraint and bounds on the variables. SIAM J. Optim. **28**(4), 2809–2838 (2018)
25. di Serafino, D., Ruggiero, V., Toraldo, G., Zanni, L.: On the steplength selection in gradient methods for unconstrained optimization. Appl. Math. Comput. **318**, 176–195 (2018). Recent Trends in Numerical Computations: Theory and Algorithms

First Order Methods in Optimization:
Theory and Applications

A Limited Memory Gradient Projection Method for Box-Constrained Quadratic Optimization Problems

Serena Crisci[1,3]([✉]) [ID], Federica Porta[2,3] [ID], Valeria Ruggiero[1,3] [ID],
and Luca Zanni[2,3] [ID]

[1] Department of Mathematics and Computer Science,
University of Ferrara, via Machiavelli 30, 44121 Ferrara, Italy
{serena.crisci,valeria.ruggiero}@unife.it
[2] Department of Physics, Informatics and Mathematics,
University of Modena and Reggio Emilia, via Campi 213/B, 41125 Modena, Italy
{federica.porta,luca.zanni}@unimore.it
[3] Member of the INdAM-GNCS Research group, Rome, Italy

Abstract. Gradient Projection (GP) methods are a very popular tool to address box-constrained quadratic problems thanks to their simple implementation and low computational cost per iteration with respect, for example, to Newton approaches. It is however possible to include, in GP schemes, some second order information about the problem by means of a clever choice of the steplength parameter which controls the decrease along the anti-gradient direction. Borrowing the analysis developed by Barzilai and Borwein (BB) for an unconstrained quadratic programming problem, in 2012 Roger Fletcher proposed a limited memory steepest descent (LMSD) method able to effectively sweep the spectrum of the Hessian matrix of the quadratic function to optimize. In this work we analyze how to extend the Fletcher's steplength selection rule to GP methods employed to solve box-constrained quadratic problems. Particularly, we suggest a way to take into account the lower and the upper bounds in the steplength definition, providing also a theoretical and numerical evaluation of our approach.

Keywords: Quadratic programming · Gradient projection methods · Steplength selection rule · Ritz-like values

1 Introduction

Let consider the following box-constrained quadratic problem (QP)

$$\min_{\ell \leq x \leq u} f(x) \equiv \frac{1}{2}x^T A x - b^T x + c, \tag{1}$$

where $A \in \mathbb{R}^{n \times n}$ is a symmetric and positive definite matrix, b, ℓ, u are vectors of $\mathbb{R}^n$, with $\ell \leq u$, and c is a scalar. Hereafter, we denote the feasible region by $\Omega = \{x \in \mathbb{R}^n : \ell \leq x \leq u\}$.

© Springer Nature Switzerland AG 2020
Y. D. Sergeyev and D. E. Kvasov (Eds.): NUMTA 2019, LNCS 11973, pp. 161–176, 2020.
https://doi.org/10.1007/978-3-030-39081-5_15

We are interested in solving problem (1) by means of the well-known Gradient Projection (GP) method combined with a linesearch strategy along the feasible direction [2, Chapter 2]. The standard GP iteration can be written as

$$
\begin{aligned}
x^{(k+1)} = x^{(k)} + \nu_k d_k &= \\
&= x^{(k)} + \nu_k \left(P_\Omega(x^{(k)} - \alpha_k g^{(k)}) - x^{(k)} \right),
\end{aligned}
\tag{2}
$$

where $P_\Omega(\cdot)$ denotes the projection operator onto the constraints set, $g^{(k)}$ stands for $\nabla f(x^{(k)})$, α_k is the steplength parameter that controls the decrease along the anti-gradient and $\nu_k \in (0, 1]$ is computed by means of a backtracking procedure, as for example the monotone or non-monotone Armijo linesearch [2,8].

Convergence results on the GP method have been obtained also in the case of general continuously differentiable objective functions, as summarized in the following theorem. For further details, see [3,10].

Theorem 1. *[3, Th 2.2] Let consider the problem*

$$
\min_{x \in \Omega} F(x),
$$

where $F : \mathbb{R}^n \to \mathbb{R}$ is a continuously differentiable function and Ω is a closed convex subset of $\mathbb{R}^n$. Let assume $\alpha_k \in [\alpha_{min}, \alpha_{max}]$, with $0 < \alpha_{min} < \alpha_{max}$, and ν_k obtained by either a monotone or a non-monotone linesearch. Then, each limit point of the sequence $\{x^k\}_{k \in \mathbb{N}}$ generated by GP method is a stationary point for the considered problem. If, in addition, F is a convex function with a Lipschitz-continuous gradient, the set of solutions is not empty and the initial level set is bounded, we have that

$$
F(x^{(k)}) - F^* = \mathcal{O}\left(\frac{1}{k}\right),
$$

where F^* is the minimum value [10, Th. 3.3].

Thanks to the very general hypothesis on α_k, a clever choice of such parameter can be exploited in order to accelerate the practical performance of the GP method. In order to understand the key principle to properly select the steplength α_k, we firstly recall some useful considerations relative to the easier case of the unconstrained minimization of the quadratic function in (1).

In this case, the GP approach (2) reduces to a standard gradient method whose iteration is given by

$$
x^{(k+1)} = x^{(k)} - \alpha_k g^{(k)},
\tag{3}
$$

and the corresponding gradient recurs according to the rule

$$
g^{(k+1)} = g^{(k)} - \alpha_k A g^{(k)}.
\tag{4}
$$

By denoting with $\{\lambda_1, \lambda_2, \ldots, \lambda_n\}$ the eigenvalues of A and with $\{v_1, v_2, \ldots, v_n\}$ a set of associated orthonormal eigenvectors, the gradient $g^{(k+1)}$ can be expressed as

$$
g^{(k+1)} = \sum_{i=1}^{n} \mu_i^{(k+1)} v_i,
\tag{5}
$$

where $\mu_i^{(k+1)} \in \mathbb{R}$ is called the i-th eigencomponent of $g^{(k+1)}$ and satisfies the following recurrence formula:

$$\mu_i^{(k+1)} = \mu_i^{(0)} \prod_{j=0}^{k}(1 - \alpha_j\lambda_i) = \mu_i^{(k)}(1 - \alpha_k\lambda_i). \tag{6}$$

Since, in absence of constraints, the gradient components at the solution are equal to zero, equality (6) highlights that if the steplength α_k is able to approximate the inverse of any eigenvalue λ_i, the corresponding eigencomponent $\mu_i^{(k+1)}$ is damped. Indeed, the most popular steplengths strategies developed for the unconstrained quadratic case (in particular those based on the Barzilai-Borwein rules [1]) are usually aimed at capturing this property, providing suitable approximations of the inverse of the eigenvalues of the Hessian matrix A and trying to reduce the gradient eigencomponents by means of an effective sweeping of the spectrum of A^{-1} (see, for example, [1,5,6,12,13]).

Among them, a promising steplength updating strategy has been suggested in [5] leading to the development of the Limited Memory Steepest Descent (LMSD) algorithm. Several numerical experiments [5] show the effectiveness of this steplength selection rule with respect to other state-of-the art strategies for choosing the steplength.

As a consequence of these encouraging results and taking into account that the convergence for the GP method (2) is guaranteed for every choice of the steplength in a closed and bounded interval, the aim of this paper is to understand how to apply the limited memory steplength approach to the GP algorithms for box-constrained quadratic problems. Particularly, we investigate if it may be still convenient to approximate the eigenvalues of the whole Hessian matrix or the spectrum of a more suitable matrix should be considered.

The paper is organized as follows. In Sect. 2 we recall the standard LMSD method and we develop its possible generalization to quadratic problems with upper and lower bounds. In Sect. 3 we present the results of different numerical experiments to evaluate the validity of the suggested generalized limited memory steplength strategy. Our conclusions are reported in Sect. 4.

Notation. In the following we denote by $A_{\mathcal{C},\mathcal{D}}$ the submatrix of A of order $\sharp\mathcal{C} \times \sharp\mathcal{D}$ given by the intersection of the rows and the columns with indices in the sets $\mathcal{C}$ and $\mathcal{D}$ respectively. Moreover, $x_{\mathcal{C}} \in \mathbb{R}^{\sharp\mathcal{C}}$ stands for the subvector of x with entries indexed in $\mathcal{C}$.

2   A Limited Memory Steplength Selection Rule for Box-Constrained Quadratic Problems

We start this section by recollecting the main features of the original LMSD algorithm for unconstrained quadratic optimization problems and then we move to the more general case of box-constrained ones.

2.1   The State-of-the Art About LMSD

The basic idea of this method is to divide up the sequence of iterations of a steepest descent scheme into groups of m iterations, referred to as *sweeps*, where m is a small positive integer, with the aim of providing, at each new sweep, m approximations of the eigenvalues of the Hessian matrix.

The starting point of such idea is that m steps of the Lanczos iterative process applied to the matrix A, with starting vector $q_1 = g^{(k)}/\|g^{(k)}\|$, provide a tridiagonal $m \times m$ matrix T whose eigenvalues, called Ritz values, are approximations of m eigenvalues of A [7]. Indeed, given an integer $m \geq 1$, the Lanczos process generates orthonormal n-vectors $\{q_1, q_2, \ldots, q_m\}$ that define a basis for the Krylov sequence

$$\left\{ g^{(k)}, Ag^{(k)}, A^2 g^{(k)}, \ldots, A^{m-1} g^{(k)} \right\} \tag{7}$$

and such that the matrix

$$T = Q^T A Q \tag{8}$$

is tridiagonal, where $Q = [q_1, q_2, \ldots, q_m]$, $Q^T Q = I$. The inverses of the m eigenvalues of the matrix T are employed as steplengths for the successive m iterations. The matrix T can be computed without involving the matrix A explicitly. To reach this goal we need the two following remarks.

1. If a limited number m of back values of the gradient vectors

$$G = [g^{(k)} \ g^{(k+1)} \ \cdots \ g^{(k+m-1)}]$$

   is stored and the $(m + 1) \times m$ matrix J containing the inverses of the corresponding last m steplengths is considered

$$J = \begin{bmatrix} \frac{1}{\alpha_k} & & \\ -\frac{1}{\alpha_k} & \ddots & \\ & \ddots & \frac{1}{\alpha_{k+m-1}} \\ & & -\frac{1}{\alpha_{k+m-1}} \end{bmatrix}$$

   then the equations arising from (4) can be rewritten in matrix form as

$$AG = [G, \ g^{(k+m)}]J. \tag{9}$$

2. Taking into account (4) and that the columns of G are in the space generated by the Krylov sequence (7), we have $G = QR$, where R is $m \times m$ upper triangular and nonsingular, assuming G is full-rank.

We remark that R can be obtained from the Cholesky factorization of $G^T G$ and the computation of Q is not required. Then from both (9) and $G^T G = R^T R$, it follows that the tridiagonal matrix T can be written as

$$T = Q^T A Q = R^{-T} G^T [G \ g^{(k+m)}]J R^{-1} = [R \ r]J R^{-1}, \tag{10}$$

where the vector r is the solution of the linear system $R^T r = G^T g^{(k+m)}$. Proper techniques can be adopted to address the case of rank-deficient G.

2.2   From the LMSD Method to the Limited Memory GP method

When we solve a quadratic box-constrained problem by means of the GP method (2), the procedure to build the matrix G using the whole gradients seems not convenient. Indeed, in this case, the relation (9) does not hold and the matrix obtained by the procedure (10) does not exhibit a tridigonal structure but it is an upper Hessenberg matrix. A first attempt at suggesting a possible way to employ the limited memory steplength selection rule in the framework of box-constrained optimization has been done in [11]. Here, the authors propose to consider a generalized matrix T, called $\tilde{T}$, defined as

$$\tilde{T} = \tilde{R}^{-T} \tilde{G}^T [\tilde{G} \ \tilde{g}^{(k+m)}] \tilde{J} \tilde{R}^{-1},$$

where the general vector $\tilde{g}^{(k)}$ is given by

$$\tilde{g}_j^{(k)} = \begin{cases} g_j^{(k)} & \text{if } \ell_j < x_j^{(k)} < u_j \\ 0 & \text{otherwise} \end{cases}, \tag{11}$$

$\tilde{G}$ is the $n \times m$ matrix $[\tilde{g}^{(k)} \ \tilde{g}^{(k+1)} \ \cdots \ \tilde{g}^{(k+m-1)}]$, $\tilde{R}$ is such that $\tilde{R}^T \tilde{R} = \tilde{G}^T \tilde{G}$ and $\tilde{J}$ is the $(m+1) \times m$ lower bidiagonal matrix

$$\tilde{J} = \begin{bmatrix} \frac{1}{\alpha_k \nu_k} & & \\ -\frac{1}{\alpha_k \nu_k} & \ddots & \\ & \ddots & \frac{1}{\alpha_{k+m-1}\nu_{k+m-1}} \\ & & -\frac{1}{\alpha_{k+m-1}\nu_{k+m-1}} \end{bmatrix}. \tag{12}$$

However, in [11] no investigation related to both the spectral properties of the steplengths generated by the suggested approach and possible extended recurrence formulas of kind (9) has been addressed. In the following, starting from an analysis aimed at clarifying the relation between subsequent gradients in a sweep, a different idea to generalize the standard limited memory procedure to GP algorithms for box-constrained quadratic problems will be developed.

At any iteration, let denote the following set of indices

$$\mathcal{F}_k = \{i \ : \ \ell_i \le x_i^{(k)} - \alpha_k g_i^{(k)} \le u_i\}$$
$$\mathcal{B}_k = \mathcal{N} \setminus \mathcal{F}_k, \quad \text{with} \quad \mathcal{N} = \{1, ..., n\}$$

in order to distinguish the cases where the projection onto the feasible set is computed from the cases where it has no effect on the components of the current iterate. The entries of the iterate $x^{(k+1)}$ generated by the GP method (2) are

$$x_i^{(k+1)} = \begin{cases} x_i^{(k)} + \nu_k \left(x_i^{(k)} - \alpha_k g_i^{(k)} - x_i^{(k)} \right) & i \in \mathcal{F}_k, \\ x_i^{(k)} + \nu_k \left(\gamma_i^{(k)} - x_i^{(k)} \right) & i \in \mathcal{B}_k, \end{cases} \tag{13}$$

where

$$\gamma_i^{(k)} = \begin{cases} \ell_i & \text{if } x_i^{(k)} - \alpha_k g_i^{(k)} < \ell_i, \\ u_i & \text{if } x_i^{(k)} - \alpha_k g_i^{(k)} > u_i. \end{cases}$$

As a consequence, for any $i = 1, \ldots, n$, the new gradient components are given by

$$g_i^{(k+1)} = \sum_{j=1}^{n} a_{ij} x_j^{(k+1)} - b_i =$$

$$= \sum_{j \in \mathcal{F}_k} a_{ij} \left(x_j^{(k)} - \nu_k \alpha_k g_j^{(k)} \right) + \sum_{j \in \mathcal{B}_k} a_{ij} \left(x_j^{(k)} - \nu_k (x_j^{(k)} - \gamma_j^{(k)}) \right) - b_i$$

$$= g_i^{(k)} - \nu_k \alpha_k \sum_{j \in \mathcal{F}_k} a_{ij} g_j^{(k)} - \nu_k \alpha_k \sum_{j \in \mathcal{B}_k} a_{ij} \frac{x_j^{(k)} - \gamma_j^{(k)}}{\alpha_k}.$$

From the previous equation we can write

$$A_{\mathcal{F}_k, \mathcal{F}_k} g_{\mathcal{F}_k}^{(k)} = \left[g_{\mathcal{F}_k}^{(k)} \; g_{\mathcal{F}_k}^{(k+1)} \right] \begin{bmatrix} \frac{1}{\alpha_k \nu_k} \\ -\frac{1}{\alpha_k \nu_k} \end{bmatrix} - A_{\mathcal{F}_k, \mathcal{N}} p^{(k)} \tag{14}$$

where $p^{(k)}$ is a vector with n entries defined as

$$p_i^{(k)} = \begin{cases} 0 & i \in \mathcal{F}_k, \\ \frac{x_i^{(k)} - \gamma_i^{(k)}}{\alpha_k} & i \in \mathcal{B}_k. \end{cases} \tag{15}$$

At the next iteration, using the same argument employed to obtain (14), we get

$$A_{\mathcal{F}_{k+1}, \mathcal{F}_{k+1}} g_{\mathcal{F}_{k+1}}^{(k+1)} = \left[g_{\mathcal{F}_{k+1}}^{(k+1)} \; g_{\mathcal{F}_{k+1}}^{(k+2)} \right] \begin{bmatrix} \frac{1}{\alpha_{k+1} \nu_{k+1}} \\ -\frac{1}{\alpha_{k+1} \nu_{k+1}} \end{bmatrix} - A_{\mathcal{F}_{k+1}, \mathcal{N}} p^{(k+1)} \tag{16}$$

with the obvious definitions for $\mathcal{F}_{k+1}$ and $\mathcal{B}_{k+1}$. At this point, under the assumption $\mathcal{F}_k \cap \mathcal{F}_{k+1} \neq \emptyset$, we consider the following subsets of indices by taking into account all the possible cases that may occur at the $(k+1)$-th iteration

$$\mathcal{F}_{(k,k+1)} := \mathcal{F}_k \cap \mathcal{F}_{k+1}$$
$$\overline{\mathcal{F}}_{(k,k+1)}^k := \mathcal{F}_k \setminus (\mathcal{F}_k \cap \mathcal{F}_{k+1})$$
$$\overline{\mathcal{F}}_{(k,k+1)}^{k+1} := \mathcal{F}_{k+1} \setminus (\mathcal{F}_k \cap \mathcal{F}_{k+1}).$$

From (14) and (16), we may write

$$A_{\mathcal{F}_{(k,k+1)}, \mathcal{F}_{(k,k+1)}} \left[g_{\mathcal{F}_{(k,k+1)}}^{(k)} \; g_{\mathcal{F}_{(k,k+1)}}^{(k+1)} \right] =$$

$$= \left[g_{\mathcal{F}_{(k,k+1)}}^{(k)} \; g_{\mathcal{F}_{(k,k+1)}}^{(k+1)} \; g_{\mathcal{F}_{(k,k+1)}}^{(k+2)} \right] \begin{bmatrix} \frac{1}{\alpha_k \nu_k} & 0 \\ -\frac{1}{\alpha_k \nu_k} & \frac{1}{\alpha_{k+1} \nu_{k+1}} \\ 0 & -\frac{1}{\alpha_{k+1} \nu_{k+1}} \end{bmatrix} - A_{\mathcal{F}_{(k,k+1)}, \mathcal{N}} \left[p^{(k)} \; p^{(k+1)} \right] +$$

$$- \left[A_{\mathcal{F}_{(k,k+1)}, \overline{\mathcal{F}}_{(k,k+1)}^k} g_{\overline{\mathcal{F}}_{(k,k+1)}^k}^{(k)} \quad A_{\mathcal{F}_{(k,k+1)}, \overline{\mathcal{F}}_{(k,k+1)}^{k+1}} g_{\overline{\mathcal{F}}_{(k,k+1)}^{k+1}}^{(k+1)} \right].$$

The argument can be generalized to a sweep of length m starting from the iteration k. By defining $\mathcal{F}_{(k,k+m-1)} := \cap_{s=k}^{k+m-1} \mathcal{F}_s \neq \emptyset$ we have

$$A_{\mathcal{F}_{(k,k+m-1)},\mathcal{F}_{(k,k+m-1)}} \left[g^{(k)}_{\mathcal{F}_{(k,k+m-1)}} \cdots g^{(k+m-1)}_{\mathcal{F}_{(k,k+m-1)}} \right] \tag{17}$$
$$= \left[g^{(k)}_{\mathcal{F}_{(k,k+m-1)}} \cdots g^{(k+m-1)}_{\mathcal{F}_{(k,k+m-1)}} \ g^{(k+m)}_{\mathcal{F}_{(k,k+m-1)}} \right] \tilde{J} +$$
$$- A_{\mathcal{F}_{(k,k+m-1)},\mathcal{N}} \left[p^{(k)} \cdots p^{(k+m-1)} \right] +$$
$$- \left[A_{\mathcal{F}_{(k,k+m-1)},\overline{\mathcal{F}}^k_{(k,k+m-1)}} \ g^{(k)}_{\overline{\mathcal{F}}^k_{(k,k+m-1)}} \cdots A_{\mathcal{F}_{(k,k+m-1)},\overline{\mathcal{F}}^{k+m-1}_{(k,k+m-1)}} \ g^{(k+m-1)}_{\overline{\mathcal{F}}^{k+m-1}_{(k,k+m-1)}} \right],$$

where $\tilde{J}$ is the $(m+1) \times m$ lower bidiagonal matrix given in (12). If $\mathcal{F}_{k+j} \subseteq \mathcal{F}_{(k,k+m-1)}$, for $j = 0, \ldots, m-1$, the term

$$A_{\mathcal{F}_{(k,k+m-1)},\overline{\mathcal{F}}^{k+j}_{k,k+m-1}} \ g^{(k+j)}_{\mathcal{F}_{(k,k+m-1)},\overline{\mathcal{F}}^{k+j}_{k,k+m-1}}$$

does not contribute to the relation (17), which can be rewritten as

$$A_{\mathcal{F}_{(k,k+m-1)},\mathcal{F}_{(k,k+m-1)}} \left[g^{(k)}_{\mathcal{F}_{(k,k+m-1)}} \cdots g^{(k+m-1)}_{\mathcal{F}_{(k,k+m-1)}} \right] \tag{18}$$
$$= \left[g^{(k)}_{\mathcal{F}_{(k,k+m-1)}} \cdots g^{(k+m-1)}_{\mathcal{F}_{(k,k+m-1)}} \ g^{(k+m)}_{\mathcal{F}_{(k,k+m-1)}} \right] \tilde{J} +$$
$$- A_{\mathcal{F}_{(k,k+m-1)},\mathcal{N}} \left[p^{(k)} \cdots p^{(k+m-1)} \right].$$

In order to preserve the validity of (18) and to correctly neglect the term $A_{\mathcal{F}_{(k,k+m-1)},\overline{\mathcal{F}}^{k+j}_{k,k+m-1}} \ g^{(k+j)}_{\mathcal{F}_{(k,k+m-1)},\overline{\mathcal{F}}^{k+j}_{k,k+m-1}}$, for $j = 0, \ldots, m-1$, we propose to interrupt a sweep and to restart the collection of new restricted gradient vectors when the condition $\mathcal{F}_{k+j} \subseteq \mathcal{F}_{(k,k+m-1)}$ is not satisfied, by developing a technique which *adaptively* controls the length of the sweep, up to the given value m. At the beginning of the iterative process, this condition does not typically hold and the sweeps have a length at most equal to 1; however, as the number of iterations increases, the components that are going to be projected onto the feasible set tend to stabilize and, as a consequence, $\mathcal{F}_{k+j} \subseteq \mathcal{F}_{(k,k+m-1)}$ occurs for a growing number of iterations. Hereafter, we suppose that $\mathcal{F}_{k+j} \subseteq \mathcal{F}_{(k,k+m-1)}$, $j = 0, \ldots, m-1$, holds. We can state that the equality (18) can be considered a possible extension of the Eq. (9) which holds in the unconstrained framework. As a consequence, in presence of box-constraints, we suggest to not store m back *whole* gradients vectors (for which no recurrence formula holds) but to consider m back gradients *restricted* to the set of indices $\mathcal{F}_{(k,k+m-1)}$. Driven by these considerations, our implementation of the limited memory steplength rule for the constrained case is based on the following generalization of the matrix G:

$$G_{(k,k+m-1)} = \left[g^{(k)}_{\mathcal{F}_{(k,k+m-1)}} \cdots g^{(k+m-1)}_{\mathcal{F}_{(k,k+m-1)}} \right]. \tag{19}$$

Given $m \geq 1$ and the $m \times m$ matrix $R_{(k,k+m-1)}$ such that $R_{(k,k+m-1)}^T$ $R_{(k,k+m-1)} = G_{(k,k+m-1)}^T G_{(k,k+m-1)}$, we propose to compute, at each new sweep, m steplengths as inverses of the eigenvalues of the symmetric matrix

$$\tilde{T}_{(k,k+m-1)} = R_{(k,k+m-1)}^{-T} G_{(k,k+m-1)}^T A_{\mathcal{F}_{(k,k+m-1)},\mathcal{F}_{(k,k+m-1)}} G_{(k,k+m-1)} R_{(k,k+m-1)}^{-1},$$

with the aim of approximating the inverses of the eigenvalues of the matrix $A_{\mathcal{F}_{(k,k+m-1)},\mathcal{F}_{(k,k+m-1)}}$. This idea mimics the approach proposed in [4] where, in the case of box constraints, novel versions of the BB rules sweeping the spectrum of a proper submatrix of A turned out to be convenient with respect to the standard ones. Indeed, in this case, under the special assumptions $m = 1$, $\mathcal{F}_{k-1} = \mathcal{F}_k$ and $\nu_{k-1} = 1$, in view of $\gamma_{\mathcal{B}_k}^{(k)} = x_{\mathcal{B}_k}^{(k)}$, the recurrence (14) can be simplified as

$$g_{\mathcal{F}_k}^{(k+1)} = g_{\mathcal{F}_k}^{(k)} - \nu_k \alpha_k A_{\mathcal{F}_k,\mathcal{F}_k} g_{\mathcal{F}_k}^{(k)}.$$

By denoting with $\{\delta_1, \ldots, \delta_r\}$ and $\{w_1, \ldots, w_r\}$ the eigenvalues and the associated orthonormal eigenvectors of $A_{\mathcal{F}_k,\mathcal{F}_k}$, respectively, where $r = \sharp \mathcal{F}_k$, and by writing $g_{\mathcal{F}_k}^{(k+1)} = \sum_{i=1}^r \bar{\mu}_i^{(k+1)} w_i$ and $g_{\mathcal{F}_k}^{(k)} = \sum_{i=1}^r \bar{\mu}_i^{(k)} w_i$, we obtain the following recurrence formula for the eigencomponents:

$$\bar{\mu}_i^{(k+1)} = \bar{\mu}_i^{(k)} (1 - \nu_k \alpha_k \delta_i).$$

This means that if the selection rule provides a good approximation of $\frac{1}{\delta_i}$, a useful reduction of $|\bar{\mu}_i^{(k+1)}|$ can be achieved. We underline that, if $m = 1$, α_k is computed in order to estimate the inverse of an eigenvalue of $A_{\mathcal{F}_{k-1},\mathcal{F}_{k-1}}$; obviously, if $\mathcal{F}_{k-1} = \mathcal{F}_k$, α_k can also provide a good approximation of one of the values $\frac{1}{\delta_i}$ and thus reduce the corresponding desired component $|\bar{\mu}_i^{(k+1)}|$.

We remark that, in view of (18), the matrix $\tilde{T}_{(k,k+m-1)}$ has the following form

$$\tilde{T}_{(k,k+m-1)} = R_{(k,k+m-1)}^{-T} G_{(k,k+m-1)}^T \left[G_{(k,k+m-1)} \, g_{\mathcal{F}_{(k,k+m-1)}}^{(k+m)} \right] \tilde{J} R_{(k,k+m-1)}^{-1} +$$

$$-R_{(k,k+m-1)}^{-T} G_{(k,k+m-1)}^T A_{\mathcal{F}_{(k,k+m-1)},\mathcal{N}} \left[p^{(k)} \cdots p^{(k+m-1)} \right] R_{(k,k+m-1)}^{-1} =$$

$$= \left[R_{(k,k+m-1)}, \; r_{(k,k+m-1)} \right] \tilde{J} R_{(k,k+m-1)}^{-1} + \tag{20}$$

$$+R_{(k,k+m-1)}^{-T} G_{(k,k+m-1)}^T A_{\mathcal{F}_{(k,k+m-1)},\mathcal{N}} \left[p^{(k)} \cdots p^{(k+m-1)} \right] R_{(k,k+m-1)}^{-1},$$

where the vector $r_{(k,k+m-1)}$ is the solution of the system $R_{(k,k+m-1)}^T r_{(k,k+m-1)} = G_{(k,k+m-1)}^T g_{\mathcal{F}_{(k,k+m-1)}}^{(k+m)}$.

Despite the carried out analysis, from the practical point of view, we want to avoid to explicitly make use of the matrix $A_{\mathcal{F}_{(k,k+m-1)},\mathcal{N}}$ and, hence, we do not consider the exact relation (18), but its inexact version where the term $A_{\mathcal{F}_{(k,k+m-1)},\mathcal{N}} \left[p^{(k)} \cdots p^{(k+m-1)} \right]$ is neglected. For this reason, we do not compute the eigenvalues of $\tilde{T}_{(k,k+m-1)}$ but the eigenvalues of the symmetric part of the matrix

$$Z_{(k,k+m-1)} = \left[R_{(k,k+m-1)}, r_{(k,k+m-1)} \right] \tilde{J} R_{(k,k+m-1)}^{-1}.$$

To explain the relation between the eigenvalues of $\tilde{T}_{(k,k+m-1)}$ and the ones of the symmetric part of $Z_{(k,k+m-1)}$, we start to clarify the details of our approach from the more understandable case of $m = 1$ where $\tilde{T}_{(k,k+m-1)}$ reduces to a scalar. In this case, at iteration $k + 1$ only one gradient is available $G_k = g_{\mathcal{F}_k}^{(k)}$. We are interested in computing

$$\tilde{T}_k = R_k^{-T} G_k^T A_{\mathcal{F}_k, \mathcal{F}_k} G_k R_k^{-1},$$

where

$$G_k^T G_k = (g_{\mathcal{F}_k}^{(k)})^T g_{\mathcal{F}_k}^{(k)} = \sqrt{(g_{\mathcal{F}_k}^{(k)})^T g_{\mathcal{F}_k}^{(k)}} \sqrt{(g_{\mathcal{F}_k}^{(k)})^T g_{\mathcal{F}_k}^{(k)}} = R_k^T R_k.$$

Then, using (14), the matrix $\tilde{T}_k$ is given by

$$R_k^{-T} G_k^T \left(\begin{bmatrix} g_{\mathcal{F}_k}^{(k)} & g_{\mathcal{F}_k}^{(k+1)} \end{bmatrix} \begin{bmatrix} \frac{1}{\alpha_k \nu_k} \\ -\frac{1}{\alpha_k \nu_k} \end{bmatrix} - A_{\mathcal{F}_k, \mathcal{N}} p^{(k)} \right) R_k^{-1} \tag{21}$$

$$= \frac{1}{\sqrt{(g_{\mathcal{F}_k}^{(k)})^T g_{\mathcal{F}_k}^{(k)}}} (g_{\mathcal{F}_k}^{(k)})^T \left(\frac{g_{\mathcal{F}_k}^{(k)} - g_{\mathcal{F}_k}^{(k+1)}}{\alpha_k \nu_k} - A_{\mathcal{F}_k, \mathcal{N}} p^{(k)} \right) \frac{1}{\sqrt{(g_{\mathcal{F}_k}^{(k)})^T g_{\mathcal{F}_k}^{(k)}}}$$

$$= \frac{(g_{\mathcal{F}_k}^{(k)})^T A_{\mathcal{F}_k, \mathcal{F}_k} g_{\mathcal{F}_k}^{(k)}}{(g_{\mathcal{F}_k}^{(k)})^T g_{\mathcal{F}_k}^{(k)}}.$$

Hence, for the special case $m = 1$, if we would consider the exact expression of $A_{\mathcal{F}_k, \mathcal{F}_k} G_k$ given by the right hand side of (14), we would obtain as unique eigenvalue of $\tilde{T}_k$ the value $\frac{(g_{\mathcal{F}_k}^{(k)})^T A_{\mathcal{F}_k, \mathcal{F}_k} g_{\mathcal{F}_k}^{(k)}}{(g_{\mathcal{F}_k}^{(k)})^T g_{\mathcal{F}_k}^{(k)}}$ which is the inverse of the Rayleigh quotient of the matrix $A_{\mathcal{F}_k, \mathcal{F}_k}$ and, hence, belongs to its spectrum.

However, in practice we compute the scalar

$$Z_k = \quad = \frac{(g_{\mathcal{F}_k}^{(k)})^T A_{\mathcal{F}_k, \mathcal{F}_k} g_{\mathcal{F}_k}^{(k)}}{(g_{\mathcal{F}_k}^{(k)})^T g_{\mathcal{F}_k}^{(k)}} + \frac{(g_{\mathcal{F}_k}^{(k)})^T A_{\mathcal{F}_k, \mathcal{N}} p^{(k)}}{(g_{\mathcal{F}_k}^{(k)})^T g_{\mathcal{F}_k}^{(k)}} \tag{22}$$

that is a value in the spectrum of $A_{\mathcal{F}_k, \mathcal{F}_k}$ affected by en error due to the presence of the second term at the right-hand side of Eq (22). An estimation of this error, at iteration $k + 1$, is given by

$$\rho_k \leq \frac{\|A_{\mathcal{F}_k, \mathcal{N}} p^{(k)}\|}{\|g_{\mathcal{F}_k}^{(k)}\|}.$$

From Eqs. (13) and (15), the following results hold

$$\|p^{(k)}\| = \|p_{\mathcal{B}_k}^{(k)}\| = \frac{\|x_{\mathcal{B}_k}^{(k+1)} - x_{\mathcal{B}_k}^{(k)}\|}{\alpha_k \nu_k} \quad \text{and} \quad \|g_{\mathcal{F}_k}^{(k)}\| = \frac{\|x_{\mathcal{F}_k}^{(k+1)} - x_{\mathcal{F}_k}^{(k)}\|}{\alpha_k \nu_k}.$$

As a consequence,

$$\rho_k \leq \frac{\|A_{\mathcal{F}_k, \mathcal{B}_k} p_{\mathcal{B}_k}^{(k)}\|}{\|g_{\mathcal{F}_k}^{(k)}\|} \leq \frac{\|A_{\mathcal{F}_k, \mathcal{B}_k}\| \|p_{\mathcal{B}_k}^{(k)}\|}{\|g_{\mathcal{F}_k}^{(k)}\|} = \frac{\|A_{\mathcal{F}_k, \mathcal{B}_k}\| \|x_{\mathcal{B}_k}^{(k+1)} - x_{\mathcal{B}_k}^{(k)}\|}{\|x_{\mathcal{F}_k}^{(k+1)} - x_{\mathcal{F}_k}^{(k)}\|}.$$

From this bound on ρ_k, we can state that, when k is sufficiently large so as the components to project onto the feasible region are almost settled, the error ρ_k is not significant and the steplength α_{k+1} is approximating the inverse of an eigenvalue of $A_{\mathcal{F}_k, \mathcal{F}_k}$. In the more general case of $m > 1$, we compute the eigenvalues of the symmetric part of $Z_{(k,k+m-1)}$ given by

$$\tilde{Z}_{(k,k+m-1)} = \frac{1}{2}\left(Z_{(k,k+m-1)} + Z_{(k,k+m-1)}^T\right). \tag{23}$$

Then, from Eq. (20) we have

$$\tilde{Z}_{(k,k+m-1)} = \tilde{T}_{(k,k+m-1)} + \tag{24}$$
$$+ \frac{1}{2}\left(R_{(k,k+m-1)}^{-T} G_{(k,k+m-1)}^T A_{\mathcal{F}_{(k,k+m-1)},\mathcal{N}}\left[p^{(k)} \cdots p^{(k+m-1)}\right] R_{(k,k+m-1)}^{-1}\right) +$$
$$+ \frac{1}{2}\left(R_{(k,k+m-1)}^{-T}\left[p^{(k)} \cdots p^{(k+m-1)}\right]^T A_{\mathcal{F}_{(k,k+m-1)},\mathcal{N}}^T G_{(k,k+m-1)} R_{(k,k+m-1)}^{-1}\right).$$

A result of perturbation matrix theory (see Corollary 6.3.4 [9]) ensures that

$$\left|\lambda_j\left(\tilde{Z}_{(k,k+m-1)}\right) - \lambda_j\left(\tilde{T}_{(k,k+m-1)}\right)\right| \leq \left\|\tilde{Z}_{(k,k+m-1)} - \tilde{T}_{(k,k+m-1)}\right\|, \tag{25}$$

where $\lambda_j(C)$ is the j-th eigenvalue of C. By denoting with $\|D\|_F$ the Frobenius norm of a matrix D, the right-hand side of (25) can be bounded from above as

$$\left\|\tilde{Z}_{(k,k+m-1)} - \tilde{T}_{(k,k+m-1)}\right\| \leq$$
$$\leq \left\|R_{(k,k+m-1)}^{-T} G_{(k,k+m-1)}^T A_{\mathcal{F}_{(k,k+m-1)},\mathcal{N}}\left[p^{(k)} \cdots p^{(k+m-1)}\right] R_{(k,k+m-1)}^{-1}\right\| \leq$$
$$\leq \left\|A_{\mathcal{F}_{(k,k+m-1)},\mathcal{N}}\left[p^{(k)} \cdots p^{(k+m-1)}\right]\right\| \|R_{(k,k+m-1)}^{-1}\| \leq$$
$$\leq \left\|A_{\mathcal{F}_{(k,k+m-1)},\mathcal{N}}\right\| \left\|\left[p^{(k)} \cdots p^{(k+m-1)}\right]\right\| \|R_{(k,k+m-1)}^{-1}\| \leq$$
$$\leq \left\|A_{\mathcal{F}_{(k,k+m-1)},\mathcal{N}}\right\| \left\|\left[p^{(k)} \cdots p^{(k+m-1)}\right]\right\|_F \|R_{(k,k+m-1)}^{-1}\| \leq$$
$$\leq \left\|A_{\mathcal{F}_{(k,k+m-1)},\mathcal{N}}\right\| \sqrt{\sum_{i=0}^{m-1}\left\|p_{\mathcal{B}_{k+i}}^{(k+i)}\right\|^2} \|R_{(k,k+m-1)}^{-1}\|,$$

where in the second inequality we use

$$\|R_{(k,k+m-1)}^{-T} G_{(k,k+m-1)}^T\| = \sqrt{\|R_{(k,k+m-1)}^{-T} G_{(k,k+m-1)}^T G_{(k,k+m-1)} R_{(k,k+m-1)}^{-1}\|} = 1.$$

We can conclude that, if m is relatively small, as the number of iterations k increases, the matrix $\tilde{Z}_{(k,k+m-1)}$ (of which we compute the eigenvalues) approaches the matrix $\tilde{T}_{(k,k+m-1)}$ (of which we would compute the eigenvalues).

3   Numerical Experiments

In this section we analyse, on some box-constrained quadratic problems, the practical behaviour of the GP method combined with

- the original limited memory (LM) steplength selection rule (which collects the whole back gradients),
- the modified LM steplength selection rule suggested in [11] (which considers the modified gradients given in (11)),
- the modified LM steplength selection rule suggested in Sect. 2.2 (which exploits the matrix $G_{(k,k+m-1)}$ defined in (19)).

Our main aim is to investigate the distribution of the inverses of the steplengths generated by the three approaches with respect to the eigenvalues of a proper submatrix of the Hessian matrix.

Since in the practical implementation of the modified LM updating strategy proposed in Sect. 2.2, we compute the steplengths as the inverses of the eigenvalues of the matrix (23), by analogy we generate the steplengths provided by both the standard LM procedure and the one proposed in [11] as the inverses of the eigenvalues of the symmetric part of the Hessenberg matrices T and $\tilde{T}$, respectively, instead of reducing them to a tridiagonal form, as made in [5] and [11]. Nevertheless, numerical experiments show that this modification does not significantly affect the results. Hereafter we denote by LMGP, Box-LMGP1 and Box-LMGP2 the GP algorithm equipped with the standard LM steplength selection rule and the modified versions developed in [11] and Sect. 2.2, respectively.

In our tests, the LMGP, Box-LMGP1 and Box-LMGP2 methods share a monotone Armijo-type linesearch procedure to select ν_k and the same stopping criterion:

$$\|\varphi(x^{(k)})\| \leq tol \|g(x^{(0)})\|, \tag{26}$$

where $\varphi(x^{(k)})$ is the projected gradient at $x^{(k)}$, i.e., the vector with entries $\varphi_i^{(k)}$, $i = 1 \ldots, n$, defined as

$$\varphi_i^{(k)} = \begin{cases} g_i^{(k)} & \text{for } \ell_i < x_i^{(k)} < u_i, \\ \max\{0, g_i^{(k)}\} & \text{for } x_i^{(k)} = u_i, \\ \min\{0, g_i^{(k)}\} & \text{for } x_i^{(k)} = \ell_i. \end{cases} \tag{27}$$

The following parameter setting is used: $tol = 10^{-8}$, $\alpha_{min} = 10^{-10}$, $\alpha_{max} = 10^6$, $\alpha_0 = (g^{(0)^T} g^{(0)})/(g^{(0)^T} A g^{(0)})$; furthermore, different values for the parameter m are used, i.e. $m = 3, 5, 7$. The feasible initial point $x^{(0)}$ is randomly generated with inactive entries.

We start to analyse the effect of the considered steplength selection rules within the GP method on a toy problem of size $n = 20$ with ten active constraints at the solution; the eigenvalues of the Hessian matrix are logarithmically distributed and the condition number is equal to 500. In Fig. 1 we report the behaviour of $\frac{1}{\alpha_k}$ (red crosses) with respect to the eigenvalues of the Hessian

Table 1. Main features of two quadratic test problems subject to lower bounds

			$n = 1000$		$na = 400$	
	$\lambda_{min}(A)$	$\lambda_{max}(A)$	Distribution of the eigenvalues of A	$\lambda_{min}(A_{\mathcal{F}^*,\mathcal{F}^*})$	$\lambda_{max}(A_{\mathcal{F}^*,\mathcal{F}^*})$
TP1	1	1000	log-spaced	3.08	753.26
TP2	9.40	1013.95	log-spaced	10	1000

matrix (green dotted lines) and the restricted Hessian submatrix (black dotted line) at each iteration k, for the case $m = 5$. In each panel of the figure, the blue lines show the maximum and the minimum eigenvalues of the whole Hessian matrix and the blue symbols "o" are used to denote the eigenvalues of the submatrix $A_{\mathcal{F}^*,\mathcal{F}^*}$. We observe that the inverses of the steplengths produced by the LMGP method may sometimes fall outside the spectrum of the restricted Hessian or even the spectrum of the whole Hessian, while the other two approaches are able to restrain this effect. In particular, the sequence $\left\{\frac{1}{\alpha_k}\right\}$ generated by the Box-LMGP2 scheme, belongs to the spectra of the current restricted Hessian matrices, providing also a reduction of the iterations needed to satisfy the stopping criterion. Indeed, the effectiveness of the Box-LMGP2 procedure allows an earlier stabilization of the active set, with respect to the other two approaches. The numerical behaviour of the considered methods was also verified on test problems of larger size. In particular, we randomly generated quadratic test problems subject to lower bounds, in which the solution, the number of active constraints at the solution, and the distribution of the eigenvalues of the dense symmetric positive definite Hessian matrix of the objective function are prefixed. For the sake of brevity, we only report here the results obtained on two test problems described in Table 1, where na is the number of active lower constraints at the solution and $\mathcal{F}^*$ denotes the set of the indices of the inactive constraints at the solution, so that $A_{\mathcal{F}^*,\mathcal{F}^*}$ is the submatrix of the Hessian matrix defined by the intersection of the rows and the columns with indices in $\mathcal{F}^*$.

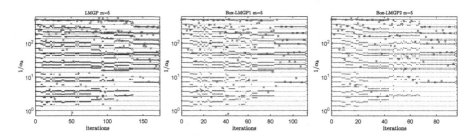

Fig. 1. Distribution of $\frac{1}{\alpha_k}$ with respect to iterations for LMGP (left panel), Box-LMGP1 (middle panel) and Box-LMGP2 (right panel) on a toy problem of size $n = 20$, for a sweep of length $m = 5$.

In Table 2 we report, for each method, the number of iterations (*iterations*) necessary to satisfy the stopping criterion, the total number of backtracking reductions (*backtr.*) on the parameter ν_k, and the final error on the objective function values, i.e. $F_k = |f(x^*) - f(x^{(iterations)})|$.

Table 2. Numerical results for problems TP1 and TP2

method	*iterations*	*backtr.*	F_k	*iterations*	*backtr.*	F_k
	TP1			**TP2**		
LMGP (m = 3)	384	131	3.96e-09	292	81	1.12e-06
Box-LMGP1 (m = 3)	233	25	4.19e-09	116	13	2.72e-07
Box-LMGP2 (m = 3)	222	23	9.31e-10	109	16	1.96e-07
LMGP (m = 5)	355	112	1.51e-08	172	25	1.04e-06
Box-LMGP1 (m = 5)	252	18	1.00e-08	114	8	9.76e-08
Box-LMGP2 (m = 5)	221	25	7.22e-09	101	8	4.82e-07
LMGP (m = 7)	375	89	1.40e-08	152	16	1.11e-06
Box-LMGP1 (m = 7)	255	16	2.03e-08	112	5	5.49e-07
Box-LMGP2 (m = 7)	233	25	7.92e-09	104	4	1.02e-07

Figure 2 enables to show the behaviour of the rules on test problem TP1. Similar results can be observed for TP2. In each panels of Fig. 2, at the k-th iteration, the black dots denote 20 eigenvalues (with linearly spaced indices, included the maximum and the minimum eigenvalues) of the submatrix of the Hessian matrix, defined by neglecting the rows and columns with indices corresponding to the active variables at the current iterate, and the red cross corresponds to the inverse of the steplength α_k. The blue lines show the maximum and the minimum eigenvalues of the whole Hessian matrix and the blue symbols "o" are used to denote 20 eigenvalues of the submatrix $A_{\mathcal{F}^*,\mathcal{F}^*}$, with linearly spaced indices (included the maximum and the minimum eigenvalues). The results shown in Figs. 1 and 2 and Table 2 allow us to make the following considerations:

- in general the steplengths generated by the Box-LMGP1 and the Box-LMGP2 methods seem to better approximate the eigenvalues of the submatrices of A restricted to the rows and columns corresponding to the inactive components of the current iterate with respect to the LMGP algorithm;
- the adaptive strategy implemented by the Box-LMGP2 approach turns out to be more effective for $m = 5$ and $m = 7$, thanks to its previously described ability in adaptively controlling the length of the sweep. Indeed such ability allows to consider shorter sweeps at the beginning of the iterative process when the sequence of the restricted Hessian matrices is not yet stabilized towards $A_{\mathcal{F}_*,\mathcal{F}_*}$ and, hence, the inverses of the steplengths generated by the limited memory strategy in a particular sweep could not provide suitable approximations of the eigenvalues of the restricted Hessian matrices involved

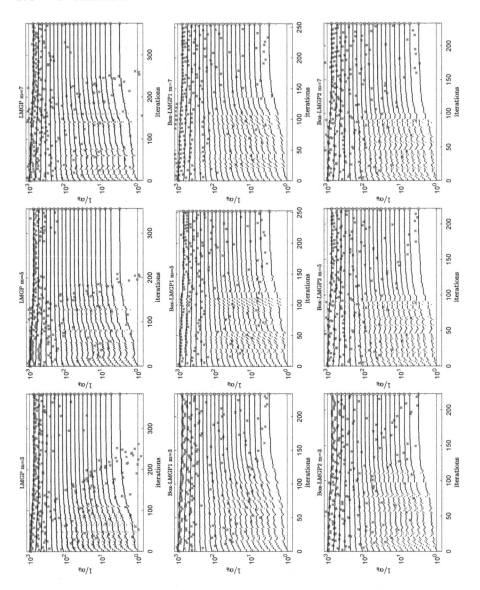

Fig. 2. Distribution of $\frac{1}{\alpha_k}$ with respect to the iterations for LMGP (first row), Box-LMGP1 (second row) and Box-LMGP2 (third row) on TP1, for different values of the length of the sweep.

in the next sweep. Longer sweeps are instead promoted with the stabilization of the final active set; furthermore, from Table 2, we may observe that the Box-LMGP2 method, compared with LMGP and Box-LMGP1, generally requires lower numbers of backtracking steps;

– within the same or higher accuracy, the effectiveness of the Box-LMGP2
  method is confirmed by the lower number of iterations with respect to the
  LMGP and Box-LMGP1 methods.

4 Conclusions

In this paper we developed an updating strategy to select the steplength in
gradient projection methods for the minimization of box-constrained quadratic
problems. In particular, we generalized a steplength selection rule proposed in
the unconstrained optimization framework and based on the storage of a limited
number of consecutive objective function gradients. By preserving the same basic
idea of exploiting stored gradient vectors, we detailed how to possibly modify the
original updating strategy in order to take into account the lower and the upper
bounds. Numerical experiments carried out on box-constrained quadratic test
problems showed that the modified procedure allowed the gradient projection
method to reach better practical performance with respect to the standard one.

Acknowledgments. We thank the anonymous reviewers for their careful reading of
our manuscript and their many insightful comments and suggestions. This work has
been partially supported by the INDAM research group GNCS.

References

1. Barzilai, J., Borwein, J.M.: Two-point step size gradient methods. IMA J. Numer.
   Anal. **8**, 141–148 (1988)
2. Bertsekas, D.P.: Nonlinear Programming, 2nd edn. Athena Scientific, Boston
   (1999)
3. Birgin, E.G., Martinez, J.M., Raydan, M.: Non-monotone spectral projected gradient methods on convex sets. SIAM J. Optim. **10**(4), 1196–1211 (2000)
4. Crisci, S., Ruggiero, V., Zanni, L.: Steplength selection in gradient projection methods for box-constrained quadratic programs. Appl. Math. Comput. **356**, 312–327
   (2019)
5. Fletcher, R.: A limited memory steepest descent method. Math. Program. Ser. A
   135, 413–436 (2012)
6. Frassoldati, G., Zanghirati, G., Zanni, L.: New adaptive stepsize selections in gradient methods. J. Ind. Manage. Optim. **4**(2), 299–312 (2008)
7. Golub, G.H., van Loan, C.F.: Matrix Computations, 3rd edn. John Hopkins University Press, Baltimore (1996)
8. Grippo, L., Lampariello, F., Lucidi, S.: A nonmonotone line-search technique for
   Newton's method. SIAM J. Numer. Anal. **23**(4), 707–716 (1986)
9. Horn, R.A., Johnson, C.R.: Matrix Analysis. Cambridge University Press, Cambridge (2012)
10. Huang, Y., Liu, H.: On the rate of convergence of projected Barzilai-Borwein meth-
    ods. Optim. Meth. Softw. **30**(4), 880–892 (2015)
11. Porta, F., Prato, M., Zanni, L.: A new steplength selection for scaled gradient
    methods with application to image deblurring. J. Sci. Comput. **65**, 895–919 (2015)

12. di Serafino, D., Ruggiero, V., Toraldo, G., Zanni, L.: On the steplength selection in gradient methods for unconstrained optimization. Appl. Math. Comput. **318**, 176–195 (2018)
13. Zhou, B., Gao, L., Dai, Y.H.: Gradient methods with adaptive step-sizes. Comput. Optim. Appl. **35**(1), 69–86 (2006)

A Gradient-Based Globalization Strategy for the Newton Method

Daniela di Serafino[1] ⓘ, Gerardo Toraldo[2] ⓘ, and Marco Viola[1(✉)] ⓘ

[1] Department of Mathematics and Physics, University of Campania "L. Vanvitelli", Caserta, Italy
{daniela.diserafino,marco.viola}@unicampania.it
[2] Department of Mathematics and Applications, University of Naples Federico II, Naples, Italy
toraldo@unina.it

Abstract. The Newton method is one of the most powerful methods for the solution of smooth unconstrained optimization problems. It has local quadratic convergence in a neighborhood of a local minimum where the Hessian is positive definite and Lipschitz continuous. Several strategies have been proposed in order to achieve global convergence. They are mainly based either on the modification of the Hessian together with a line search or on the adoption of a restricted-step strategy. We propose a globalization technique that combines the Newton and gradient directions, producing a descent direction on which a backtracking Armijo line search is performed. Our work is motivated by the effectiveness of gradient methods using suitable spectral step-length selection rules. We prove global convergence of the resulting algorithm, and quadratic rate of convergence under suitable second-order optimality conditions. A numerical comparison with a modified Newton method exploiting Hessian modifications shows the effectiveness of our approach.

Keywords: Newton method · Gradient method · Global convergence

1 Introduction

We consider the following unconstrained minimization problem

$$\min f(\mathbf{x}), \quad \mathbf{x} \in \mathbb{R}^n, \tag{1}$$

where f is a twice continuously differentiable function. In the following, the gradient and the Hessian of f are denoted by $\mathbf{g}(\mathbf{x})$ and $H(\mathbf{x})$, respectively.

The Newton method for problem (1) generates a sequence $\{\mathbf{x}_k\}$ by using the iterative scheme

$$\mathbf{x}_{k+1} = \mathbf{x}_k + \mathbf{d}_k^N, \quad k = 0, 1, 2, ..., \tag{2}$$

This work was partially supported by Gruppo Nazionale per il Calcolo Scientifico - Istituto Nazionale di Alta Matematica (GNCS-INdAM). Marco Viola was also supported by the MOD_CELL_DEV Project - Programma di finanziamento della Ricerca di Ateneo, University of Naples Federico II.

© Springer Nature Switzerland AG 2020
Y. D. Sergeyev and D. E. Kvasov (Eds.): NUMTA 2019, LNCS 11973, pp. 177–185, 2020.
https://doi.org/10.1007/978-3-030-39081-5_16

where $\mathbf{d}_k^N$ solves the linear system

$$H_k \mathbf{d} = -\mathbf{g}_k, \tag{3}$$

with $H_k = H(\mathbf{x}_k)$ and $\mathbf{g}_k = \mathbf{g}(\mathbf{x}_k)$. If $\mathbf{x}^*$ is a solution of (1), with $H(\mathbf{x}^*) \succ 0$ and $H(\mathbf{x})$ Lipschitz continuous in a neighborhood of $\mathbf{x}^*$, then the Newton method is locally quadratically convergent to $\mathbf{x}^*$ [11]. Global convergence holds for convex problems, provided some form of reduction on the pure Newton step $\mathbf{d}_k^N$ is allowed (see, e.g., the damped Newton method by Nesterov [16]). For nonconvex optimization the Newton direction computed through (3) may not be a descent direction for f at $\mathbf{x}_k$. In order to deal with nonconvex problems, Modified Newton (MN) methods have been proposed. The basic idea is to replace H_k in (3) by a matrix $\widetilde{H}_k = H_k + E_k$, where E_k is such that $\widetilde{H}_k$ is positive definite [14], so that the solution $\mathbf{d}_k^{MN}$ of the linear system $\widetilde{H}_k \mathbf{d} = -\mathbf{g}_k$ is a descent direction at $\mathbf{x}_k$, and $\mathbf{x}_{k+1}$ can be computed by performing a line search along $\mathbf{d}_k^{MN}$. Global convergence can be proved for MN methods, provided that the matrices $\widetilde{H}_k$ have uniformly bounded condition numbers, i.e.,

$$\kappa(\widetilde{H}_k) \leq \zeta$$

for some $\zeta > 0$ independent of k. Effective ways to compute $\widetilde{H}_k$ have been proposed, such as those in [10,12], based on modified Cholesky factorizations which automatically compute the Cholesky factorization of a positive definite matrix $H_k + E_k$, where E_k is a nonnegative diagonal matrix which is zero if H_k is "sufficiently" positive definite. Another popular choice is to take $E_k = \lambda I$, where I is the identity matrix and λ is a suitable positive constant, so that

$$\mathbf{d}_k^{MN} = -(H_k + \lambda I)^{-1}\mathbf{g}_k. \tag{4}$$

Let us recall that the Newton method computes $\mathbf{x}_{k+1}$ by using the second-order Taylor approximation to f at $\mathbf{x}_k$:

$$\psi_k(\mathbf{x}) = f(\mathbf{x}_k) + \mathbf{g}_k^T(\mathbf{x} - \mathbf{x}_k) + \frac{1}{2}(\mathbf{x} - \mathbf{x}_k)^T H_k(\mathbf{x} - \mathbf{x}_k). \tag{5}$$

In MN methods, the quadratic model at $\mathbf{x}_k$ is forced to be convex, in order to provide a descent direction. Because of that, if f is nonconvex in a neighborhood of $\mathbf{x}_k$, the quadratic model may be inadequate. Starting from this consideration, we propose a globalization strategy for the Newton method in which the search direction is a combination of the Newton and Steepest Descent (SD) directions, such as

$$\mathbf{d}_k = \mathbf{d}_k^N - \lambda \mathbf{g}_k = -(H_k^{-1} + \lambda I)\mathbf{g}_k, \tag{6}$$

where λ is chosen so that $\mathbf{d}_k$ satisfies some descent condition at $\mathbf{x}_k$. The directions (4) and (6) are somehow related to the Trust Region (TR) approach, also known as restricted-step approach, which represents the most popular globalization strategy for the Newton method. In a TR approach, given $\mathbf{x}_k$, the iterate $\mathbf{x}_{k+1}$ is computed as the (possibly approximate) solution of the problem

$$\min \{\psi_k(\mathbf{x}) : \|\mathbf{x} - \mathbf{x}_k\| \leq \Delta_k\}, \tag{7}$$

where Δ_k is the radius of the ball where the quadratic model is trusted to be a reliable approximation of $f(\mathbf{x})$. This radius is updated at each iteration, according to some rule. Moré and Sorensen (see [14] and references therein) show that $\widetilde{\mathbf{x}}_k$ solves (7) if and only if $\mathbf{d}_k = \widetilde{\mathbf{x}}_k - \mathbf{x}_k$ satisfies

$$
\begin{aligned}
(H_k + \lambda I)\,\mathbf{d} &= -\mathbf{g}_k, \\
\lambda\,(\Delta_k - \|\mathbf{d}\|) &= 0,
\end{aligned}
\tag{8}
$$

for some $\lambda > 0$ such that $H_k + \lambda I$ is positive semidefinite; therefore, the MN direction (4) can be seen as a TR direction. A search direction of the form (6) reminds us of the so-called dogleg method, and more generally of the two-dimensional subspace minimization, which are designed to give an approximate solution to the TR subproblem (7). These strategies make a search in the two-dimensional space spanned by the steepest descent direction and the Newton one.

A search direction of the form (6) is a way to combine a locally superlinearly convergent strategy (Newton) with a globally convergent strategy (SD) in such a way that the latter brings the iterates close to a solution, namely in a so-called basin of attraction of the Newton method. The effectiveness of the approach obviously relies on a suitable choice of λ in (6). Although appearing a quite natural strategy, we found the use of the SD direction in globalizing Newton or quasi-Newton methods (see, e.g., [13] and [3, Section 1.4.4]) to be much less popular than we expected. This is probably due to lack of confidence in the SD methods, which have long been considered rather ineffective because of their slow convergence rate and their oscillatory behavior. However, starting from the publication of the Barzilai and Borwein (BB) method [2], it has become more and more evident that suitable choices of the step length in gradient methods may lead to effective algorithms [4, 6, 7], which have also shown good performance in solving problems arising in several application fields [1, 5, 9, 19].

The remainder of this paper is organized as follows. In Sect. 2 we describe our globalization strategy, proving global convergence of the corresponding Newton method. In Sect. 3 we report results of numerical experiments carried out with our algorithm, including a comparison with an MN method. We provide some conclusions in Sect. 4.

2   A Globalized Newton Method

A line-search method for solving (1) uses the iterative scheme

$$
\mathbf{x}_{k+1} = \mathbf{x}_k + \alpha_k \mathbf{d}_k, \quad k = 0, 1, 2, \ldots,
$$

where $\mathbf{d}_k$ is the search direction and $\alpha_k > 0$ is the step length. Decreasing line-search methods, i.e., such that $\{f(\mathbf{x}_k)\}$ is a strictly decreasing sequence, require $\mathbf{d}_k$ to be a direction of strict descent, i.e., a direction such that

$$
\mathbf{d}_k^T \mathbf{g}_k < 0.
\tag{9}
$$

Algorithm 1. NSD method

1: Let $\mathbf{x}_0 \in \mathbb{R}^n$, $\varepsilon \in (0,1)$, $\sigma \in (0,1/2)$ and $\rho \in (0,1)$.
2: **for** $k = 0,1,2,\dots$ **do**
3:     $\widetilde{\mathbf{g}}_k = -\mathbf{g}_k/\|\mathbf{g}_k\|$;
4:     **if** H_k is nonsingular **then**
5:         $\mathbf{d}_k^N = -H_k^{-1}\mathbf{g}_k$;
6:         $\widetilde{\mathbf{d}}_k^N = \mathbf{d}_k^N/\|\mathbf{d}_k^N\|$;
7:         $\beta_k = \rho^h$, where $h = \min\left\{s \in \mathbb{N}_0 : \cos\langle\rho^s\widetilde{\mathbf{d}}_k^N + (1-\rho^s)\widetilde{\mathbf{g}}_k, -\mathbf{g}_k\rangle > \varepsilon\right\}$;
8:     **else**
9:         $\beta_k = 0$;
10:    **end if**
11:    $\mathbf{d}_k = \beta_k\widetilde{\mathbf{d}}_k^N + (1-\beta_k)\widetilde{\mathbf{g}}_k$;
12:    $\overline{\alpha}_k = \beta_k\|\mathbf{d}_k^N\| + (1-\beta_k)\xi_k\|\mathbf{g}_k\|$ for a suitably chosen $\xi_k > 0$;
13:    select α_k satisfying (11) by a backtracking procedure starting from $\overline{\alpha}_k$;
14:    $\mathbf{x}_{k+1} = \mathbf{x}_k + \alpha_k\mathbf{d}_k$;
15: **end for**

The convergence of the line-search method depends on the choice of $\mathbf{d}_k$ and α_k. The descent condition (9) is not sufficient to guarantee convergence, and a key requirement for $\mathbf{d}_k$ is to satisfy the following angle criterion:

$$\cos\langle-\mathbf{g}_k, \mathbf{d}_k\rangle = \frac{-\mathbf{g}_k^T\mathbf{d}_k}{\|\mathbf{g}_k\|\|\mathbf{d}_k\|} > \varepsilon \tag{10}$$

for some $\varepsilon > 0$ independent of k. About the step length, the Armijo condition

$$f(\mathbf{x}_k + \alpha_k\mathbf{d}_k) - f(\mathbf{x}_k) \leq \sigma\alpha_k\mathbf{d}_k^T\mathbf{g}_k, \quad \sigma \in (0,1), \tag{11}$$

is a common choice. We note that (11) does not prevent the method from taking too small steps. Such a drawback can be overcome if a backtracking line-search procedure is adopted to choose α_k (see [17, page 37]).

In our globalization strategy for the Newton method, the search direction is defined as a convex combination of the normalized Newton and steepest descent directions. A general description of our approach, referred to as NSD method, is given in Algorithm 1. We note that if H_k is singular β_k is set to 0 and $\mathbf{d}_k$ to $\widetilde{\mathbf{g}}_k$, i.e., the iteration becomes a gradient step. The parameter ξ_k is chosen using a spectral step-length selection rule for the gradient method [6].

The following theorem shows that the limit points of the sequence generated by the NSD method are stationary.

Theorem 1. *Let $f \in C^2(\mathbb{R}^n)$. Then the NSD method is well defined and, for any choice of $\mathbf{x}_0$, every limit point of the sequence $\{\mathbf{x}_k\}$ is a stationary point.*

Proof. We note that the strategy used to choose β_k in Algorithm 1 ensures that a direction $\mathbf{d}_k$ satisfying condition (10) can be found. Indeed, if H_k is singular, then (10) trivially holds. If H_k is nonsingular and $\widetilde{\mathbf{d}}_k^N$ satisfies (10), then $\beta_k = \rho^0 = 1$ and $\mathbf{d}_k = \widetilde{\mathbf{d}}_k^N$; otherwise, since $\cos\langle\widetilde{\mathbf{g}}_k, -\mathbf{g}_k\rangle = 1$ and $\mathbf{d}_k \to \widetilde{\mathbf{g}}_k$

as $s \to \infty$, a value of β_k such that (10) holds will be computed in a finite number of steps. Since α_k is obtained through a backtracking strategy to satisfy the Armijo rule (11), Proposition 1.2.1 in [3] can be applied to conclude the proof.

It is well known that Newton-type methods are superlinearly convergent when started from an initial point sufficiently close to a local minimum. One would hopefully expect a globalization strategy to be able to preserve the convergence rate of the pure Newton method. This is the case of the NSD method, as shown by the following theorem.

Theorem 2. *Let $f \in C^2(\mathbb{R}^n)$ and let $\{x_k\}$ be the sequence generated by the NSD method. Suppose that there exists a limit point $\hat{x}$ of $\{x_k\}$ where $H(\hat{x}) \succ 0$ and that $H(x)$ is Lipschitz continuous in a neighborhood of $\hat{x}$. Then $\{x_k\}$ converges to $\hat{x}$ provided that the value of ε used in Algorithm 1 is sufficiently small. Furthermore, the rate of convergence is quadratic.*

Proof. Let $\mathcal{K} \subset \mathbb{N}$ be the index set of the subsequence of $\{x_k\}$ converging to $\hat{x}$ and let λ_1 and λ_n be the smallest and the largest eigenvalue of $H(\hat{x})$, respectively. If we choose $\mu_1, \mu_2 > 0$ such that $0 < \mu_1 < \lambda_1 \leq \lambda_n < \mu_2$, the continuity of $H(x)$ implies that there exists a scalar $\delta > 0$ such that $H(x) \succ 0$ and its eigenvalues lie in (μ_1, μ_2) for each $x \in B_\delta(\hat{x}) = \{x \in \mathbb{R}^n : \|x - \hat{x}\| < \delta\}$. Without loss of generality, we can choose δ such that, starting from a point in $B_\delta(\hat{x})$, the iterates of the Newton method remain in $B_\delta(\hat{x})$ (see [14, Theorem 2.3]). We can also assume that $H(x)$ is Lipschitz continuous $B_\delta(\hat{x})$. Let $k_\delta \in \mathcal{K}$ be such that $x_{k_\delta} \in B_\delta(\hat{x})$. We have that

$$\cos\langle d_{k_\delta}^N, -g_{k_\delta} \rangle = \frac{g_{k_\delta}^T H_{k_\delta}^{-1} g_{k_\delta}}{\|H_{k_\delta}^{-1} g_{k_\delta}\| \|g_{k_\delta}\|} > \frac{\mu_1}{\mu_2}.$$

Therefore, if $\varepsilon < \frac{\mu_1}{\mu_2}$, the backtracking strategy at line 7 of Algorithm 1 selects $\beta_{k_\delta} = 1$ and the $(k_\delta + 1)$-st iterate is obtained by a line search over the Newton direction. By reasoning as in the proof of part (b) in [18, Theorem 1.4.9], we have that if k_δ is large enough, then $\|g_{k_\delta}\|$ is small and the step length $\alpha_k = 1$ satisfies the Armijo condition (11), because $H(x)$ is Lipschitz continuous in $B_\delta(\hat{x})$ and $\sigma < \frac{1}{2}$. By the choice of δ we have that $x_{k_\delta+1} \in B_\delta(\hat{x})$. Then Algorithm 1 becomes the Newton method and the sequence $\{x_k\}$ converges to $\hat{x}$ with quadratic rate.

3 Numerical Experiments

We developed a MATLAB implementation of the NSD method and compared it with a MATLAB implementation of the MN method available from https://github.com/hrfang/mchol, which exploits the modified Cholesky factorizations described in [10]. In the NSD method, we set $\xi_k = \max\{\xi_k^{BB2}, \nu\}$, where ξ_k^{BB2} is computed by using the Barzilai-Borwein step-length selection rule defined

Table 1. Performance of the NSD and MN methods in the solution of the test problems (the mark "—" indicates that the required accuracy has not been satisfied within 1000 iterations). Note that we set to 0 all the values below the machine epsilon.

prob	NSD			MN		
	fval	# its	# evals	fval	# its	# evals
1	0.0000 e+00	20	49	0.0000 e+00	18	96
3	1.1279 e−08	3	5	1.1279 e−08	3	5
4	3.4891 e−10	438	3859	3.7800 e−05	—	2007
6	0.0000 e+00	11	21	0.0000 e+00	11	21
7	6.9588 e−02	9	17	6.9588 e−02	9	17
8	2.2500 e−05	37	83	2.2500 e−05	37	83
9	9.3763 e−06	140	354	9.3763 e−06	138	344
10	0.0000 e+00	6	12	2.4571 e−13	967	1933
11	8.5822 e+04	10	19	8.5822 e+04	10	19
12	0.0000 e+00	8	16	0.0000 e+00	23	58
13	3.0282 e−04	16	35	3.0282 e−04	9	37
14	0.0000 e+00	25	57	0.0000 e+00	25	57
15	1.7085 e−10	18	35	1.7085 e−10	18	35
16	0.0000 e+00	10	22	4.5201 e−01	—	2137
17	0.0000 e+00	456	1022	0.0000 e+00	59	169
18	0.0000 e+00	7	14	0.0000 e+00	7	14
19	0.0000 e+00	33	78	0.0000 e+00	33	78
20	-4.0000 e+00	—	21981	−4.0000 e+00	—	21981
21	0.0000 e+00	2	3	0.0000 e+00	2	3
22	0.0000 e+00	2	3	0.0000 e+00	2	3
23	6.9492 e−15	29	67	6.9492 e−15	29	67
25	3.0227 e−10	18	35	3.0227 e−10	18	35
27	4.8254 e−01	8	15	5.0000 e−01	2	20
28	3.4102 e+00	4	8	8.8077 e+00	9	54
30	3.9789 e−01	5	9	3.9789 e−01	5	29
31	−1.0153 e+01	10	20	−1.0153 e+01	13	26
32	−1.0402 e+01	9	28	−4.9728 e−06	10	39
33	−3.8351 e+00	9	20	−1.9733 e+00	8	20
34	−2.1546 e−01	7	14	−2.1546 e−01	7	14
35	−1.3803 e+01	6	12	−1.4427 e+01	5	9
37	−1.0000 e+00	11	23	0.0000 e+00	2	23
38	2.2875 e+00	6	13	2.2875 e+00	6	13
39	2.1831 e−01	7	16	2.1831 e−01	6	21
40	5.1001 e−01	6	13	−4.6516 e−01	7	33
41	0.0000 e+00	40	101	0.0000 e+00	28	78
43	3.7532 e−16	6	13	0.0000 e+00	20	79

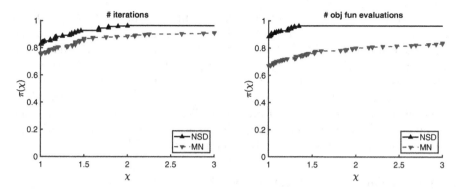

Fig. 1. Performance profiles of the NSD and MN methods in the solution of the test problems with 10 starting points. We consider the number of iterations (left) and the number of objective function evaluations (right).

in [2, equation (5)] and $\nu = 10^{-5}$. Moreover, we set $\rho = 0.9$ and $\varepsilon = 10^{-6}$. In the MN method, we chose the modified Cholesky factorization named GMW-II, which minimizes the 2-norm of the matrix E_k in the modified Hessian. Both methods used an Armijo backtracking line search with quadratic and cubic interpolation (see [17, Section 3.5]) and $\sigma = 10^{-4}$. The methods were stopped as soon as $\|\mathbf{g}_k\| < 10^{-6}$ or a maximum number of 1000 iterations was achieved.

The two algorithms were run on 36 problems from the collection available at https://people.sc.fsu.edu/~jburkardt/m_src/test_opt/test_opt.html, which includes MATLAB implementations of the well-known Moré-Garbow-Hillstrom benchmark problems [15] and other unconstrained optimization problems. All the experiments were carried out using MATLAB R2018b on a 64-bit Intel i7-6500 (3.10 GHz) processor, with 12 GB of RAM and 4 MB of cache memory and the Windows 10 operating system.

We first tested the two methods using the starting points provided with the test problems. For each method and each problem, we report in Table 1 the objective function value at the computed solution (fval), the number of iterations (# its) and the number of objective function evaluations (# evals). These results show that the NSD method was generally able to obtain objective function values at the computed solutions smaller than or equal to those reached by the MN method. Furthermore, we verified that in most of the cases where the two algorithms computed the same solution, NSD performed a number of iterations (each requiring the solution of a linear system) that is comparable with the number of iterations of the MN method. However, NSD generally required a smaller number of objective function evaluations, indicating a smaller number of line-search steps per iteration.

In order to further assess the performance of NSD, we ran tests using multiple starting points. Besides the original starting point $\mathbf{x}_0$, for each problem we considered 9 more points obtained by adding to each entry $(\mathbf{x}_0)_i$ a value γ_i randomly chosen in $[-10^{-2}a_i, 10^{-2}a_i]$, where $a_i = |(\mathbf{x}_0)_i|$. This allowed us to compare the two algorithms on a total of 360 problem instances. Again NSD was generally able to achieve objective function values smaller than or comparable with those computed by MN. The results of these tests are summarized by the performance profiles [8] in Fig. 1, which compare the two algorithms in terms of iterations and objective function evaluations on the 273 instances (76% of the problems) where they computed the same optimal solution up to the third significant digit. We see that NSD was more efficient in terms of both iterations (i.e., number of linear system solved) and objective function evaluations (i.e., line-search steps), and appeared to be more robust than MN.

4   Conclusions and Future Work

We proposed a globally convergent algorithm for unconstrained minimization, in which the Newton and the steepest descent directions are combined to produce a descent direction. Although simple and straightforward, this idea has been little explored to globalize the Newton method. Preliminary computational experiments show that the proposed strategy achieves promising results compared with a well-established modified Newton approach. Future work will include extensive testing of the proposed globalization strategy and its application to quasi-Newton methods.

References

1. Antonelli, L., De Simone, V., di Serafino, D.: On the application of the spectral projected gradient method in image segmentation. J. Math. Imaging and Vis. **54**(1), 106–116 (2016)
2. Barzilai, J., Borwein, J.M.: Two-point step size gradient methods. IMA J. Numer. Anal. **8**(1), 141–148 (1988)
3. Bertsekas, D.P.: Nonlinear Programming, 3rd edn. Athena Scientific, Belmont (2016)
4. Crisci, S., Ruggiero, V., Zanni, L.: Steplength selection in gradient projection methods for box-constrained quadratic programs. Appl. Math. Comput. **356**, 312–327 (2019)
5. De Asmundis, R., di Serafino, D., Landi, G.: On the regularizing behavior of the SDA and SDC gradient methods in the solution of linear ill-posed problems. J. Comput. Appl. Math. **302**, 81–93 (2016)
6. di Serafino, D., Ruggiero, V., Toraldo, G., Zanni, L.: On the steplength selection in gradient methods for unconstrained optimization. Appl. Math. Comput. **318**, 176–195 (2018)
7. di Serafino, D., Toraldo, G., Viola, M., Barlow, J.: A two-phase gradient method for quadratic programming problems with a single linear constraint and bounds on the variables. SIAM J. Optim. **28**(4), 2809–2838 (2018)
8. Dolan, E.D., Moré, J.J.: Benchmarking optimization software with performance profiles. Math. Program. Ser. B **91**(2), 201–213 (2002)
9. Dostál, Z., Toraldo, G., Viola, M., Vlach, O.: Proportionality-based gradient methods with applications in contact mechanics. In: Kozubek, T., et al. (eds.) HPCSE 2017. LNCS, vol. 11087, pp. 47–58. Springer, Cham (2018). https://doi.org/10.1007/978-3-319-97136-0_4

10. Fang, H., O'Leary, D.P.: Modified Cholesky algorithms: a catalog with new approaches. Math. Program. **115**(2), 319–349 (2008)
11. Fletcher, R.: Practical Methods of Optimization. Wiley, New York (2013)
12. Gill, P., Murray, W.: Newton-type methods for unconstrained and linearly constrained optimization. Math. Program. **7**(1), 311–350 (1974)
13. Han, L., Neumann, M.: Combining quasi-Newton and Cauchy directions. Int. J. Appl. Math. **22**(2), 167–191 (2003)
14. Moré, J.J., Sorensen, D.C.: Newton's method. In: Golub, G. (ed.) Studies in Numerical Analysis, pp. 29–82. The Mathematical Association of America, Providence (1984)
15. Moré, J.J., Garbow, B.S., Hillstrom, K.E.: Algorithm 566: FORTRAN subroutines for testing unconstrained optimization software [C5], [E4]. ACM Trans. Math. Softw. **7**(1), 136–140 (1981)
16. Nesterov, Y.: Introductory Lectures on Convex Optimization: A Basic Course, vol. 87. Springer, Boston (2013). https://doi.org/10.1007/978-1-4419-8853-9
17. Nocedal, J., Wright, S.: Numerical Optimization. Springer Series in Operations Research, 2nd edn. Springer, New York (2006). https://doi.org/10.1007/978-0-387-40065-5
18. Polak, E.: Optimization: Algorithms and Consistent Approximations, vol. 124. Springer, New York (2012). https://doi.org/10.1007/978-1-4612-0663-7
19. Zanella, R., Boccacci, P., Zanni, L., Bertero, M.: Efficient gradient projection methods for edge-preserving removal of Poisson noise. Inverse Prob. **25**(4), 045010 (2009)

On the Steplength Selection in Stochastic Gradient Methods

Giorgia Franchini[1,2,3]($\boxtimes$) iD, Valeria Ruggiero[2,3] iD, and Luca Zanni[1,3] iD

[1] Department of Physics, Informatics and Mathematics, University of Modena
and Reggio Emilia, Modena, Italy
{giorgia.franchini,luca.zanni}@unimore.it
[2] Department of Mathematics and Computer Science, University of Ferrara,
Ferrara, Italy
valeria.ruggiero@unife.it
[3] INdAM Research Group GNCS, Rome, Italy

Abstract. This paper deals with the steplength selection in stochastic gradient methods for large scale optimization problems arising in machine learning. We introduce an adaptive steplength selection derived by tailoring a limited memory steplength rule, recently developed in the deterministic context, to the stochastic gradient approach. The proposed steplength rule provides values within an interval, whose bounds need to be prefixed by the user. A suitable choice of the interval bounds allows to perform similarly to the standard stochastic gradient method equipped with the best-tuned steplength. Since the setting of the bounds slightly affects the performance, the new rule makes the tuning of the parameters less expensive with respect to the choice of the optimal prefixed steplength in the standard stochastic gradient method. We evaluate the behaviour of the proposed steplength selection in training binary classifiers on well known data sets and by using different loss functions.

Keywords: Stochastic gradient methods · Steplength selection rule · Ritz-like values · Machine learning

1 Introduction

One of the pillars of machine learning is the development of optimization methods for the numerical computation of parameters of a system designed to make decisions based on yet unseen data. Supported on currently available data or examples, these parameters are chosen to be optimal with respect to a loss function [3], measuring some *cost* associated with the prediction of an event. The problem we consider is the unconstrained minimization of the form

$$\min_{x\mathbb{R}^d} F(x) \equiv \mathbb{E}[f(x,\xi)] \tag{1}$$

where ξ is a multi-value random variable and f represents the cost function.

© Springer Nature Switzerland AG 2020
Y. D. Sergeyev and D. E. Kvasov (Eds.): NUMTA 2019, LNCS 11973, pp. 186–197, 2020.
https://doi.org/10.1007/978-3-030-39081-5_17

While it may be desirable to minimize (1), such a goal is untenable when one does not have complete information about the probability distribution of ξ. In practice, we seek the solution of a problem that involves an estimate of the objective function $F(x)$. In particular, we minimize the sum of cost functions depending on a finite training set, composed by sample data ξ_i, $i \in \{1 \ldots n\}$:

$$\min_{x \in \mathbb{R}^d} F_n(x) \equiv \frac{1}{n} \sum_{i=1}^{n} f(x, \xi_i) \equiv \frac{1}{n} \sum_{i=1}^{n} f_i(x), \tag{2}$$

where n is the number of samples and each $f_i(x) \equiv f(x, \xi_i)$ denotes the cost function related to the instance ξ_i of the training set elements. For very large training set, the computation of $\nabla F_n(x)$ is too expansive and it results inapposite for an online training with a growing amount of samples; then, exploiting the redundancy of the data, Stochastic Gradient (SG) method and its variants, requiring only the gradient of one or few terms of $F_n(x)$ at each iteration, have been chosen as the main approaches for addressing the problem (2). In Algorithm 1 we resume the main steps of a generalized SG method [3]. Each k-th iteration of SG requires a realization ξ_k of the random variable ξ, a strategy to devise a stochastic gradient vector $g(x_k, \xi_k) \in \mathbb{R}^d$ at the current iterate x_k and an updating rule for the steplength or scalar learning rate $\eta_k > 0$.

Algorithm 1. Stochastic Gradient (SG) method

1: Choose an initial iterate x_1.
2: **for** $k = 1, 2, \ldots$ **do**
3:     Generate a realization of the random variable ξ_k.
4:     Compute a stochastic vector $g(x_k, \xi_k)$.
5:     Choose a learning rate $\eta_k > 0$.
6:     Set the new iterate as $x_{k+1} \leftarrow x_k - \eta_k g(x_k, \xi_k)$.
7: **end for**

In particular, we point out two different strategies for the choices of ξ_k and $g(x_k, \xi_k)$:

- a realization of ξ_k may be given by the choice of a single training sample, or, in other words, a random index i_k is chosen from $\{1, 2, \ldots, n\}$ and the stochastic gradient is defined as

$$g(x_k, \xi_k) = \nabla f_{i_k}(x_k), \tag{3}$$

where $\nabla f_{i_k}(x_k)$ denotes the gradient of the i_k-th component function of (2) at x_k;

- the random variable ξ_k may represent a small subset $S_k \subset \{1, ..., n\}$ of samples, randomly chosen at each iteration, so that the stochastic gradient is defined as

$$g(x_k, \xi_k) = \frac{1}{|S_k|} \sum_{i \in S_k} \nabla f_i(x_k) \tag{4}$$

where $|S_k|$ denotes the number of elements of the set S_k.

The described approaches give rise to the standard simple SG and its mini-batch version, respectively.

The convergence results of Algorithm 1 equipped with the rule (3) or (4) apply to both the objective functions $F(x)$ and $F_n(x)$. In [3] the theoretical properties of generalized SG schemes are proved for strongly convex and nonconvex loss functions. In particular, we refer to the case of a strongly convex loss function, such as the ones used in the numerical results section. We recall that a function F is said to be strongly convex with parameter c when for all x, y and $a \in [0, 1]$ we have

$$F(ax + (1 - a)y) \leq aF(x) + (1 - a)F(y) - \frac{1}{2}ca(1 - a)\|x - y\|^2,$$

where c is a positive scalar.

Standard assumptions for the analysis of SG methods are that the gradient of F is L-Lipschitz continuous and the first and second moments of the stochastic directions $\{g(x_k, \xi_k)\}$ satisfy the following inequalities:

$$\mu_G \parallel \nabla F(x_k) \parallel_2^2 \geq \nabla F(x_k)^T \mathbb{E}_{\xi_k}[g(x_k, \xi_k)] \geq \mu \parallel \nabla F(\omega_k) \parallel_2^2$$
$$\mathbb{V}_{\xi_k}[g(x_k, \xi_k)] \leq M + M_V \parallel \nabla F(x_k) \parallel_2^2,$$

where μ, μ_G, M, M_V are suitable positive scalars. Thus, when the variable steplength η_k is in the interval $[\overline{\eta}_{min}, \overline{\eta}_{max}]$, with $\overline{\eta}_{min} > 0$ and $\overline{\eta}_{max} \leq \frac{\mu}{LM_G}$, the expected optimality gap related to the SG iteration satisfies the following asymptotic condition [3, Th. 4.6]

$$\mathbb{E}[F(x_k) - F_*] \leq \xrightarrow{k \to \infty} \overline{\eta}_{max} \frac{LM}{2c\mu},$$

where F_* is the required minimum value. This result shows that if the steplength is sufficiently small, then the expected objective values will converge to a neighborhood of the optimal value. In practice, since the constants related to the assumptions, such as the Lipschitz parameter of ∇F, or the parameters involved in the bounds of the moments of the stochastic directions, are unknown and not easy to approximate, the steplength is selected as a fixed small value η. We observe that there is no guidance on the specific choice of this value, which, however, plays a crucial role in the effectiveness of the method. Indeed, a too small steplength can give rise to a very slow learning process. For this reason, in some recent papers (see, for example, [8,9]), rules for an adaptive selection of the steplength have been proposed. In this work, we tailor the limited memory steplength selection rule proposed in [4] to the SG framework.

2   Selections Based on the Ritz-like Values

In the deterministic framework, a very effective approach for the steplength selection in the gradient methods is proposed in [4] for unconstrained quadratic programming problems and then extended to general nonlinear problems. In order to capture some second order information of the considered problem, the steplengths are defined as the inverse of suitable approximations of the eigenvalues of the Hessian matrix, given by its Ritz values. The key point is to obtain the Ritz values in an inexpensive way.

Let assume we have to solve the quadratic programming problem $\min_x \phi(x) \equiv \frac{1}{2}x^T A x - b^T x$ by means of the gradient method. The basic idea in [4] is to divide the sequence of iterations into groups of m_R iterations referred to as *sweeps*, where m_R is a small positive integer, and, for each sweep, to set the steplengths as the inverse of some Ritz values of the Hessian matrix A, computed by exploiting the gradients of the previous sweep. In particular, at the iteration $k \geq m_R$, we denote by G and J the matrices obtained collecting m_R gradient vectors computed at previous iterates and the related steplengths:

$$G = [g_{k-m_R}, \ldots, g_{k-1}], \quad J = \begin{pmatrix} \eta_{k-m_R}^{-1} & & & \\ -\eta_{k-m_R}^{-1} & \ddots & & \\ & \ddots & \eta_{k-1}^{-1} & \\ & & -\eta_{k-1}^{-1} \end{pmatrix},$$

where we use the notation $g_i \equiv \nabla\phi(x_i)$. In view of the recurrence formula linking the gradients of $\phi(x)$ at two successive iterates

$$g_i = g_{i-1} - \eta_{i-1} A g_{i-1}, \quad i > 0,$$

we can write

$$AG = [G, g_k]J. \tag{5}$$

This equation is useful to compute the tridiagonal matrix T obtained from the application of m_R iterations of the Lanczos process to the matrix A, with starting vector $q_1 = g_{k-m_R}/\| g_{k-m_R} \|$; this procedure generates an orthogonal matrix $Q = [q_1, \ldots, q_{m_R}]$, whose columns are a basis for the Krylov subspace $\{g_{k-m_R}, A g_{k-m_R}, A^2 g_{k-m_R}, \ldots, A^{m_R-1} g_{k-m_R}\}$, such that

$$T = Q^T A Q. \tag{6}$$

The steplengths for the next m_R gradient iterations are defined as the inverse of the eigenvalues θ_i of T, that are the so-called Ritz values:

$$\eta_{k-1+i} = \frac{1}{\theta_i}, \quad i = 1, \ldots, m_R. \tag{7}$$

The explicit computation of the matrix Q can be avoided, by observing that $G = QR$, where R is upper triangular, and R can be obtained from the Cholesky factorization of $G^T G$, that is $G^T G = R^T R$. Then the matrix T can be computed from equation (6) as follows:

$$T = R^{-T}G^T A G R^{-1} = R^{-T}G^T[G, g_k]JR^{-1} = [R, r]JR^{-1}, \tag{8}$$

where the vector r is the solution of the linear system $R^T r = G^T g_k$.

For a non-quadratic objective function, the recurrence (5) does not hold and the Eq. (8) provides an Hessenberg matrix; nevertheless, we can compute the symmetric tridiagonal matrix $\bar{T}$ by replacing the strictly upper triangle of T by the transpose of its strictly lower triangle (in Matlab notation $\bar{T} = tril(T) + tril(T, -1)'$); the eigenvalues θ_i of $\bar{T}$ tend to approximate m_R eigenvalues of the Hessian matrix [4,7].

In the stochastic context, we propose to introduce in the SG methods a selection rule for the steplength based on the just described Ritz-like approach; in this case, the implementation of this technique involves some important differences. The main difference is the use of stochastic gradients instead of full gradients in the construction of the matrix G:

$$G = [g_{k-m_R}(x_{k-m_R}, \xi_{k-m_R}), \dots, g_{k-1}(x_{k-1}, \xi_{k-1})];$$

we observe that the stochastic gradients are related to different samples of the data. By means of this matrix G, the matrix T of (8) can be computed; we propose to approximate second order information with the eigenvalues of its symmetric part $\tilde{T} = (T + T^T)/2$. Another key point is that, among the m_R eigenvalues θ_i of $\tilde{T}$, only the Ritz values belonging to an appropriate range $[\eta_{min}, \eta_{max}]$ have to be considered. As a consequence, the steplengths in a new sweep are defined in the following way:

$$\eta_{k-i+1} = \max\left\{\min\left\{\eta_{max}, \frac{1}{\theta_i}\right\}, \eta_{min}\right\}, \ i = 1, \dots, m_R. \tag{9}$$

Moreover, a further value $\eta_{ini} \in [\eta_{min}, \eta_{max}]$ is introduced for setting the starting sweep: $\eta_i = \eta_{ini}, i = 0, \dots, m_R - 1$. This reference value is also used as steplength in a recovery procedure, when all the eigenvalues of $\tilde{T}$ are negative and they have to be discarded.

The proposed steplength approach depends on the chosen interval $[\eta_{min}, \eta_{max}]$ and on η_{ini}. However, the effectiveness of the corresponding SG methods is weakly affected by variations of these parameters. This behaviour introduces greater flexibility with respect to the choice of a fixed small scalar, that must be carefully best-tuned. In particular, the numerical results of the next section highlight that the version of SG equipped with the Ritz-like selection rule for the steplengths appears to be more robust than that with a constant steplength and it provides numerical results with a comparable accuracy.

3 Numerical Experiments

In order to evaluate the effectiveness of the proposed steplength rule for SG methods, we consider the optimization problems arising in training binary classifiers for two well known data-sets:

- the *MNIST* data-set of handwritten digits (http://yann.lecun.com/exdb/mnist), commonly used for testing different systems that process images; the

images are in gray-scale $[0, 255]$, in our case normalized in the interval $[0, 1]$, centered in a box of 28×28 pixels; from the whole data-set of $60,000$ images, $11,800$ images were extracted exclusively relating to digits 8 and 9;
- the web data-set *w8a* downloadable from https://www.csie.ntu.edu.tw/~cjlin/libsvmtools/datasets/binary.html, containing 49,749 examples; each example is described by 300 binary features.

We built linear classifiers corresponding to three different loss functions; in all cases, a regularization term was added to avoid overfitting. Thus the minimization problem has the form

$$\min_{x \in \mathbb{R}^d} F_n(x) + \frac{\lambda}{2} \|x\|_2^2, \tag{10}$$

where $\lambda > 0$ is a regularization parameter. By denoting as $a_i \in \mathbb{R}^d$ and $b_i \in \{1, -1\}$ the feature vector and the class label of the i-th sample, respectively, the loss function $F_n(x)$ assumes one of the following form:

- logistic regression loss:

$$F_n(x) = \frac{1}{n} \sum_{i=1}^n \log \left[1 + e^{(-b_i a_i^T x)} \right];$$

- square loss:

$$F_n(x) = \frac{1}{n} \sum_{i=1}^n (1 - b_i a_i^T x)^2;$$

- smooth hinge loss:

$$F_n(x) = \frac{1}{n} \sum_{i=1}^n \begin{cases} \frac{1}{2} - b_i a_i^T x, & \text{if } b_i a_i^T x \leq 0 \\ \frac{1}{2}(1 - b_i a_i^T x)^2, & \text{if } 0 < b_i a_i^T x < 1 \\ 0, & \text{if } b_i a_i^T x \geq 1. \end{cases}$$

We compare the effectiveness of the following schemes:

- simple SG with fixed steplength and SG equipped with the Ritz-like steplength rule (9), named SG FR;
- SG with a fixed mini-batch size in the version with fixed steplength, denoted by SG mini-batch, and the one equipped with the Ritz-like steplength rule (9), named SG FR mini-batch.

3.1 Numerical Results

In all the numerical experiments we use the following setting:

- the regularization parameter λ is equal to 10^{-8};
- the size of the mini-batch is $|S| = 20$;
- in the FR methods, the length of the sweep is $m_R = 3$;

- each method is stopped after 15 epochs, i.e., after a time interval equivalent to 15 evaluations of a full gradient of F_n; in this way we compare the behaviour of the methods in a time equivalent to 15 iterations of a full gradient method applied to $F_n(x)$.

In the following tables we report the results obtained by the considered methods on the *MNIST* and *w8a* data-sets, by using the three loss functions (logistic regression, square and smooth hinge functions). For any test problem, we perform 10 runs of each method and we report the following results:

- the average value of the optimality gap $F_n(\bar{x}) - F_*$, where $\bar{x}$ is the iterate obtained at the end of the 15 epochs and F_* is an estimate of the optimal objective value obtained by a full gradient method with a large number of iterations;
- the related average accuracy $A(\bar{x})$ with respect to the training set employed for training the binary classifier, that is the percentage of well-classified examples.

We carried out the 10 different simulations with the same parameters, but leaving the possibility to the random number generator to vary. Indeed, due to the stochastic nature of the methods, the average values in different simulations provide more reliable results.

First of all, we describe the numerical results obtained by the different versions of SG related to the best-tuned setting of the parameters. In Table 1, we report the value of the fixed steplength η_{OPT} corresponding to the best performance of SG in 15 epochs. The steplength of SG mini-batch is set as $|S| \cdot \eta_{OPT}$.

Table 1. Values of the best-tuned steplength η_{OPT} in 15 epochs for the SG method in the case of the two data-sets and the three loss functions.

	MNIST			w8a		
Loss function	Logistic regression	Square	Smooth hinge	Logistic regression	Square	Smooth hinge
η_{OPT}	10^{-2}	10^{-4}	10^{-2}	10^{-1}	10^{-3}	$5\ 10^{-2}$

In the FR case, the following setting provides the best results:

- in SG FR, for both *MNIST* and *w8a*, we set $[\eta_{min}, \eta_{max}] = [10^{-4}\eta_{OPT}, 5\eta_{OPT}]$ and $\eta_{ini} = 0.1\ \eta_{OPT}$;
- in SG FR mini-batch, for both *MNIST* and *w8a*, we set $[\eta_{min}, \eta_{max}] = [10^{-8}\eta_{OPT}, 50\eta_{OPT}]$ and $\eta_{ini} = 0.1\ \eta_{OPT}$. We can observe that, in the mini-batch version, the method allows to choose the steplengths within a greater interval, showing more robustness.

In Tables 2, 3 and 4, we show the results obtained for the logistic regression, square and smooth hinge loss functions, respectively.

Table 2. Numerical results of the considered methods with $F_n(x)$ given by the logistic regression after 15 epochs.

	MNIST		w8a	
Method	$F_n(\overline{x}) - F_*$	$A(\overline{x})$	$F_n(\overline{x}) - F_*$	$A(\overline{x})$
SG	0.0107	0.987	0.00419	0.903
SG FR	0.0210	0.984	0.0251	0.899
SG mini-batch	0.0103	0.987	0.0141	0.901
SG FR mini-batch	0.0129	0.986	0.00441	0.903

Table 3. Numerical results of the considered methods with $F_n(x)$ given by the square loss after 15 epochs.

	MNIST		w8a	
Method	$F_n(\overline{x}) - F_*$	$A(\overline{x})$	$F_n(\overline{x}) - F_*$	$A(\overline{x})$
SG	0.00557	0.978	0.00229	0.889
SG FR	0.00729	0.978	0.00417	0.888
SG mini-batch	0.00585	0.977	0.00145	0.890
SG FR mini-batch	0.00647	0.977	0.00871	0.887

We observe that the results obtained with the FR adaptive steplength rule are well comparable with the ones obtained with the standard SG method equipped with the best-tuned steplength.

However, the considerable numerical experimentation carried out to obtain the above tables allows to remark that the optimal steplength search process for the SG method is often computationally long and expensive, since it requires a trial-and-error approach, while the FR adaptive rule appears to be weakly affected by the values η_{min} and η_{max} defining its working interval. In the following figures, we highlight the differences between the two approaches with respect to different settings of parameters in terms of the behavior of the optimality gap $F_n(\overline{x}) - F_*$ in 15 epochs. In particular, in Figs. 1 and 2, we show the behaviour

Table 4. Numerical results of the considered methods with $F_n(x)$ given by the smooth hinge loss after 15 epochs.

	MNIST		w8a	
Method	$F_n(\overline{x}) - F_*$	$A(\overline{x})$	$F_n(\overline{x}) - F_*$	$A(\overline{x})$
SG	0.00607	0.989	0.00136	0.904
SG FR	0.00847	0.987	0.0127	0.898
SG mini-batch	0.00754	0.987	0.000656	0.904
SG FR mini-batch	0.00793	0.987	0.00361	0.902

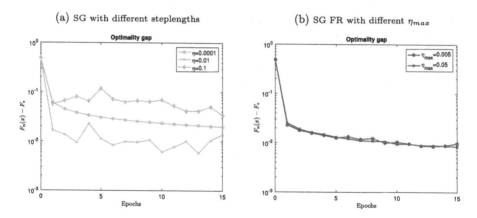

Fig. 1. Behaviour of SG and SG FR in 15 epochs on the *MNIST* data-set; test problem with smooth hinge loss function.

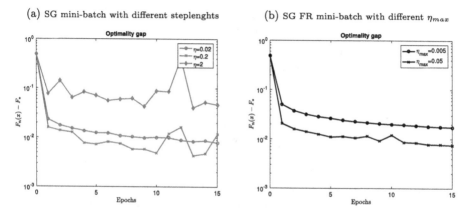

Fig. 2. Behaviour of SG mini-batch and SG FR mini-batch in 15 epochs on the *MNIST* data-set; test problem with smooth hinge loss function.

of the considered methods when the smooth hinge is used as loss function on the *MNIST* data-set, while the results obtained with the logistic regression loss function on the *w8a* data-set are reported in Figs. 3 and 4. On the left panels of the figures, the behaviour of SG and SG mini-batch methods equipped with different values of the steplength η in 15 epochs is shown; the right panels report the results obtained when the methods SG FR and SG FR mini-batch are executed with the same values for η_{ini} and η_{min} and different values of η_{max}. We observe that the adaptive steplength rule FR seems to be slightly dependent on the value of η_{max}, making the choice of a suitable value of this parameter a less difficult task with respect to the setting of η in the SG and SG mini-batch schemes.

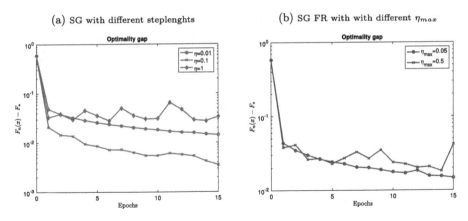

Fig. 3. Behaviour of SG and SG FR in 15 epochs on the *w8a* data-set; test problem with logistic regression loss function.

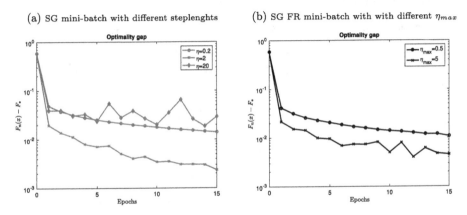

Fig. 4. Behaviour of SG mini-batch and SG FR mini-batch in 15 epochs on the *w8a* data-set; test problem with logistic regression loss function.

Finally, the last two figures compare the accuracy obtained in 15 epochs by the two approaches, SG with constant steplength and SG equipped by FR adaptive steplength rule, when the parameters η and η_{max} are not set at the best-tuned values, as in the experiments related to the previous tables. In particular, in Figs. 5 and 6, we show a comparison between SG or SG mini-batch with a prefixed non-optimal steplength η or $|S|\eta$ respectively, and the corresponding versions equipped with the adaptive steplength rule based on the Ritz-like values. In these figures, for the FR case we set $\eta_{min} = 10^{-4}\eta$, $\eta_{max} = 5\eta$, $\eta_{ini} = 10^{-1}\eta$ and for FR mini-batch case we set $\eta_{min} = 10^{-8}\eta$, $\eta_{max} = 50\eta$, $\eta_{ini} = 10^{-1}\eta$. In practice, in order to obtain the numerical results shown in Fig. 5 (*w8a* data-set and logistic regression loss function), the best-tuned values have been multiplied by a factor 10^{-3}, while in the case of Fig. 6 (*MNIST* data-set and smooth hinge loss function), the parameters are set equal to 10^{-2} times the best-tuned values.

(a) $\eta = 0.0001$ in SG, $\eta_{max} = 0.0005$ in SG FR   (b) $|S| = 20$; $\eta_{max} = 0.005$

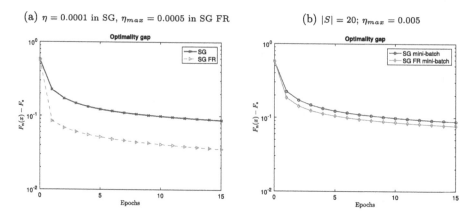

Fig. 5. Comparison between SG with respect to SG FR (on the left) and between SG mini-batch and SG FR mini-batch (on the right) in 15 epochs on the *w8a* data-set; test problem with logistic regression loss function.

(a) $\eta = 0.0001$ in SG. $\eta_{max} = 0.0005$ in SG FR   (b) $|S| = 20$; $\eta_{max} = 0.005$

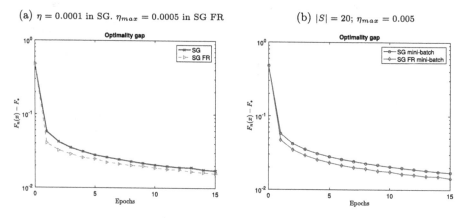

Fig. 6. Comparison between SG with respect to SG FR (on the left) and between SG mini-batch and SG FR mini-batch (on the right) in 15 epochs on the *MNIST* data-set; test problem with smooth hinge loss function.

As can be seen in the two figures, the selected steplength values guarantee the convergence of SG and SG mini-batch, but they are too small and produce a slow descent of the optimality gap; on the other hand, the FR approach appears less dependent on an optimal setting of the parameters and it enables us to obtain smaller optimality gap values after the same number of epochs exploited by SG.

4   Conclusions

In this work we propose to tailor the steplength selection rule based on the Ritz-like values, used successfully in the deterministic gradient schemes, to the SG methods. The numerical experimentation highlights that this strategy enables to achieve the performance of SG with fixed best-tuned steplength. Although, also in this case, it is necessary to carefully select a thresholding range for the steplengths, the proposed strategy appears slightly dependent on the parameters defining the bounds for the steplengths, making the parameters setting less expensive with respect to the SG framework. Furthermore, we remark that the mini-batch version of the proposed approach, allows more flexibility, since it needs of less restricted bounds for the steplengths range. In conclusion, the proposed technique provides a guidance on the learning rate selection and it allows to perform similarly to the SG approach equipped with the best-tuned steplength.

Future works will involve the introduction of the proposed technique in the variance reductions methods and its validation on other loss functions. Following the suggestions in [1,2,5,6], a very interesting analysis will concern the possibility of combining the proposed steplength selection rule with inexact line search techniques used in SG methods.

References

1. Bellavia, S., Krejic, N., Krklec Jerinkic, N.: Subsampled inexact Newton methods for minimizing large sums of convex functions. arXiv:1811.05730 (2018)
2. Bollapragada, R., Byrd, R., Nocedal, J.: Adaptive sampling strategies for stochastic optimization. SIAM J. Optim. **28**(4), 3312–3343 (2018)
3. Bottou, L., Curtis, F.E., Nocedal, J.: Optimization methods for large-scale machine learning. SIAM Rev. **60**(2), 223–311 (2018)
4. Fletcher, R.: A limited memory steepest descent method. Math. Program. Ser. A **135**, 413–436 (2012)
5. Krejic, N., Krklec Jerinki, N.: Nonmonotone line search methods with variable sample size. Numer. Algorithms **68**, 711–739 (2015)
6. Paquette, C., Scheinberg, K.: A stochastic line search method with convergence rate analysis. arXiv:1807.07994v1 (2018)
7. di Serafino, D., Ruggiero, V., Toraldo, G., Zanni, L.: On the steplength selection in gradient methods for unconstrained optimization. Appl. Math. Comput. **318**, 176–195 (2018)
8. Sopyla, K., Drozda, P.: SGD with BB update step for SVM. Inf. Sci. Inform. Comput. Sci. Intell. Syst. Appl. **316**(C), 218–233 (2015)
9. Tan, C., Ma, S., Dai, Y., Qian, Y.: BB step size for SGD. In: Lee, D., Sugiyama, M., Luxburg, U., Guyon, I., Garnett, R. (eds.) Advances in Neural Information Processing Systems (NIPS 2016), vol. 29 (2016)

Efficient Block Coordinate Methods for Blind Cauchy Denoising

Simone Rebegoldi[1]([⊠]) [iD], Silvia Bonettini[2] [iD], and Marco Prato[2] [iD]

[1] Dipartimento di Scienze Biomediche, Metaboliche e Neuroscienze,
Università di Modena e Reggio Emilia, Via Campi 287, 41125 Modena, Italy
`simone.rebegoldi@unimore.it`
[2] Dipartimento di Scienze Fisiche, Informatiche e Matematiche,
Università di Modena e Reggio Emilia, Via Campi 213/b, 41125 Modena, Italy
`{silvia.bonettini,marco.prato}@unimore.it`

Abstract. This paper deals with the problem of image blind deconvolution in presence of Cauchy noise, a type of non-Gaussian, impulsive degradation which frequently appears in engineering and biomedical applications. We consider a regularized version of the corresponding data fidelity function, by adding the total variation regularizer on the image and a Tikhonov term on the point spread function (PSF). The resulting objective function is nonconvex with respect to both the image and PSF block, which leads to the presence of several uninteresting local minima. We propose to tackle such challenging problem by means of a block coordinate linesearch based forward backward algorithm suited for nonsmooth nonconvex optimization. The proposed method allows performing multiple forward-backward steps on each block of variables, as well as adopting variable steplengths and scaling matrices to accelerate the progress towards a stationary point. The convergence of the scheme is guaranteed by imposing a linesearch procedure at each inner step of the algorithm. We provide some practical sound rules to adaptively choose both the variable metric parameters and the number of inner iterations on each block. Numerical experiments show how the proposed approach delivers better performances in terms of efficiency and accuracy if compared to a more standard block coordinate strategy.

Keywords: Blind deconvolution · Cauchy noise · Nonconvex optimization

1 Problem Formulation

In image restoration, the vast majority of the literature relies on the assumption that the observed data is corrupted by additive white Gaussian noise [9,24]. Such assumption, combined with a maximum a posteriori approach, leads to the minimization of a penalized least squares functional, which is usually convex and hence easy to minimize. However, in several real applications, the most adequate noise model may not be the Gaussian one. Typical examples include

© Springer Nature Switzerland AG 2020
Y. D. Sergeyev and D. E. Kvasov (Eds.): NUMTA 2019, LNCS 11973, pp. 198–211, 2020.
https://doi.org/10.1007/978-3-030-39081-5_18

Poisson noise in photon counting imaging [4], impulse noise due to analogue-to-digital conversion errors [19] or multiplicative noise in radar imagery [1].

Recently, much attention has been dedicated to Cauchy noise [14, 16, 18, 25], a kind of impulsive noise which corrupts synthetic aperture radar (SAR) images, underwater acoustic signals, low-frequency atmospheric signals and multiple-access in wireless communication systems [3, 20, 25]. The probability density function associated to the "zero-centered" Cauchy distribution is given by

$$p(v) = \frac{1}{\pi} \frac{\gamma}{\gamma^2 + v^2} \tag{1}$$

where $\gamma > 0$ is a scale parameter. We observe that (1) is even, bell-shaped and exhibits a thicker tail than the Gaussian bell curve, which tells us that an image corrupted by Cauchy noise is more likely to contain major noise peaks than the same image corrupted by white Gaussian noise.

Assume that a digital image $g \in \mathbb{R}^n$ has been acquired according to the model $g = \omega \otimes x + v$, where $x \in \mathbb{R}^n$ is the true image, $\omega \in \mathbb{R}^n$ is the point spread function (PSF) describing the blurring process, $v \in \mathbb{R}^n$ is the random noise vector following a Cauchy distribution and $\otimes$ denotes the convolution operation under periodic boundary conditions. Then, according to the maximum a posteriori approach [25], it is possible to derive a variational model to restore the corrupted image, where the corresponding fit-to-data term is given by

$$J_C(x; \omega) = \frac{1}{2} \sum_{i=1}^{n} \log\big(\gamma^2 + ((\omega \otimes x)_i - g_i)^2\big). \tag{2}$$

The image restoration is then performed by minimizing, with respect to the unknown x, the sum of J_C plus some appropriate convex regularizers, such as the total variation functional [24]. Note that the data fidelity term (2) is non-convex, thus several local minima may exist and any optimization method may get stuck in one of them. The authors in [25] address the nonconvexity of the problem by adding a quadratic penalty term to the objective function, whereas the works [7, 18] directly address the problem by means of first order splitting algorithms, for which convergence guarantees hold even in the nonconvex case.

In this paper, we extend the previously described model to blind deconvolution, namely to the restoration of images corrupted by Cauchy noise when the blur is unknown. Blind deconvolution is a severely ill-posed problem which has been widely treated in the literature, especially in the case of Gaussian and Poisson noise [2, 17, 21–23]. However, to the best of our knowledge, Cauchy noise has not yet been considered in the blind deconvolution framework. In this case, the problem amounts to minimizing the data fidelity (2) with respect to the coupled unknowns (x, ω), possibly adding regularizers and constraints on both x and ω in order to prevent the model from favouring trivial solutions [17, 21]. Here we choose to add the total variation function on the image and a zero-order Tikhonov term on the PSF, while imposing nonnegativity on both unknowns and a normalization constraint on the PSF only. Therefore, we obtain the regularized problem

$$\underset{x\in\mathbb{R}^n,\ \omega\in\mathbb{R}^n}{\operatorname{argmin}}\quad J(x;\omega) \equiv J_C(x;\omega) + \rho_x TV(x) + \iota_{\Omega_x}(x) + \rho_\omega \|\omega\|^2 + \iota_{\Omega_\omega}(\omega) \quad (3)$$

where $TV(x) = \sum_{i=1}^n \|\nabla_i x\|$ is the total variation regularizer, being $\nabla_i \in \mathbb{R}^{2\times n}$ the discrete gradient operator at pixel i, $\rho_x, \rho_\omega > 0$ are the regularization parameters, $\Omega_x = \{x \in \mathbb{R}^n : x \geq 0\}$ and $\Omega_\omega = \{\omega \in \mathbb{R}^m : \omega \geq 0, \sum_{i=1}^m \omega_i = 1\}$ are the constraints sets, and ι_{Ω_x} (resp. ι_{Ω_ω}) denotes the indicator function of Ω_x (resp. Ω_ω). Problem (3) is nonsmooth and nonconvex, not only globally, but also with respect to the variables blocks x and ω; therefore it is crucial to devise efficient optimization tools able to avoid uninteresting local minima and speed up the convergence to a meaningful stationary point.

The aim of this paper is to propose an efficient alternating minimization strategy to solve problem (3). The proposed method applies a specific variable metric forward–backward algorithm to each block of variables, and ensures global convergence towards a stationary point by performing a backtracking procedure along the descent direction. Unlike other standard approaches, it allows selecting the metric parameters in a variable manner; this could be helpful in order to get better reconstructions of both the image x and the PSF ω with reduced computational times. We support this remark by reporting a numerical experience on some test images corrupted by Cauchy noise, where we show the validity of the proposed approach in terms of both the quality of the reconstructed objects and the convergence speed towards the limit point.

2   The Proposed Algorithm

In its standard version, the alternating minimization strategy [13] consists in cyclically minimizing the objective function J with respect to a single block of variables, while keeping the other one fixed. Starting from an initial guess $(x^{(0)}, \omega^{(0)})$, the resulting iterative sequence $\{(x^{(k+1)}, \omega^{(k+1)})\}_{k\in\mathbb{N}}$ is given by

$$\begin{cases} x^{(k+1)} = \underset{x\in\mathbb{R}^n}{\operatorname{argmin}}\ J(x;\omega^{(k)}) \\ \omega^{(k+1)} = \underset{\omega\in\mathbb{R}^n}{\operatorname{argmin}}\ J(x^{(k+1)};\omega) \end{cases} \quad (4)$$

It is known that solving exactly the subproblems in (4) could be too costly or generate oscillating sequences in the absence of strict convexity assumptions [13]. Since the objective function in (3) can be decomposed into the sum

$$J(x;\omega) = J_0(x;\omega) + J_1(x) + J_2(\omega)$$

where $J_0(x; \omega) = J_C(x; \omega) + \rho_\omega \|\omega\|^2$ is continuously differentiable, $J_1(x) = \rho_x TV(x) + \iota_{\Omega_x}(x)$ and $J_2(\omega) = \iota_{\Omega_\omega}(\omega)$ are continuous on their domains and convex, it is much more convenient to adopt a proximal–linearized version of the previous alternating scheme. To this aim, we recall that the proximal operator of a convex function g is defined as [9, Sect. 3.4]

$$\text{prox}_g^{\alpha, D}(x) = \underset{u \in \mathbb{R}^n}{\text{argmin}} \; g(u) + \frac{1}{2\alpha} \|u - x\|_D^2 \tag{5}$$

where $\alpha > 0$, $D \in \mathbb{R}^{n \times n}$ is a symmetric positive definite matrix and $\|y\|_D = \sqrt{y^T D y}$ denotes the norm induced by D of a vector $y \in \mathbb{R}^n$. Then one can consider the following inexact scheme

$$\begin{cases} x^{(k+1)} = \text{prox}_{\rho_x TV + \iota_{\Omega_x}}^{\alpha_x^{(k)}, D_x^{(k)}} \left(x^{(k)} - \alpha_x^{(k)} (D_x^{(k)})^{-1} \nabla_x J_0(x^{(k)}; \omega^{(k)}) \right) \\ \omega^{(k+1)} = \text{prox}_{\iota_{\Omega_\omega}}^{\alpha_\omega^{(k)}, D_\omega^{(k)}} \left(\omega^{(k)} - \alpha_\omega^{(k)} (D_\omega^{(k)})^{-1} \nabla_\omega J_0(x^{(k+1)}; \omega^{(k)}) \right) \end{cases} \tag{6}$$

where $\alpha_x^{(k)}, \alpha_\omega^{(k)} > 0$ are the steplength parameters, $D_x^{(k)}, D_\omega^{(k)} \in \mathbb{R}^{n \times n}$ are the scaling matrices and $\nabla_x J_0(x; \omega)$, $\nabla_\omega J_0(x; \omega)$ denote the partial gradients of J_0 at point (x, ω) with respect to the two blocks of variables.

The alternating scheme (6) is also denominated block coordinate forward–backward algorithm [5, 8, 10], since it alternates, on each block of variables, a forward (gradient) step on the differentiable part with a backward (proximal) step on the convex part. A standard strategy for the parameters selection in (6) consists in choosing $D_x^{(k)}$ (resp. $D_\omega^{(k)}$) as the identity matrix, while taking $\alpha_x^{(k)}$ (resp. $\alpha_\omega^{(k)}$) as the inverse of the Lipschitz constant of the partial gradient $\nabla_x J_0(x; \omega^{(k)})$ (resp. $\nabla_\omega J_0(x^{(k+1)}; \omega)$), see for instance [5]. However, this choice is not applicable whenever the partial gradients fail to be Lipschitz continuous and it might generate sequences of extremely small steplengths. Furthermore, computing the proximal operators appearing in (6) could still be computationally expensive. Indeed, even though $\text{prox}_{\iota_{\Omega_\omega}}^{\alpha_\omega^{(k)}, D_\omega^{(k)}}$ reduces to the projection onto the set Ω_ω, which can be computed by means of linear-time algorithms [6], the evaluation of $\text{prox}_{\rho_x TV + \iota_{\Omega_x}}^{\alpha_x^{(k)}, D_x^{(k)}}$ requires an inner routine which could heavily slow down the iterations of the outer loop, if an excessive precision in the evaluation is required.

In order to overcome such limitations, the authors in [8] propose a modification of the inexact scheme (6) which can be seen as a block coordinate extension of the variable metric linesearch based forward–backward algorithm (VMILA) [7]. At each outer iteration $k \in \mathbb{N}$, the proposed block-VMILA algorithm applied to problem (3) generates two sequences of inner iterates $\{x^{(k,\ell)}\}$ and $\{\omega^{(k,\ell)}\}$, by performing $N_x^{(k)}$ VMILA steps on the image block and $N_\omega^{(k)}$ VMILA steps on the PSF block. The simplified outline of the algorithm is reported below.

$$
\begin{cases}
x^{(k,0)} = x^{(k)} \\
\text{FOR } \ell = 0,1,\ldots,N_x^{(k)} - 1 \\
\quad u_x^{(k,\ell)} \approx_\eta \operatorname{prox}_{\rho_x TV + \iota_{\Omega_x}}^{\alpha_x^{(k,\ell)}, D_x^{(k,\ell)}} \left(x^{(k,\ell)} - \alpha_x^{(k,\ell)}(D_x^{(k,\ell)})^{-1} \nabla_x J_0(x^{(k,\ell)}; \omega^{(k)}) \right) \\
\quad x^{(k,\ell+1)} = x^{(k,\ell)} + \lambda_x^{(k,\ell)}(u_x^{(k,\ell)} - x^{(k,\ell)}) \\
x^{(k+1)} = x^{(k,N_x^{(k)})} \\
\\
\omega^{(k,0)} = \omega^{(k)} \\
\text{FOR } \ell = 0,1,\ldots,N_\omega^{(k)} - 1 \\
\quad u_\omega^{(k,\ell)} = \operatorname{prox}_{\iota_{\Omega_\omega}}^{\alpha_\omega^{(k,\ell)}, D_\omega^{(k,\ell)}} \left(\omega^{(k,\ell)} - \alpha_\omega^{(k,\ell)}(D_\omega^{(k,\ell)})^{-1} \nabla_\omega J_0(x^{(k+1)}; \omega^{(k,\ell)}) \right) \\
\quad \omega^{(k,\ell+1)} = \omega^{(k,\ell)} + \lambda_\omega^{(k,\ell)}(u_\omega^{(k,\ell)} - \omega^{(k,\ell)}) \\
\omega^{(k+1)} = \omega^{(k,N_\omega^{(k)})}
\end{cases}
\tag{7}
$$

We now clarify the meaning of the iterates and parameters involved in (7).

- The point $u_x^{(k,\ell)} \in \mathbb{R}^n$ is a suitable approximation of the forward–backward step performed on the image block, where $\alpha_x^{(k,\ell)} > 0$ is a steplength parameter belonging to the interval $[\alpha_{\min}, \alpha_{\max}]$ and $D_x^{(k,\ell)} \in \mathcal{D}_\mu$ is a symmetric positive definite scaling matrix with eigenvalues in $[\frac{1}{\mu}, \mu]$, $\mu > 1$. In order to compute $u_x^{(k,\ell)}$, we consider the convex minimum problem associated to the proximal evaluation at the gradient step $z_x^{(k,\ell)} = x^{(k,\ell)} - \alpha_x^{(k,\ell)}(D_x^{(k,\ell)})^{-1}\nabla_x J_0(x^{(k,\ell)}; \omega^{(k)})$, namely

$$
\min_{u \in \mathbb{R}^n} h_x^{(k,\ell)}(u) \equiv \frac{\|u - z_x^{(k,\ell)}\|_{D_x^{(k,\ell)}}^2}{2\alpha_x^{(k,\ell)}} + \rho_x TV(u) + \iota_{\Omega_x}(u) + C_x^{(k,\ell)}
$$

where $C_x^{(k,\ell)} \in \mathbb{R}$ is a suitable constant, and then we reformulate it as the equivalent dual problem [7, Sect. 3]

$$
\max_{v \in \mathbb{R}^{3n}} \Psi_x^{(k,\ell)}(v) \equiv -\frac{\|\alpha_x^{(k,\ell)}(D_x^{(k,\ell)})^{-1}A^T v - z_x^{(k,\ell)}\|_{D_x^{(k,\ell)}}^2}{2\alpha_x^{(k,\ell)}} - \iota_{\Omega_x^*}(v) + C_x^{(k,\ell)}
$$

where $A = (\nabla_1^T \cdots \nabla_n^T \ I)^T \in \mathbb{R}^{3n \times n}$ and $\Omega_x^* = \{v \in \mathbb{R}^{3n} : \|(v_{2i-1}, v_{2i})\|_2 \leq \rho_x, \ i = 1,\ldots,n, \ v_i \leq 0, \ i = 2n+1,\ldots,3n\}$. Chosen a parameter $\eta \in (0,1)$, we look for a primal point $u_x^{(k,\ell)} \in \Omega_x$ and a dual variable $v^{(k,\ell)} \in \Omega_x^*$ satisfying [7, Eq. 15]

$$
h_x^{(k,\ell)}(u_x^{(k,\ell)}) \leq \eta \Psi_x^{(k,\ell)}(v_x^{(k,\ell)}).
\tag{8}
$$

The primal-dual pair $(u_x^{(k,\ell)}, v_x^{(k,\ell)})$ can be computed by generating a sequence of dual variables converging to the solution of the dual problem and then stopping the dual iterates when condition (8) is achieved.

- The point $u_\omega^{(k,\ell)}$ is the projection of the gradient step onto the nonnegativity orthant plus a normalization constraint. As described in [6], this problem is equivalent to solving a one-dimensional root finding problem for a piecewise linear monotonically nondecreasing function, which can be tackled by specialized linear time algorithms. In the numerical experiments, we will adopt the same secant-based method used in [6,11].
- The linesearch parameters $\lambda_x^{(k,\ell)}, \lambda_\omega^{(k,\ell)} \in (0,1]$ are chosen in such a way that the sufficient decrease conditions (9)-(10) are imposed on the image and PSF block, respectively. This is done by performing a backtracking procedure along the descent directions defined by $d_x^{(k,\ell)} = u_x^{(k,\ell)} - x^{(k,\ell)}$ and $d_\omega^{(k,\ell)} = u_\omega^{(k,\ell)} - \omega^{(k,\ell)}$. Note that the stopping criterion (8) and the projection operation onto Ω_ω imply that $h_x^{(k,\ell)}(u_x^{(k,\ell)}) < 0$ and $h_\omega^{(k,\ell)}(u_\omega^{(k,\ell)}) < 0$, so that the two inequalities (9)-(10) can be actually considered as descent conditions.

The block-VMILA scheme is entirely reported in Algorithm 1. Under mild assumptions on the objective function and provided that the proximal operators are computed with increasing accuracy, it is possible to prove that each limit point (if any) of the sequence generated by the block-VMILA scheme is stationary for the objective function [8, Theorem 1]. In particular, Algorithm 1 applied to problem (3) satisfies the required assumptions for ensuring the stationarity of the limit points. We report the corresponding theoretical result below. For the sake of clarity, we recall that the subdifferential of a convex function g at point y is given by $\partial g(y) = \{v \in \mathbb{R}^n : g(z) \geq g(y) + v^T(z-y), \forall z \in \mathbb{R}^n\}$, whereas the normal cone to a convex set Ω at point y is defined as $N_\Omega(y) = \{v \in \mathbb{R}^n : v^T(z-y) \leq 0, \forall z \in \mathbb{R}^n\}$.

Theorem 1. *Let $J(x; \omega)$ be defined as in problem (3). Any limit point $(\bar{x}, \bar{\omega})$ of the sequence $\{(x^{(k)}, \omega^{(k)})\}_{k \in \mathbb{N}}$ generated by Algorithm 1 is stationary, that is*

$$0 \in \nabla J_0(\bar{x}; \bar{\omega}) + (\rho_x \partial TV(\bar{x}) + N_{\Omega_x}(\bar{x})) \times N_{\Omega_\omega}(\bar{\omega}).$$

The convergence of the iterates $\{(x^{(k)}, \omega^{(k)})\}_{k \in \mathbb{N}}$ to a stationary point can be proved when the proximal operators are computed exactly and by assuming that the objective function satisfies the so-called Kurdyka–Lojasiewicz property at each of its stationary points [5, Definition 3]. The proof of this stronger convergence result for the block-VMILA scheme can be found in [8, Theorem 2].

3   Numerical Experience

For our numerical experiments, we consider a dataset of five 256×256 grayscale images and assume that the true PSF ω is associated to a Gaussian kernel with window size 9×9 and standard deviation equal to 2. The scale parameter γ has been set equal to 0.02 as in [25], whereas the regularization parameters ρ_x and ρ_ω have been manually tuned in order to provide the most visually satisfactory reconstructions for both the image and the PSF. The blurred and noisy images have been obtained by convolving the true objects with ω and then

---

Algorithm 1. Block-VMILA for blind Cauchy denoising

Choose the initial guesses $x^{(0)} \in \Omega_x$, $\omega^{(0)} \in \Omega_\omega$, the real numbers $0 < \alpha_{\min} \leq \alpha_{\max}$, $\mu \geq 1$, $\delta, \beta, \eta \in (0, 1)$ and the nonnegative integers $\bar{N}_x$, $\bar{N}_\omega$.

FOR $k = 0, 1, 2, \ldots$

STEP 1. 1. Set $x^{(k,0)} = x^{(k)}$.
2. Choose the number of inner iterations $N_x^{(k)} \leq \bar{N}_x$.
3. FOR $\ell = 0, 1, \ldots, N_x^{(k)} - 1$

1. Choose $\alpha_x^{(k,\ell)} \in [\alpha_{\min}, \alpha_{\max}]$, $D_x^{(k,\ell)} \in \mathcal{D}_\mu$.
2. Set $z_x^{(k,\ell)} = x^{(k,\ell)} - \alpha_x^{(k,\ell)} (D_x^{(k,\ell)})^{-1} \nabla_x J_0(x^{(k,\ell)}; \omega^{(k)})$.
3. Compute $u_x^{(k,\ell)} \approx_\eta \text{prox}_{\rho_x TV + \iota_{\Omega_x}}^{\alpha_x^{(k,\ell)}, D_x^{(k,\ell)}} (z_x^{(k,\ell)})$ according to (8).
4. Set $d_x^{(k,\ell)} = u_x^{(k,\ell)} - x^{(k,\ell)}$.
5. Compute the smallest nonnegative integer m such that

$$J(x^{(k,\ell)} + \delta^m d_x^{(k,\ell)}; \omega^{(k)}) \leq J(x^{(k,\ell)}; \omega^{(k)}) + \beta \delta^m h_x^{(k,\ell)}(u_x^{(k,\ell)}). \quad (9)$$

6. Compute $x^{(k,\ell+1)} = x^{(k,\ell)} + \delta^m d_x^{(k,\ell)}$.

4. Set $x^{(k+1)} = x^{(k,N_x)}$.

STEP 2. 1. Set $\omega^{(k,0)} = \omega^{(k)}$.
2. Choose the number of inner iterations $N_\omega^{(k)} \leq \bar{N}_\omega$.
3. FOR $\ell = 0, 1, \ldots, N_\omega^{(k)} - 1$

1. Choose $\alpha_\omega^{(k,\ell)} \in [\alpha_{\min}, \alpha_{\max}]$, $D_\omega^{(k,\ell)} \in \mathcal{D}_\mu$.
2. Set $z_\omega^{(k,\ell)} = \omega^{(k,\ell)} - \alpha_\omega^{(k,\ell)} (D_\omega^{(k,\ell)})^{-1} \nabla_\omega J_0(x^{(k+1)}; \omega^{(k,\ell)})$.
3. Compute $u_\omega^{(k,\ell)} = \text{prox}_{\iota_{\Omega_\omega}}^{\alpha_\omega^{(k,\ell)}, D_\omega^{(k,\ell)}} (z_\omega^{(k,\ell)})$.
4. Set $d_\omega^{(k,\ell)} = u_\omega^{(k,\ell)} - \omega^{(k,\ell)}$.
5. Compute the smallest nonnegative integer m such that

$$J(x^{(k+1)}; \omega^{(k,\ell)} + \delta^m d_\omega^{(k,\ell)}) \leq J(x^{(k+1)}; \omega^{(k,\ell)}) + \beta \delta^m h_\omega^{(k,\ell)}(u_\omega^{(k,\ell)}). \quad (10)$$

6. Compute $\omega^{(k,\ell+1)} = \omega^{(k,\ell)} + \delta^m d_\omega^{(k,\ell)}$.

4. Set $\omega^{(k+1)} = \omega^{(k,N_\omega)}$.

---

adding Cauchy noise, which has been generated by dividing two independent realizations of a normal random variable and then multiplying the result by the parameter γ (see [25, Sect. 5.1] for more details).

We consider the following two alternative choices for the scaling matrices $D_x^{(k,\ell)}$ and $D_\omega^{(k,\ell)}$ appearing in Algorithm 1.

– Split Gradient (SG): according to the strategy proposed in [15], we first decompose the partial gradients $\nabla_x J_0(x^{(k,\ell)}; \omega^{(k)})$ and $\nabla_\omega J_0(x^{(k+1)}; \omega^{(k,\ell)})$

into the difference of a positive and a nonnegative part, and then we choose $D_x^{(k,\ell)}$ and $D_\omega^{(k,\ell)}$ as diagonal matrices whose diagonal elements are given by

$$\left(D_x^{(k,\ell)}\right)_{jj}^{-1} = \max\left\{\min\left\{\frac{x_j^{(k,\ell)}}{V_x^{(k)}(x^{(k,\ell)})}, \mu_k\right\}, \frac{1}{\mu_k}\right\}$$

$$\left(D_\omega^{(k,\ell)}\right)_{jj}^{-1} = \max\left\{\min\left\{\frac{\omega_j^{(k,\ell)}}{V_\omega^{(k)}(\omega^{(k,\ell)})}, \mu_k\right\}, \frac{1}{\mu_k}\right\}$$

where $V_x^{(k)}(x) = \Omega^{(k)T}\frac{\omega^{(k)}\otimes x - g}{\gamma^2 + (\omega^{(k)}\otimes x - g)^2}$ and $V_\omega^{(k)}(\omega) = X^{(k)T}\frac{\omega\otimes x^{(k+1)} - g}{\gamma^2 + (\omega\otimes x^{(k+1)} - g)^2}$ are the positive parts coming from the decompositions of the partial gradients, being $\Omega^{(k)}, X^{(k)} \in \mathbb{R}^{n\times n}$ the two block circulant matrices with circulant blocks such that $\omega^{(k)} \otimes x = \Omega^{(k)}x$ and $\omega \otimes x^{(k+1)} = X^{(k)}\omega$, whereas $\{\mu_k\}_{k\in\mathbb{N}}$ is a sequence of positive thresholds chosen as $\mu_k = \sqrt{1 + P/k^p}$ with $p = 2$, $P = 10^{10}$, so that the scaling matrices are gradually converging to the identity matrix.
- Identity Matrix (I): in this case, we let $D_x^{(k,\ell)} = D_\omega^{(k,\ell)} = I_n$ for all k, ℓ.

For both choices of the scaling matrices, the steplength parameters $\alpha_x^{(k,\ell)}$ and $\alpha_\omega^{(k,\ell)}$ are computed by appropriately alternating the two scaled Barzilai-Borwein (BB) rules [6]. If we focus on the image block (the reasoning follows identically for the PSF block), each scaled BB steplength is obtained by imposing a quasi-Newton property on the matrix $B(\alpha) = \alpha^{-1}D_x^{(k,\ell)}$, thus obtaining

$$\alpha_x^{BB1} = \frac{s_x^{(k,\ell)T}D_x^{(k,\ell)}D_x^{(k,\ell)}s_x^{(k,\ell)}}{s_x^{(k,\ell)T}D_x^{(k,\ell)}y_x^{(k,\ell)}}, \qquad \alpha_x^{BB2} = \frac{s_x^{(k,\ell)T}\left(D_x^{(k,\ell)}\right)^{-1}y_x^{(k,\ell)}}{y_x^{(k,\ell)T}\left(D_x^{(k,\ell)}\right)^{-1}\left(D_x^{(k,\ell)}\right)^{-1}y_x^{(k,\ell)}}$$

where $s_x^{(k,\ell)} = x^{(k,\ell)} - x^{(k,\ell-1)}$ is the difference between two consecutive inner iterates and $y_x^{(k,\ell)} = \nabla_x J_0(x^{(k,\ell)}; \omega^{(k)}) - \nabla_x J_0(x^{(k,\ell-1)}; \omega^{(k)})$ is the difference between two consecutive partial gradients. At each inner iterate, one of the two scaled BB steplengths is chosen according to the alternation strategy described in [6, Sect. 3.3]. The chosen value is then constrained within the interval $[\alpha_{min}, \alpha_{max}]$ with $\alpha_{min} = 10^{-10}$ and $\alpha_{max} = 10^2$.

It is well known that, in the quadratic case, the BB rules well approximate the reciprocals of some eigenvalues of the Hessian matrix, which is a desirable property in order to ensure fast convergence of gradient methods [12]. There is some evidence that such good behaviour still holds for non-quadratic problems and it can be further enhanced when the BB rules are combined with the split gradient strategy, as confirmed by several numerical experiments in previous works [6–8]. Therefore it seems reasonable to expect the same accelerated behaviour for the block-VMILA algorithm.

According to Algorithm 1, the number of inner steps $N_x^{(k)}$ and $N_\omega^{(k)}$ may vary at each outer iteration k, provided that they do not exceed the a priori fixed upper bounds $\bar{N}_x$ and $\bar{N}_\omega$. Then following the strategy adopted in [8], we stop the inner iterates $x^{(k,\ell)}$ (resp. $\omega^{(k,\ell)}$) when either the maximum number of inner iterations is achieved or when the quantity $h_x^{(k,\ell)}(u_x^{(k,\ell)})$ (resp. $h_\omega^{(k,\ell)}(u_\omega^{(k,\ell)})$) is sufficiently small. This is reasonable, since $h_x^{(k,\ell)}(u_x^{(k,\ell)})$ (resp. $h_\omega^{(k,\ell)}(u_\omega^{(k,\ell)})$) is zero if and only the inner iterate is stationary for the objective function restricted to the single block of variables (see [7,8] and references therein). In conclusion, the inner iterations numbers are set as

$$N_x^{(k)} = \min\{\min\{\ell \in \mathbb{N} : |h_x^{(k,\ell)}(u_x^{(k,\ell)})| \leq \epsilon_x^{(k)}\}, \bar{N}_x\}$$
$$N_\omega^{(k)} = \min\{\min\{\ell \in \mathbb{N} : |h_\omega^{(k,\ell)}(u_\omega^{(k,\ell)})| \leq \epsilon_\omega^{(k)}\}, \bar{N}_\omega\}$$

where $\bar{N}_x = \bar{N}_\omega = 5$ and the tolerance parameters $\epsilon_x^{(k)}$ and $\epsilon_\omega^{(k)}$ are halved whenever the inner routine does not perform more than one inner iteration.

In addition to the two variants of Algorithm 1 described above, we implement also the more standard block coordinate forward–backward scheme (6). For this method, the scaling matrices $D_x^{(k)}$ and $D_\omega^{(k)}$ are chosen equal to the identity matrix, whereas the steplength parameters are selected as $\alpha_x^{(k)} = 1/L_x^{(k)}$ and $\alpha_\omega^{(k)} = 1/L_\omega^{(k)}$, being $L_x^{(k)} = \gamma^{-2}\|\Omega^{(k)}\|_1\|\Omega^{(k)}\|_\infty$, $L_\omega^{(k)} = \gamma^{-2}\|X^{(k)}\|_1\|X^{(k)}\|_\infty$ two upper estimates for the Lipschitz constants of the partial gradients $\nabla_x J_0(x^{(k,\ell)};\omega^{(k)})$ and $\nabla_\omega J_0(x^{(k+1)};\omega^{(k,\ell)})$. The proximal operator of the total variation term plus the indicator function of Ω_x is computed inexactly, by using the same stopping criterion (8) adopted for Algorithm 1. The resulting method can be considered as an inexact version of the proximal alternating linearized minimization (PALM) algorithm devised in [5], which does not originally include the possibility of computing inexactly the proximal operators of the convex terms involved in the minimization problem. For that reason, we will refer to this implementation of the block scheme (6) as the inexact PALM (inePALM in short).

For all algorithms, the initial guess $x^{(0)}$ has been chosen equal to the observed image g, whereas the initial PSF $\omega^{(0)}$ has been set as either the constant image satisfying the normalization constraint or a Gaussian function with standard deviation equal to 1. In Fig. 1, we observe that the block-VMILA scheme is able to provide accurate reconstructions of the ground truth, whereas the inePALM estimated images still look blurry, as if no information on the PSF had been retrieved by the algorithm. This is confirmed by looking at Figs. 4-5, where we report the plots of the relative root-mean-square error (RMSE) with respect to time on both the reconstructed image $x^{(k)}$ and PSF $\omega^{(k)}$, i.e.

$$RMSE(x^{(k)}) = \frac{\|x^{(k)} - \bar{x}\|}{\|\bar{x}\|}, \quad RMSE(\omega^{(k)}) = \frac{\|\omega^{(k)} - \bar{\omega}\|}{\|\bar{\omega}\|}$$

being $\bar{x}$, $\bar{\omega}$ the true image and PSF, respectively. We see that the inePALM algorithm does not move away from the initial PSF, whatever the test problem considered. Consequently, the algorithm fails in deblurring the observed images and gets stuck in stationary points which provide bigger values than the ones identified by block-VMILA, as we can deduce from the function plots reported in Figs. 2-3. The inePALM failure is due to the excessively small steplengths $\alpha_\omega^{(k)}$,

Fig. 1. From left to right: blurred and noisy images, VMILA-SG and inexact PALM reconstructions obtained by initializing the point spread function with a Gaussian blur.

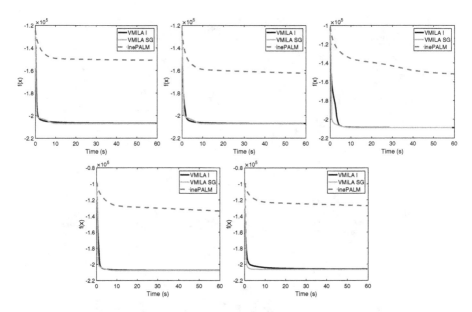

Fig. 2. Decrease of the objective function values vs time obtained by initializing the point spread function with a constant image. From left to right: baboon, boat and cameraman (top row), parrot and peppers (bottom row).

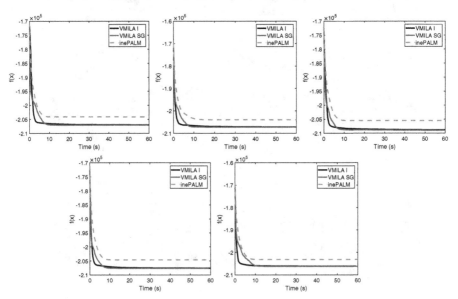

Fig. 3. Decrease of the objective function values vs time obtained by initializing the point spread function with a Gaussian function. From left to right: baboon, boat and cameraman (top row), parrot and peppers (bottom row).

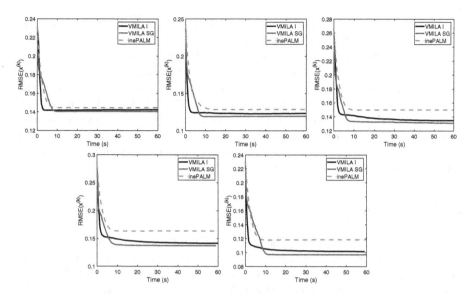

Fig. 4. Relative root-mean-square error (RMSE) on the image vs time obtained by initializing the point spread function with a Gaussian function. From left to right: baboon, boat and cameraman (top row), parrot and peppers (bottom row).

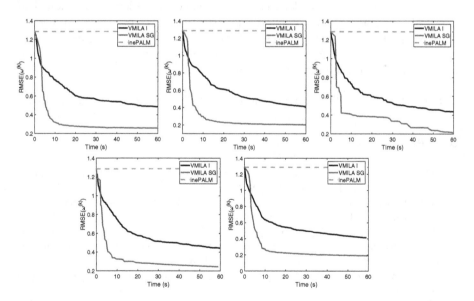

Fig. 5. Relative root-mean-square error (RMSE) on the PSF vs time obtained by initializing the point spread function with a Gaussian function. From left to right: baboon, boat and cameraman (top row), parrot and peppers (bottom row).

which in turn depend on the extremely large Lipschitz constants $L_\omega^{(k)}$. This issue is completely avoided by the block-VMILA scheme, thanks to the use of larger steplengths in the computation of the forward–backward step.

From the plots in Figs. 4-5, it is also evident that the block-VMILA algorithm performs better when equipped with the split gradient strategy for the scaling matrix. Indeed, for each test problem, the VMILA-SG variant provides a reconstructed PSF with relative error smaller than 50% in less than 10 seconds on a laptop equipped with a 2.60 GHz Intel(R) Core(TM) i7-4510U and 8 GB of RAM, whereas the non-scaled VMILA-I variant takes in average sextuple the time to get to the same accuracy level.

4   Conclusions

In this paper, we have addressed the problem of blind Cauchy denoising by means of a block coordinate linesearch based forward backward method. The proposed approach allows the user to freely select the metric parameters, while imposing a backtracking procedure along the feasible direction in order to guarantee the convergence towards a stationary point. The scheme is accelerated in practice by making use of very well-known adaptive strategies for the selection of the parameters, such as the Barzilai-Borwein rules or the split-gradient strategy, which capture some second order local information of the objective function. The numerical experiments here presented demonstrate that the proposed approach is highly competitive with respect to standard block coordinate methods, avoiding uninteresting stationary points and providing more accurate reconstructions of both the image and the point-spread-function.

Acknowledgements. The authors are members of the INDAM research group GNCS.

References

1. Aubert, G., Ajoul, J.-F.: A variational approach to removing multiplicative noise. SIAM J. Appl. Math. **68**(4), 925–946 (2008)
2. Ayers, G.R., Dainty, J.C.: Iterative blind deconvolution method and its applications. Opt. Lett. **13**(7), 547–549 (1988)
3. Banerjee, S., Agrawal, M.: Underwater acoustic communication in the presence of heavy-tailed impulsive noise with bi-parameter Cauchy-Gaussian mixture model. In: Ocean Electronics (SYMPOL), pp. 1–7 (2013)
4. Bertero, M., Boccacci, P., Ruggiero, V.: Inverse Imaging with Poisson Data, pp. 2053–2563. IOP Publishing, Bristol (2018)
5. Bolte, J., Sabach, S., Teboulle, M.: Proximal alternating linearized minimization for nonconvex and nonsmooth problems. Math. Program. **146**(1–2), 459–494 (2014)
6. Bonettini, S., Zanella, R., Zanni, L.: A scaled gradient projection method for constrained image deblurring. Inverse Problems **25**(1), 015002 (2009)
7. Bonettini, S., Loris, I., Porta, F., Prato, M., Rebegoldi, S.: On the convergence of a linesearch based proximal-gradient method for nonconvex optimization. Inverse Problems **33**(5), 055005 (2017)

8. Bonettini, S., Prato, M., Rebegoldi, S.: A block coordinate variable metric line-search based proximal gradient method. Comput. Optim. Appl. **71**(1), 5–52 (2018)
9. Chambolle, A., Pock, T.: An introduction to continuous optimization for imaging. Acta Numerica **25**, 161–319 (2016)
10. Chouzenoux, E., Pesquet, J.C., Repetti, A.: A block coordinate variable metric forward-backward algorithm. J. Glob. Optim. **66**(3), 457–485 (2016)
11. Dai, Y.H., Fletcher, R.: New algorithms for singly linearly constrained quadratic programming problems subject to lower and upper bounds. Math. Program. **106**, 403–421 (2006)
12. Frassoldati, G., Zanghirati, G., Zanni, L.: New adaptive stepsize selections in gradient methods. J. Ind. Manag. Optim. **4**(2), 299–312 (2008)
13. Grippo, L., Sciandrone, M.: On the convergence of the block nonlinear Gauss–Seidel method under convex constraints. Oper. Res. Lett. **26**(3), 127–136 (2000)
14. Idan, M., Speyer, J.: Cauchy estimation for linear scalar systems. IEEE Trans. Autom. Control **55**(6), 1329–1342 (2010)
15. Lantéri, H., Roche, M., Cuevas, O., Aime, C.: A general method to devise maximum likelihood signal restoration multiplicative algorithms with non-negativity constraints. Signal Process. **81**(5), 945–974 (2001)
16. Laus, F., Pierre, F., Steidl, G.: Nonlocal myriad filters for Cauchy noise removal. J. Math. Imaging Vision **60**(8), 1324–1354 (2018)
17. Levin, A., Weiss, Y., Durand, F., Freeman, W.T.: Understanding blind deconvolution algorithms. IEEE Trans. Pattern Anal. Mach. Intell. **33**(12), 2354–2367 (2011)
18. Mei, J.-J., Dong, Y., Huang, T.-Z.: Cauchy noise removal by nonconvex ADMM with convergence guarantees. J. Sci. Comput. **74**(2), 743–766 (2018)
19. Nikolova, M.: A variational approach to remove outliers and impulse noise. J. Math. Imaging Vision **20**(1–2), 99–120 (2004)
20. Peng, Y., Chen, J., Xu, X., Pu, F.: SAR images statistical modeling and classification based on the mixture of alpha-stable distributions. Remote Sens. **5**, 2145–2163 (2013)
21. Perrone, D., Favaro, P.: A clearer picture of total variation blind deconvolution. IEEE Trans. Pattern Anal. Mach. Intell. **38**(6), 1041–1055 (2016)
22. Prato, M., La Camera, A., Bonettini, S., Bertero, M.: A convergent blind deconvolution method for post-adaptive-optics astronomical imaging. Inverse Problems **29**, 065017 (2013)
23. Prato, M., La Camera, A., Bonettini, S., Rebegoldi, S., Bertero, M., Boccacci, P.: A blind deconvolution method for ground based telescopes and Fizeau interferometers. New Astron. **40**, 1–13 (2015)
24. Rudin, L.I., Osher, S., Fatemi, E.: Nonlinear total variation based noise removal algorithms. J. Phys. D. **60**(1–4), 259–268 (1992)
25. Sciacchitano, F., Dong, Y., Zeng, T.: Variational approach for restoring blurred images with Cauchy noise. SIAM J. Imaging Sci. **8**(3), 1894–1922 (2015)

High Performance Computing in Modelling and Simulation

A Parallel Software Platform for Pathway Enrichment

Giuseppe Agapito$^{(\boxtimes)}$ and Mario Cannataro

University of Catanzaro, 88100 Catanzaro, Italy
{agapito,cannataro}@unicz.it

Abstract. Biological pathways are complex networks able to provide a view on the interactions among bio-molecules inside the cell. They are represented as a network, where the nodes are the bio-molecules, and the edges represent the interactions between two biomolecules. Main online repositories of pathways information include KEGG that is a repository of metabolic pathways, SIGNOR that comprises primarily signaling pathways, and Reactome that contains information about metabolic and signal transduction pathways. Pathways enrichment analysis is employed to help the researchers to discriminate relevant proteins involved in the development of both simple and complex diseases, and is performed with several software tools. The main limitation of the current enrichment tools are: (*i*) each tool can use only a single pathway source to compute the enrichment; (*ii*) researchers have to repeat the enrichment analysis several times with different tools (able to get pathway data from different data sources); (*iii*) enrichment results have to be manually merged by the user, a tedious and error-prone task even for a computer scientist. To face this issues, we propose a parallel enrichment tool named Parallel Enrichment Analysis (PEA) ables to retrieve at the same time pathways information from KEGG, Reactome, and SIGNOR databases, with which to automatically perform pathway enrichment analysis, allowing to reduce the computational time of some order of magnitude, as well as the automatic merging of the results.

Keywords: Pathways · Pathway database · Parallel computing · Pathway enrichment analysis

1 Introduction

Biological pathways are human representations of the existent interactions among biomolecules, that regulate how cellular functions are carried out both in healthy and diseased state and how cells can interact with the external environment. Biological Pathways can be classified into three categories: **Signalling Pathways**, **Metabolic Pathways**, and **Regulatory Pathways**. Several online databases store, represent, and share different types of pathways. For example, Reactome and KEGG store all three types of pathways while SIGNOR

© Springer Nature Switzerland AG 2020
Y. D. Sergeyev and D. E. Kvasov (Eds.): NUMTA 2019, LNCS 11973, pp. 215–222, 2020.
https://doi.org/10.1007/978-3-030-39081-5_19

includes only signaling pathways and Metacyc comprises only metabolic pathways. Although, databases containing the same kinds of pathways (like KEGG and Reactome) show minimal overlap on the number of pathways and gene coverage as described in [9]. Each database has its own representation conventions and data access methods, making data integration from multiple databases a considerable challenge. This calls for a need to define a unique file format that makes it possible to standardize data coming from various data sources.

Pathways are represented using structured file formats (i.e., XML, OWL, RDF, XML-based, and psimi-xml) or unstructured text files. The most used structured file format is the Biological Pathway eXchange format (BIOPAX), categorized in - (LEVELS 1,2,3) [2]. Thus, the search, comparison, and identification of similar data types from different sources is often difficult. To fill this gap, in this work we present a parallel software algorithm named Parallel Enrichment Analysis (PEA) with which to simply and effectively manage information contained in several online databases based on BioPAX and XML formats. Pathway enrichment tools such as CePa [5], SPIA [10], and TPEA [7] can perform pathway enrichment analysis, exploiting data coming from a single specific data source. PEA instead, can perform pathway enrichment analysis, exploiting data coming from several data source into the same analysis (in this version, PEA can retrive information from Reactome, KEGG, and SIGNOR databases, as well as each data compliant with the BioPAX Level 3 format). PEA is implemented using the cross platform language *Java 8*, using a multi-threads solution, where threads are mapped on the available physical cores. Retrieving data from multiple databases is an easily parallelizable task, because there is no need to share information among threads to retrieve information from independent data sources. As a result, the enrichment results coming from different data source are automatically merged together, allowing thus to obtain more informative results without be necessary to use multiple software tools. Pathways enrichment analysis in PEA is implemented as a customized version of the *Hypergeometric* distribution function.

The remaining part of the manuscript is organized as follows: Sect. 2 presents the state of the art of pathway databases, along with some well-known enrichment tools. Section 3 introduces the PEA software platform and its capability, whereas in Sect. 4 the PEA's performance are evaluated. Finally, Sect. 5 concludes the paper and delineate some possible future works and extensions.

2   Related Works

The number of pathway databases is growing quickly in recent years. This is advantageous because biologists often need to use information from many sources to support their research. Here we report a short list of well-known pathway databases.

- Reactome is an open source pathway database [3,4]. Currently Reactome contains the whole known pathways coming from 22 different organisms including human. Pathway can be download in different formats comprising SBML, BioPAX and other graphical formats.

- KEGG is a collection of 19 interconnected databases, including genomic, chemical, pathway and phenotypic information [6]. KEGG stores pathways from several organisms, including human. KEGG data can be accessed using the KEGG API or KEGG FTP, allowing users to download each pathway in the KGML format (KEGG XML format).
- SIGnaling Network Open Resource (SIGNOR) is a collection of approximately 12,000 manually-annotated causal relationships and about 2800 human proteins participating in signal pathways [8]. Pathways are manually curated, and available for the download from the Download page.
- PathwayCommons [1] is a collection of public pathway databases, providing an access point for a collection of public databases. PathwayCommons provides a web interface to browse pathways, as well as a web service API for automatic access to the data. Also, PSI-MI and BioPAX, formats are supported for the data download.

On the other hand, each database has its own representation conventions, pathway are encoded using different type of file such as: *BIOPAX -(LEVEL 1,2,3) -*, *SIF*, *SBML*, *XML*, and flat text files. BIOPAX -Biological Pathway Exchange-, is a meta language defined in OWL and is represented in the RDF/XML format. Simple Interaction Format *(SIF)* is a data exchange format used to represent molecular interactions and extended to represent biological models. Systems Biology Markup Language (SBML) is focused on describing biological networks. Here we introduce some available pathway enrichment tools and their functionalities.

- Topology-based Pathway Enrichment Analysis (TPEA) [7] is a pathway enrichment tool based on graph topology. Pathways enrichment analysis in TPEA is obtained by using the KEGG database. TPEA is available as a R package at https://cran.r-project.org/web/packages/TPEA/.
- CePa [5] performs enrichment pathways analysis based on topological information in addition to gene-set information. Enrichment analysis in CePa is obtained exploiting the pathways contained in KEGG database. CePa is available as R package, at https://cran.r-project.org/web/packages/CePa/.
- Signaling Pathway Impact Analysis (SPIA) [10] combines the evidence obtained from the classical enrichment analysis with the measure of the perturbation on pathways under a given condition. Pathways are obtained from KEGG. SPIA is an R package available at https://bioconductor.org/packages/SPIA/.

All the listed pathway enrichment tools can perform enrichment analysis exploiting the information available into a single data source. Due to this limitation, researchers have to repeat several time the enrichment analysis by using the opportune software tool compatibles with the chosen data source. In many cases, software tools are available only for well-known data repository such as KEGG or Reactome, limiting the number of data source usable from the researchers. This scenario, calls for the need to define software tools can dig with many databases at the same time. To overcome the limitation of the existent software tools, we propose Parallel Enrichment Analysis (PEA) software tools. PEA preforms parallel pathway enrichment analysis from different data sources, providing more relevant results, allowing researchers to foster a greater understanding of the underlying biology. PEA can analyze data coming from KEGG, Reactome, and SIGNOR in parallel as well as, to perform pathways enrichment using any pathway data coded using the BioPAX-Level 3 format.

3 Parallel Enrichment Analysis Platform

Parallel Enrichment Analysis (PEA) is a software platform developed to perform pathway enrichment analysis in parallel, using data coming from different data sources. PEA is developed by using the Java 8 cross-platform programming language, making it compatible with all the operating systems that support Java. To efficiently manage concurrency we employed the built-in support for multithreaded programming. PEA is based on a modular architecture that provides a two-fold advantage, first each module can handle a specific type of data, and second it is possible to add new modules making PEA easily compatible with new data sources. The PEA's architecture is depicted in Fig. 1.

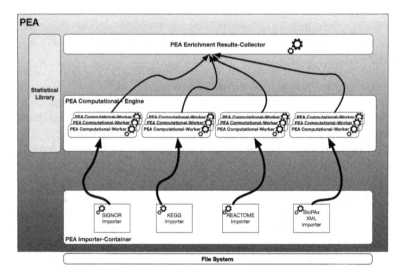

Fig. 1. PEA architecture.

The key modules of PEA are: *Importer-Container*, *Enrichment-Computational-Engine*, *Enrichment-Results-Collector*, and *Statistical-Library*. In the current version of PEA the *Importer-Container* comprises KEGG, SIGNOR, Reactome, and the General BioPAX and XML - Importer. The *Importer-Container* provides built-in functionalities, allowing to include additional modules with a minimum programming effort, simulating the Java plug-in importer.

The available importer-modules are independent, property that would enable PEA to retrieve, and preprocess data in parallel making the import process scalable. Preprocessing is mandatory to convert data in a suitable format for the subsequent enrichment analysis. Conversely, from the other enrichment tools, PEA provides an outlet with which to get data from multiple data sources i.e., Reactome, KEGG, and SIGNOR. In addition, it provides a generic BioPAX and XML importer. Traditional pathways enrichment tools cannot use a generic pathway, i.e., a new pathway coded in BioPAX or XML by the user, with which to perform pathway enrichment analysis. PEA, introducing a general BioPAX-XML importer, can also use generic pathways to perform pathway enrichment analysis. The *Enrichment-Computational-Engine* module, allows to carry out the enrichment analysis quickly, efficiently, and automatically from all the available data source in the same experiment. The enrichment analysis is done in parallel for each data source. When an importer finishes its job, it sends the preprocessed data to the *Enrichment-Computational-Engine* that in parallel starts to perform the enrichment per each single data source. The enrichment results are handled from the *Enrichment-Results-Collector* that merges together the results obtained from the multiple enrichment in an single result file. Pathway enrichment results are sorted in a increasing probability value (p-value) of statistical relevance, and only the results for which the computed p-value is greater than a threshold defined by the user (default p-value is 0.05) are stored and provided to the users. In addition, in PEA the FDR (false discovery rate) and Bonferroni methods of multiple statistical test adjustment are available. Finally, the *Statistical-Library* module is a collection of statistical measures for the automatic analysis of pathways coming from multiple data sources.

4   Performance Evaluation

This Section presents the performance evaluation of PEA currently implemented as a multi-threaded Java application. The experiments were run on a single cluster's node made available by the ICAR-CNR Institute, Italy, and composed by 24 computing nodes (CN) each one containing 132 GBytes of RAM, 1 Tbytes hard disk, and 16 Intel(R) Xeon(R) CPU E5–2670 at (2.60GHz) with 16 cores each. Since Intel uses the hyper-threading technology, each CN has 16 CPUs and 32 virtual cores. The operating system is Linux Red Hat version 4.4.6-4 (x86_64), whereas the versions of JRE and JDK used are, OpenJDK Runtime Environment, and OpenJDK 64-Bit Server VM both at 64 bit.

For our experiments, we used Reactome, KEGG, and SIGNOR pathway databases, the data sources from which to retrieve the information and perform the multiple pathway enrichment analysis. To compute pathway enrichment analysis, we simulated an up-regulate genes dataset, including about 50,000 distinct genes. PEA parallel engine presents a multi-level of parallelism: the first one concerns the capability to load in parallel the pathways data from the three databases. In PEA, pathway data are loaded through ad-hoc Pipes. Pipes are independent, making it possible to retrieve data in parallel from all the available databases. In the current implementation, pipes are implemented as single-thread, to avoid that performance decreases significantly due to the delay introduced by I/O operations performed from additional threads per single database. In the current version of PEA, the downloaded information from the three databases have been stored into three separate files. The second level of parallelism concerns with the enrichment. In PEA, to speed up the enrichment analysis, the enrichment-engine splits the enrichment genes dataset among the available physical cores, to balance the workload, speeding-up the time necessary to compute the enrichment. In the experiments, PEA has run on one computation node of the cluster varying the number of active threads in the enrichment analysis. Mapping each thread on a physical core, i.e., changing the number of parallel worker between 2, 4, 8, and 12. Each thread is executed on a core for a maximum of 12 cores for single data source (to avoid that PEA-Threads saturate the available phisical cores). The sequential PEA has run on just one core, running each data importer and enrichment analysis sequentially, employing only a single thread for the whole analysis process. We measured the response time of PEA (i.e., the execution times of the preprocess and enrichment). For each experiment configuration, we made 10 runs and measured the mean of execution times. The execution times are reported in Table 1.

Table 1. Table shows the execution times obtained by PEA using 2, 4, 6, 8, 10, 12 cores, computing pathway enrichment analysis from all the available databases. Loading is done in parallel, using one thread per database. EA refers to Enrichment Analysis.

#*Cores*	Prep times (ms)	EA Times (ms)
1	236970	36313
2	192739	7850
4	192228	7590
6	191624	6600
8	191624	6240
10	192228	5760
12	192225	5180

Figure 2 conveys the PEA's speed-up for loading the input data and for the pathway enrichment analysis, respectively.

The PEA's enrichment analysis speed-up (black line in Fig. 2) is super-linear until 4 cores, and start to decrease when passing to 6, 8, 10 and 12 cores for each databases. This behavior can be explained because, until 4 threads per single data-source the available physical cores are enough to directly handle 12 threads (the CPU has 16 physical cores). The speed-up starts to decrease when, to map the threads per single data-source, in addition to the physical cores it is necessary to use the Hyper-Threading technology available in the Intel's CPUs. This knowledge can be used, to automatically set the maximum number of threads to

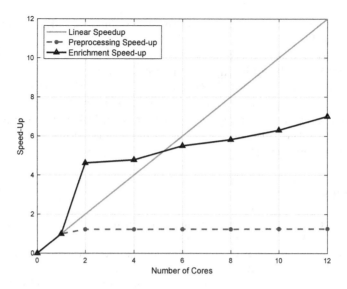

Fig. 2. PEA speed up.

use in the enrichment, avoiding to decrease the performance of the system due
to bad resource management. The PEA's pre-processing speed-up (dashed blue
line in Fig. 2) is sub-linear due to disk readings. On the other hand, reading in
parallel the three dataset allow to save time, respect the sequential read (let see
Table 1). Using more threads to read a single file, it results in a decreasing of the
system performance. In fact, in the current version of PEA data from the three
datasets are stored as single huge files, making multi-thread reading ineffective.
Thus, using a redundant array of independent disks (RAID), and splitting data
on more small files, could contribute to increase the overall performance using
more threads to load pathway data. To alleviate the computational impact of
pre-processing on the performance of PEA, we are planning to load the databases
only at the startup, keeping them in the main memory.

5   Conclusion

In this article, we have proposed PEA a parallel framework that can manage information from multiple pathway databases, to produce more informative
pathway enrichment analysis. Most of the existing pathway enrichment analysis
tools are designed only to use data coming from a single database. PEA instead,
is the first attempt to perform pathway enrichment analysis exploiting information coming from different pathway databases in a single experiment. In the
current version of PEA, we incorporated three pathway databases: KEGG, Reactome, SIGNOR, and generic pathways coded in BioPAX (Level 3), with which
to perform enrichment analysis. In our simulation studies, we demonstrated that

the proposed parallel approach outperforms the sequential pathway enrichment analysis methods.

Future work will regard the possibility to make PEA compatible with other pathway databases. In addition, we will provide an engine able to store the pathway data on multiple files, contributing to increase the performance of the loading task as well as the overall performance of PEA. Finally, PEA can be implemented using the distributed computing model, based on message passing to better exploit the performance of distributed architecture as well as multi CPUs or multi-cores architectures.

References

1. Cerami, E.G., et al.: Pathway commons, a web resource for biological pathway data. Nucleic Acids Res. **39**, D685–D690 (2010)
2. Demir, E., et al.: The BioPAX community standard for pathway data sharing. Nat. Biotechnol. **28**(9), 935 (2010)
3. Fabregat, A., et al.: The reactome pathway knowledgebase. Nucleic Acids Res. **46**(D1), D649–D655 (2017)
4. Fabregat, A., et al.: Reactome pathwayanalysis: a high-performance in-memory approach. BMC Bioinformatics **18**(1), 142 (2017). https://doi.org/10.1186/s12859-017-1559-2
5. Gu, Z., Wang, J.: CePa: an R package for finding significant pathways weighted by multiple network centralities. Bioinformatics **29**(5), 658–660 (2013)
6. Kanehisa, M., Goto, S.: KEGG: kyoto encyclopedia of genes and genomes. Nucleic Acids Res. **28**(1), 27–30 (2000)
7. Liu, D., et al.: Pathway enrichment analysis approach based on topological structure and updated annotation of pathway. Briefings in Bioinf. **20**(1), 168–177 (2017)
8. Perfetto, L., et al.: SIGNOR: a database of causal relationships between biological entities. Nucleic Acids Res. **44**(D1), D548–D554 (2015)
9. Rahmati, S., Abovsky, M., Pastrello, C., Jurisica, I.: pathDIP: an annotated resource for known and predicted human gene-pathway associations and pathway enrichment analysis. Nucleic Acids Res. **45**(D1), D419–D426 (2016)
10. Tarca, A.L., et al.: A novel signaling pathway impact analysis. Bioinformatics **25**(1), 75–82 (2009). https://doi.org/10.1093/bioinformatics/btn577. https://www.ncbi.nlm.nih.gov/pubmed/18990722

Hierarchical Clustering of Spatial Urban Data

Eugenio Cesario[1,2] , Andrea Vinci[2](✉) , and Xiaotian Zhu[1]

[1] Monmouth University, West Long Branch, NJ, USA
{ecesario,s1201934}@monmouth.edu
[2] ICAR-CNR, Rende, Italy
vinci@icar.cnr.it

Abstract. The growth of data volume collected in urban contexts opens up to their exploitation for improving citizens' quality-of-life and city management issues, like resource planning (water, electricity), traffic, air and water quality, public policy and public safety services. Moreover, due to the large-scale diffusion of GPS and scanning devices, most of the available data are geo-referenced. Considering such an abundance of data, a very desirable and common task is to identify homogeneous regions in spatial data by partitioning a city into uniform regions based on pollution density, mobility spikes, crimes, or on other characteristics. Density-based clustering algorithms have been shown to be very suitable to detect density-based regions, i.e. areas in which urban events occur with higher density than the remainder of the dataset. Nevertheless, an important issue of such algorithms is that, due to the adoption of global parameters, they fail to identify clusters with varied densities, unless the clusters are clearly separated by sparse regions. In this paper we provide a preliminary analysis about how hierarchical clustering can be used to discover spatial clusters of different densities, in spatial urban data. The algorithm can automatically estimate the area of data having different densities, it can automatically estimate parameters for each cluster so as to reduce the requirement for human intervention or domain knowledge.

1 Introduction

In our days, we are experiencing the most rapid growth in urbanization in history. The population which lives in cities is growing from 2.86 billion in 2000 to 4.98 billion in 2030, according to a United Nations report. Thus, the 60% of people in the world is going to live in cities by 2030. Such rapid urbanization is bringing significant environmental, economic and social changes, and also raises new issues in city management, related to public policy, safety services, resource management (electricity, water) and air pollution.

Moreover, a large-scale diffusion of scanning devices, gps, and image processing leads to an abundance of geo-referenced data. Furthermore, more and more Point of Interest (POI) databases are created which annotate spatial objects with categories, e.g. buildings are identified as restaurants, and systems, such as

© Springer Nature Switzerland AG 2020
Y. D. Sergeyev and D. E. Kvasov (Eds.): NUMTA 2019, LNCS 11973, pp. 223–231, 2020.
https://doi.org/10.1007/978-3-030-39081-5_20

Google Earth, already fully support the visualization of POI objects on maps. Considering such an abundance of data, it is becoming crucial their acquisition, integration, and analysis of big and heterogeneous urban information to tackle the major issues that cities face today, including air pollution, energy consumption, traffic flows, human mobility, environmental preservation, commercial activities and savings in public spending [6].

As more and more data become available for a spatial area, it is desirable to identify different functions and roles which different parts of this spatial area play; in particular, a very desirable and common task is to identify homogeneous regions in spatial data and to describe their characteristics, creating high-level summaries for spatial datasets which are valuable for planners, scientists, and policy makers. For example, environmental scientists are interested in partitioning a city into uniform regions based on pollution density and on other environmental characteristics. Similarly, city planners might be interested in identifying uniform regions of a city with respect to the functions they serve for the people who live in or visit this part of a city. Furthermore, policy officers are interested in detecting high crime density areas (or crime hotspots), to better control the city territory in terms of public safety.

Among several spatial analysis approaches, density-based clustering algorithms have been shown to be very suitable to detect density-based regions, i.e. areas in which urban events (i.e., pollution peaks, traffic spikes, crimes) occur with higher density than the remainder of the dataset. In fact, they can detect dense regions within a given geographical area, where shapes of the detected regions are automatically traced by the algorithm without any pre-fixed division in areas. Also, they can find arbitrarily shaped and differently sized clusters, which are considered to be the dense regions separated by low-density regions. Moreover, density-based clustering requires no prior information regarding the number of clusters, which is another positive benefit that makes such methodology suitable for these cases.

An important issue of such density-based algorithms is that, due to the adoption of global parameters, they fail to identify clusters with varied densities, unless the clusters are clearly separated by sparse regions. In fact, they can result in the discovery of several small non significant clusters that actually do no represent dense regions, or they can discover a few large regions that actually are no longer dense as well. In this paper we provide a preliminary analysis about how hierarchical clustering can be used to discover spatial clusters of different densities, in spatial urban data. The algorithm can automatically estimate the area of the data having different densities, it can automatically estimate parameters for each cluster so as to reduce the requirement for human intervention or domain knowledge.

The rest of the paper is organized as follows. Section 2 reports the most important approaches in spatial clustering literature exploiting different densities and the most representative projects in that field of research. Section 3 presents the proposed algorithm by describing its main steps. Section 4 describes a preliminary experimental evaluation, performed on a real-world case study. Finally, Sect. 5 concludes the paper and plans future research works.

2   Related Works

In urban datasets the detection of areas in which events occur with higher density than the remainder of the dataset is becoming a more and more desirable and common task [2,3,5,9,10]. To this purpose, several density-based approaches have been used and we briefly describe here some of the most representative ones.

DBSCAN [7] is the classic density-based algorithm proposed in literature, which builds clusters by grouping data points that are sufficiently dense, where the density associated with a point is obtained by counting the number of points in a region of specified radius ϵ around this point. The main issue of this algorithm is that it does not perform well under multi-density circumstance and requires much subjective intervention in the parameter estimation.

OPTICS [1] (Ordering Points to Identify the Clustering Structure) is an enhanced method upon DBSCAN, which creates an ordering of the objects augmented by reachability distance and makes a reachability-plot out of this ordering. The reachability-plot, which contains information about the intrinsic clustering structure, is the basis for interactive cluster analysis, but it does not produce the clustering result explicitly.

DBSCAN-DLP (Multi-density DBSCAN based on Density Levels Partitioning) [11] partitions a dataset into different density level sets by analyzing the statistical characteristics of its density variation, and then estimates ϵ for each density level set. Finally, DBSCAN clustering is performed on each density level set with corresponding ϵ to get clustering results.

GADAC (A new density-based scheme for clustering based on genetic algorithm) has been proposed in [8] to determine appropriate parameters for DBSCAN. It exploits a genetic algorithm to find clusters of varied densities, by selecting several radius values.

The VDBSCAN [9] (Varied Density Based Spatial Clustering of Applications with Noise) is an algorithm which detects clustering models at different values of densities. Specifically, the approach computes k-dist value for each object (i.e., th minimum distance such that k points are included in the object's neighborhood) and sorts them in ascending order, then make a visualization of the sorted values. The sharp changes in the k-dist plot correspond to a list of radiuses for different density varied clusters.

KDDClus [10] is another density-based clustering approach for an automatic estimation of parameters, to detect clusters of varied densities implementing a bottom-up approach. It estimates the density of a pattern by averaging the distances of all its k-nearest neighbors, and uses 1-dimension clustering on these density values to get a partition of different levels. Then, the algorithm detects radiuses for different densities, and finally DBSCAN is run to find out clusters of varied densities.

3   A Framework for Hierarchical Clustering of Urban Spatial Data

This section describes the main steps of the algorithm we exploit to perform *hierarchical density-based clustering* on spatial urban data. Let be $\mathscr{D}$ a dataset collecting spatial urban data instances, $\mathscr{D} = < x_1, x_2, \ldots, x_N >$, where each x_i is a data tuple described by $< latitude, longitude >$, i.e. coordinates of the place event occurs. The meta-code of the approach, based on the algorithm proposed in [11], is reported in Fig. 1. The algorithm receives in input the dataset $\mathscr{D}$, and returns the discovered knowledge models, i.e., a set of spatial clusters $\mathscr{DR} = \{DR_1, \ldots, DR_K\}$ of different densities.

The algorithm begins by computing, for each point x_i, the *k-nearest neighbor distance* of x_i, given a certain k. This is performed by the COMPUTE-K-DIST$(K, \mathscr{D})$ method, which computes the distance between each $x_i \in \mathscr{D}$ and its kth-nearest neighbor (line L1). It is worth noting that the *k-nearest neighbor distance* value of a certain point x_i can indicate its density appropriately (for more details, see [11]): higher such a distance, lower is the density of points around x_i. As soon as this step is completed, the COMPUTE-DENSITY-VARIATION(K-Dist-List) method computes the *density variation* of each point p_i with respect to p_j (with $i \neq j$) and returns the *density variation list* (line L2). On the basis of the computed density variation values, the PARTITION-DENSITY-VARIATION($Density - Variation - List, \mathscr{D}$) method creates a list of *density level sets* (line L3): a *density level set* consists of points whose densities are approximately the same ([11]), that is, density variations of data points within the same density level set are lower than τ, where τ is a density variation threshold. Doing this, a multi-density dataset can be divided into several density level sets, each of which stands for a density distribution. At this point, the COMPUTE-EPS-VALUES() method computes *coefficient of variability* values (which are used to compare the dispersion of two sample sets) and scatterness values, which are suitable to compute the level-turning line for the ϵ values (line L4). Such values are stored and returned in the $\epsilon - list$, i.e., a list of ϵ values that are estimated as the best values with respect to the different densities in the data ([11]). Finally, for each ϵ in the ϵ-list, the clustering algorithm is executed and the discovered clusters are added to the final cluster set (lines L5–L8). All non-marked points are recognized as noise. The final result consists in a set of spatial clusters, each one representing a event-dense urban region, detected by different ϵ-value settings (i.e., by different densities).

4   Experimental Results

This section describes a preliminary experimental evaluation of the approach described above. To do that, we executed different tests on a real-world dataset collecting geo-referenced crime events occurred in an urban area of Chicago. The goal of our analysis is to perform a comparative analysis of spatial clustering to

```
MDC(𝒟)
  Input:  𝒟 = {x₁,...,xₙ}: a dataset of points;
  Output: 𝒟ℛ = {DR₁,...,DRₖ}: a set of K dense regions;
  L1:  K-Dist-List ← COMPUTE-K-DIST(K, 𝒟)
  L2:  Density-Variation-List ← COMPUTE-DENSITY-VARIATION(K-Dist-List)
  L3:  Density-Level-Sets-List  ←  PARTITION-DENSITY-VARIATION(Density −
       Variation − List, 𝒟)
  L4:  ε-list ← COMPUTE-EPS-VALUES(Density − Level − Sets − List)
  L5:  for each εᵢ = ε₁,...,εₓ in ε-list do
  L6:     CS_{εᵢ} ← DBSCAN(𝒦, εᵢ, 𝒟ℒ𝒮, 𝒟)
  L7:     𝒟ℛ ← 𝒟ℛ ∪ CS_{εᵢ}
  L8:  end for
       return (𝒟ℛ)
```

Fig. 1. Multi-density clustering algorithm

detect dense regions of geo-localized urban events, when a unique global density parameter or multi-density parameters are used.

The data that we used to train the models and perform the experimental evaluation is housed on Plenario [4], a publicly available data search and exploration platform that was developed (and currently managed) by the University of Chicago's Urban Center for Computational and Data. The platform hosts several data sets regarding various city events, i.e., traffic crashes, food inspections, crime events, etc. For the sake of our experimental evaluation, the analysis has been performed on the *'Crimes - 2001 to present'* dataset, a collection of crime events occurred in a large area of Chicago on 2012. The selected area includes different zones of the city, some growing in terms of population, others in terms of business activities, with different crime-densities over their territory (so making it interesting for a comparison between single-density and multi-density spatial analysis). Its perimeter is about $52\ Km$ and its area is approximately $135\ Km^2$. The total number of collected crimes is 100 K, while the average number of crimes per week is 2,275. The total size of this data set is 123 MB.

Figure 2 shows the number of data points having at least min_pts neighbors within the reachability radius ϵ, varying the specific density (pts/m^2) (in log scale) and the min_pts parameter. Given the min_pts and $density$ values, the ϵ parameter is computed as $density = min\_pts/(\pi * \epsilon^2)$. In other words, the chart shows how many core points are counted considering the chosen ϵ and min_pts parameters. The chart shows how the points density is sensitive to the min_pts parameter, and how min_pts and ϵ parameter values impact on the results of the density-based clustering algorithm. In particular two phenomena can be observed: (i) the number of data points having a specific density decreases as min_pts increases and (ii) for each min_pts value, there is a density limit beyond which the number of core points does not decrease. The first phenomenon shows how the min_pts value has to be carefully chosen to make valuable the results of the clustering algorithm. The second phenomenon is due to the nature of the exploited dataset, where the granularity of spatial coordinates is of 1 foot (0.3048 m), thus having data points which share the same coordinates, and having neighbors at a zero distance. The same phenomena can be observed also in Fig. 3, which depicts the number of core points with respect to ϵ and for several values of min_pts.

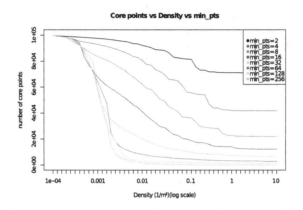

Fig. 2. Number of core points, w.r.t. the specific density (pts/m^2) and min_pts.

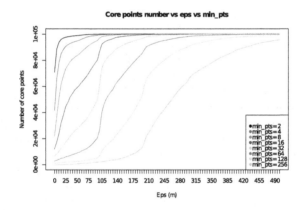

Fig. 3. Number of core points, w.r.t. the ϵ and min_pts parameters.

Figure 4 shows a preliminary result of our analysis, where dense regions are discovered exploiting global or local density parameter values. In particular, Fig. 4(a) shows the results achieved by fixing a global $\epsilon = 150$ m and $min\_pts = 64$, where each region is represented by a different colour. Interestingly, the algorithm detects a set of significant regions clearly recognizable through different colours: a large region (in red) in the central part of the area along with seven smaller areas (in green, blue and light-blue) on the left and right side, corresponding to zones with the highest concentration of geo-localized

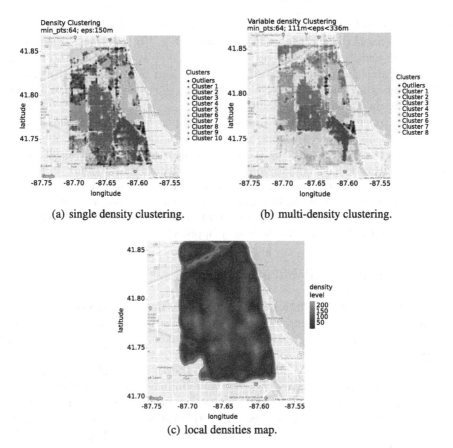

(a) single density clustering.　　　　　(b) multi-density clustering.

(c) local densities map.

Fig. 4. Dense regions discovered in an area of Chicago, detected exploiting global or variable density parameter values. (Color figure online)

events. Many other smaller regions representing very local high-density zones are distributed in the whole area. When a unique global parameter is used, the density-based algorithm can fail to identify clusters with varied densities, unless the clusters are clearly separated by sparse regions. In fact, they can result in the discovery of several small non significant clusters that actually do no represent dense regions, or they can discover a few large regions that actually are no longer dense as well. On the other side, Fig. 4(b) shows the result achieved when different density parameter values are used. Interestingly, the algorithm detects a set of eight significant regions clearly recognizable through different colors. Comparing the two images, we can observe a large red region of the central part of Fig. 4(a); on the other side, in Fig. 4(b) such a region is split in three clusters, each one with different characteristics and different densities.

Figure 4(c) shows a density map of the same dataset. There is a set of several significant regions clearly recognizable through different color intensities, where darker (lighter) intensities of blue represent lower (higher) densities of geo-localized events in the selected area. Moreover, in Fig. 4(b) and (c) we can observe several high-density areas (in red, blue, cyan, yellow and green in Fig. 4(b); in the central area, in low-density areas at the top and bottom in Fig. 4(c)), that group points that can be analyzed as an homogeneous set, and that are not detected by the global density algorithm.

We can conclude that, when a unique global parameter is used, the density-based algorithm can fail to identify clusters with varied densities, unless the clusters are clearly separated by sparse regions. In fact, they can result in the discovery of several small (and non significant) clusters that actually do no represent dense regions, or they can discover a few large regions that actually are no longer dense as well. A multi-density approach seems to be more suitable for clustering and analyzing urban spatial data, as it can detect regions that are more meaningful w.r.t. the global density approach.

5   Conclusions and Future Work

Spatial analysis of urban data is becoming a very desirable and common task, aimed at describing building high-level summaries for spatial datasets which are valuable for planners, scientists, and policy makers. Density-based clustering algorithms have been shown to be very suitable to detect density-based regions, but due to the adoption of global parameters, they fail to identify clusters with varied densities. This paper has provided a preliminary analysis about a hierarchical multi-density clustering algorithm that can be used to discover clusters in spatial urban data. A preliminary experimental evaluation of the approach, performed on a real-world dataset, has shown a comparative analysis of the results achieved when a unique global density parameter and multi-density parameters are used. The initial results reported in the paper are encouraging and show several benefits when a multi-density approach is used. As future work, we will investigate this issue more in detail and we will perform a more extensive experimental evaluation.

References

1. Ankerst, M., Breunig, M.M., Kriegel, H.P., Sander, J.: OPTICS: ordering points to identify the clustering structure. ACM Sigmod Rec. **28**, 49–60 (1999)
2. Catlett, C., Cesario, E., Talia, D., Vinci, A.: A data-driven approach for spatio-temporal crime predictions in smart cities. In: 2018 IEEE International Conference on Smart Computing (SMARTCOMP), pp. 17–24. IEEE (2018)
3. Catlett, C., Cesario, E., Talia, D., Vinci, A.: Spatio-temporal crime predictions in smart cities: a data-driven approach and experiments. Pervasive Mob. Comput, **53**, 62–74 (2019)
4. Catlett, C., et al.: Plenario: an open data discovery and exploration platform for urban science. IEEE Data Eng. Bull. **37**(4), 27–42 (2014)

5. Cesario, E., Talia, D.: Distributed data mining patterns and services: an architecture and experiments. Concurrency Comput. Pract. Experience **24**(15), 1751–1774 (2012)
6. Cicirelli, F., Guerrieri, A., Mastroianni, C., Spezzano, G., Vinci, A. (eds.): The Internet of Things for Smart Urban Ecosystems. IT. Springer, Cham (2019). https://doi.org/10.1007/978-3-319-96550-5
7. Ester, M., Kriegel, H.P., Sander, J., Xu, X., et al.: A density-based algorithm for discovering clusters in large spatial databases with noise. In: KDD, vol. 96, pp. 226–231 (1996)
8. Lin, C.Y., Chang, C.C., Lin, C.C.: A new density-based scheme for clustering based on genetic algorithm. Fundamenta Informaticae **68**(4), 315–331 (2005)
9. Liu, P., Zhou, D., Wu, N.: VDBSCAN: varied density based spatial clustering of applications with noise. In: 2007 International Conference on Service Systems and Service Management, pp. 1–4. IEEE (2007)
10. Mitra, S., Nandy, J.: KDDClus: a simple method for multi-density clustering. In: Proceedings of International Workshop on Soft Computing Applications and Knowledge Discovery (SCAKD 2011), Moscow, Russia, pp. 72–76. Citeseer (2011)
11. Xiong, Z., Chen, R., Zhang, Y., Zhang, X.: Multi-density dbscan algorithm based on density levels partitioning. J. Inf. Comput. Sci. **9**(10), 2739–2749 (2012)

Improving Efficiency in Parallel Computing Leveraging Local Synchronization

Franco Cicirelli⦿, Andrea Giordano$^{(\boxtimes)}$⦿, and Carlo Mastroianni⦿

ICAR-CNR, via P. Bucci, cubo 8/9 C, Rende, CS, Italy
`{franco.cicirelli,andrea.giordano,carlo.mastroianni}@icar.cnr.it`

Abstract. In a parallel computing scenario, a complex task is typically split among many computing nodes, which are engaged to perform portions of the task in a parallel fashion. Except for a very limited class of application, computing nodes need to coordinate with each other in order to carry out the parallel execution in a consistent way. As a consequence, a synchronization overhead arises, which can significantly impair the overall execution performance. Typically, synchronization is achieved by adopting a centralized synchronization barrier involving all the computing nodes. In many application domains, though, such kind of global synchronization can be relaxed and a lean synchronization schema, namely local synchronization, can be exploited. By using local synchronization, each computing node needs to synchronize only with a subset of the other computing nodes. In this work, we evaluate the performance of the local synchronization mechanism when compared to the global synchronization approach. As a key performance indicator, the efficiency index is considered, which is defined as the ratio between useful computation time and total computation time, including the synchronization overhead. The efficiency trend is evaluated both analytically and through numerical simulation.

Keywords: Parallel computing · Efficiency · Synchronization · Max-Plus Algebra

1 Introduction

In order to parallelize the computation needed to solve a problem, different portions of the problem are assigned to different computing nodes which process data in parallel. Important application fields in which parallel computing is of outmost important to achieve significant improvements in terms of execution and efficiency are: biology, geology, hydrology, logistics and transportation, social sciences, smart electrical grids (see [1–4]). An interesting application field where parallel computing is gaining importance is the urban-computing one: in this context, it is necessary to analyze as different aspects as the mobility of people or vehicles, the air quality, the consumption of water and electricity, and so on.

© Springer Nature Switzerland AG 2020
Y. D. Sergeyev and D. E. Kvasov (Eds.): NUMTA 2019, LNCS 11973, pp. 232–242, 2020.
https://doi.org/10.1007/978-3-030-39081-5_21

The objective is to improve the quality of the services offered to the citizens (see [5–7]). Two further emerging fields are the "Internet of Things" (IoT) [8,9] and some alternatives to the classic paradigm of Cloud Computing, i.e., the Fog Computing and Edge Computing [10,11], where the computation is naturally parallel and is brought closer to the user and to the data.

A common classification of parallel computing includes the notion of "embarrassingly parallel" [12]. In this case, computation at the single nodes can be performed in isolation. A more common case is when parallel tasks exchange data during computation, and therefore the advancement of execution at the different nodes must be coordinated, or synchronized. Synchronization [13] means that, at certain time instants, one node must wait for the data coming from other nodes before proceeding to the next piece of computation. In this paper, we consider the very common case of step-based computation: the computation is split in work units called "steps", and synchronization occurs at the end of each step.

It is important to define how and when the computing nodes synchronize with each other. A useful distinction is between *global* and *local* synchronization [14]. Synchronization is global, or *all–to–all*, when each node can start the execution at a given step only after the node itself and all the other nodes have completed their execution of the step before. In many application domains, though, such kind of global synchronization can be relaxed and another synchronization schema, namely local synchronization, can be exploited. With local synchronization, each computing node needs to synchronize only with a subset of the other computing nodes, which are referred to as "adjacent" or "neighbor" nodes.

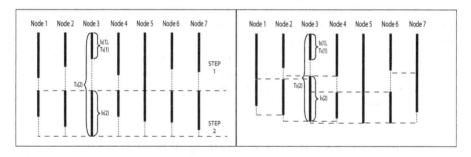

Fig. 1. Dynamics of seven nodes for two steps using global synchronization (left) and local synchronization (right). The solid vertical lines represent the execution times, the dashed vertical lines are the waiting times and the horizontal dashed lines represent the synchronization points.

Figure 1 shows an example of global and local synchronization for a system composed of seven nodes, for two consecutive steps. In the case of global synchronization, at each step, all the nodes must wait for the slowest one before advancing to the next step (see figure on the left). More in detail, node 5 is the slowest at step 1 while node 3 is the slowest at step 2. The scenario of local synchronization is depicted in the right part of Fig. 1, in the case that each node has

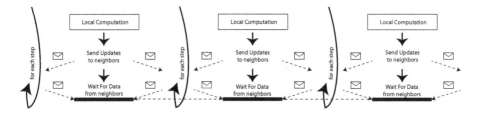

Fig. 2. Execution loop under local synchronization.

two neighbors, named left and right neighbors. Each node advances to the next step of computation when the node itself and its two neighbors have completed the execution of the current step.

Figure 2 shows the loop executed by each computing node, at each step, when adopting the local synchronization pattern. The loop in composed of three phases: (i) the node executes the local computation, related to the specific region for the current step; (ii) the node sends data to its neighbour nodes; (iii) the node waits for the analogous data coming from its neighbours.

The rest of the paper is organized as follows. Section 2 presents a mathematical model for global and local synchronization, which allows to define the computation time and the overhead time spent for synchronization. For the sake of simplicity, the model of local synchronization considers the case in which each node synchronizes only with two neighbors. Anyway, the achieved results are general and hold also with a greater number of neighbors provided it is a constant value. Section 3 presents an asymptotic analysis that shows that the mean computation time per step is bounded with local synchronization and unbounded with global synchronization. Section 4 reports an analysis of efficiency, defined as the ratio between useful computation time and total computation time, including the overhead time due to synchronization. Section 5 reports some numerical results obtained with simulation, which confirm the provided analytical assessment. Finally, Sect. 6 concludes the paper.

2   A Model for Local and Global Synchronziation

Let us denote by N the number of nodes, by $l_i(k)$ the time needed by node i, $1 \leq i \leq N$, to execute the local computation at the step k, and by $T_i(k)$ the time elapsed at node i from the beginning of the computation (i.e., start of the step 1) until the end of the step k. As an example, in Fig. 1, assuming that the values of $l_i(k)$ are the same for the cases of global and local synchronization, we show the corresponding execution advancements. It can be seen that the times $T_i(k)$ tend to be shorter in the case of local synchronization.

For the sake of ease of exposition, we consider a simple case of the model for which communication times are negligible and each node i needs to synchronize with two neighbors, indexed as $i - 1$ and $i + 1$, except for nodes 1 and N that have only one neighbor. In addition, we assume that the computation time $l_i(k)$

depends only on the node i but does not depend on the step k, and that $l_i(k) = l_i$, $1 \leq i \leq N$, are i.i.d. random variables with average $\bar{l}_i$.

In the case of global synchronization, each step k begins contemporarily at each node when the slowest node at the step $k-1$ has completed its computation. So we have:

$$T_i(k) = \sum_{j=1}^{k-1} \max(l_1, \ldots, l_N) + l_i, \ k \geq 1, \tag{1}$$

In the case of local synchronization we have the recursive equation:

$$T_i(k) = \max(T_i(k-1), T_{i-1}(k-1), T_{i+1}(k-1)) + l_i, \ k \geq 1, \ 1 \leq i \leq N. \tag{2}$$

in which we assume $T_i(0) = 0$, $1 \leq i \leq N$ and $T_0(k) = T_{N+1}(k) = 0$, $k \geq 0$.

In both cases, the random sequence $\{\boldsymbol{T}(k)/k, \ k \geq 1\}$, where $\{\boldsymbol{T}(k) = (T_1(k), \ldots T_N(k))\}$ falls into the framework of stochastic equations as described in [15]. Using the results from [15] it can be shown that when N is finite and $\{\boldsymbol{l}(k) = (l_1(k), \ldots, l_N(k)), \ k \geq 1\}$ are independent, the sequence $\{\boldsymbol{T}(k)/k, \ k \geq 1\}$ is ergodic and stable.

2.1   Computation and Overhead Times

In the following analysis, it is appropriate to distinguish between the computation time (the time spent by nodes for useful computation) and the overhead time (the waiting time needed for synchronization).

The computation time at step k, defined as the sum of computation times of all the nodes at step k, and denoted as $t_S(k)$, is equal to:

$$t_S(k) = l_1(k) + l_2(k) + \cdots + l_N(k) \tag{3}$$

The average value of the computation step is:

$$\bar{t}_S(k) = \bar{l}_1 + \bar{l}_2 + \cdots + \bar{l}_N = N \cdot \bar{l} \tag{4}$$

where $\bar{l}$ is the average of the average computation times $\bar{l}_i$ of the N nodes. With the assumptions specified above, $\bar{t}_S(k)$ is the same at each step.

The overhead time at step k is defined as the sum of the times $o_i(k)$ spent by each node for synchronization, after completing the computation at step $k-1$ and before starting the computation at step k. It is equal to:

$$t_O(k) = o_1(k) + o_2(k) + \cdots + o_N(k) \tag{5}$$

In the case of *global synchronization*, the expression for the overhead time of node i at step k, $o_i(k)$, must take into account the fact that node i needs to wait until all the other nodes have finished their computation at step $k-1$. Since all the nodes start their computation precisely at the same time, the overhead experienced by node i at step k is equal to:

$$o_i(k) = \max_{j=1..N} (l_j) - l_i \tag{6}$$

and the expression for the overall overhead time is:

$$to(k) = \sum_{i=1}^{N} o_i(k) = N \cdot \left(\max_{j=1..N} l_j \right) - t_S(k) \tag{7}$$

In the case of *local synchronization*, $o_i(k)$ is the time that node i needs to wait until its two neighbors have completed their computation of step $k-1$, and is equal to:

$$o_i(k) = max\left(T_{i-1}(k-1), T_i(k-1), T_{i+1}(k-1)\right) - T_i(k-1) \tag{8}$$

Now that we have defined the computation and overhead time for both global and local synchronization, we define as $t_L(k) = t_S(k) + t_O(k)$ the full length of the step k, including both the actual computation and the overhead. Recalling the definition of $T_i(k)$, it follows that the sum of the full lengths of computation $t_L(k)$, from step 1 up to step k, is equal to the sum of $T_i(k)$ for all the nodes i:

$$\sum_{i=1}^{N} T_i(k) = \sum_{j=1}^{k} t_L(j) = \sum_{j=1}^{k-1} t_L(j) + t_L(k) = \sum_{i=1}^{N} T_i(k-1) + t_L(k) \tag{9}$$

From the above expression we find another expression for $t_L(k)$:

$$t_L(k) = \sum_{i=1}^{N} T_i(k) - \sum_{i=1}^{N} T_i(k-1) \tag{10}$$

3   Asymptotic Analysis of Global and Local Synchronization

In the following, analytical results are given showing that with global synchronization the mean computation time per step is unbounded when the number of nodes increases. Conversely, with local synchronization the mean computation time per step is bounded irrespective of the number of nodes. In the following, the mean computation time per step is also referred to as the cycle time, $T_{cycle} = \lim_{k \to \infty} \frac{1}{k} \mathbf{E}(T_i(k))$.

3.1   Analysis of Global Synchronization

With global synchronization, represented in (1), the analysis can be performed by exploiting the well-known results coming from order statistics and extreme value theory [16–18]. In the case that the random variable l_i has a continuous distribution with the support[1] on a semi-infinite interval, we can state that, as

---

[1] In the case that l_i are independent (not necessarily identically distributed) random variables with a continuous distribution having support on a bounded interval, the mean computation time $\lim_{k \to \infty} \mathbf{E}(T_i(k))/k$ is always a constant, irrespective of the number of nodes N.

the number of nodes N grows, the cycle time $\lim_{k \to \infty} \frac{1}{k} \mathbf{E}(T_i(k))$ grows as well, and in the limit as $N \to \infty$ we have that:

$$\lim_{k \to \infty} \frac{1}{k} \mathbf{E}(T_i(k)) = \mathbf{E}(\max(l_1, \ldots, l_N)) \to \infty.$$

For example, in the case that l_i are i.i.d random variables distributed exponentially, having parameter λ and average $1/\lambda$, for sufficiently large N we have[2]:

$$\lim_{k \to \infty} \frac{1}{k} \mathbf{E}(T_i(k)) \approx (1/\lambda) (\ln N + 0.5772)$$

Similar expressions are not available in the general case, but hold for a large class of distributions. Such a class includes the distributions [19] having pure-exponential and non-pure exponential tails, like the gamma distribution.

With global synchronization, if the computation times are random with unbounded support, then as the number of nodes N increases, on average, the time to complete the next computation step also becomes unbounded.

3.2   Analysis of Local Synchronization

With local synchronization, a node can start the next computation step after a finite number of neighbors finish their computations of the current step. In this scenario, a different conclusion is obtained: the cycle time $\lim_{k \to \infty} \frac{1}{k} \mathbf{E}(T_i(k))$ becomes bounded irrespective of the number of nodes N. In what follow, for simplicity we consider the case of synchronization with only two neighbors, i.e., the scenario described in (2). It is convenient to describe the model (2) by adopting the framework of discrete event dynamic systems, and the evolution can be captured in terms of the Max-Plus algebra. Subsequently, the well-known results of the Max-Plus theory can be used to study the cycle time $\lim_{k \to \infty} \frac{1}{k} \mathbf{E}(T_i(k))$.

The Max-Plus theory defines an algebra over the vector space $\mathbb{R} \cup \{-\infty\}$ with the operation $x \oplus y$ defined as $\max(x, y)$ and the operation $x \otimes y$ defined as $x + y$.

The vector $\mathbf{T}(k) = (T_1(k), \ldots, T_N(k))$ collects the values of $T_i(k)$ for each node i at the step k, where $\mathbf{T}(0)$ is the column-vector of zeros. The $N \times N$ matrix $M(k) = [M(k)]_{ij}$, $1 \leq i, j \leq N$, is defined as:

$$\mathbf{M}(k) = \begin{pmatrix} l_1 & l_1 & -\infty & -\infty & \ldots & -\infty & -\infty \\ l_2 & l_2 & l_2 & -\infty & \ldots & -\infty & -\infty \\ -\infty & l_3 & l_3 & l_3 & \ldots & -\infty & -\infty \\ -\infty & -\infty & l_4 & l_4 & \ldots & -\infty & -\infty \\ \vdots & \vdots & \vdots & \vdots & \ddots & \ldots & \ddots \\ -\infty & -\infty & -\infty & -\infty & \ldots & l_{N-1} & l_{N-1} \\ -\infty & -\infty & -\infty & -\infty & \ldots & l_N & l_N \end{pmatrix}.$$

---

[2] It is well-known that the right part is the approximation of the expected maximum of N i.i.d. random variables with exponential distribution equal to $\mu \sum_{i=1}^{N} i^{-1}$, with $\sum_{i=1}^{N} i^{-1}$ being the Nth harmonic number.

With this notation, equation (2) can be rewritten as

$$T(k+1) = \mathbf{M}(k) \otimes T(k), \ k \geq 0, \tag{11}$$

where the matrix-vector product is defined by:

$$[\mathbf{M}(k) \otimes T(k)]_i = \max_{1 \leq j \leq N} ([\mathbf{M}(k)]_{ij} + T_j(k))$$

Indeed, this expression equals to:

$$[T(k+1)]_i = [\mathbf{M}(k) \otimes T(k)]_i = l_i \otimes T_{i-1}(k) \oplus l_i \otimes T_i(k) \oplus l_i \otimes T_{i+1}(k) =$$
$$max(T_{i-1}(k) + l_i, T_i(k) + l_i, T_{i+1}(k) + l_i) = max(T_{i-1}(k), T_i(k), T_{i+1}(k)) + l_i$$

where $T_0(k) = T_{N+1}(k) = 0$.

Now we can make use of the well-known asymptotic results from the Max-Plus theory (see, for example, [20]). The matrix $\mathbf{T}(k)$ has at least one finite entry on each row, which is the necessary and sufficient condition for $T_j(k)$ to be finite. From [20, Lemma 6.1] we find that there exists $\gamma > 0$ such that $\lim_{k \to \infty} \frac{1}{k} \mathbf{E}(T_i(k)) = \gamma$, $1 \leq i \leq N$. In the case when l_i are stochastically bounded by a single variable (say L), having moment generating function (say $L(s)$), the upper bound for the value of γ is (see [20, Proposition 6.2]):

$$\gamma \leq \inf\{x > \mathbf{E}(L) \text{ such that } M(x) > \ln 3\}, \tag{12}$$

where $M(x) = \sup_{\theta \in \mathbb{R}}(\theta x - \ln L(\theta))$ and $\mathbf{E}(L) = L'(0)$.

This result tells us that in the case of local synchronization the mean computation time remains finite for any number of nodes N.

4   Estimation of Efficiency

The aim of this section is to show that global and local synchronizations exhibit very different behaviors in terms of efficiency, where the efficiency is the ratio of the useful computation and the total time (useful computation time plus overhead time).

To show this let us start defining the overall computation and overall overhead times from the beginning of execution up to step k:

$$T_S(k) = \sum_{j=1}^{k} t_S(j) \qquad T_O(k) = \sum_{j=1}^{k} t_O(j) \tag{13}$$

where $t_S(k)$ and $t_S(k)$ are defined in (3) and (5). The efficiency is defined as:

$$E_f(k) = \frac{\bar{T}_S(k)}{\bar{T}_S(k) + \bar{T}_O(k)} = \frac{1}{1 + \bar{T}_O(k)/\bar{T}_S(k)} \tag{14}$$

We are interested in the efficiency when both k and N grow indefinitely. In the following subsections we see that:

– with global synchronization, the efficiency tends to zero as the number of nodes grows;
– with local synchronization, the efficiency has a non-zero lower bound.

The consequences of these different trends are remarkable not only with very large values of N, but also for ordinary values of N, which means that the use of local synchronization leads to significant improvements in practical scenarios.

Efficiency with Global Synchronization. With global synchronization, the ratio $\bar{T}_O(k)/\bar{T}_S(k)$ can be derived from (4), (7) and (13) and is equal to:

$$\frac{\bar{T}_O(k)}{\bar{T}_S(k)} = \frac{k \cdot N \cdot E\left[\max_{j=1..N} \bar{l}_j\right] - \bar{T}_S(k)}{\bar{T}_S(k)} \tag{15}$$

From (4) we see that $\bar{T}_S(k) = k \cdot N \cdot \bar{l}$, and the above expression becomes:

$$\frac{\bar{T}_O(k)}{\bar{T}_S(k)} = \frac{E\left[\max_{j=1..N} \bar{l}_j\right]}{\bar{l}} - 1 \tag{16}$$

The value of the denominator is clearly a bounded value, for each value of N, since it is the average of N finite values. On the other hand, the value of the numerator is unbounded, as N increases, for a wide set of probability distributions that can be assumed to characterize the computation time l_i. It follows that the ratio (16) is unbounded and that the efficiency – see expression (14) – tends to zero when increasing the number of nodes. In other words, when N increases, the overhead becomes prevalent with respect to the actual computation time.

For example, if the l_i are i.i.d. and have an exponential distribution, with parameter λ and average $1/\lambda$, the maximum of N such distributions is known to be approximated by the expression $1/\lambda \cdot H(N) \simeq 1/\lambda \cdot ln(N)$, where $H(N)$ is the harmonic number $\sum_{i=1..N}(1/i)$.

In this case, the efficiency can be expressed as:

$$E_f(k) \simeq \frac{1}{\ln(N)} \tag{17}$$

This expression means that the efficiency decreases and tends to zero as the number of nodes increases. When the local computation time is distributed with any other distribution of the exponential family, the expression for the maximum of a set of i.i.d random variables will be different but the trend will still be logarithmic.

Efficiency with Local Synchronization. In the case of local synchronization, the value of efficiency for $k \to \infty$ can be computed as:

$$E_f = \lim_{k \to \infty} E_f(k) = \lim_{k \to \infty} \frac{\bar{T}_S(k)}{\bar{T}_O(k) + \bar{T}_S(k)} = \tag{18}$$

$$\lim_{k \to \infty} \frac{1/k \cdot \sum_{j=1}^{k} \bar{t}_S(j)}{1/k \cdot \sum_{i=1}^{N} \mathbf{E}(T_i(k))} = \tag{19}$$

$$\lim_{k \to \infty} \frac{1/k \cdot k \cdot N \cdot \bar{l}}{\sum_{i=1}^{N} \mathbf{E}(T_i(k))/k} = \tag{20}$$

$$\frac{N \cdot \bar{l}}{\sum_{i=1}^{N} \lim_{k \to \infty} \mathbf{E}(T_i(k))/k} = \tag{21}$$

$$\frac{N \cdot \bar{l}}{N \cdot T_{cycle}} \geq \frac{\bar{l}}{\gamma} \tag{22}$$

The expression (19) comes by applying expression (9) at the denominator, and then multiplying the numerator and the denominator by $1/k$. To obtain expression (20) we apply (4), and in (21) we exchange the limit and the summation at the denominator. To obtain the left side of (22) we rely on Max-Plus that ensures that the value of the limit, the cycle time, is the same at each node, see Sect. 3.2. The final inequality applies because we know by the Max-Plus theory that the cycle time T_{cycle} is finite and has un upper bound γ, i.e., $T_{cycle} \leq \gamma$.

The conclusion is that, in the case of local synchronization, the efficiency has a non-zero lower bound, irrespective of the number of nodes.

5 Numerical Results

In the following we report some numerical results both for global and local synchronizations under negligible communication times and i.i.d. computation times l_i. We used Matlab to simulate the computational behavior modeled by (1) and (2) with different values of the number of nodes N. For l_i we considered many types of distributions that are well-suited workload models in parallel computing systems [21] (exponential, hyper-gamma, lognormal, etc.). We report here the results obtained with the exponential distribution. However, we found that the behavior described in the following applies also with the other distributions.

The performance is assessed by computing the efficiency using a single simulation run with $k = 10000$ and the batch-means method. We consider the case in which l_i has an exponential distribution with the mean equal to 1.0. Figure 3 shows the values of the achieved efficiency versus the number of nodes N in the case of global and local synchronization. The figure also reports the bound on the efficiency obtained with the Max-Plus algebra (see Expression (22)) and the value of efficiency obtained for $N = 1000$ nodes.

The experimental values are consistent with the theoretical bounds discussed in the previous sections, and it can be seen that the use of local synchronization allows the efficiency to be notably increased with respect to global synchronization. In particular, it is confirmed that with global synchronization the value of efficiency decreases towards zero, while with local synchronization the efficiency is always higher than the theoretical bound, which in this case is 0.3.

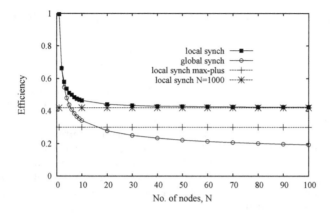

Fig. 3. Values of efficiency as function of N under local and global synchronization. The local computation times have exponential distribution. The plot shows the bound obtained with Max-Plus algebra and the numerical value with $N = 1000$.

From the figure it can be seen that a "practical" bound equal to about 0.41 is already reached when the number of nodes is 50 or larger. It is also interesting to notice that the bound on efficiency derived with Max-Plus algebra, although underestimated, is still larger than the value of efficiency obtained with global synchronization with 20 or more nodes.

From the unreported experiments performed by using different distributions of the computation time we have seen that the advantage of local synchronization increases with the variance of the adopted distribution.

6   Conclusions

In this paper, we analyzed and compared the performance of two different approaches for synchronizing the execution of parallel computation, namely local synchronization and global synchronization. The efficiency, which is given by the ratio of useful computation time and the total computation time, including the overhead, is used as a key performance indicator. We found that, as the number of nodes increases, the efficiency tends to zero in the case of global synchronization, while it has a non-zero lower bound in the case of local synchronization. This result was achieved by using the Max-Plus algebra, and it was confirmed experimentally though numerical simulation.

References

1. Cicirelli, F., Forestiero, A., Giordano, A., Mastroianni, C.: Transparent and efficient parallelization of swarm algorithms. ACM Trans. Auton. Adapt. Syst. **11**(2), 14 (2016)
2. Cicirelli, F., Forestiero, A., Giordano, A., Mastroianni, C.: Parallelization of space-aware applications: modeling and performance analysis. J. Netw. Comput. Appl. **122**, 115–127 (2018)

3. Gong, Z., Tang, W., Bennett, D.A., Thill, J.C.: Parallel agent-based simulation of individual-level spatial interactions within a multicore computing environment. Int. J. Geogr. Inf. Sci. **27**(6), 1152–1170 (2013)
4. Tang, W., Bennett, D.A., Wang, S.: A parallel agent-based model of land use opinions. J. Land Use Sci. **6**(2–3), 121–135 (2011)
5. Mastroianni, C., Cesario, E., Giordano, A.: Efficient and scalable execution of smart city parallel applications. Concurrency Comput. Pract. Experience **30**(20), e4258 (2018)
6. Zheng, Y., Capra, L., Wolfson, O., Yang, H.: Urban computing: concepts, methodologies, and applications. ACM Trans. Intell. Syst. Technol. **5**(3), 1–55 (2014)
7. Blecic, I., Cecchini, A., Trunfio, G.A., Verigos, E.: Urban cellular automata with irregular space of proximities. J. Cell. Automata **9**(2–3), 241–256 (2014)
8. Atzori, L., Iera, A., Morabito, G.: The internet of things: a survey. Comput. Netw. **54**(15), 2787–2805 (2010)
9. Lee, I., Lee, K.: The Internet of Things (IoT): applications, investments, and challenges for enterprises. Bus. Horiz. **58**(4), 431–440 (2015)
10. Krishnan, Y.N., Bhagwat, C.N., Utpat, A.P.: Fog computing–network based cloud computing. In: 2nd IEEE International Conference on Electronics and Communication Systems (ICECS), pp. 250–251 (2015)
11. Hu, P., Dhelim, S., Ning, H., Qiu, T.: Survey on fog computing: architecture, key technologies, applications and open issues. J. Netw. Comput. Appl. **98**, 27–42 (2017)
12. Ekanayake, J., Fox, G.: High performance parallel computing with clouds and cloud technologies. In: Avresky, D.R., Diaz, M., Bode, A., Ciciani, B., Dekel, E. (eds.) CloudComp 2009. LNICSSTE, vol. 34, pp. 20–38. Springer, Heidelberg (2010). https://doi.org/10.1007/978-3-642-12636-9_2
13. Fujimoto, R.: Parallel and Distributed Simulation Systems. Wiley, New York (2000)
14. Cicirelli, F., Forestiero, A., Giordano, A., Mastroianni, C., Razumchik, R.: Global and local synchronization in parallel space-aware applications. In: 32nd European Conference on Modelling and Simulation, ECMS 2018, Wilhelmshaven, Germany, May 2018
15. Borovkov, A.A.: Ergodicity and stability theorems for a class of stochastic equations and their applications. Theory Probab. Appl. **23**(2), 227–247 (1979)
16. David, H.A., Nagaraja, H.N.: Order Statistics, 3rd edn. Wiley, New York (2003)
17. Ang, A.S., Tang, W.: Probability Concepts in Engineering Planning and Design, vol. II. Rainbow Bridge (1984)
18. Madala, S., Sinclair, J.B.: Performance of synchronous parallel algorithms with regular structures. IEEE Trans. Parallel Distrib. Syst. **2**(1), 105–116 (1991)
19. Whitt, W., Crow, C., Goldberg, D.: Two-moment approximations for maxima. Oper. Res. **55**(3), 532–548 (2007)
20. Baccelli, F., Konstantopoulos, P.: Estimates of cycle times in stochastic petri nets. In: Karatzas, I., Ocone, D. (eds.) Applied Stochastic Analysis, pp. 1–20. Springer, Heidelberg (1992). https://doi.org/10.1007/BFb0007044
21. Lublin, U., Feitelson, D.G.: The workload on parallel supercomputers: modeling the characteristics of rigid jobs. J. Parallel Distrib. Comput. **63**(11), 1105–1122 (2003)

A General Computational Formalism
for Networks of Structured Grids

Donato D'Ambrosio[1]([⊠])[iD], Paola Arcuri[1][iD], Mario D'Onghia[1][iD],
Marco Oliverio[1][iD], Rocco Rongo[1][iD], William Spataro[1][iD], Andrea Giordano[2][iD],
Davide Spataro[3][iD], Alessio De Rango[4][iD], Giuseppe Mendicino[4][iD],
Salvatore Straface[4][iD], and Alfonso Senatore[4][iD]

[1] DEMACS, University of Calabria, Rende, Italy
{donato.dambrosio,rocco.rongo,william.spataro}@unical.it
arcuripaola93@gmail.com, mariodonghia93@gmail.com, marcopecos@gmail.com
[2] ICAR-CNR, Rende, Italy
giordano@icar.cnr.it
[3] ASML, Eindhoven, The Netherlands
davide.spataro@asml.com
[4] DIATIC, University of Calabria, Rende, Italy
{alessio.derango,giuseppe.mendicino,
salvatore.straface,alfonso.senatore}@unical.it

Abstract. Extended Cellular Automata (XCA) represent one of the most known parallel computational paradigm for the modeling and simulation of complex systems on stenciled structured grids. However, the formalism does not perfectly lend itself to the modeling of multiple automata were two or more models co-evolve by interchanging information and by synchronizing during the dynamic evolution of the system. Here we propose the Extended Cellular Automata Network (XCAN) formalism, an extension of the original XCA paradigm in which different automata are described by means of a graph, with vertices representing automata and inter-relations modeled by a set of edges. The formalism is applied to the modeling of a theoretical 2D/3D coupled system, where space/time variance and synchronization aspects are pointed out.

Keywords: Extended Cellular Automata Network · Modeling · Stenciled structured grid · Direct acyclic graph · Space/time granularity · Synchronization

1 Introduction

Mesh algorithms are widely used in Science and Engineering for modeling and simulating a wide variety of complex phenomena, e.g. in Computational Fluid Dynamics [2,17]. In this field, even if unstructured grids are gaining popularity (see e.g. [38,43]), structured grids are still the most utilized since they generally require less computing power and memory, often allow for better convergence,

© Springer Nature Switzerland AG 2020
Y. D. Sergeyev and D. E. Kvasov (Eds.): NUMTA 2019, LNCS 11973, pp. 243–255, 2020.
https://doi.org/10.1007/978-3-030-39081-5_22

besides being more suitable for the execution on modern many-core accelerators like GPUs due to data locality [3, 7–9, 28, 39].

Cellular Automata probably represent one of the most known example of parallel computational paradigms for *stenciled structured grids* (i.e., for grids on which a spatial *neighborhood relation* is defined). Originally proposed by John von Neumann to study self-reproduction in biological systems [34], the model is based on an infinite grid of cells with integer coordinates in which each cell changes state based on a local law of evolution (*transition function*). Cellular Automata are universal computational models [4, 10, 42], widely adopted in both theoretical studies [29, 35, 44], and applications [1, 21, 27, 31]. Nevertheless, they do not straightforwardly lend themselves to the modeling of some complex systems like, for instance, lava flows [11]. As a matter of fact, in this case it is not straightforward to simulate lava feeding to vents in terms of *local interactions*, or to perform evaluations regarding the global state of the system, for instance for evaluating if the phenomenon evolution is terminated (i.e., if lava has solidified everywhere). For this reason, Di Gregorio and Serra proposed an extension of the original von Neumann's cellular model, known as Multi-component, Complex or even Extended Cellular Automata (XCA) [26]. Even if computationally equivalent to the original CA, XCA permit to relax the locality constraint when it is advantageous. In addition, it permits to split both the cell state in *substates*, and the transition function in *elementary processes* or *kernels*. For this reason, they were widely adopted for the simulation of different complex natural phenomena [13, 14, 16, 18, 19, 25, 30, 36, 40, 40]. Furthermore, the formalism is general and other numerical methods, like for instance Explicit Finite Differences, can be seen as a particular case of XCA (see e.g. [6, 12, 20, 23, 33]). Different software systems were also developed from the XCA formalism. Among them, well known examples are Camelot [15], libAuToti [41], OpenCAL [12, 37], and vinoAC [24][1]. Cellular Automata Networks were also proposed [5], albeit they essentially represent an alternative formulation of XCA, where the network defines the order of application of the considered local/global laws.

However, though powerful and highly flexible, the computational paradigms cited above are not straightforwardly applicable for modeling the co-evolution of automata of different dimensions and space/time granularity, as pointed out in [32], where automaton-automaton interactions were explicitly defined and implemented. Specifically, they considered a coupled system composed by a two-dimensional model simulating water runoff on a topographic surface, and a three-dimensional automaton modeling water infiltration and groundwater propagation, respectively. In such a case, besides the different dimensions, the automata could have different spacial granularity due to the data sampling methods adopted. For instance, a LIDAR technique could be used for the high-resolution sampling of the upper topographic surface elevations, while core drilling could be adopted to (roughly) characterize the underground, resulting in a finer description (and therefore in a smaller grid cell) for the two-dimensional automaton. In a similar context, a base time step variance could also be observed (with

---

[1] The vinoAC acronym does not explicitly appear in the text.

the finer-grained spacial model generally needing a smaller time step), and a higher level structure required for coordinating the dynamical co-evolution of the automata [22].

In order to address the above issues, here we propose the Extended Cellular Automata Network (XCAN) formalism for modeling a network of co-evolving automata. XCAN defines the interfaces between interconnected automata, manages the dynamical evolution of each model belonging to the network and their synchronization, by taking into account the space/time granularity variance within the network. The paper is organized as follows: Sect. 2 provides an alternative formulation of the XCA computational paradigm, by introducing some features that are considered useful in the modeling of different complex systems; Sect. 3 introduces XCAN by providing the formal definition; Sect. 4 describes a simplified, hypothetical, example of application; Sect. 5 concludes the paper with a general discussion and future developments.

2 Extended Cellular Automata

An Extended Cellular Automata (XCA) is defined as a tuple:

$$A = \langle (D, S, I), X, P, Q, (\Sigma, \Phi), R, \Gamma, (T, t), \omega \rangle$$

where the round parentheses are used to group related objects, and the D, X, Q, and (Σ, Φ) objects are mandatory. The meaning of each model object is described below.

- $D = [\iota_{1_{min}}, \iota_{1_{max}}] \times \ldots \times [\iota_{d_{min}}, \iota_{d_{max}}] \subset \mathbb{Z}^d$ is the d-dimensional discrete (integer coordinates) computational domain.
- $S = \{x = (x_1, x_2, \ldots, x_d) \in \mathbb{R}^d \mid \mu : D \to \mathbb{R}^d, \ \mu(\iota) = x\}$ is the d-dimensional (real coordinates) domain corresponding to the discrete domain D under the function μ, this latter associating cells in D to points in $\mathbb{R}^d$.
- $I = \{I_i = [\zeta_{i_{1_{min}}}, \zeta_{i_{1_{max}}}] \times \ldots \times [\zeta_{i_{d_{min}}}, \zeta_{i_{d_{max}}}] \subseteq D \mid \iota_{i_{j_{min}}} \leq \zeta_{i_{j_{min}}}, \zeta_{i_{j_{max}}} \leq \iota_{i_{j_{max}}} \mid \forall j = 1, 2, \ldots, d\}$ is the set of domain interfaces. It can be used to provide input to the system (e.g., to set boundary conditions), or for reduction/generalization (or aggregation/disaggregation) operations.
- $X = \{X_i = \{\xi_{i_1}, \xi_{i_2}, \ldots, \xi_{i_{|X_i|}}\} \mid i = 1, 2, \ldots, |X|, \ \xi_{i_j} \in \mathbb{Z}^d \ \forall j = 1, 2, \ldots, |X_i|\}$ is the set of neighborhood stencils. The neighborhood of a cell $\iota \in D$ under the stencil X_i is defined as:

$$V(\iota, X_i) = \{\iota + \xi_{i_1}, \iota + \xi_{i_2}, \ldots, \iota + \xi_{i_{|X_i|}}\}$$

The *neighborhood radius* along the jth direction is defined as:

$$r_j = (r_j^-, r_j^+) = \left(\min(0, \min_{i=1}^{|X|} \xi_{i_j}), \max(0, \max_{i=1}^{|X|} \xi_{i_j}) \right) \tag{1}$$

while the neighborhood radius is defined as $r = \max_{j=1}^d \{r_j^-, r_j^+\}$.

- $P = \{P_i = \{p_{i_1}, p_{i_2}, \ldots, p_{i_{|P_i|}}\} \mid i = 1, 2, \ldots, |P|, \ p_{i_j} \in Q_P \ \forall j = 1, 2, \ldots, |P_i|\}$ is the set of parameters. It can be used to define space-invariant values, as well as to store information resulting from non-local operations (see below).
- $Q = Q_1 \times \ldots \times Q_{|Q|}$ is the set of cell states, expressed as the Cartesian product of the Q_i *substate sets*. Accordingly, a state $q \in Q$ is expressed as:

$$q = (q_1, q_2, \ldots, q_{|Q|}) \mid q_i \in Q_i \ \forall i = 1, 2, \ldots, |Q|$$

and $q_i \in Q_i$ is referred as the ith *substate*. The *cell state function* defines cell state assignment:

$$c : D \to Q \mid \iota \in D \mapsto q \in Q$$

Eventually, an automaton configuration is defined as the state assignment to each domain cell:

$$C = \{(\iota, q) \mid q = c(\iota), \ \iota \in D, \ q \in Q\}$$

- $\Sigma = \{\sigma_1, \sigma_2, \ldots, \sigma_{|\Sigma|} \mid \sigma_i : Q_P^{|P_j|} \times Q^{|X_k|} \to Q, \ j \in \{1, 2, \ldots, |P|\}\}$ is the set of local transition functions. The input is given by the states of the neighboring cells involved into the local transition rule, besides the values of zero or more parameters. A local transition is classified as an *internal transition* if the states of the neighboring cells, except the state of cell itself, are irrelevant, i.e., if $\sigma_i : Q_P^{|P_j|} \times Q \to Q$.
- $\Phi = \{\phi_1, \phi_2, \ldots, \phi_{|\Sigma|} | \phi_i : Q_P^{|P_j|} \times Q^{|I_k^+|} \to Q^{|I_k|}, \ \phi_i|_{\iota \in I_k} = \sigma_i(P_j, c(V(\iota, X))), j \in \{1, 2, \ldots, |P|\}, \ k \in \{1, 2, \ldots, |I|\}\}$. I_k^+ represents the set resulting from the union of I_k and its adjacent borders along each direction, where the border width along each direction is defined by Eq. 1:

$$I_k^+ = [\zeta_{k_{1_{min}}} - r_1^-, \zeta_{k_{1_{max}}} + r_1^+] \times \ldots \times [\zeta_{k_{d_{min}}} - r_d^-, \zeta_{k_{d_{max}}} + r_d^+]$$

ϕ_i applies the corresponding local function σ_i to the non-local domain I_k.
- $R = \{\rho_1, \rho_2, \ldots, \rho_{|R|} \mid \rho_i : Q_P^{|P_j|} \times Q^{|I_k|} \to Q_P, \ j \in \{1, 2, \ldots, |P|\}, \ k \in \{1, 2, \ldots, |I|\}\}$ is the set of reduction (or aggregation) functions.
- $\Gamma = \{\gamma_1, \gamma_2, \ldots, \gamma_{|\Gamma|} \mid \gamma_i : Q_P^{|P_j|} \times Q \to Q^{|I_k|}, \ j \in \{1, 2, \ldots, |P|\}, \ k \in \{1, 2, \ldots, |I|\}\}$ is the set of generalization (or disaggregation) functions.
- $T = \circ_{i=1}^{|\Phi|+|R|+|\Gamma|} \tau_i \mid \tau_i \in \Phi \cup R \cup \Gamma$ is the function determining the automaton global transition.
- $t \in \mathbb{R}^+$ is the time corresponding to state transition of the system (i.e., to the application of the T function).
- $\omega : \{C \mid C$ is a configuration of $A\} \to \{0, 1\}$ is the *quiescent function*. If ω is *true*, A is in quiescent state.

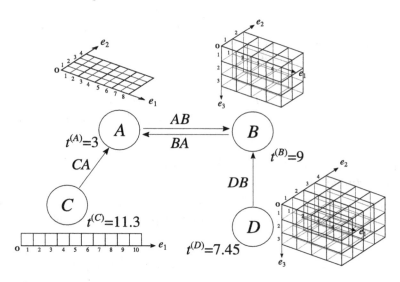

Fig. 1. Example of a network composed by four XCA models, namely A, B, C, and D, each of them characterized by a dimension and a space/time granularity. One-, two and three-dimensional models are here considered, with different spacial cardinality and time clock. The edges CA, AB, BA, and DB define dependence relationships of A from C, of B from A, of A from B, and of b from D, respectively. Please, note that the edges AB and BA define a two-sided relationship between the automata A and B.

3   Extended Cellular Automata Network

An Extended Cellular Automata Network (XCAN) is a direct acyclic graph (DAG), where nodes represent XCA models, while edges connections between couples of automata. An edge AB connecting the automaton A to the automaton B specifies that the state of B (also) depends on the state of A. In the case a dependence of A from B have to be modeled, a further edge BA can be considered. Additional components are considered to define topological/functional relations between connected vertices and to characterize the dynamical evolution of the network. An example is shown in Fig. 1. An XCAN is formally defined as:

$$N = <V, E, (I, \pi), \tau, \tau_{\mathbb{N}}, \omega>$$

where

- $V = \{A_i \mid i = 1, 2, \ldots, |V|, A_i \text{ is a XCA}\}$ is the set of XCA models belonging to the network.
- $E = \{AB \mid A, B \in V\}$ is the set of edges. The notation AB denotes an edge from A to B, meaning that the state of B is affected by the state of A.
- $I = \{\{I_i^{(A)}, I_j^{(B)}\} \mid AB \in E, I_i^{(A)} \in I^{(A)}, I_j^{(B)} \in I^{(B)}\}$ is the set of interfaces of connected automata. The notation $I^{(A)}$ denotes that the set I belongs to the automaton A.

- $\pi = \{\pi_{AB} : \mathcal{P}(I_i^{(A)}) \to \mathcal{P}(I_j^{(B)}) \mid AB \in E\}$ is the set of partition functions that define a bijection between interface cells of one automaton and interface subsets of the other. Specifically:

$$\pi_{AB} = \begin{cases} \pi_{AB,\gamma} : I_i^{(A)} \to \mathcal{P}(I_j^{(B)}) \mid \iota \in I_i^{(A)} \mapsto H \subseteq I_j^{(B)} & \text{if } |I_i^{(A)}| < |I_j^{(B)}| \\ \pi_{AB,\sigma} : I_i^{(A)} \to I_j^{(B)} \mid \iota \in I_i^{(A)} \mapsto \iota \in I_j^{(B)} & \text{if } |I_i^{(A)}| = |I_j^{(B)}| \\ \pi_{AB,\rho} = \pi_{BA,\gamma} \mid \iota \in I_i^{(B)} \mapsto H \subseteq I_j^{(A)} & \text{if } |I_i^{(A)}| > |I_j^{(B)}| \end{cases}$$

where H is the generic *macrocell* (i.e., a set of one or more cells) of the *macropartitioned* interface.

- $\tau = \{\tau_{AB} : Q^{(A)|H^{(A)}|} \times Q^{(B)|H^{(B)}|} \to Q^{(B)|H^{(B)}|} \mid AB \in E\}$ is the set of *induction* (or interaction) functions that, depending on the interface partitioning, are defined as:

$$\tau_{AB} = \begin{cases} \tau_{AB,\gamma} : Q^{(A)} \times Q^{(B)|H^{(B)}|} \to Q^{(B)|H^{(B)}|} & \text{if } |I_i^{(A)}| < |I_j^{(B)}| \\ \tau_{AB,\sigma} : Q^{(A)} \times Q^{(B)} \to Q^{(B)} & \text{if } |I_i^{(A)}| = |I_j^{(B)}| \\ \tau_{AB,\rho} : Q^{(A)|H^{(A)}|} \times Q^{(B)} \to Q^{(B)} & \text{if } |I_i^{(A)}| > |I_j^{(B)}| \end{cases}$$

- $\tau_{\mathbb{N}} : \mathbb{N} \times C^{(N)} \to C^{(N)}$ is the network *control unit*. It determines the network configuration update by executing the base steps reported in the ordered list below at discrete iterations:
  1. Evaluates the elapsed time that each automaton $A \in V$ would have if evolved for one more step;
  2. Evolves the automata having the minimum elapsed time, as evaluated in the previous base-step, by applying the corresponding global transition function T;
  3. Applies the induction functions τ_{AB}, being B evolved in the current step, A any other automaton s.t. $AB \in E$.
- $\omega : V \to \{true, false\}$ is the termination function. When ω is *true* the network computation halts.

At step $n = 0$, the network is in the initial state $C_0^{(N)}$. $\tau_{\mathbb{N}}$ is then applied to allow the network evolve, by producing a sequence of configurations $C_1^{(N)}, C_2^{(N)}, \cdots$. After each step, the ω function is applied. Possible halt criteria could be the following:
- A predefined number of iterations have been computed;
- The analyzed process/phenomenon is completed;
- All the automata are in the quiescent state.

4   An Example of Extended Cellular Automata Network

As an example of application, here we consider a conceptual, oversimplified network composed by two XCA, namely a two-dimensional model A and three-dimensional model B, interconnected by a single edge AB, defining a dependence of B from A (Fig. 2). Interaction between A and B occurs at domain

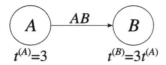

Fig. 2. Example of a simplified network composed by two XCA models, namely A and B. The single edge AB connects the two automata by defining a dependence of B from A. The time corresponding to a discrete step of the two automata is shown, being 3 physical time units for A, and 9 time units for B.

interfaces. Different spacial granularity here induces a macrocell partitioning on the finer-grained interface, which is that of A (Fig. 3). Moreover, the different temporal granularity causes that the finer-grained automaton, that is A again, is updated more frequently. Automata interaction exclusively occurs when B evolves. In the following, the above outlined conceptual models and the interconnection network are defined, by ignoring components and specifications that are not assumed necessary to characterize the network dynamical behavior. The _ symbol is used to mark missing or unspecified components.

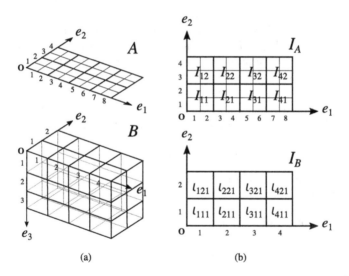

(a)   (b)

Fig. 3. Example of a coupled 2D/3D system. (a) The domain of the 2D automaton A is represented by a 8×4 structured grid of square cells, while the domain of the 3D automaton B by a $4 \times 2 \times 3$ structured grid of cubic cells. (b) A couple of interfaces, $I_A = A$ and $I_B = \{e_3 = 0\}$ on A and B, respectively, are defined to permit synchronization (i.e., information exchange during the dynamical evolution of the system) between the two automata. In particular, the finer-grained interface I_A is partitioned in macrocells $(I_{jk}$, with $j, k = 1, 2, 3, 4)$ to math the coarse grained interface I_B, where the original partitioning in cells $(\iota_{jk}$, with $j, k = 1, 2, 3, 4)$ is maintained.

The automaton A is defined as

$$A = \langle (D, \_, I), \_, \_, Q, (\_, \_), \rho, \_, (\_, t), \_ \rangle$$

where:

- $D = [1, 8] \times [1, 4] \subset \mathbb{Z}^2$ is the two-dimensional discrete domain.
- $I = [1, 8] \times [1, 4] = D$ is the interface with the B automaton.
- $Q = \mathbb{R}$ is the set of states for the cell.
- $\rho : Q^4 \to \mathbb{R}$ is a reduction function, which takes the states of four cells and provides a reduced value, e.g., average. In such a case, we would have:

$$(q_1, q_2, q_3, q_4) \mapsto (q_1 + q_2 + q_3 + q_4)/4$$

- $t = 3$ is the time corresponding to the state transition (i.e., to the application of the τ function - here omitted).

Similarly, the automaton B is defined as

$$B = \langle (D, \_, I), \_, P, Q, (\sigma, \_), \_, \_, (\_, t), \_ \rangle$$

where:

- $D = [1, 4] \times [1, 2] \times [1, 3] \subset \mathbb{Z}^3$ is the three-dimensional discrete domain.
- $I = [1, 4] \times [1, 2] \times \{1\} \subset D^{(B)}$ is the interface with the automaton A.
- $P = \{p\}$ is the set of parameters, here composed by the single parameter p.
- $Q = \mathbb{R}$ is the set of states for the cell.
- $\sigma : P \times Q \to Q$ is the internal transition that takes a real value and a cell state and provides a new state for the cell. For instance, the function could simply add the value of the parameter to the cell state. In such a case, we would have:

$$(p, q) \mapsto q' = p + q$$

- $t = 9$ is the time corresponding to the state transition (i.e., to the application of the τ function - here omitted).

The related network is defined as

$$N = <V, E, (I, \pi), \tau, \tau_{\mathbb{N}}, \_>$$

- $V = \{A, B\}$ is the set of XCA belonging to the network.
- $E = \{AB\}$ is the set of edges, composed by the only edge AB defining the dependence of B from A.
- $I = \{I^{(A)}, I^{(B)}\}$ is the couple of interfaces of connected automata.
- $\pi = \pi_{AB} : I^{(B)} \to \mathcal{P}(I^{(A)})$ is the partition function defining the bijection between each cell of $I^{(B)}$ and a square regions (macrocells) of 2×2 cells of $I^{(A)}$ due to the different interface granularity (see Fig. 3b). In the specific case, the following function could be adopted:

$$(i_1, i_2, 1) \mapsto [2 \cdot (i_1 - 1) + 1, 2 \cdot (i_1 - 1) + 2] \times [2 \cdot (i_2 - 1) + 1, 2 \cdot (i_2 - 1) + 2]$$

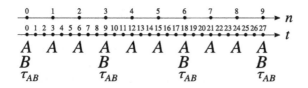

Fig. 4. Sequence of the first nine steps of evolution for the network. Both the step (n) and the time (t) axis are considered for reference. Note that, since the edge AB is defined, the τ_{AB} function is executed each time the automaton B executes a state transition.

- $\tau = \tau_{AB,\rho} : Q^{(A)^4} \times Q^{(B)} \to Q^{(B)}$ is the induction function from A to B that takes in input the states of a macrocell on $I^{(A)}$ and the state of a cell on $I^{(B)}$ and gives the new state for the cell on $I^{(B)}$

$$((q_1, q_2, q_3, q_4)^{(A)}, q^{(B)}) \mapsto q'^{(B)}$$

By considering the definitions of the automata A and B, it can be defined as the composition of a reduction on A and an internal transition on B:

$$\tau_{AB,\rho}((q_1, q_2, q_3, q_4)^{(A)}, q^{(B)}) = \sigma^{(B)}\left(\rho^{(A)}((q_1, q_2, q_3, q_4)^{(A)}), q^{(B)}\right)$$
$$= (q_1 + q_2 + q_3 + q_4)^{(A)}/4 + q^{(B)}$$

- $\tau_{\mathbb{N}} : \mathbb{N} \times C^{(N)} \to C^{(N)}$ is the network *control unit*, as defined in Sect. 3.

At step $n = 0$, the network is in the initial state. $\tau_{\mathbb{N}}$ is then applied to let the network evolve, by producing a sequence of configurations, as depicted in Fig. 4. Note that, while A evolves independently from B, each time B evolves, the τ_{AB} function is applied to take into consideration the dependence of B from A.

5   Conclusions

In this paper we have introduced the Extended Cellular Automata Network formal computational paradigm. With respect to XCA, the network provides a higher level of abstraction, by allowing for the co-evolution of different extended cellular automata models, each one characterized by its own dimension, and space/time granularity. Indeed, the network permits to define the hierarchical relationship among automata by adopting a direct acyclic graph in which the vertices represent the automata, and the edges the inter-relations between them. The network also defines automaton-automaton communication interfaces and external functions, which model effects that occur between connected automata.

Preliminarily, we have provided an alternative formulation of Extended Cellular Automata that is more suitable for defining the new XCAN model. Moreover, we have provided a first theoretical example of a network composed by one two-dimensional and one three-dimensional automaton, each one whit different space/time granularity. In particular, we have pointed out aspects related to the definition of communication interfaces, and have defined a simple external function to account for the influence that the two-dimensional automaton has on the three-dimensional one. Eventually, we have shown a possible co-evolution process in which the automata evolution is coordinated by the network.

As future development of this work, we intend to define an application program interface (API) based on XCAN and to provide an implementation of such an API. The software will be developed with the purpose of both test different parallelization scenarios, such as shared and distributed memory systems, and strategies concerning, for instance, load balancing and halos exchanging optimizations. In addition, different co-evolution strategies will be explored to ensure the network evolve consistently. Eventually, the system will be applied to the modeling and simulation of a real system, in order to evaluate correctness and efficiency.

References

1. Aidun, C., Clausen, J.: Lattice-Boltzmann method for complex flows. Annu. Rev. Fluid Mech. **42**, 439–472 (2010)
2. Andersson, B., et al.: Computational Fluid Dynamics for Engineers. Cambridge University Press, Cambridge (2011). https://doi.org/10.1017/CBO9781139093590
3. Belli, G., et al.: A unified model for the optimal management of electrical and thermal equipment of a prosumer in a DR environment. IEEE Trans. Smart Grid **10**(2), 1791–1800 (2019). https://doi.org/10.1109/TSG.2017.2778021
4. Burks, A.W.: Programming and the theory of automata. In: Burks, A.W. (ed.) Essays on Cellular Automata, chap. 2, pp. 65–83. University of Illinois Press, Urbana (1970)
5. Calidonna, C.R., Naddeo, A., Trunfio, G.A., Di Gregorio, S.: CANv2: a hybrid CA model by micro and macro-dynamics examples. In: Bandini, S., Manzoni, S., Umeo, H., Vizzari, G. (eds.) ACRI 2010. LNCS, vol. 6350, pp. 128–137. Springer, Heidelberg (2010). https://doi.org/10.1007/978-3-642-15979-4_13
6. Cervarolo, G., Mendicino, G., Senatore, A.: A coupled ecohydrological-three-dimensional unsaturated flow model describing energy, H2O and CO2 fluxes. Ecohydrology **3**(2), 205–225 (2010)
7. Cicirelli, F., Furfaro, A., Giordano, A., Nigro, L.: An agent infrastructure for distributed simulations over HLA and a case study using unmanned aerial vehicles. In: 40th Annual Simulation Symposium (ANSS 2007), pp. 231–238, March 2007. https://doi.org/10.1109/ANSS.2007.10
8. Cicirelli, F., Giordano, A., Nigro, L.: Distributed simulation of situated multi-agent systems. In: 2011 IEEE/ACM 15th International Symposium on Distributed Simulation and Real Time Applications, pp. 28–35, September 2011. https://doi.org/10.1109/DS-RT.2011.11
9. Cicirelli, F., Forestiero, A., Giordano, A., Mastroianni, C.: Transparent and efficient parallelization of swarm algorithms. ACM Trans. Auton. Adapt. Syst. (TAAS) **11**(2), 14 (2016)
10. Cook, M.: Universality in elementary cellular automata. Complex Syst. **15**(1), 1–40 (2004)

11. Crisci, G.M., et al.: Predicting the impact of lava flows at Mount Etna. Italy. J. Geophys. Res. Solid Earth **115**(B4), 1–14 (2010)
12. D'Ambrosio, D., et al.: The open computing abstraction layer for parallel complex systems modeling on many-core systems. J. Parallel Distrib. Comput. **121**, 53–70 (2018). https://doi.org/10.1016/j.jpdc.2018.07.005
13. D'Ambrosio, D., Rongo, R., Spataro, W., Trunfio, G.A.: Meta-model assisted evolutionary optimization of cellular automata: an application to the SCIARA model. In: Wyrzykowski, R., Dongarra, J., Karczewski, K., Waśniewski, J. (eds.) PPAM 2011. LNCS, vol. 7204, pp. 533–542. Springer, Heidelberg (2012). https://doi.org/10.1007/978-3-642-31500-8_55
14. D'Ambrosio, D., Rongo, R., Spataro, W., Trunfio, G.A.: Optimizing cellular automata through a meta-model assisted memetic algorithm. In: Coello, C.A.C., Cutello, V., Deb, K., Forrest, S., Nicosia, G., Pavone, M. (eds.) PPSN 2012. LNCS, vol. 7492, pp. 317–326. Springer, Heidelberg (2012). https://doi.org/10.1007/978-3-642-32964-7_32
15. Dattilo, G., Spezzano, G.: Simulation of a cellular landslide model with CAMELOT on high performance computers. Parallel Comput. **29**(10), 1403–1418 (2003)
16. De Rango, A., Napoli, P., D'Ambrosio, D., Spataro, W., Di Renzo, A., Di Maio, F.: Structured grid-based parallel simulation of a simple DEM model on heterogeneous systems. In: 2018 26th Euromicro International Conference on Parallel, Distributed and Network-based Processing (PDP), pp. 588–595, March 2018. https://doi.org/10.1109/PDP2018.2018.00099
17. Deng, X., Min, Y., Mao, M., Liu, H., Tu, G., Zhang, H.: Further studies on geometric conservation law and applications to high-order finite difference schemes with stationary grids. J. Comput. Phys. **239**, 90–111 (2013)
18. Filippone, G., Spataro, W., D'Ambrosio, D., Spataro, D., Marocco, D., Trunfio, G.: CUDA dynamic active thread list strategy to accelerate debris flow simulations. In: Proceedings - 23rd Euromicro International Conference on Parallel, Distributed, and Network-Based Processing, PDP 2015, pp. 316–320 (2015)
19. Filippone, G., D'Ambrosio, D., Marocco, D., Spataro, W.: Morphological coevolution for fluid dynamical-related risk mitigation. ACM Trans. Model. Comput. Simul. (TOMACS) **26**(3), 18 (2016)
20. Folino, G., Mendicino, G., Senatore, A., Spezzano, G., Straface, S.: A model based on cellular automata for the parallel simulation of 3D unsaturated flow. Parallel Comput. **32**(5), 357–376 (2006)
21. Frish, U., Hasslacher, B., Pomeau, Y.: Lattice gas automata for the Navier–Stokes equation. Phys. Rev. Lett. **56**(14), 1505–1508 (1986)
22. Fujimoto, R.M.: Parallel and Distribution Simulation Systems, 1st edn. Wiley, New York (1999)
23. Cervarolo, G., Mendicino, G., Senatore, A.: Coupled vegetation and soil moisture dynamics modeling in heterogeneous and sloping terrains. Vadose Zone J. **10**, 206–225 (2011)
24. Giordano, A., De Rango, A., D'Ambrosio, D., Rongo, R., Spataro, W.: Strategies for parallel execution of cellular automata in distributed memory architectures. In: 2019 27th Euromicro International Conference on Parallel, Distributed and Network-Based Processing (PDP), pp. 406–413, February 2019. https://doi.org/10.1109/EMPDP.2019.8671639
25. Giordano, A., et al.: Parallel execution of cellular automata through space partitioning: the landslide simulation Sciddicas3-Hex case study. In: 2017 25th Euromicro International Conference on Parallel, Distributed and Network-based Processing (PDP), pp. 505–510, March 2017. https://doi.org/10.1109/PDP.2017.84

26. Di Gregorio, S., Serra, R.: An empirical method for modelling and simulating some complex macroscopic phenomena by cellular automata. Future Gen. Comput. Syst. **16**, 259–271 (1999)
27. Higuera, F., Jimenez, J.: Boltzmann approach to lattice gas simulations. Europhys. Lett. **9**(7), 663–668 (1989)
28. Jammy, S., Mudalige, G., Reguly, I., Sandham, N., Giles, M.: Block-structured compressible Navier–Stokes solution using the OPS high-level abstraction. Int. J. Comput. Fluid Dyn. **30**(6), 450–454 (2016)
29. Langton, C.: Computation at the edge of chaos: phase transition and emergent computation. Physica D **42**, 12–37 (1990)
30. Lucà, F., D'Ambrosio, D., Robustelli, G., Rongo, R., Spataro, W.: Integrating geomorphology, statistic and numerical simulations for landslide invasion hazard scenarios mapping: an example in the Sorrento Peninsula (Italy). Comput. Geosci. 67(1811), 163–172 (2014)
31. McNamara, G., Zanetti, G.: Use of the Boltzmann equation to simulate lattice-gas automata. Phys. Rev. Lett. **61**, 2332–2335 (1988)
32. Mendicino, G., Pedace, J., Senatore, A.: Stability of an overland flow scheme in the framework of a fully coupled eco-hydrological model based on the macroscopic cellular automata approach. Commun. Nonlinear Sci. Numer. Simul. **21**(1–3), 128–146 (2015)
33. Mendicino, G., Senatore, A., Spezzano, G., Straface, S.: Three-dimensional unsaturated flow modeling using cellular automata. Water Resour. Res. **42**(11), W11419 (2006)
34. von Neumann, J.: Theory of Self-Reproducing Automata. University of Illinois Press, Champaign (1966)
35. Ninagawa, S.: Dynamics of universal computation and 1/f noise in elementary cellular automata. Chaos Solitons Fractals **70**(1), 42–48 (2015)
36. Oliverio, M., Spataro, W., D'Ambrosio, D., Rongo, R., Spingola, G., Trunfio, G.: OpenMP parallelization of the SCIARA cellular automata lava flow model: performance analysis on shared-memory computers. Procedia Comput. Sci. **4**, 271–280 (2011)
37. De Rango, A., Spataro, D., Spataro, W., D'Ambrosio, D.: A first multi-GPU/multi-node implementation of the open computing abstraction layer. J. Comput. Sci. **32**, 115–124 (2019). https://doi.org/10.1016/j.jocs.2018.09.012
38. Reguly, I., et al.: Acceleration of a full-scale industrial CFD application with OP2. IEEE Trans. Parallel Distrib. Syst. **27**(5), 1265–1278 (2016). https://doi.org/10.1109/TPDS.2015.2453972
39. Reguly, I., Mudalige, G., Giles, M., Curran, D., McIntosh-Smith, S.: The OPS domain specific abstraction for multi-block structured grid computations. In: Proceedings of WOLFHPC 2014: 4th International Workshop on Domain-Specific Languages and High-Level Frameworks for High Performance Computing - Held in Conjunction with SC 2014: The International Conference for High Performance Computing, Networking, Storage and Analysis, pp. 58–67 (2014). https://doi.org/10.1109/WOLFHPC.2014.7
40. Spataro, D., D'Ambrosio, D., Filippone, G., Rongo, R., Spataro, W., Marocco, D.: The new SCIARA-fv3 numerical model and acceleration by GPGPU strategies. Int. J. High Perform. Comput. Appl. **31**(2), 163–176 (2017). https://doi.org/10.1177/1094342015584520
41. Spingola, G., D'Ambrosio, D., Spataro, W., Rongo, R., Zito, G.: Modeling complex natural phenomena with the libAuToti cellular automata library: an example of

application to lava flows simulation. In: PDPTA - International Conference on Parallel and Distributed Processing Techniques and Applications, pp. 277–283 (2008)

42. Thatcher, J.W.: Universality in the von Neumann cellular model. In: Burks, A.W. (ed.) Essays on Cellular Automata, chap. 5, pp. 132–186. University of Illinois Press, Urbana (1970)

43. Tie, B.: Some comparisons and analyses of time or space discontinuous Galerkin methods applied to elastic wave propagation in anisotropic and heterogeneous media. Adv. Model. Simul. Eng. Sci. **6**(1) (2019). https://doi.org/10.1186/s40323-019-0127-x

44. Wolfram, S.: A New Kind of Science. Wolfram Media Inc., Champaign (2002)

Preliminary Model of Saturated Flow
Using Cellular Automata

Alessio De Rango[1]([⊠]) ⓘ, Luca Furnari[1] ⓘ, Andrea Giordano[3] ⓘ,
Alfonso Senatore[1] ⓘ, Donato D'Ambrosio[2] ⓘ, Salvatore Straface[1] ⓘ,
and Giuseppe Mendicino[1] ⓘ

[1] DIATIC, University of Calabria, Rende, Italy
alessio.derango@unical.it
[2] DeMACS, University of Calabria, Rende, Italy
[3] ICAR-CNR, Rende, Italy

Abstract. A fully-coupled from surface to groundwater hydrological model is being developed based on the Extended Cellular Automata formalism (XCA), which proves to be very suitable for high performance computing. In this paper, a preliminary module related to three-dimensional saturated flow in porous media is presented and implemented by using the OpenCAL parallel software library. This allows to exploit distributed systems with heterogeneous computational devices. The proposed model is evaluated in terms of both accuracy and precision of modeling results and computational performance, using single layered three-dimensional test cases at different resolutions (from local to regional scale), simulating pumping from one or more wells, river-groundwater interactions and varying soil hydraulic properties. Model accuracy is compared with analytic, when available, or numerical (MOD-FLOW 2005) solution, while the computational performance is evaluated using an Intel Xeon CPU socket. Overall, the XCA-based model proves to be accurate and, mainly, computationally very efficient thanks to the many options and tools available with the OpenCAL library.

Keywords: High-performance computing · Extended Cellular Automata · Computational fluid dynamics

1 Introduction

In a context of climate change, water resources management will become a key factor to research sustainable development [14]. The development of more and more efficient hydrological models is an essential element to study superficial and subsurface water dynamics. Many models have been recently proposed to predict these phenomena (cf. [20]). Most of them use PDEs (Partial Difference Equation) to study and analyze the real event, approximating the solution by adopting a specific numerical solver as, for instance, Finite Differences, Finite Elements, Finite Volume Methods, etc. (cf. [4,9]). In most cases, an explicit

© Springer Nature Switzerland AG 2020
Y. D. Sergeyev and D. E. Kvasov (Eds.): NUMTA 2019, LNCS 11973, pp. 256–268, 2020.
https://doi.org/10.1007/978-3-030-39081-5_23

recurrence relation is obtained, which expresses the next state of the generic (discrete) system element as a function of the current states of a limited number of neighboring elements. For this reason, most explicit schemes can be formalized in terms of Cellular Automata (CA) [18], which is one of the most known and utilized decentralized discrete parallel computational models (cf. [2,3,11,15,16,21]). Nevertheless, a kind of global control on the system to be simulated is often useful to make the modeling of certain processes more straightforward. Therefore, as pointed out in [6], both decentralized transitions and steering operations can be adopted to model complex systems.

Different parallel integrated development environments (IDEs) and software libraries were proposed to efficiently model complex systems formalized in terms of Cellular Automata. Among them, Camelot represents a (commercial) example of CA IDE, accounting for both local interactions and global steering, able to accelerate the computation on clusters of distributed nodes interconnected via network [1]. The libAuToti library [22] is essentially feature-equivalent to Camelot, even released as an Open Source software library. Unfortunately, both Camelot and libAutoti are no longer developed. However, the open source Open-CAL [5] software library has been recently proposed as a modern alternative to the aforementioned simulation software. Specifically, OpenCAL allows to exploit heterogeneous computational devices like classical CPU sockets and modern accelerators like GPUs on clusters of interconnected workstations. It was successfully applied to the simulation of different systems, including a debris flow evolving on real topographic surface, graphics convolutional filters, fractal generation, as well as particle systems based on the Discrete Element Method [7,8].

In this paper we developed a new, preliminary, groundwater model based on the discretization of the Darcy law, by adopting an explicit Finite Difference scheme to obtain a discrete formulation [12,17]. We then implemented it by using OpenCAL, and therefore applied the model to simulate two different cases of study, namely a constant pumping rate and an idealized case of interaction between aquifer and river. We evaluated the model in terms of both accuracy and computational efficiency. As regards accuracy, we compared the outcomes with the analytical solutions, when available, and the MODFLOW 2005 model [13] while, regarding computational performances, they were preliminary evaluated by considering an Intel Xeon-based workstation.

The paper is organized as follows. Section 2 formalizes the groundwater flow model, while Sect. 3 briefly presents the OpenCAL parallel software library. Section 4 illustrates the considered case studies, together outcomes and accuracy evaluation, while Sect. 5 describes the obtained computational performances. Finally, Sect. 6 concludes the paper with a general discussion envisaging possible future developments.

2   Preliminary Model of Saturated Flow

Groundwater phenomena is governed by the continuity equation that expresses the mass balance for each cell. In particular, during a fixed time interval, the sum of all flows into and out of the cell must be equal to the rate of change in storage within the cell. The discrete governing equation is defined as follows:

$$\sum Q_i = \frac{\Delta h}{\Delta t} \cdot \Delta x \cdot \Delta y \cdot S_y \tag{1}$$

Where Q is the flow rate into the cell [m³ s⁻¹], Δh is the head change [m] over a time interval Δt [s], Δx and Δy identify the cell dimensions [m] and S_y is the specific yield [.].

In the considered preliminary test cases, a two-dimensional space, composed of regular cells, is adopted. In order to estimate the hydraulic head change within the cell, the four cells at the cardinal points are considered. In particular, the flows from the central cell are represented as positive inflow or negative outflow. As shown in Fig. 1.

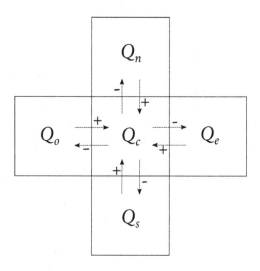

Fig. 1. Interaction between neighbor cells.

Therefore, the summation of flows is calculated as:

$$\sum Q_i = Q_n + Q_s + Q_e + Q_o + Q_c \tag{2}$$

Applying the Darcy law,

$$Q_i = K_{s_i} \cdot A_{transv} \cdot \frac{h_i - h_c}{\Delta x} = T_i \cdot (h_i - h_c) \tag{3}$$

Where K_{s_i} is the saturated conductivity [m s⁻¹], A_{transv} is the area of the cell [m²], h is the hydraulic head [m] and T_i is hydraulic transmissivity [m² s⁻¹], calculated as $T_i = K_{s_i} \cdot b$ where b is the aquifer thickness [m].

Eventually, the explicit formula is obtained as follows:

$$h^{t+1} = h^t + \frac{\Delta t}{S_y \cdot \Delta x^2} \cdot \sum Q_i \qquad (4)$$

The Cellular Automata formalism is used in order to solve the saturated flow equation. Cellular Automata (CA) are computational models whose evolution is governed by laws which are purely local. In their essential definition, a CA can be described as a d–dimensional space composed of regular cells. At time $t = 0$, cells are in an arbitrary state and the CA evolves by changing the states of the cells in discrete steps of time and by applying simultaneously to each of them the same law of evolution, or *transition function*. Input for each cell is given by the states of neighboring cells and the neighborhood conditions are determined by a geometrical pattern, which is invariant in time and space. Despite their simple definition, CA may give rise to extremely complex behavior at a macroscopic level. In fact, even if the local laws that regulate the dynamics of the system are known, the global behavior of the system can be very hard to be predicted.

The Cellular Automata formulation of the aforementioned preliminary model is defined as:

$$SF_{CA} = <E^d, X, S, P, \sigma>$$

- $E^d = \{(x,y)|x,y \in N, 0 \le x \le l_x, 0 \le y \le l_y\}$ is the set of square cells covering the bi-dimensional finite region where the phenomenon evolves;
- $X = \{(0,0),(-1,0),(0,1),(0,-1),(0,1)\}$ identifies the pattern of cells (Von Neumann neighbourhood) that influence the cell state change;
- $S = S_h \times S_{K_s} \times S_y \times S_w$ is the finite set of states considered as Cartesian product of substates where S_h is the head state, S_k identifies the saturated conductivity, S_y identifies the specific yield and S_w identifies the source term;
- $P = \{Dim_x, Dim_y, b\}$ is the finite set of parameters (invariant in time and space) which affect the transition function used for the first case study. In addition, for the second case study, further parameters are considered such as: $\{K_{sb}, M, H_w, W, R_b\}$;
- σ : identifies the transition function applied to each cell at every time step, which describes water dynamics inside the aquifer by applying Eq. 4.

3   The OpenCAL Software System and Implementation Details

OpenCAL (Open Computation Abstraction Layer) is an open source parallel computing abstraction layer for scientific computing and is based on the Extended Cellular Automata general formalism [10] as a Domain Specific Language. Moreover, Cellular Automata, Finite Difference Method and other structured grid-based methods can be straightforwardly supported.

The library permits to abstract the implementation phase by providing the building blocks of the XCA computational paradigm through its API structures and functions. In particular, the computational domain, substates, and neighborhood are formalized at the higher level of abstraction, as well as local interactions and global operations (e.g., reductions operations), which are commonly required by structured grid-based applications. Moreover, the initial conditions of the system, and a termination criterion to halt the system evolution can be defined. In addition, OpenCAL provides embedded optimization algorithms and allows for a fine grained control over the simulation [5]. Furthermore, starting from an OpenCAL-based serial implementation of a model, different parallel versions can be obtained with the minimum effort, including those for multi- and many-core shared memory devices, as well as for distributed memory systems. For this purpose, the OpenCAL implementation level is transparently exploited, whose components are based on the OpenMP, OpenCL, and MPI APIs.

As regards the implementation of the proposed underground model, we considered two different versions in this preliminary work, namely a straightforward basic implementation, that does not consider any OpenCAL specific optimizations, and an explicit implementation, that in most cases permits to avoid unneeded memory access operations [5]. More in detail, in the first version, the *implicit* update policy transparently updates the whole set of model substates at the end of each execution of each elementary process composing the transition function. Conversely, the *explicit* scheme permits to selectively update substates, depending if they are actually modified.

Note that only the serial and OpenMP-based components of the library were considered in this work, namely OpenCAL (which actually refers to both the serial implementation component, and the name of the whole software library) and OpenCAL-OMP, respectively.

4   Case Studies

In this preliminary development phase of the model, two different transient case studies where conducted, considering an unconfined aquifer, following [19]. The first one is a standard hydrology problem, the water table drawdown caused by a well with a constant pumping rate. The second one is the aquifer response to a stream-stage variation. In particular, in these case studies the Q_c term from the Eq. (2) is referred to the source term, which is equal to zero when well or river cells are not considered. A square aquifer of $1\,\text{km}^2$ is considered, in order to test the accuracy and computational performance of the model. For both case studies, two different meshes, with different sizes, are adopted. For the first test Δx is fixed to $10\,\text{m}$, for a total of $100 \times 100 = 10^4$ cells, in the second test $\Delta x = 1\,\text{m}$, for a total of $1000 \times 1000 = 10^6$ cells.

Hydraulic saturated conductivity K_s is set equal to $1.25 \cdot 10^{-5}\,\text{m s}^{-1}$, and the specific yield S_y is set 0.1. The aquifer thickness is $50\,\text{m}$, temporal step size Δt is calculated using the Courant-Friedrichs-Lewy (CFL) condition $\Delta t = \frac{\Delta x^2 \cdot S_y}{4T}$ and is set to $4000\,\text{s}$ and $40\,\text{s}$, for the first and second test cases, respectively.

Both cases studies are compared with the widely used groundwater model MODFLOW 2005 [13]. Its solutions are achieved, only on the less dense mesh, using the Preconditioned Conjugate-Gradient(PCG) solver set with the same Δt used in the SF_{CA} model (4000 s).

4.1 Drawdown Made by a Well with Constant Pumping Rate

This problem could be resolved using the analytic solution (5) obtained by Theis [23,24], which is valid for confined aquifer with full penetrating well.

$$s = \frac{Q}{4\pi T} W(u) \tag{5}$$

$$W(u) = \int_u^\infty \frac{e^{-u}}{u} du \tag{6}$$

$$u = \frac{r^2 \cdot S_y}{4tT} \tag{7}$$

Where s is drawdown [m] (Fig. 2b), Q is the pumping rate [m^3 s^{-1}], t is the time since pumping starts [s], T is the hydraulic transmissivity [m^2 s^{-1}], $W(u)$ is called *well function* and could be approximated using a numerical solution, r is the radial distance from the well [m] and S_y is specific yield. This solution could be used also for unconfined aquifer, if the drawdown is relatively small compared to the saturated thickness.

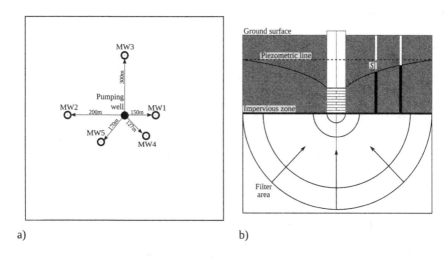

a)   b)

Fig. 2. (a) Scheme of the position of the monitoring wells and the pumping well, (b) vertical and horizontal section of the pumping well, s is the drawdown made by the pumping in respect to the piezometric line which represents the undisturbed condition.

In this specific case study, a well is placed in the center of the domain with a constant pumping rate of 0.001 m^3 s^{-1}. The initial head is equal to 50 m all over the domain and the Dirichlet condition is used for the boundaries, fixing the hydraulic head to a constant value.

Monitoring wells are placed, according to [19], at distances equal to 150 m, 200 m and 300 m from the pumping well on one cardinal direction. Moreover, two further monitoring wells are placed at distances of 127 m and 170 m on the 45° direction (Fig. 2a), to verify the quality of the Von Neumann neighborhood.

The results of a simulation with 10 m mesh size (100 × 100 cells) over 12 days are shown in Fig. 3.

The lines represent the analytical solutions obtained by Theis, the red squares are the numerical solutions obtained by MODFLOW 2005 and the blue crosses are the numerical solutions obtained by SF_{CA}. This visual comparison shows a very good fit by the SF_{CA} model and confirm that Von Neumann neighborhood does not generate numerical distortions even in diagonal directions.

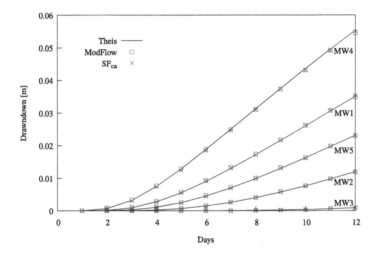

Fig. 3. Drawdown obtained by analytical solution, MODFLOW 2005 and SF_{CA} at each monitoring well (please refer to Table 1 for labels).

Table 1. RMSE [m] referred to analytical solution.

Well	Distance [m]	MODFLOW	SF_{CA}
MW1	150	2,10E−04	1,49E−04
MW2	200	8,00E−05	3,37E−05
MW3	300	4,71E−05	3,35E−06
MW4	127	1,30E−04	3,24E−04
MW5	170	3,13E−04	6,74E−05

Also Table 1 confirms a very got fit by SF_{CA} compared to the analytical solution, even better than MODFLOW 2005 results. In particular, the Root Mean Square Error (RMSE) calculated using SF_{CA} is always lower than the MODFLOW RMSE, unless for MW4, which points out a better performance by MODFLOW 2005.

Further tests are also executed with the denser grid (1000×1000 cells) with similar results.

4.2   Interaction Between Aquifer and River

One of the most relevant problems in alluvial hydrology is the interaction between the aquifer system and the river stage. Modeling this phenomenon requires that two different behaviors are considered, being in one case the aquifer a source for the river (therefore, aquifer head will decrease in the proximity of the river) and, in the opposite case, the river a source for the aquifer (therefore, the aquifer head will increase in the proximity of the river).

To discriminate these two different behaviors, the river package in MODFLOW 2005 uses the bottom of streambed (R_b) as a control variable. Specifically, if the aquifer hydraulic head (h) is above the bottom of streambed ($h > R_b$), it acts as a source. Otherwise ($h \leq R_b$), the river acts as a source:

$$Q_C = \begin{cases} \frac{K_{sb} \cdot \Delta x \cdot W}{M} \cdot (h_w - h) & if \quad h > R_b \\ \frac{K_{sb} \cdot \Delta x \cdot W}{M} \cdot (H_w + M) & if \quad h \leq R_b \end{cases} \tag{8}$$

Equation 8 describes how the river source is modeled both in MODFLOW 2005 and in the SF_{CA} model, where K_{sb} is the streambed hydraulic conductivity, set to $1.0 \cdot 10^{-5}$ m s^{-1}, W is the river width fixed to 5 m, M is the riverbed thickness set to 0.5 m, the bottom of streambed R_b is set to 46.5 m and H_w is the water level in the stream above the surface of the streambed. h_w is the hydraulic head in the stream and in this case study it is assumed to change accordingly to a triangular law during the simulations (Fig. 5). The relative schematization is illustrated in Fig. 4b.

According to [19], river is located at a distance of 250 m from the west border of the domain and flows from north to south. Neumann boundary conditions of no flows are set in the northern and southern borders, while Dirichlet conditions are imposed to the western and eastern borders. The initial head is fixed equal to 50 m all over the domain.

Simulations lasted 30 days. During the first 10 days, H_w is set to 48 m, between 10 and 15 days it linearly increases up to 50 m, during the next 5 days it linearly decreases again to 48 m keeping this constant value until the end of the experiment (Fig. 5). Therefore, three different behaviors can be identified: initially, the aquifer recharges the river, then the river recharges the aquifer and, finally, the aquifer again recharges the river.

Monitoring wells are placed at increasing distance from the west border. Specifically, MW100 is placed at 100 m distance, MW350 at 350 m, MW450 at 450 m, MW 550 at 550 m and MW650 at 650 m (Fig. 4a). The closest well to

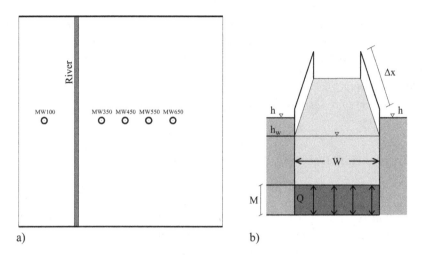

a)   b)

Fig. 4. (a) Overview of the position of the monitoring wells and the river, (b) Cross-section of the case study identifying the interaction between the river and the aquifer.

the river is MW350, it is expected that in this monitoring well the influence of the river stage variation is mostly amplified. On the other hand, MW650 is the farthest well, where the head variation during the simulation should be the lowest.

The results of the simulations, with 10 m mesh size, are represented in Fig. 5. The red line represents the H_w variation, black lines the MODFLOW 2005 solutions and blue crosses the SF_{CA} model solutions. Also in this second case study, the SF_{CA} model shows an excellent fit and confirms the trend expected by physical considerations.

5   Computational Performance

In order to evaluate the computational performance of the two different Open-CAL implementations of the SF_{CA} groundwater flow model here proposed (cf. Sect. 3), the simulation of the test studies reported in Sects. 4.1 and 4.2 were performed over a computational domain of 1000×1000 cells of 1 m side. Both OpenCAL and SF_{CA} were compiled by considering the gcc GNU C compiler version 8.3.0 with the -O3 optimization flag. All serial and multi-thread tests were performed on a hyper-threading 8-core Intel Xeon 2.0 GHz E5-2650 CPU based workstation running Arch Linux by considering 1, 2, 4, 8 and 16 threads. Here, the OpenMP static scheduler for parallel loop work-sharing was considered for the OpenCAL-OMP benchmarks.

The speed-up analysis results of the first test case related to OpenCAL and OpenCAL-OMP versions are shown in Fig. 6, whereas the results of the second test case are shown in Fig. 7. A speed-up value of about 12 was obtained by considering the explicit update scheme for both the considered test cases, while

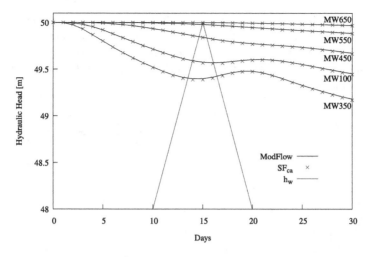

Fig. 5. Hydraulic head in the stream H_w and hydraulic head at different monitoring wells obtained by MODFLOW 2005 and SF_{CA} models. (Color figure online)

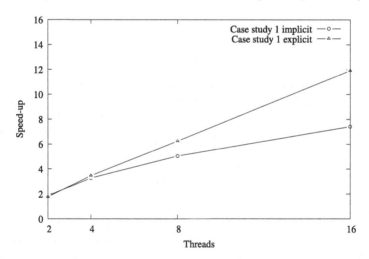

Fig. 6. Speed-up achieved by the OpenCAL-OMP simulation of the test study described in Sect. 4.1. Results of both the implicit and explicit implementations are shown. Speed-up values were evaluated by considering the corresponding serial execution time that was 1587 and 1304 s for the implicit and explicit OpenCAL schemes, respectively.

speed-up of about 7.4 and 4.8 were registered for the first and second test cases, respectively, when the implicit scheme was adopted.

The result obtained in the case of the implicit version is explained by considering that the whole (global) state of the automaton is transparently updated, even if only the S_h substate is actually modified by the transition function (cf.

Sect. 2). Nevertheless, the implicit scheme implementation represents the first fundamental step in the OpenCAL development process that permits to rapidly obtain a non optimized working version of the model in order to test its correctness. It is worth to note that no in-depth knowledge of OpenCAL details is required in this phase.

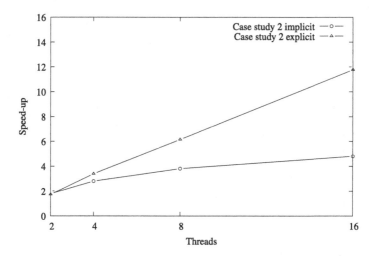

Fig. 7. Speed-up achieved by the OpenCAL-OMP simulation of the test study described in Sect. 4.2. Results of both the implicit and explicit implementations are shown. Speed-up values were evaluated by considering the corresponding serial execution time that was 4476 and 3227 s for the implicit and explicit OpenCAL schemes, respectively.

6   Conclusions

This paper reports the first implementation of the SF_{CA} model for the simulation of groundwater phenomena, using the OpenCAL parallel computational library. Two case studies are considered, one related to a drawdown made by a well with constant pumping rate and the second regarding the interaction between an aquifer and a river. The parallel implementation of the model is carried out by considering the OpenMP component of OpenCAL. Several tests are carried out to validate the accuracy and computational performance of the model. The MODFLOW 2005 model and the analytical solution, obtained by Theis, are used to compare the results of the developed model. The results show good accuracy of the model for both case studies. In particular, for the first case study, the model achieved a better agreement with the analytical solution than MODFLOW 2005 for most of the monitoring wells. Regarding computational performances, the tests pointed out overall good performance and scalability, achieving a speed-up of about 12 by considering 16 threads for both test cases.

Future work will regard the extension of the model to the non-saturated case and performance improvements. In particular, regarding the performance, other compilers will be tested in spite of `gcc`, e.g. `clang`. Moreover, OpenCAL will be better exploited to improve performance. For instance, the embedded quantization optimization could be considered that, based on a set of one or more thresholds, is able to exclude cells classified as *stationary* from the computation. Eventually, other OpenCAL components will be taken into account in order to exploit modern accelerators like GPUs.

Acknowledgements. L. Furnari acknowledges for the program "POR Calabria FSE/FESR 2014/2020 - Mobilitá internazionale di Dottorandi a Assegni di ricerca/Ricercatori di Tipo A" Actions 10.5.6 and 10.5.12.

References

1. Cannataro, M., Di Gregorio, S., Rongo, R., Spataro, W., Spezzano, G., Talia, D.: A parallel cellular automata environment on multicomputers for computational science. Parallel Comput. **21**(5), 803–823 (1995)
2. Cervarolo, G., Mendicino, G., Senatore, A.: A coupled ecohydrological-three-dimensional unsaturated flow model describing energy, H_2O and CO_2 fluxes. Ecohydrology **3**(2), 205–225 (2010)
3. Cervarolo, G., Mendicino, G., Senatore, A.: Coupled vegetation and soil moisture dynamics modeling in heterogeneous and sloping terrains. Vadose Zone J. **10**, 206–225 (2011)
4. Chang, K.S., Song, C.J.: Interactive vortex shedding from a pair of circular cylinders in a transverse arrangement. Int. J. Numer. Methods Fluids **11**(3), 317–329 (1990)
5. D'Ambrosio, D., et al.: The open computing abstraction layer for parallel complex systems modeling on many-core systems. J. Parallel Distrib. Comput. **121**, 53–70 (2018)
6. Dattilo, G., Spezzano, G.: Simulation of a cellular landslide model with camelot on high performance computers. Parallel Comput. **29**(10), 1403–1418 (2003)
7. De Rango, A., Napoli, P., D'Ambrosio, D., Spataro, W., Di Renzo, A., Di Maio, F.: Structured grid-based parallel simulation of a simple DEM model on heterogeneous systems, pp. 588–595 (2018)
8. De Rango, A., Spataro, D., Spataro, W., D'Ambrosio, D.: A first multi-GPU/multi-node implementation of the open computing abstraction layer. J. Comput. Sci. **32**, 115–124 (2019)
9. Deng, X., Min, Y., Mao, M., Liu, H., Tu, G., Zhang, H.: Further studies on geometric conservation law and applications to high-order finite difference schemes with stationary grids. J. Comput. Phys. **239**, 90–111 (2013)
10. Di Gregorio, S., Serra, R.: An empirical method for modelling and simulating some complex macroscopic phenomena by cellular automata. Futur. Gener. Comput. Syst. **16**, 259–271 (1999)
11. Filippone, G., D'Ambrosio, D., Marocco, D., Spataro, W.: Morphological coevolution for fluid dynamical-related risk mitigation. ACM Trans. Model. Comput. Simul. (TOMACS) **26**(3), 18 (2016)

12. Folino, G., Mendicino, G., Senatore, A., Spezzano, G., Straface, S.: A model based on cellular automata for the parallel simulation of 3D unsaturated flow. Parallel Comput. **32**(5), 357–376 (2006)
13. Harbaugh, A.: MODFLOW-2005, the U.S. geological survey modular ground-water model-the ground-water flow process. U.S. Geological Survey (2005)
14. Kundzewicz, Z.W., et al.: The implications of projected climate change for freshwater resources and their management. Hydrol. Sci. J. **53**(1), 3–10 (2008)
15. Lucá, F., D'Ambrosio, D., Robustelli, G., Rongo, R., Spataro, W.: Integrating geomorphology, statistic and numerical simulations for landslide invasion hazard scenarios mapping: an example in the Sorrento Peninsula (Italy). Comput. Geosci. **67**(1811), 163–172 (2014)
16. Mendicino, G., Pedace, J., Senatore, A.: Stability of an overland flow scheme in the framework of a fully coupled eco-hydrological model based on the Macroscopic Cellular Automata approach. Commun. Nonlinear Sci. Numer. Simul. **21**(1–3), 128–146 (2015)
17. Mendicino, G., Senatore, A., Spezzano, G., Straface, S.: Three-dimensional unsaturated flow modeling using cellular automata. Water Resour. Res. **42**(11), 2332–2335 (2006)
18. von Neumann, J.: Theory of Self-Reproducing Automata. University of Illinois Press, Champaign (1966)
19. Ravazzani, G., Rametta, D., Mancini, M.: Macroscopic cellular automata for groundwater modelling: a first approach. Environ. Model Softw. **26**(5), 634–643 (2011)
20. Senatore, A., Mendicino, G., Smiatek, G., Kunstmann, H.: Regional climate change projections and hydrological impact analysis for a Mediterranean basin in Southern Italy. J. Hydrol. **399**(1), 70–92 (2011)
21. Spataro, D., D'Ambrosio, D., Filippone, G., Rongo, R., Spataro, W., Marocco, D.: The new SCIARA-fv3 numerical model and acceleration by GPGPU strategies. Int. J. High Perform. Comput. Appl. **31**(2), 163–176 (2017)
22. Spingola, G., D'Ambrosio, D., Spataro, W., Rongo, R., Zito, G.: Modeling complex natural phenomena with the libAuToti cellular automata library: an example of application to lava flows simulation. In: PDPTA - International Conference on Parallel and Distributed Processing Techniques and Applications, pp. 277–283 (2008)
23. Straface, S.: Estimation of transmissivity and storage coefficient by means of a derivative method using the early-time drawdown. Hydrol. J. **17**(7), 1679 (2009)
24. Theis, C.V.: The relation between the lowering of the piezometric surface and the rate and duration of discharge of a well using ground-water storage. Trans. Am. Geophys. Union **16**, 519–524 (1935)

A Cybersecurity Framework for Classifying Non Stationary Data Streams Exploiting Genetic Programming and Ensemble Learning

Gianluigi Folino$^{(\boxtimes)}$ (ID), Francesco Sergio Pisani, and Luigi Pontieri (ID)

ICAR-CNR, Rende, CS, Italy
{gianluigi.folino,fspisani,luigi.pontieri}@icar.cnr.it

Abstract. Intrusion detection systems have to cope with many challenging problems, such as unbalanced datasets, fast data streams and frequent changes in the nature of the attacks (concept drift). To this aim, here, a distributed genetic programming (GP) tool is used to generate the combiner function of an ensemble; this tool does not need a heavy additional training phase, once the classifiers composing the ensemble have been trained, and it can hence answer quickly to concept drifts, also in the case of fast-changing data streams. The above-described approach is integrated into a novel cybersecurity framework for classifying non stationary and unbalanced data streams. The framework provides mechanisms for detecting drifts and for replacing classifiers, which permits to build the ensemble in an incremental way. Tests conducted on real data have shown that the framework is effective in both detecting attacks and reacting quickly to concept drifts.

Keywords: Cybersecurity · Intrusion detection · Genetic programming

1 Introduction

The problem of classifying attacks in the cybersecurity field involves many issues, such as the need of dealing with fast data streams, the non-stationary nature of attacks (concept drift) and the uneven distribution of the classes. Classical data mining algorithms usually are not able to handle all these issues. Ensemble-based algorithms [5] fit well this challenging scenario, as they are incremental, robust to noise, scalable and operate well on unbalanced datasets. Data-driven strategies for combining the classifiers composing the ensemble have proven to be more effective than classic combination schemes relying on non-trainable aggregation functions (such as weighted voting of the base models' predictions). However, since it may be necessary to re-train (part of) the ensemble when new data become available, this phase should not be computationally expensive. In this work, we describe an approach to the detection of attacks in fast data

© Springer Nature Switzerland AG 2020
Y. D. Sergeyev and D. E. Kvasov (Eds.): NUMTA 2019, LNCS 11973, pp. 269–277, 2020.
https://doi.org/10.1007/978-3-030-39081-5_24

streams, which can incrementally update an ensemble of classifiers when changes in the data are detected. This approach, named *CAGE-MetaCombiner* (*CMC* for short), is a distributed intrusion detection framework, which exploits the distributed GP tool CAGE [4] to generate the combiner function of the ensemble by mixing up a collection of non-trainable functions. A major advantage of using non-trainable functions as building blocks is that they can be evolved without any extra phase of training and, thus, they are particularly apt to handle concept drifts, also in the case of stringent processing-time constraints. More details on the ensemble-based algorithm and an experimentation on artificial datasets can be found in [7], while here we describe the general architecture and the usage of a real intrusion detection dataset. The rest of the paper is structured as follows. Section 2 gives some background information about Genetic Programming. Section 3 describes the proposed approach and system architecture. Section 4 illustrates experiments conducted to validate the approach and to compare it with related ones. Finally, Sect. 5 draws some conclusions and discusses a number of open issues and relevant lines of future work.

2   Background: Genetic Programming

Genetic Programming (GP), inspired by Darwin's evolutionary theories, evolves a population of solutions (individuals) to a problem for a number of generations. It can be used to learn both the structure and parameters of the model; the individuals (chromosomes) of typical GP approaches are trees. The internal nodes of the tree are functions and the leaves are typically the problem variables, constants or random numbers. The initial population of GP is a set of trees generated randomly. During the evolutionary process, the individuals are evolved until they reach the optimal solution (or a good approximation of it) of the problem, or until a maximum number of generations is reached. The evolution is driven by a function of fitness, which is chosen for the particular problem to be solved and represents the goodness of a solution of the problem. Similar to other evolutionary algorithms, for each generation, two genetic operators (crossover and mutation) are performed on some individuals, chosen randomly on the basis of their fitness: individuals with better fitness have more chance to be chosen.

The crossover operator swaps two random subtrees of two individuals (parents) and generates two new individuals (children). Moreover, the mutation operator is performed on a single individual and mutates a random subtree and generate a new individual. Figure 1, show an example of the crossover and mutation operator. By these two operators GP can search the problem landscape in order to find the optimal solution. Then, the new generated individuals are added to the populations and compete with other individuals based on their fitness, i.e., the better individuals have more chance to survive. This process leads to find better solutions during the evolution of the process.

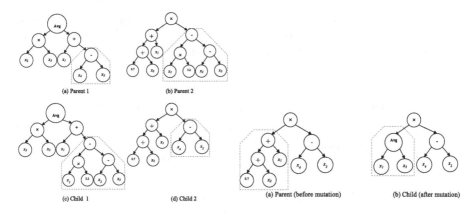

Fig. 1. An example of GP crossover (left) and mutation (right). In GP crossover two random subtrees of the parents are selected and swapped with each other and generates two new individuals. Here, the function set contains Avg, +, and ×, and the terminal set contains problem variables and some random numbers. In GP mutation, a random subtree of the parent is selected and substituted with a new random subtree.

3 Proposed Framework: Approach and System Architecture

This section describes our Intrusion Detection (ID) approach, and an IDS architecture supporting the main tasks (e.g., drift detection, models' induction, etc.) involved in it.

3.1 Stream-Oriented Ensemble Discovery and Prediction Approach

The proposed approach relies on analysing a stream $D = d_0, d_1, \ldots$ of log data, containing both labelled and unlabelled data tuples, with the help of an ensemble model E, which consists of two components: *(i)* a list $base(E)$ of base classifiers, say $h_1, \ldots, h_k$ for some k in $\mathbb{N}$, such that each h_j encodes a function mapping any data tuple d of D to an anomaly score $h_j(d) \in [0, 1]$ (the higher the score, the more likely d is estimated to be an intrusion attack); *(ii)* a combiner function ϕ^E, which maps any data tuple d of D to an overall anomaly score $\phi^E(d) \in [0, 1]$, derived from the predictions $h_1(d), \ldots, h_k(d)$.

Both the base classifiers and the ensemble E are built through a continuous learning and prediction scheme, where incoming labelled tuples are used as training examples, while unlabelled one are classified, as either malicious or normal, using E, as they arrive. The approach relies on a window-based processing strategy, where the input data stream D is split into a sequence $D_0, D_1, \ldots, D_i, D_{i+1}, \ldots$ of non-overlapping fixed-length windows, consisting each of n temporally contiguous tuples.

The main computation steps of our continuous learning and prediction approach are summarised below.

As soon as a new window D_i has been gathered, the approach tries to assess whether some kind of concept drift has occurred, by suitably comparing the data distribution of D_i with that of the previous window(s).

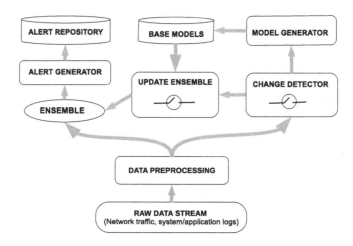

Fig. 2. A general architecture for classifying large streams of IDS data.

In case a drift is detected, the labelled tuples of D_i are split into two sets: a training subset T_i and a validation set V_i. The former is used to train α different base classifiers, by resorting to different induction algorithms, like those described in the experimental section. Among the discovered classifiers, the l ones with the highest accuracy scores are chosen and added to $base(E)$. For efficiency reasons, at most m base classifiers are kept in E. Hence, whenever $|base(E)|$ overcomes m, $|base(E)| - m$ classifiers must be removed from $base(E)$, using some suitable selection strategy. At this point, a novel combiner function ψ^E is derived for E by running a GP procedure that tries to maximise the accuracy of ψ^E over V_i. More details on this respect are given in the next subsection.

As a last processing step for the current window D_i, the ensemble E is applied to the tuples of D_i, in order to associate the unlabelled ones with a class label and to update the statistics needed for concept-drift detection.

3.2   Conceptual Architecture and Implementation Details

Figure 2 shows the architecture of the intrusion-detection system that was implemented to validate the approach described above. The architecture features several functional modules, which correspond to the main kinds of data-stream processing tasks that are involved in the approach. We next briefly describe these modules and some details on how their current implementation in the prototype system.

Change Detector. This module analyses the data stream in search of possible changes in the distribution of the target classes (i.e. normal vs attacks), representing evidence for a concept drift. For the sake of efficiency (and suitability for an online ID setting), such analysis relies on the incremental computation of statistics for every new incoming data window D_i.

The current prototype system resorts to a classic change-detection method of this kind, namely ADaptative WINDdowing (ADWIN) [1] (using a confidence level δ of 0.002). However, notice that our approach can adopt alternative drift detection methods—see [9] for a survey on this topic.

Model Generator. This module is responsible for inducing a collection of base models from given training tuples, and it is called when building an initial version of the ensemble and whenever a concept drift is detected on a window D_i (to extract a collection of new base models that capture well the emerging concept). Currently, this module implements a number of classic classifier-induction methods, including decision tree learning, Bayesian methods, logistic regression and k-NN (see Sect. 4 for details).

Update Ensemble. This module is meant to produce an updated version of the ensemble E, starting from an updated collection of base models. To this end, it must face two main tasks: *(i)* discard some of the base models, in case they are more than the maximum threshold m, and *(ii)* generate a combiner function for E via evolutionary computation.

As to the first task, many strategies were defined in the literature [8] for discarding inaccurate/redundant/obsolete base models, like the one used in the experimentation and described in the next section, which could be integrated in our framework.

In order to define an optimal combiner function for the selected base models, a GP strategy is adopted in the framework, briefly described below. Basically, any candidate combiner (function) is encoded as a GP tree where each non-leaf node can be associated with a non-trainable aggregation function (i.e. average, weighted average, multiplication, maximum and median, replicated with a different arity, from 2 to 5), while each leaf refers to one of the base models. The fitness score for evaluating the quality of such a tree w.r.t. the current data window D_i is simply computed as the prediction error made by the resulting ensemble on the validation set V_i – needing no extra computation on other training data. Further details on how the combiner function of the ensemble is evolved are reported in [6]. The tool used to evolve the combiner functions is a distributed GP implementation of a fine-grained cellular model, named CAGE [4], which can run on both distributed-memory parallel computers and distributed environments.

Other Modules. In principle, the stream of input log data may come from different kinds of sources, such as network-traffic logs, system/application logs, etc. Thus, suitable data preprocessing and feature extraction methods must be implemented, in the *Data Preprocessing* module, to turn these data into a homogenous collection of data tuples, in order to make them undergo our analysis approach.

Finally, the *Alert Generation* module is devoted to classify the incoming unlabelled tuples, using the ensemble model E, and to generate and register suitable alerts for each detected intrusion attack.

4   Experimental Section

In this section, we verified the quality of the proposed approach on the ISCX IDS dataset [12]. This dataset was created by capturing seven days of network traffic in a controlled testbed made of a subnetwork placed behind a firewall. Normal traffic was generated with the aid of agents that simulated normal requests of human users following some probability distributions extrapolated from real traffic. Attack were generated with the aid of human operators. The result is a fully labelled dataset containing realistic traffic scenarios (see Table 1). Different days contain different attack scenarios, ranging from HTTP Denial of Service, DDos, Brute Force SSH and attempts of infiltrating the subnetwork from the inside.

A Linux cluster with 16 Itanium2 1.4 GHz nodes (2 GBytes of RAM), connected by a Myrinet high performance network, was employed to conduct the experiments. The GP framework was run by using the same parameters as in the original paper, since in this work we are no interested in tuning the parameters to improve the performance. In particular, the probability of crossover was set to 0.7, the probability of mutation was set to 0.1, a population of 120 individuals (each one with a maximum depth set to 7) was used for 500 generations. All the results were obtained by averaging 30 runs.

Among the many metrics for evaluating classifier systems, in this paper we choose recall and precision, because they give an idea of the capability of the system to individuate the attacks and to reduce the number of false alarms. We remind that the recall indicates the ratio between the correctly predicted attacks to the total number of the attacks (a value of 100% means that all the attacks were detected). The precision indicates the ratio between the number of correctly predicted attacks and the total number of predicted attacks (a value of 100% means that no false alarms were signaled).

The AUC metric quantifies the area under the ROC curve. The ROC curve is computed comparing the false positive rate (i.e., recall) and the true positive rate (i.e., the ratio between the false alarm signaled above all normal connections processed). It is evident that an AUC close to 1 means an optimal recognition rate.

In order to train the classifiers of the ensemble, each window of the stream is divided into two equal parts (each on of 50%): the first part is used to train the base classifiers, and the other part is used as a validation set for evolving the combination function through the evolutionary algorithm. The base ensemble is composed of 10 classifiers, while the maximum number of classifiers is 20.

CAGE-MetaCombiner (CMC) adopts many different learners as base classifiers, all taken from the well-known WEKA tool[1]. In more detail, the classifiers

---

[1] http://www.cs.waikato.ac.nz/ml/weka.

are the following ones: J48 (decision trees), JRIP rule learner (Ripper rule learning algorithm), NBTree (Naive Bayes tree), Naive Bayes, 1R classifier, logistic model trees, logistic regression, decision stumps and 1BK (k-nearest neighbor algorithm).

Table 1. Main characteristics of the ISCX IDS dataset.

Day	Description	Pcap file' size (GB)	No. of flows	Perc. of attacks
Day 1	*Normal traffic without malicious activities*	16.1	359,673	0.000%
Day 2	*Normal traffic with some malicious activities*	4.22	134,752	1.545%
Day 3	*Infiltrating the network from the inside &* Normal traffic	3.95	153,409	6.395%
Day 4	*HTTP Denial of Service & Normal traffic*	6.85	178,825	1.855%
Day 5	*Distributed Denial of Service using an IRC Botnet*	23.4	554,659	6.686%
Day 6	*Normal traffic without malicious activities*	17.6	505,057	0.000%
Day 7	*Brute Force SSH + Normal activities*	12.3	344,245	1.435%

Table 1 illustrates the different distributions of the attacks, which usually are grouped in a small range of windows; in addition, for different days, different types of attack are detected. This characteristic is very useful in testing the capability of our algorithm to handle drifts, as, when a new type of attack is present in the data, usually a drift can be observed.

We compared our approach with the HoeffdingTree algorithm (classic and boosted version) for different width of the windows (1,000, 2,000 and 5,000 of tuples). The results are reported in Table 2.

Differently from our approach, the HoeffdingTree algorithm updates more frequently its model; therefore, for small windows (1k) it preforms better in terms of precision and AUC. However, as the size of the window increases (2k and 5k), our approach performs sensibly better. Furthermore, while for both the versions of the HoeffdingTree, a performance degradation of the ensemble-based algorithm can be observed when the width is increased, this behavior does not affect our approach; probably, it is due to the fact that, in our case, when a drift is detected, the algorithm updates/replaces the models and re-weights the classifiers (by recomputing the combination function).

Table 2. Precision, rEcall and AUC for the comparison among our approach, the Hoeffding tree (classical and boosted version) on the ISCX dataset.

	Precision			Recall			AUC		
	1k	2k	5k	1k	2k	5k	1k	2k	5k
CMC	83.46±0.28	87.59±0.03	92.42±2.20	88.39±0.69	88.25±1.08	82.44±0.69	0.84±.022	0.87±.003	0.89±.001
HT boosted	84.72±2.67	81.79±2.82	79.50±0.14	85.20±1.04	81.35±2.14	70.62±1.15	0.82±.021	0.79±.007	0.75±.009
HoeffdingTree	89.22±1.67	87.51±2.23	87.23±1.80	92.66±0.92	87.76±2.86	79.69±0.53	0.89±.014	0.88±.001	0.85±.012

5   Conclusion, Discussion and Future Work

An ensemble-based framework for detecting intrusions in streaming logs has been presented. The framework features a drift detection mechanism, allowing it to cope with the non-stationary nature of intrusion attacks, and a wheel strategy for removing inaccurate base models when such drift occur. The ensemble's combiner function is updated by using a GP method, meant to maximise the ensemble's accuracy on a small validation set. Preliminary test results on public data confirmed that the framework is a viable and effective solution for detecting intrusions in real-life application scenarios. Several issues major lines of extension are open for this work, which are summarised below.

Integrating Outlier-Oriented Models. The base models currently used in our ensemble approach are learnt by only reusing classifier-induction algorithms, which need to be trained with labelled example tuples covering two classes (at least): normal behaviour vs. intrusion attacks. To allow the framework to react more promptly to novel attack types, we plan to also incorporate (unsupervised) outlier detection models, and evaluate the ability of our ensemble learning scheme to effectively combine the anomaly scores returned by such base models—along the recent research line of outlier ensembles [11].

Using More Expressive Combiners. In order to allow our framework to benefit from the discovery of expressive combiner functions, one possible solution is to enlarge the set of non-trainable functions that can be used as building block in the GP search. To this end, in addition to the aggregation functions considered in this work, we will investigate on using fuzzy-logics functions (e.g., t-norms, t-corms), as well as *generalised mixture functions* [2], which compute a weighted combination of their input variables where the weights are determined dynamically, based on the actual values of the variables.

An even more flexible "context-aware" ensemble scheme could be obtained when allowing the combiner function to possibly change the logics for combining the predictions made by the base models on a given test instance, according to the characteristics of the instance itself. This is in line with the classic area of Mixture of Experts models [10] and the more recent one of Dynamic Ensemble Selection/Weighting [3] approaches, where the final overall prediction returned for a test instance x is (mainly) grounded on the predictions made by the base classifiers that look more competent for (the region of the instance space that contain) x. How to efficiently and effectively extend our approach is a challenging matter of study.

Other Work. A further direction of future work concerns adapting the fitness score (used to guide the evolutionary search of the ensemble's combiner function) in a way that the degree of diversity among the base models is taken into account. Indeed, if having a high level of diversity is a desideratum for an ensemble classifier in general, it may become a key feature for making it robust enough towards the dynamically changing nature of intrusion detection scenarios.

References

1. Bifet, A., Gavalda, R.: Learning from time-changing data with adaptive windowing. In: SDM, vol. 7, pp. 443–448. SIAM (2007)
2. Costa, V.S., Farias, A.D.S., Bedregal, B., Santiago, R.H., Canuto, A.M.P.: Combining multiple algorithms in classifier ensembles using generalized mixture functions. Neurocomputing **313**, 402–414 (2018)
3. Cruz, R.M., Sabourin, R., Cavalcanti, G.D.: Dynamic classifier selection: recent advances and perspectives. Inf. Fusion **41**, 195–216 (2018)
4. Folino, G., Pizzuti, C., Spezzano, G.: A scalable cellular implementation of parallel genetic programming. IEEE Trans. Evol. Comput. **7**(1), 37–53 (2003)
5. Folino, G., Pisani, F.S.: Combining ensemble of classifiers by using genetic programming for cyber security applications. In: Mora, A.M., Squillero, G. (eds.) EvoApplications 2015. LNCS, vol. 9028, pp. 54–66. Springer, Cham (2015). https://doi.org/10.1007/978-3-319-16549-3_5
6. Folino, G., Pisani, F.S., Sabatino, P.: A distributed intrusion detection framework based on evolved specialized ensembles of classifiers. In: Squillero, G., Burelli, P. (eds.) EvoApplications 2016. LNCS, vol. 9597, pp. 315–331. Springer, Cham (2016). https://doi.org/10.1007/978-3-319-31204-0_21
7. Folino, G., Pisani, F.S., Sabatino, P.: An incremental ensemble evolved by using genetic programming to efficiently detect drifts in cyber security datasets. In: Genetic and Evolutionary Computation Conference, Companion Material Proceedings, GECCO 2016, Denver, CO, USA, 20–24 July 2016, pp. 1103–1110 (2016)
8. Gama, J., Žliobaitė, I., Bifet, A., Pechenizkiy, M., Bouchachia, A.: A survey on concept drift adaptation. ACM Comput. Surv. (CSUR) **46**(4), 44 (2014)
9. Gonçalves Jr., P.M., de Carvalho Santos, S.G., Barros, R.S., Vieira, D.C.: A comparative study on concept drift detectors. Expert Syst. Appl. **41**(18), 8144–8156 (2014)
10. Masoudnia, S., Ebrahimpour, R.: Mixture of experts: a literature survey. Artif. Intell. Rev. **42**(2), 275–293 (2014)
11. Micenková, B., McWilliams, B., Assent, I.: Learning outlier ensembles: the best of both worlds-supervised and unsupervised. In: ACM SIGKDD 2014 Workshop ODD2 (2014)
12. Shiravi, A., Shiravi, H., Tavallaee, M., Ghorbani, A.A.: Toward developing a systematic approach to generate benchmark datasets for intrusion detection. Comput. Secur. **31**(3), 357–374 (2012)

A Dynamic Load Balancing Technique for Parallel Execution of Structured Grid Models

Andrea Giordano[1] , Alessio De Rango[2] , Rocco Rongo[3] ,
Donato D'Ambrosio[3] , and William Spataro[3](✉)

[1] ICAR-CNR, Rende, Italy
giordano@icar.cnr.it
[2] DIATIC, University of Calabria, Rende, Italy
alessio.derango@unical.it
[3] DEMACS, University of Calabria, Rende, Italy
{rocco.rongo,donato.dambrosio,william.spataro}@unical.it

Abstract. Partitioning computational load over different processing elements is a crucial issue in parallel computing. This is particularly relevant in the case of parallel execution of structured grid computational models, such as Cellular Automata (CA), where the domain space is partitioned in regions assigned to the parallel computing nodes. In this work, we present a dynamic load balancing technique that provides for performance improvements in structured grid model execution on distributed memory architectures. First tests implemented using the MPI technology have shown the goodness of the proposed technique in sensibly reducing execution times with respect to not-balanced parallel versions.

Keywords: Parallel computing · Parallel software tools · Load balancing · Cellular Automata

1 Introduction

The computational demands of complex systems simulation, such as in Computational Fluid Dynamic (CFD), are in general very compute intensive and can be satisfied only thanks to the support of advanced parallel computer systems. In the field of Modeling and Simulation (M&S), approximate numerical solutions of differential equations which rule a physical system (e.g., Navier-Stokes equations) are obtained by using parallel computers [4]. Classical approaches based on calculus (e.g., Partial Differential Equations - PDEs) often fail to solve these kinds of equations analytically, making a numerical computer-based methodology mandatory in case of solutions for real situations. Discretization methods, such as the Finite Element Method (FEM) or Finite Difference Method (FDM) (c.f., [7,9,14,16]), which estimate values at points over the considered domain, are often adopted to obtain approximate numerical solutions of the partial differential equations describing the system. Among these discrete numerical methodologies, Cellular Automata (CA) have proven to be particularly adequate for

© Springer Nature Switzerland AG 2020
Y. D. Sergeyev and D. E. Kvasov (Eds.): NUMTA 2019, LNCS 11973, pp. 278–290, 2020.
https://doi.org/10.1007/978-3-030-39081-5_25

systems whose behavior can be described in terms of local interactions. Originally studied by John von Neumann for studying self-reproducing issues [27], CA models have been developed by numerous researchers and applied in both theoretical and scientific fields (c.f., [2,10,13,15,21,24–26,29,30,38]).

Thanks to their rule locality nature and independently from the adopted formal paradigm, complex systems simulation based on the above numerical methodologies can be straightforwardly implemented on parallel machines. Typically, parallelizations based on OpenMP, MPI and the more recent GPGPU approach (e.g., with CUDA or OpenCL) have proven to be valuable solutions for efficient implementations of computational models (e.g., [1,3,18,19,31,32,36,37]). Nevertheless, computational layers based on higher level abstractions (i.e, by using a Domain Specific Language - DSL) of the adopted paradigm have been applied with success for simulating complex systems (e.g., [9,22,34]).

Parallel Computing applications require the best distribution of computational load over processing elements, in order to better exploit resources [23]. For instance, this is the case of CA which model the dynamics of topologically connected spatial systems (e.g., lava or debris flows), in which the evolution initially develops within a confined sub-region of the cellular space, and further expanding on the basis of the topographic conditions and source location(s) [6]. Indeed, when a data-parallel spatial decomposition is utilized, most computation can take place where the so-called "active" cells are located, with the risk of overloading the involved processing element(s). In this case, an effective load-balancing technique can mitigate this issue.

Load Balancing (LB) consists in partitioning the computation between processing elements of a parallel computer, to obtain optimal resource utilization, with the aim of reducing the overall execution time. In general, the particular parallel programming paradigm which is adopted and the interactions that occur among the concurrent task determine the suitability of a *static* or *dynamic* LB technique. Static LB occurs when tasks are distributed to the processing elements before execution, while dynamic LB refers to the case when the workload is dynamically distributed during the execution of the algorithm. More specifically, if the computation requirements of a task are known *a priori* and do not change during computation, a static mapping could represent the most appropriate choice. Nevertheless, if these are unknown before execution, a static mapping can lead to a critical imbalance, resulting in dynamic load balancing being more efficient. However, if the amount of data to be exchanged by processors produces elevated communication times (e.g., due to overhead introduced system monitoring, process migration and/or extra communication), a static strategy may outperform the advantages of the dynamic mapping, providing for the necessity of a static load balancing.

The simulation of complex physical phenomena implies the handling of an important amount of data. Several examples of both static and dynamic LB approaches can be find in literature for structured grid based models parallel implementations. In CAMEL [11] authors adopt a static load balancing strategy based on the scattered decomposition technique [28] which effectively can reduce the number of non-active cells per processor. A dynamic load-balancing algorithm is adopted in P-CAM [35]. P-CAM is a simulation environment based on a computational framework of interconnected cells which are arranged in graphs defining the cell interconnections and interactions. When a simulation starts, a decomposition of these graphs on a parallel machine is generated and a migration of cells occur thanks to a load balancing strategy. Another notable example based on a dynamic LB technique is presented in [39], and applied to an open-source parallel library for implementing Cellular Automata models, and provides for meaningful performance improvements for the simulation of topologically connected phenomena. Other recent examples can be found in [12] referred to multi-physics simulations centered around the lattice Boltzmann method, and in [17] where authors study LB issues in decentralized multi-agent systems by adopting a sandpile cellular automaton approach.

Accordingly, we here present an automatic domain detection feature that dynamically (and optimally) balances computational load among processing elements of a distributed memory parallel computer during a simulation. The paper is organized as follows. In Sect. 2 parallel decomposition techniques for CA models in distributed memory environments is discussed; subsequently, in Sect. 3, the proposed dynamic load balancing technique for CA is presented while preliminary experimental results are reported in Sect. 4. Eventually, a brief discussion and future developments section concludes the paper.

2   Parallel Execution of Cellular Automata in Distributed Memory Architectures

As anticipated, the Cellular Automata (CA) computational paradigm is particularly suitable for describing some complex systems, whose dynamics may be expressed in terms of local laws of evolution. In particular, CA can be fruitfully adopted to model and simulate complex systems that are characterized by an elevated number of interacting elementary components. Nevertheless, thanks to their implicit parallel nature, CA can be fruitfully parallelized on different parallel machines to scale and speed up their execution. A CA can be considered as a d-dimensional space (i.e., the cellular space), partitioned into elementary uniform units called cells, representing a finite automaton (f_a). Input for each f_a is given by the cell's state and the state of a geometrical pattern, invariant in time and space over the cellular space, called the cell's neighbourhood. Two well-known examples of two-dimensional neighbourhoods are shown Fig. 1. At time $t = 0$, the CA initial configuration is defined by the f_a's states and the CA subsequently evolves by applying the f_a's *transition function*, which is simultaneously applied to each f_a at each computational step. Despite their simplicity, CA can produce complex global behavior and are equivalent to Turing Machines from a computational viewpoint.

The CA execution on both sequential or parallel computers consists evaluating the transition functions over the cellular space. In this case, at step t, the

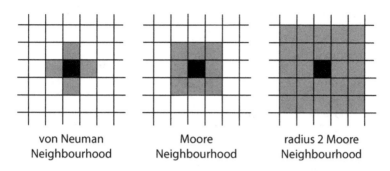

Fig. 1. CA neighbourhoods: von Neumann and Moore neighbourhoods are the left ones, respectively, both with visibility radius equal to 1, while a Moore neighbourhood with visibility radius equal to 2 is shown on the right.

evaluation of a cell transition function takes as input the state of neighbouring cells at step $t-1$. This requires that the cell states at step $t-1$ have to be stored in memory when executing step t, obtaining the so-called "maximum parallelism" behaviour. This issue can be straightforwardly implemented by using two matrices for the space state: the *read* and the *write* matrix. When executing of a generic step, the evaluation of the transition function is obtained by reading states from the read matrix and by then writing results to the write matrix. When all of the transition functions of cells have been evaluated, the matrices are swapped and the following CA step can take place.

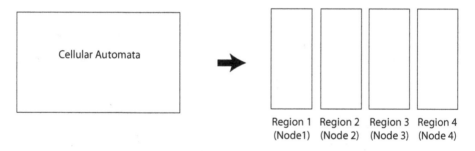

Fig. 2. Cellular space partitioning in regions that are assigned to different parallel computing nodes.

Being parallel computational models, CA execution can be efficiently parallelized by adopting a typical data-parallel approach, consisting in partitioning the cellular space in regions (or territories) and assigning each of then to a specific computing element [8,20] as shown in Fig. 2. Each region is assigned to a different computing element (node), which is in responsible of executing the transition function of all the cells belonging to that region. Since the computation of

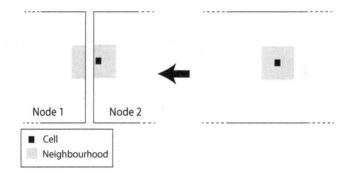

Fig. 3. The neighbourhood of a edge cell overlapping more regions

a transition function of a cell is based on the states of the cell's neighbourhood, this can overlap two regions for cells located at the edge part, as seen in Fig. 3.

Therefore, the transition function execution of cells in this area needs information that belongs to a adjacent computing node. As a consequence, in order to keep the parallel execution consistent, the states of these border cells (or *halo* cells) need to be exchanged among neighboring nodes at each computing step. In addition, the border area of a region (the *halo* cells) is divided in two different sub-areas: the *local border* and the *mirror border* (see Fig. 4). The local border is handled by the local node and its content replicated in the mirror border of the adjacent node.

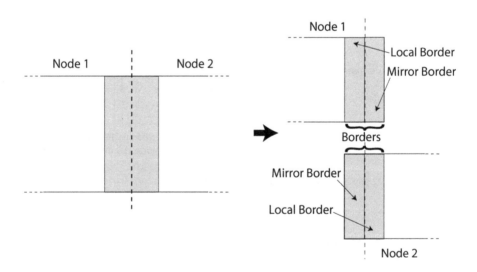

Fig. 4. Border areas of two adjacent nodes

The parallel execution of a CA in a distributed memory architecture consists in each node executing the loop illustrated in Algorithm 1. At each iteration, the node sends its halo borders to neighbour nodes, receiving in turn the corresponding halo cells (lines 3–4). Afterwards, the transition function is applied by reading input values of cells from the read matrix and updating the new result to the write matrix (i.e., recall the "maximum parallelism" behaviour described above). After all cells have been have updated, a swap between the read and write matrices occurs and the next CA step can be taken into account (line 7).

Algorithm 1. CA execution

---

1 while !StopCriterion()   // Loop until CA stop criterion met
2 do
3 | SendBorderToNeighbours() // Send halo borders to left and right
  | neighbour nodes
4 | ReceiveBorderFromNeighbours() // Receive halo borders from left
  | and right neighbour nodes
5 | ComputeTransitionFunction() // Read from read matrix and write
  | to write matrix
6 | SwapReadWriteMatrices() // Swap read and write matrices
7 | $step \leftarrow step + 1$ // Next CA step

---

3   A Dynamic Load Balancing Technique for Partitioned Cellular Automata

The proposed LB algorithm consists in dynamically computing the optimal workload that each processing element has to take into account for achieving uniform execution times over a simulation. The approach presented in this paper relies on exchanging computation load among computing nodes at given steps on the basis of a specific criterion. For example, the balancing of the load can occur at a predefined rate of CA time steps, or when the execution times among nodes is particularly unbalanced, and so on. The load balancing is actually achieved by means of CA columns exchange among nodes. In particular, each node exchanges columns with its neighbour nodes in a parallel fashion by exploiting the same communication channel already in use for the halo exchange[1]. During a simulation, the execution times experienced by the nodes for executing the current step are retrieved and stored in each node. When the LB phase has to be executed, each node sends its step time to a specific "master" node, which is in charge of establishing a suitable columns exchanges that nodes must perform in order to achieve a balanced workload.

---

[1] For the sake of simplicity, in this first implementation of the LB procedure the exchange of columns is *not toroidal*, i.e. it is not possible to exchange columns between the rightmost node and the leftmost node.

Algorithm 2. CA execution with Dynamic Load Balancing

```
1  while !StopCriterion()                  // Loop until CA stop criterion met
2  do
3  |  if LoadBalancingCriterion()          // Load Balancing step reached
4  |  then
5  |  |  SendMyExecInfo()                   // send to Master my size and timings
6  |  |  SequenceOfFLows = ReceiveSeqFlows()       // Receive sequence of
   |  |     flows from Master
7  |  |  if IamLBMaster then
8  |  |  |  LBInfo=ReceiveExecInfo()        // Receive sizes and timings
   |  |  |     from all nodes
9  |  |  |  newRegionSizes = LoadBalance(LBInfo)   // Determine new node
   |  |  |     sizes
10 |  |  |  allSequenceFlows = ComputeFlows(NewRegionSizes)
   |  |  |     // Determine sequence of flows for all nodes
11 |  |  |  SendFlows(allSequenceFlows)     // Send sequence of flows to
   |  |  |     nodes
12 |  |  forall flows ∈ SequenceOfFLows do
13 |  |  |  ExchangeLBFlows(flows)
   |
   |     // Back to normal CA execution
14 |  SendBorderToNeighbours()
15 |  ReceiveBorderFromNeighbours()
16 |  ComputeTransitionFunction()
17 |  SwapReadWriteMatrices()
18 |  step ← step + 1                       // Next CA step
```

The new CA loop containing the code implementing the load balancing is described in Algorithm 2. Please note that the normal CA execution takes place at every step as reported at lines 14–18. As mentioned before, the LB is executed when some conditions are met, i.e., when LoadBalancingCriterion() is true (line 3). At line 5, each node sends its CA space size (i.e., the number of columns), along with the elapsed time required for computing the last CA step, to the master node (SendMyExecInfo()), and waits for receiving information from this latter on columns to be exchanged to achieve load balancing (line 6). In particular, this information (SequenceOfFlows) consists in a *sequence* of columns to be exchanged with the left and the right neighbours. The reason behind why a sequence of columns exchange is necessary, rather than just an simple exchange of a given number of columns to left and right nodes, will be clarified in the following. At line 12–13 the actual columns exchange takes place. In particular, each flows of the sequence, i.e, the columns to be exchanged with the left and right neighbour, is in practice applied through ExchangeLBFlows().

Let us now summarize the master behaviour (lines 8–11). At line 8, the master receives information about the nodes state (i.e., space size and elapsed times) and determines, at line 9 (LoadBalance()), the new region sizes for the nodes, which minimize the unbalancing of the workload. On the basis of the new region sizes, the determination of the flows of columns that must be exchanged among nodes can be straightforwardly computed. However, it can be possible that some of the determined flows may exceed the columns availability of a given node. For instance, let us assume there are 3 nodes N1, N2 and N3, each having a 100 column CA space size right before the LB phase evaluation. Let us also assume that the LoadBalance() function computes new region sizes as 70, 20 and 210. In this case, in order to achieve 210 columns for N3, a flow of 110 columns should be sent from N2 to N3 (recall that it is not possible to exchange columns between N3 and N1, as reported in Sect. 2), though N2 hosts only 100 columns. In this example, this issue can be addressed by simply considering 2 exchange phases. In the first step, all the 100 columns between N2 to N3 are exchanged, while the remaining 10 columns are exchanged in a second phase. Note that in the second phase N2 hosts 30 columns, having received them from N1 in the first phase, and so is now able to send the 10 columns to N3. The aforementioned sequence of flows are computed by the master in the ComputeFlows() function (line 10) and thus sent to all the nodes by the SendFlows() function (line 11).

It is worth to note that Algorithm 2 represents a general framework for achieving a dynamic load balancing during a simulation. Most of the functions seen in the above pseudo-code do not require further specifications, except for the two methods: LoadBalancingCriterion() and LoadBalance(). The implementation of these two functions determines *when* the LB should take place and *how* to resize regions so as to achieve the "optimal" load balancing. In our preliminary implementation, the load balancing occurs at a predefined rate of CA steps while the LoadBalance() function follows an heuristic based on resizing the regions taking into account the region time differences normalized with respect to their old sizes. However, other strategies can be considered and *linked* to Algorithm 2 by implementing specific versions of the two methods just described.

4   Experimental Results

Preliminary experiments were carried out for testing the performance of the proposed LB algorithm on the SciddicaT CA debris flow model [5]. The testbed is composed by a grid of 296 columns × 420 rows, representing the DEM (Digital Elevation Model) of the Tessina landslide, occurred in Northern Italy in 1992. In order to create an initial unbalanced condition among processing nodes, landslide sources were located in the lower rows of the morphology, corresponding to higher topographic elevations (see Fig. 5). As the simulation develops, the landslide expands to lower topographic altitudes, thus progressively interesting other processing elements. The simulation was run for 4000 computational steps, corresponding to the full termination of the landslide event. Other parallelizations (e.g., multi-node and multi-GPGPU implementations) performed on SciddicaT on the same data set can be found in [9] and [33].

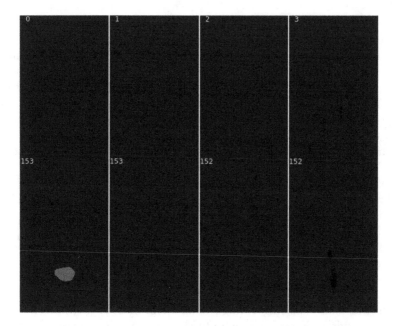

Fig. 5. Initial node partitioning for the Tessina landslide simulation. The initial landslide source in indicated in red. The upper numbering indicates the core id, the middle indicates the node partitioning as number of columns. (Color figure online)

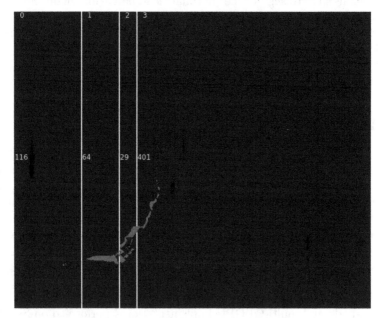

Fig. 6. Final load balanced configuration referred to the last step of the simulation, with the landslide that has evolved towards the upper-right side of the CA space. Please note the new node partitioning corresponding to a balanced configuration.

4.1   Performances

The preliminary tests were performed on a 4-core i7 Linux-based PC with 8 GB RAM. The opensource C++ OpenMPI 2.0 version of MPI was used for message passing among processes. Besides a normal not-balanced execution, three different load balancing tests were considered by considering different LB steps applications, i.e., the `LoadBalancingCriterion()` is TRUE each 250, 400 and 800 steps, respectively, for each of the three LB experiments. Table 1 summarizes the obtained results.

Table 1. Execution times of preliminary tests executed for assessing the performance of the Load Balancing algorithm on the SCIDDICA CA debris-flow model. Four different tests are reported, referred to Normal (i.e., not load balanced) execution, LB executed each 250 steps, LB each 400 steps and LB each 800 steps, respectively.

Test	Times (s)	Improvement (%)
No LB	675	-
LB each 250 steps	503	25%
LB each 400 steps	501	26%
LB each 800 steps	484	28%

As seen, the application of the LB algorithm permitted an improvement up to 28% of the overall execution time (about 484 s versus 675 s of the not-balanced algorithm). Furthermore, for these preliminary tests, improvements are noted gradually as the number of the LB application step increases, proving that in these experiments the LB algorithm indeed provides benefits in reducing execution times, though introducing, as expected, some degree of overhead which is however counterbalanced by the aforementioned absolute time decrease. Eventually, Fig. 6 shows the last step of the simulation referred to the 800-step based LB experiments. As noted, starting from an initial uniform node partitioning, the simulation ends with a new node partitioning corresponding to a balanced execution time node configuration.

5   Conclusions

We here present a dynamic load balancing feature that exploits computational resources to reduce overall execution times in parallel executions of CA models on distributed memory architectures. Specifically, the algorithm executes load balancing among processors to reduce processor timings at regular intervals, based to an algorithm which computes the optimal distribution load exchange among adjacent nodes. Preliminary experiments, considering the SciddicaT CA landslide model and executed for assessing the advantage of the dynamically load balanced version with respect the non-balanced one, resulted in good improvements on a standard 4-core i7 based PC. In particular, improvements up to 28%

were obtained when the LB algorithm is applied for the simulation of the 1992 Tessina landslide (Italy).

Further experiments will be carried out in order to compute most favorable LB parameters (e.g., other LB steps), besides testing the algorithm with other LB strategies as, for instance, considering a LB criterion only when elapsed times between nodes are significant, and other heuristics to compute the new node workload. For instance, automated optimization techniques such as evolutionary algorithms or other heuristics, will be considered for calibrating LB parameters referred to the particular parallel system and the adopted simulation model.

Future developments will regard the application of the LB algorithm on other CA models, besides the extension on two-dimensional node partitioning, thus permitting the application of the LB technique also on more complex network topologies (i.e., meshes, hypercubes, etc.).

Acknowledgments. Authors thank BS student Rodolfo Calabrò from University of Calabria for helping in code implementation and testing phases.

References

1. Abraham, M.J., et al.: GROMACS: high performance molecular simulations through multi-level parallelism from laptops to supercomputers. SoftwareX **1**, 19–25 (2015)
2. Aidun, C., Clausen, J.: Lattice-Boltzmann method for complex flows. Annu. Rev. Fluid Mech. **42**, 439–472 (2010)
3. Amritkar, A., Deb, S., Tafti, D.: Efficient parallel CFD-DEM simulations using OpenMP. J. Comput. Phys. **256**, 501–519 (2014)
4. Andersson, B., Andersson, R., Håkansson, L., Mortensen, M., Sudiyo, R., van Wachem, B.: Computational Fluid Dynamics for Engineers. Cambridge University Press (2011). https://doi.org/10.1017/CBO9781139093590
5. Avolio, M., et al.: Simulation of the 1992 Tessina landslide by a cellular automata model and future hazard scenarios. Int. J. Appl. Earth Obs. Geoinf. **2**(1), 41–50 (2000)
6. Cannataro, M., Di Gregorio, S., Rongo, R., Spataro, W., Spezzano, G., Talia, D.: A parallel cellular automata environment on multicomputers for computational science. Parallel Comput. **21**(5), 803–823 (1995)
7. Cervarolo, G., Mendicino, G., Senatore, A.: A coupled ecohydrological-three-dimensional unsaturated flow model describing energy, H_2O and CO_2 fluxes. Ecohydrology **3**(2), 205–225 (2010)
8. Cicirelli, F., Forestiero, A., Giordano, A., Mastroianni, C.: Parallelization of space-aware applications: modeling and performance analysis. J. Netw. Comput. Appl. **122**, 115–127 (2018)
9. D'Ambrosio, D., et al.: The open computing abstraction layer for parallel complex systems modeling on many-core systems. J. Parallel Distrib. Comput. **121**, 53–70 (2018). https://doi.org/10.1016/j.jpdc.2018.07.005
10. Di Gregorio, S., Filippone, G., Spataro, W., Trunfio, G.: Accelerating wildfire susceptibility mapping through GPGPU. J. Parallel Distrib. Comput. **73**(8), 1183–1194 (2013)

11. Di Gregorio, S., Rongo, R., Spataro, W., Spezzano, G., Talia, D.: High performance scientific computing by a parallel cellular environment. Future Gener. Comput. Syst. **12**(5), 357–369 (1997)
12. Duchateau, J., Rousselle, F., Maquignon, N., Roussel, G., Renaud, C.: An out-of-core method for physical simulations on a multi-GPU architecture using lattice Boltzmann method. In: 2016 International IEEE Conferences on Ubiquitous Intelligence & Computing, Advanced and Trusted Computing, Scalable Computing and Communications, Cloud and Big Data Computing, Internet of People, and Smart World Congress, pp. 581–588. IEEE (2016)
13. Filippone, G., D'ambrosio, D., Marocco, D., Spataro, W.: Morphological coevolution for fluid dynamical-related risk mitigation. ACM Trans. Model. Comput. Simul. (ToMACS) **26**(3), 18 (2016)
14. Folino, G., Mendicino, G., Senatore, A., Spezzano, G., Straface, S.: A model based on cellular automata for the parallel simulation of 3D unsaturated flow. Parallel Comput. **32**(5), 357–376 (2006)
15. Frish, U., Hasslacher, B., Pomeau, Y.: Lattice gas automata for the Navier-Stokes equation. Phys. Rev. Lett. **56**(14), 1505–1508 (1986)
16. Cervarolo, G., Mendicino, G., Senatore, A.: Coupled vegetation and soil moisture dynamics modeling in heterogeneous and sloping terrains. Vadose Zone J. **10**, 206–225 (2011)
17. Gasior, J., Seredynski, F.: A cellular automata-like scheduler and load balancer. In: El Yacoubi, S., Wąs, J., Bandini, S. (eds.) ACRI 2016. LNCS, vol. 9863, pp. 238–247. Springer, Cham (2016). https://doi.org/10.1007/978-3-319-44365-2_24
18. Gerakakis, I., Gavriilidis, P., Dourvas, N.I., Georgoudas, I.G., Trunfio, G.A., Sirakoulis, G.C.: Accelerating fuzzy cellular automata for modeling crowd dynamics. J. Comput. Sci. **32**, 125–140 (2019)
19. Giordano, A., De Rango, A., D'Ambrosio, D., Rongo, R., Spataro, W.: Strategies for parallel execution of cellular automata in distributed memory architectures, pp. 406–413, February 2019. https://doi.org/10.1109/EMPDP.2019.8671639
20. Giordano, A., et al.: Parallel execution of cellular automata through space partitioning: the landslide simulation Sciddicas3-Hex case study, pp. 505–510, February 2017
21. Higuera, F., Jimenez, J.: Boltzmann approach to lattice gas simulations. Europhys. Lett. **9**(7), 663–668 (1989)
22. Jammy, S., Mudalige, G., Reguly, I., Sandham, N., Giles, M.: Block-structured compressible Navier-Stokes solution using the ops high-level abstraction. Int. J. Comput. Fluid Dyn. **30**(6), 450–454 (2016)
23. Kumar, V.: Introduction to Parallel Computing, 2nd edn. Addison-Wesley Longman Publishing Co., Inc., Boston (2002)
24. Langton, C.: Computation at the edge of chaos: phase transition and emergent computation. Physica D **42**, 12–37 (1990)
25. Lucà, F., D'Ambrosio, D., Robustelli, G., Rongo, R., Spataro, W.: Integrating geomorphology, statistic and numerical simulations for landslide invasion hazard scenarios mapping: an example in the Sorrento Peninsula (Italy). Comput. Geosci. **67**(1811), 163–172 (2014)
26. McNamara, G., Zanetti, G.: Use of the Boltzmann equation to simulate lattice-gas automata. Phys. Rev. Lett. **61**, 2332–2335 (1988)
27. von Neumann, J.: Theory of Self-Reproducing Automata. University of Illinois Press, Champaign (1966)
28. Nicol, D.M., Saltz, J.H.: An analysis of scatter decomposition. IEEE Trans. Comput. **39**(11), 1337–1345 (1990)

29. Ninagawa, S.: Dynamics of universal computation and 1/f noise in elementary cellular automata. Chaos, Solitons Fractals **70**(1), 42–48 (2015)
30. Ntinas, V., Moutafis, B., Trunfio, G., Sirakoulis, G.: Parallel fuzzy cellular automata for data-driven simulation of wildfire spreading. J. Comput. Sci. **21**, 469–485 (2016)
31. Oliverio, M., Spataro, W., D'Ambrosio, D., Rongo, R., Spingola, G., Trunfio, G.: OpenMP parallelization of the SCIARA Cellular Automata lava flow model: performance analysis on shared-memory computers. Procedia Comput. Sci. **4**, 271–280 (2011)
32. Procacci, P.: Hybrid MPI/OpenMP implementation of the ORAC molecular dynamics program for generalized ensemble and fast switching alchemical simulations (2016)
33. Rango, A.D., Spataro, D., Spataro, W., D'Ambrosio, D.: A first multi-GPU/multi-node implementation of the open computing abstraction layer. J. Comput. Sci. **32**, 115–124 (2019). https://doi.org/10.1016/j.jocs.2018.09.012, http://www.sciencedirect.com/science/article/pii/S1877750318303922
34. Reguly, I., et al.: Acceleration of a full-scale industrial CFD application with OP2. IEEE Trans. Parallel Distrib. Syst. **27**(5), 1265–1278 (2016). https://doi.org/10.1109/TPDS.2015.2453972
35. Schoneveld, A., de Ronde, J.F.: P-CAM: a framework for parallel complex systems simulations. Future Gener. Comput. Syst. **16**(2–3), 217–234 (1999)
36. Spataro, D., D'Ambrosio, D., Filippone, G., Rongo, R., Spataro, W., Marocco, D.: The new SCIARA-fv3 numerical model and acceleration by GPGPU strategies. Int. J. High Perform. Comput. Appl. **31**(2), 163–176 (2017). https://doi.org/10.1177/1094342015584520
37. Was, J., Mróz, H., Topa, P.: Gpgpu computing for microscopic simulations of crowd dynamics. Comput. Inform. **34**(6), 1418–1434 (2016)
38. Wolfram, S.: A New Kind of Science. Wolfram Media Inc., Champaign (2002)
39. Zito, G., D'Ambrosio, D., Spataro, W., Spingola, G., Rongo, R., Avolio, M.V.: A dynamically load balanced cellular automata library for scientific computing. In: CSC, pp. 322–328 (2009)

Final Sediment Outcome
from Meteorological Flood
Events: A Multi-modelling Approach

Valeria Lupiano[1], Claudia R. Calidonna[2]([⊠]), Elenio Avolio[2],
Salvatore La Rosa[4], Giuseppe Cianflone[4,6], Antonio Viscomi[4],
Rosanna De Rosa[4], Rocco Dominici[4,6], Ines Alberico[3], Nicola Pelosi[3],
Fabrizio Lirer[3], and Salvatore Di Gregorio[2,5]

[1] IRPI UOS, National Research Council, Rende, CS, Italy
[2] ISAC UOS, National Research Council, Lamezia Terme, CZ, Italy
cr.calidonna@isac.cnr.it
[3] ISMAR UOS, National Research Council, Napoli, Italy
[4] DIBEST, University of Calabria, 87036 Rende, CS, Italy
[5] DeMaCS, University of Calabria, 87036 Rende, CS, Italy
[6] EalCUBO, 87036 Arcavacata di Rende, CS, Italy

Abstract. Coastal areas are more and more exposed to the effects of climatic change. Intense local rainfalls increases the frequency of flash floods and/or flow-like subaerial and afterward submarine landslides. The overall phenomenon of flash flood is complex and involves different phases strongly connected: heavy precipitations in a short period of time, soil erosion, fan deltas forming at mouth and hyperpycnal flows and/or landslides occurrence. Such interrelated phases were separately modelled for simulation purposes by different computational models: Partial Differential Equations methods for weather forecasts and sediment production estimation and Cellular Automata for soil erosion by rainfall and subaerial sediment transport and deposit. Our aim is to complete the model for the last phase of final sediment outcome. This research starts from the results of the previous models and introduces the processes concerning the demolition of fan deltas by sea waves during a sea-storm and the subsequent transport of and sediments in suspension by current at the sea-storm end and their deposition and eventual flowing on the sea bed. A first reduced implementation of the new model SCIDDICA-ss2/w&c1 was applied on the partial reconstruction of the 2016 Bagnara case regarding the meteorological conditions and the flattening of Sfalassà's fan delta.

Keywords: Extreme event · Flood · Sediment transport and deposition · Subaerial and subaqueous flow-like landslide · Modelling and simulation

© Springer Nature Switzerland AG 2020
Y. D. Sergeyev and D. E. Kvasov (Eds.): NUMTA 2019, LNCS 11973, pp. 291–306, 2020.
https://doi.org/10.1007/978-3-030-39081-5_26

1   Introduction

Climatic change in coastal areas increases dramatically the frequency of extreme meteorological occurrences, which induce a chain of disastrous events: extremely intense local rainfalls determine heavy soil erosion and can trigger flash floods or debris/mud/granular flows, which, on the shore, can continue as subaqueous debris flows or hyperpycnal flows at mouth of watercourses with the possibility to evolve offshore in density currents. This study is interested in the last part of this composite phenomenon, its evolution from the shoreline toward the open sea.

Different scenarios may be considered: the flood (or the debris flow) that reaches the sea as a hyperpycnal stream that flows on the sea bed and originates a density current; a delta fan, which was produced by the flood, is demolished by one or more sea-storm events of diverse intensity forming detrital submarine flows, particular meteorological conditions could give rise to the suspension of detrital matter and subsequent deposition, when the waves energy decreases. The overall phenomenon is very complex, its modelling and computer simulation (M&S) can be important in order to forecast the natural hazard and manage risk situations.

Different computational approach were adopted for M&S of the diverse inter-related phenomenological components, e.g. classic PDE approximation methods for weather forecasting [1–3] and sediment production estimation [4]; an alternative computational paradigm, the Cellular Automata, was utilized for M&S of "surface flows" [5], a CA methodology for M&S of complex systems was developed [6–8]. The most interesting models related to the phenomenological aspects of final sediment outcome are treated in several CA studies of M&S: SCIDDICA-ss2 and SCIDDICA-ss3 (Simulations through Computational Innovative methods for the Detection of Debris flow path using Interactive Cellular Automata for subaerial, subaqueous and mixed subaerial and subsequent subaqueous landslides) [9–13], M&S of density currents of Salles et al. [14] which is partially derived by a previous SCIDDICA version, SCAVATU (Simulation by Cellular Automata for the erosion of VAst Territorial Units by rainfall) [15], RUSICA (RUdimental SImulation of Coastal erosion by cellular Automata) [16], M&S of hot mudflows [17]; M&S of long-term soil redistribution by tillage [18]; M&S of soil surface degradation by rainfall [19]. The CA for surface flows [6] may be regarded as a two-dimensions space, partitioned in hexagonal cells of uniform size, the cell corresponds usually to a portion of surface; each characteristic, relevant to the evolution of the system and relative to the surface portion corresponding to the cell, is individuated as a sub-state, the third dimension (the height) features, e.g., the altitude, may be included among the sub-states of the cell; each cells embeds an identical computing device, the elementary automaton (ea), whose input is given by the sub-states of the six adjacent cells, the CA evolves changing the state at discrete times simultaneously, according to the transition function of the ea. The transition function accounts for the dynamics of the system and is compound by a sequence of "elementary" processes.

SCIDDICA-ss2 was selected as a significant base for developing in incremental way a model able to account for the physical processes regarding the final sediment outcome that are not present in SCIDDICA-ss2, but may opportunely be imported by the other models SCAVATU and RUSICA. SCIDDICA-ss2 was validated on the well-known 1997 Albano lake debris flow [10,11], this model accounted for the following processes both in subaerial zones and in subaqueous ones, summing up: determination of debris outflows towards the adjacent cells, altitude, kinetic head, debris thickness variation by detrital cover erosion, kinetic head variation by turbulence dissipation, air-water interface effects on the outflows. SCIDDICA-ss2 is able to reproduce the phenomenology of the debris flows, which overcome the coastline and continue as submarine detrital flows, while the following processes, concerning sea-storm events and their consequences have to be introduced: energy transmission from subaqueous currents and waves to granular matter in suspension, on the sea bed, in the flooded area of the shoreline, processes of transport, suspension and sedimentation of granular matter. SCIDDICA, RUSICA and SCAVATU can consider different type of flowing matter: debris, mud, sand, particles; in order to avoid confusion, the general term granular matter will be adopted, granulometries will be specified if necessary. A partial implementation of this extended model SCIDDICA-ss2/w1 was performed, the 2016 Bagnara flood event was considered for simulation in its very final parts, when the Sfalassà's fan delta [20,21] produced by the flood was demolished during successive sea-storms. A partial reconstruction of this last part, starting from the fan delta positioning and structure together with the meteorological conditions during the sea-storms, permitted to "play" with many possible scenarios in order to examine some possible final sediment outcomes. The paper is organised as follows: next section introduces to the new version of SCIDDICA, Sect. 3 presents the SCIDDICA-ss2/w&c1 partial implementation and the geological setting of the study area, the most interesting simulation results are reported, and finally conclusions are discussed.

2   The Model SCIDDICAss2/w&c1

An outline of the model SCIDDICA-ss2/w&c1 is reported in this section, starting from the original model SCIDDICA-ss2 new sub-states and new procedures are introduced in order to account also for the phenomenology of dispersion of sediments.

2.1   An Outline of the Original Model SCIDDICA-ss2

SCIDDICA-ss2 wants to capture the phenomenology of "surface flow" of granular type, debris, mud and the like in terms of complexity emergence both in subaerial areas and in subaqueous ones, it is able to model flow parting and confluence of flows according to the morphology, soil erosion and so on, it is a two-dimensions hexagonal CA model specified by the quintuple: $<R, X, S, P, \tau>$ where:

- $R = \{(x,y)|(x,y \in \mathcal{Z}) \wedge (0 \leq x \leq l_x) \wedge (0 \leq y \leq l_y)\}$ is the finite surface with the regular hexagonal tessellation, that cover the region, where the phenomenon evolves.
- $X = (0,0), (0,1), (0,-1), (1,0), (-1,0), (1,1), (-1,-1)$ identifies the geometrical pattern of cells, which influence any state change of the generic cell, identified as the central cell: X includes the central cell (index 0) and the six adjacent cells (indexes 1, ..., 6).
- S is the set of the ea states, they are specified in terms of sub-states of type ground and flow sub-states.
  - Ground sub-states: A is the cell altitude, D is the depth of soil erodable stratum that could be transformed by erosion in granular matter.
  - Granular matter sub-states: T is the average thickness of granular matter of the cell, X and Y are the coordinates of its barycenter with reference to the cell center, and K is its kinetic head.
  - Flow sub-states: ^iE is the part of outflow, the so called "external flow" (normalised to a thickness), that penetrates the adjacent cell i, $1 \leq i \leq 6$, from central cell, ^iX_E and ^iY_E are the coordinates of its barycenter with reference to the adjacent cell center, ^iK_E is its kinetic head, (six components for each sub-state); ^iI is the part of outflow toward the adjacent cell, the so called "internal flow", (normalized to a thickness) that remains inside the central cell, ^iX_I and ^iY_I are the coordinates of its barycenter with reference to the central cell center, ^iK_I is its kinetic head, (six components for all the sub-states).
- P is the set of the global physical and empirical parameters of the phenomenon, they are enumerated in the following list and are better explicated in next section: p_a is the cell apothem; p_t is the temporal correspondence of a CA step; p_{adhw}, p_{adha} are the air/water adhesion values, i.e. the landslide matter thickness, that may not be removed; p_{fcw}, p_{fca} are the air/water friction coefficient for the granular matter outflows; $p_{tdw}, p_{tda}, p_{fcw}, p_{edw}, p_{eda}$ are air/water parameters for energy dissipation by turbulence, air/water parameters for energy dissipation by erosion; p_{ml} is the matter loss in percentage when the landslide matter enters into water; p_{mtw}, p_{mta} are the air/water activation thresholds of the mobilization; p_{erw}, p_{era} are the air/water progressive erosion parameters; p_{wr} is the water resistance parameter.
- $\tau : S^7 \rightarrow S$ is the deterministic state transition for the cells in R, basic processes of the transition function are here sketched, note that ΔQ means variation of the ground sub-state Q, ground sub-state Q of the adjacent cell i, $1 \leq i \leq 6$, is specified as Q_i; the subscripts "w" and "a" in parameter names are omitted, when the formula is considered valid both in water and air.
- Mobilization Effects. When the kinetic head value overcomes an opportune threshold $(K > p_{mt})$, depending on the soil features and its saturation state, a mobilization of the detrital cover occurs proportionally to the quantity overcoming the threshold: $p_{er}(K - p_{mt}) = \Delta T = -\Delta D = -\Delta A$ (the detrital cover depth diminishes as the granular matter thickness increases), $-\Delta K = p_{ed}(K - p_{mt})$ calculates the kinetic head loss.

- Turbulence Effect. The effect of the turbulence is modelled by a proportional kinetic head loss at each SCIDDICA step: $-\Delta K = p_{td}K$.
- Granular Matter Outflows. The computation of the outflows f_i, $0 \leq i \leq 6$ from the central cell toward the cell i (f_0, is the quantity that does not flow) is performed in two steps: determination of the outflows minimizing the differences of "heights" h_i, $0 \leq i \leq 6$ in the neighborhood by the "algorithm of minimization of differences" [5, 6], and computation of the shift of the outflows with subsequent determination of external and internal outflow sub-states.
  First step: The quantity d to be distributed from the central cell: $d = T - p_{adh} = \Sigma_{0 \leq i \leq 6} f_i$. "Heights" are specified as: $h_0 = A + K + p_{adh}$; $h_i = A_i + T_i$, $0 \leq t \leq 6$; the values of flows are obtained in order to minimize $\Sigma_{0 \leq i < j \leq 6}(|(h_i + f_i) - (h_j + f_j)|)$.
  Second step: The s_i shift of f_i with slope θ_i related to heights h_i, h_0 $0 \leq i \leq 6$ is modelled as the body barycentre's movement on the slope:
  $s_i = v_i + p_t + g(\sin\theta_i - p_{fca}\cos\theta_i)p_t^2$, with g the gravity acceleration and initial velocity $v_i = \sqrt{2gH}$ in the subaerial case;
  $s_i = (1 - exp(-p_{wr} \cdot p_t))v_i \cdot p_t + g\prime(\sin\theta_i - p_{fcw} \cdot \cos\theta_i \ p_{wr})) + g\prime(\sin\theta_i - p_{fca} \cdot \cos\theta_i)p_t/p_{wr}$, the water resistance is considered for the subaqueous case, a modified Stokes equations, it is adopted with a form factor proportional to mass and $g\prime < g$, accounting for buoyancy.
- Flows Composition. The new values of the ground and granular matter sub-states are calculated according to the new values of flow sub-states.
- Air-Water Interface. An external flow from an air cell (altitude higher than water level) to a water cell (altitude lower than water level) can imply a loss of matter (water inside debris and fine grains) proportional to debris mass, specified by p_{ml}; it implies a correspondent loss of kinetic energy, determined by kinetic head decrease.

At the beginning of the simulation, we specify the states of the cells in R, defining the initial CA configuration. The initial values of the sub-states are accordingly initialized. In particular, A assumes the morphology values except for the detachment area, where the thickness of the landslide mass is subtracted from the morphology value; T is zero everywhere except for the detachment area, where the thickness of landslide mass is specified; D assumes initial values corresponding to the maximum depth of the mantle of soil cover, which can be eroded. All the values related to the remaining sub-states are zero everywhere. At each next step, the function τ is applied to all the cells in R, so that the configuration changes in time and the CA evolution is obtained.

2.2   Outline of the Preliminary Version of SCIDDICA-ss2/w&c1

SCIDDICA-ss2/w&c1 is a two-dimension hexagonal CA model, extension of SCIDDICA-ss2, which is specified by the septuplet:

$<R_{w\&c1}, G_{w\&c1}, X_{w\&c1}, S_{w\&c1}, P_{w\&c1}, \tau_{w\&c1}, \gamma_{w\&c1}>$ where:

- $R_{w\&cl} = R$
- $G_{w\&cl}$ is the set of cells, which undergo to the influences of the "external world"; in this case, they are the "underwater" cells, which are exposed to the effect of the waves.
- $X_{w\&cl} = X$
- $S_{w\&cl}$ is the set of the ea states, they are specified in terms of sub-states of type ground, granular matter and flow sub-states. This model has to account layers of matter of different granularity (n layers) and sub-states, related to waves and currents.
  - Ground sub-states are the same as in SCIDDICA-ss2.
  - Granular matter sub-states for each layer j, $1 \leq j \leq n$: T_j is the average thickness of granular matter of the cell, X_j and Y_j are the coordinates of its barycenter with reference to the cell center, K_j is its kinetic head; S_j is the granular matter of layer j in suspension in the cell, normalized to a thickness.
  - Flow sub-states for each layer j, $1 \leq j \leq n$: ^iE_j is the part of outflow, the so called "external flow" (normalized to a thickness), that penetrates the adjacent cell i, $1 \leq i \leq n$, from central cell, $^iX_{Ej}$ and $^iY_{Ej}$ are the coordinates of its barycenter with reference to the adjacent cell center, $^iK_{Ej}$ is its kinetic head, (six components for each sub-state); ^iI_j is the part of outflow toward the adjacent cell, the so called "internal flow", (normalized to a thickness) that remains inside the central cell, $^iX_{Ij}$ and $^iY_{Ej}$ are the coordinates of its barycenter with reference to the central cell center, $^iK_{Ij}$ is its kinetic head, (six components for all the sub-states); ^iS_j is the part of suspended matter outflow (normalized to a thickness), that penetrates the adjacent cell i, $1 \leq i \leq 6$, from central cell, $^iX_{Sj}$ and $^iY_{Sj}$ are the coordinates of its barycenter with reference to the adjacent cell center.
  - Wave and current sub-states: A_w, the wave amplitude; L_w, the wave length; X_w, Y_w, component $x - y$ of wave direction; the $C_x, C_y, x - y$ speed components of the surface current.
- $P_{w\&cl}$ is the set of the global physical and empirical parameters of the phenomenon, there are the same parameter of P except p_{adhw}, p_{adha} the air/water adhesion values and p_{adhw}, p_{adha} the air/water friction coefficient for the granular matter outflows; they are multiplied because take a different value for each layer j, $1 \leq j \leq n$: $^jp_{adhw}$, $^jp_{adha}$. Parameters, regarding the wave demolition of the layers, are the activation thresholds of the mobilization $^jp_{mt}$ ($^jp_{mts}$ for the suspension dynamics at the shoreline) and the progressive erosion parameters $^jp_{er}$ ($^jp_{ers}$ for the suspension dynamics at the shoreline) for each layer j, $1 \leq j \leq n$.
- $\tau_{w\&cl}$, contains all the elementary processes of τ, they are applied to each layer according to proper sub-states and parameters; furthermore the following processes are considered with only a type of granulometry for simplicity sake in the exposition:
  - Suspension by erosion for cells at the shoreline. When the wave amplitude overcomes an opportune threshold ($A_w > {}^1p_{mts}$), depending on the layer

features, a mobilization of the layer occurs proportionally to the quantity overcoming the threshold: $^1p_{ers}(A_w - {}^1p_{mts}) = \Delta S_1 = -\Delta T_1$ (the layer's thickness diminishes as the matter in suspension increases). Note that deposit process can follow immediately to the suspension process at the shoreline.

- Suspension, deposit and deposit mobilization processes. These processes derive from the model RUSICA [16,21], but they are adapted to the phenomenological conditions of the sea storm: matter in suspension can be increased by deposit erosion depending on the energy on the bottom sea: $^1p_{ers}(e_b - {}^1p_{mts}) = \Delta S_1 = -\Delta T_1$, where e_b is the energy at the bottom sea; the wave energy inside a cell e_c can maintain granular matter in suspension until a maximum density $d_{mx} = p_{mt}(e_c - p_{mne})/(-A)$ depending on granulometry of the matter, otherwise precipitation of a part of suspended matter occurs until the equilibrium is reached: $\Delta T_1 = -\Delta S_1 = S_1 - Ad_{mx}$.

- Diffusion of the granular matter in suspension during a sea storm. A diffusion mechanism, using the "algorithm of minimization of differences" [5,6], is adopted, it is similar to that of the previous section, a value of K_1 depending on the wave sub-states is introduced and A (value of central cell) is considered constant in the neighborhood, because there is no slope for matter in suspension, but only differences in density.

- Transport of the granular matter in suspension by sea currents. The shift s is specified as $s = p_t\sqrt{(C_x^2 + C_y^2)}$, originating for the granular matter in suspension, the outflows $^iS_1$ toward the adjacent cells according the Θ angle of the current such that $\Sigma^i_{1\le i\le 6}S_1 = S_1 \cdot s/p_a$. The slow deposition process of granular matter transported by current is not considered.

- $\gamma_{w\&cl}$ is the external influence function, that returns step by step ($\mathcal{Z}$) the significant values regarding waves and currents, it is split in the following functions:

- $\gamma_w : \mathcal{Z} \times G_{w\&cl} \to A_w \times L_w \times X_w \times Y_w$ for generation of wave values cell by cell of $G_{w\&cl}$ according to the observations or meteorological forecasting.

- $\gamma_c : \mathcal{Z} \times G_{w\&cl} \to C_x \times C_y$ for generation of current values cell by cell of $G_{w\&cl}$ according to the observations or meteorological forecasting.

SCIDDICA-ss2 and RUSICA coalesce with strong simplifications in this preliminary version w&cl; layers of different granulometry are here considered separately, but phenomenological conditions of their possible innermost mixing are tackled superficially and need simulations of real events in order to improve the model by an accurate parametrization or introducing new sub-states, parameters and elementary processes according to the methodology of incremental CA modelling, furthermore the model is "cut" for reduced applications.

3   Examples of Simulation of Demolition Process of Fan Delta by Sea-Storm and Sediment Outcome: The 2015–16 Sfalassà Case

A summary of observation of demolition of fan delta by sea storms precedes in this chapter the section specifying the partial implementation of SCIDDICA-ss2/w&c1; the simulations of the demolition of the 2015 fan delta follow, at the end, an hypothetical case of transport in suspension was simulated.

3.1   Summary of Observations Related to the Sfalassà's Area and Meteorological Events

The interested area is characterised as follows: the wide of continental shelf varies from 120 m north of Marturano Promontory to 50 m at Cacili promontory. It reaches the maximum extension, about 300 m, at south of Marturano Promontory and narrows to about 100 m in the area close to Sfalassà's mouth to become again wider (about 180 m) southward (Calabria, Italy). The continental shelf has a slope $\leq 7°$, it continues with a continental slope that between 20 m and 50 m depth has a slope range of $20°–30°$, that becomes of $10°–20°$ at higher depth. The submarine retrogressive scars confined at canyon head-walls coincide with both Sfalassà and Favazzina fiumara mouth. The emerged beach has a height varying between 0 and 3.5 m a.s.l. and a mean slope of 6° in accordance with the slope of continental shelf. The volume of fan delta, grown in consequence of flood event occurred on 2 November 2015 is of about 25.000 m^3 this value is in agreement with the volume of fan delta formed in consequence of flood events and the evaluation of yield sediment derived from EPM model [24].

From a meteorological point of view, the flood event was analyzed in detail in a recent work [23], where different configurations of a mesoscale atmospheric model were tested and several sensitivity tests were carried out. The precipitation affected principally the eastern side of the Calabria, although huge quantities of rain have been recorded on the southerly Tyrrhenian areas. An upper level trough over Sicily, synergistically to a low-pressure area at the surface, favored the advection of moist and warm air masses from the Ionian Sea towards the eastern side of the Calabria. Considering the observed precipitation collected by regional network of the "Centro Funzionale Multirischi" (http://www.cfd.calabria.it), it can be seen how several rain gauges recorded precipitation >500 mm for the whole event, some of which just located on the western slopes of the Aspromonte Mountain, upstream to the Sfalassà's area. The rain gauge of Bagnara Calabra recorded 96 mm on 31 October and 112 mm on 1 November, demonstrating the extraordinariness of the event. Consequently to flood a significant fan delta was formed at the mouth of Sfalassà watercourse by detrital matter consisting of badly classified sediments from coarse sand 0.75 mm to blocks of 1.0 m, an average value of 150 mm may be considered. The thickness of this fan delta is approximately 5 cm in average, surface of the underwater detrital cover, until the original coastline, was estimated 9250 m^2. Meteo-marine

Fig. 1. Interested area of Bagnara site and sequence of photographs showing effect of flash-flood and sea storms

events (at least three strong sea storms) occurred during the period 2015.11.25 to 2016.01.28, they brought to destroy almost completely the fun delta. The first sea storm (general information in Table 1st event) took away a thin strip of the delta fan parallel to the beach, the stronger second sea storm (general information in Table 2nd event) took away a larger strip of the delta fan parallel to the beach, the last strongest sea storm (general information in Table 3rd event) demolished the delta fan.

Regarding the storms, a marine-waves analysis was carried out. For the case we considered high-resolution ($0.125° \times 0.125°$) simulations carried out with the model WAM (WAve Model), developed by ECMWF (European Centre for Medium-Range Weather Forecasts). The sea storms are classified reporting in the tables the most significant marine parameters:

SWH (Significant wave height [m]); MWP (Mean wave period [s]); MWD (Mean wave direction [degrees]); $PP1D$ (Peak period of 1D spectra [s]); $HMAX$ (Maximum individual wave height [m]); $TMAX$ (Period corresponding to maximum individual wave height [s]).

Table 1. Extreme events after flash flood occurred ranging November 2015 January 2016 here statistic values regarding model output WAM for the three entire period

	ESWH (m)	MWP [s]	MWD [degrees]	PP1D [s]	HMAX [m]	TMAX [s]
25–28/11/2015						
MAX	1,85	6,52	297,92	7,43	3,59	5,90
MIN	0,98	5,80	266,58	6,19	1,88	4,80
AVE	1,32	6,24	283,30	6,62	2,55	5,62
std	0,26	0,18	10,88	0,40	0,53	0,22
03–08/01/2016						
MAX	2,11	7,43	294,29	8,73	4,01	6,74
MIX	1,00	4,75	265,76	6,13	1,93	4,33
AVE	1,48	6,64	283,30	7,41	2,85	5,94
std	0,40	0,33	4,53	0,99	0,76	0,31
12–18/01/2016						
MAX	2,43	7,62	317,07	9,06	4,63	6,82
MIX	0,89	5,32	275,50	6,17	1,69	5,07
AVE	1,53	6,60	294,97	7,71	2,92	5,93
std	0,43	0,56	12,17	0,87	0,81	0,43

The values are extrapolated by the full gridded output of WAM, in particular at an offshore position located about 3 km away from the coastline. Such values opportunely simulated with near-shore model amplify the effect especially for the waves height. These parameters, for the different sea storms, are taken into account for the SCIDDICA simulations.

The Table 1 shows considered events, according periods, reporting statistical values for the entire storm.

The effect of sea storms may be deduced roughly by a series of photographs of the area after the sea storms (see Fig. 1).

The most realistic scenario of the event considers that the strength of the waves did not allow for suspension the matter, that constituted the fan delta because of its granulometry, so granular matter, that was eroded by the strength of the waves, flowed on the seabed without be significantly influenced by the currents [22].

3.2   Partial Implementation of SCIDDICA-ss2/w&c1 and First Simulations

Implementation of SCIDDICA-ss2/w&c1 was performed on the previous one of SCIDDICA-ss2 in language C++, introducing partially and adapting sub-states, parameters and processes of RUSICA. An important limit regards the granulometry of granular matter, there is only a layer, corresponding to the fan delta, whose granulometry is an "equivalent" granulometry of a mixture of deposits.

This could be a solution in some cases, but it is not always satisfying, particularly when a very heterogeneous granulometry could involve both suspension and flowing on the sea bed. The aim of this first version is to test the model according phenomenological view-point, i.e. if the elementary processes account for the main mechanisms of the overall phenomenon. Two cases are considered, the real case of demolition of the fan delta by successive sea storms of different intensity and the very hypothetical case of a sea storm in the same area, but with suspension: if the granular matter would be very fine, the sea storm could cause diffusion in suspension, then it could be transported by the currents at the end of the sea storm. Note that the model provides for an adherence effect, i.e. a smallest part of matter remains always in the cell (less than 1 mm) and cannot be transported outside the cell; it permits to mark the flows.

3.3 Simulation of the Demolition of the 2015 Fan Delta of Watercourse Sfalassà

Three sea storms are considered after the 2015 flash flood in the Bagnara area that formed the fan delta Fig. 1. The first sea storm in the Table 1 is the shortest lasting and lesser sea storm. It does not weight much, only a thin part facing the sea of the fan delta is eroded and a small flow of the eroded part is channelized toward the offshore depression; only a minimum quantity reaches it (Fig. 2b).

Simulation of the second sea storm starts from the final conditions immediately after the first sea storm. The considered second storm 1 is longer and stronger than the first one, it effects the central area of the fan delta; a subaqueous flow of granular matter begins to reach the depression (Fig. 2c).

Simulation of the third sea storm in Table 1 starts from the final conditions immediately after the second sea storm. This sea storm is the longest and strongest, the fan delta is destroyed except two small parts on the right and on the left (Fig. 2d). Almost all the matter, that formed initially the fan delta, reached the sea depression. The global evolution of the system in the simulation reproduces significantly the real event: the erosion propagates progressively from the water line to the internal fringe area.

3.4 A Very Hypothetical Case of Transport in Suspension

In Fig. 2 simulation steps of erosion of the Sfalassà's fan delta and resulting subaqueous flows of the eroded granular matter, thickness of granular matter is reported: (a) initial erosion of the fan delta (initial conditions), (b) effect at the end of the first sea storm corresponding approximately to 3 days and 6 h, (c) effect at end of the second sea storm approximately with duration of 5 days and 12 h, (d) third storm (duration of 6 days and 21 h) dismantling delta fun i.e. area until original coastline.

Regarding the calibration phase of parameters, the main effort regarded this new part of model, i.e., the elementary process (suspension, deposit and deposit mobilization) concerning the demolition of the fan delta; a simple trial and error

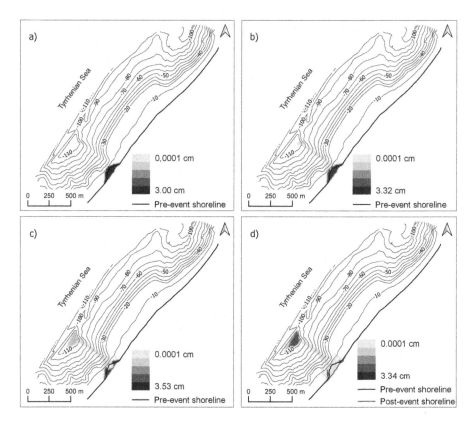

Fig. 2. Simulation steps of erosion of the Sfalassà's fan delta and resulting subaqueous flows thickness: (a) step 1, initial erosion, (b) step 1350: end of the first sea storm, (c) step 3550: end of the second sea storm approximately, (d) step 6300: the third sea storm dismantling of the delta fun.

method was possible to be applied with satisfying results in comparison with the partial reconstruction of the real case.

The second case consider a hypothetical sea storm on a not-realistic fan delta as far composition in comparison with real case, its granulometry permits the suspension after the delta demolition, after the sea storm, a constant current intercepts the granular matter and transported it.

Initial position of the fan delta is the same as in the previous case, of course with a different thin granulometry. The effect of the sea storm is just a diffusion of the eroded granular matter in suspension (step 1000 of the simulation, (Fig. 3a). The sea storm ceases at the step 2500, the further erosion at the step 2000 is reported in the Fig. 3b).

After the sea storm, a hypothetical (and not realistic) current effected the suspended granular matter and channelizes it in its direction, toward NE. The transport path is reported clearly in (Fig. 3a) (step 3000, 500 steps of the trans-

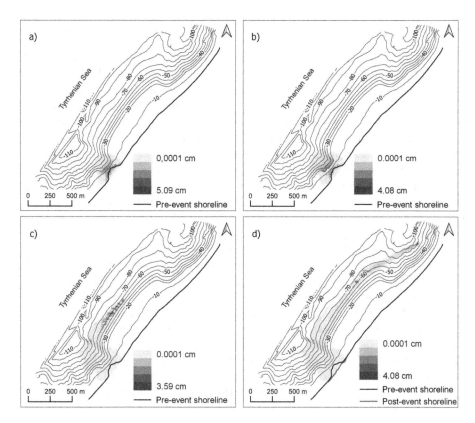

Fig. 3. Simulation steps of hypothetical diffusion in suspension of the eroded granular matter of the fan delta: (a) step 1000, (b) step 2000, (c) step 3000, (d) step 4000.

port in suspension). The last (Fig. 3d) step 4000 shows as the path continues to be channelized in NE direction. Such a current does not exist and its distance from the coast is improbable, but bathymetry data at disposal don't permit to see the evolution of the simulation for a sufficient space in order to evaluate the model in the case of another direction of current; that is the reason so unnatural distance from the coast has been hypothesized.

Regarding the calibration phase of parameters, a simple trial and error method was satisfying for the elementary processes concerning diffusion, suspension and transport of granular matter for a hypothetical (not real) case.

The main effort regarded this new part of model, i.e., the elementary process (suspension, deposit and deposit mobilization) concerning the demolition of the fan delta; a simple trial and error method was possible to be applied with satisfying results.

4   Conclusions

SCIDDICA-ss2/w&c1 was defined on the base of two models CA model SCID-DICA-ss2, that is a well-founded model, very successful for a large range of applications and RUSICA still in a phase of development. Such a model is very complex and our effort was the inclusion and refinement of elementary processes for harmonizing the two models and introducing new features according the incremental modelling method for macroscopic CA. Therefore, the implementation of the model was partial in order to focus itself on the phenomenology in a first rough way and understand the factors that permit a correct emergence of the overall phenomenon. This partial and rough implementation of SCIDDICA-ss2/w&c1 was applied on a real case, whose data for simulation are incomplete and approximate and on a hypothesized case, which isn't realistic, considering the geological characteristics of the same area of the first case, but interesting for a comparison of the different behaviours in the same initial context. The development of the simulated event in the first case may be considered successful in the limits of the implementation if we consider the demolition of the fan delta by the succession of the sea storms thanks to a good knowledge of meteorological data. The final outcome of the sediments is correct but times for a complete deposit in the depression could be different. About, the second case, the initial diffusion in suspension of eroded matter of fan delta and the transport by current was simulated satisfactory obviously only from a phenomenological viewpoint, a real case with enough precise data is necessary for correcting and improving the model. This is just a preliminary step. Further work and effort has to be pursued in order to better outperform the model and its results due to presence of several parameters necessary to describe a macro complex system such as this and with a huge and long time involved area. In these case it is necessary to obtain detailed and long term data to define better the parameters values interval and consequently obtain a more precise pattern of sediments distribution.

References

1. Pielke Sr., R.A.: Mesoscale Meteorological Modeling, 3rd edn. Academic Press, Boston (2013)
2. Kalnay, E.: Atmospheric Modeling, Data Assimilation and Predictability. Cambridge University Press, Cambridge (2003)
3. Richardson, L.F.: Weather Prediction by Numerical Process. Cambridge University Press, Cambridge (1922)
4. Larosa, S., et al.: A PyQGIS plug-in for the sediments production calculation based on the Erosion Potential Method. In: Gao, W. (eds.) Frontiers of Earth Science (2019, Subm.). ISSN 2095-0195 (print version), ISSN 2095-0209 (el. version)
5. Avolio, M.V., Di Gregorio, S., Spataro, W., Trunfio, G.A.: A theorem about the algorithm of minimization of differences for multicomponent cellular automata. In: Sirakoulis, G.C., Bandini, S. (eds.) ACRI 2012. LNCS, vol. 7495, pp. 289–298. Springer, Heidelberg (2012). https://doi.org/10.1007/978-3-642-33350-7_30

6. Di Gregorio, S., Serra, R.: An empirical method for modelling and simulating some complex macroscopic phenomena by cellular automata. Futur. Gener. Comput. Syst. **16**, 259–271 (1999)

7. Calidonna, C.R., Di Gregorio, S., Mango Furnari, M.: Mapping applications of cellular automata into applications of cellular automata networks. Comput. Phys. Commun. **147**, 724–728 (2002)

8. Calidonna, C.R., Naddeo, A., Trunfio, G.A., Di Gregorio, S.: From classical infinite space-time CA to a hybrid CA model for natural sciences modeling. Appl. Math. Comput. **218**, 8137–8150 (2012)

9. Calidonna, C.R., Di Napoli, C., Giordano, M., Mango Furnari, M., Di Gregorio, S.: A network of cellular automata for a landslide simulation. In: ICS 2001, pp. 419–426 (2001)

10. Mazzanti, P., Bozzano, F., Avolio, M.V., Lupiano, V., Di Gregorio, S.: 3D numerical modelling of submerged and coastal landslides propagation. In: Mosher, D.C., et al. (eds.) Submarine Mass Movements and Their Consequences. NTHR, vol. 28, pp. 127–139. Springer, Dordrecht (2010). https://doi.org/10.1007/978-90-481-3071-9_11

11. Avolio, M.V., Lupiano, V., Mazzanti, P., Di Gregorio, S.: An advanced cellular model for flow-type landslide with simulations of subaerial and subaqueous cases. In: Proceedings of the EnviroInfo 2009, vol. 501, pp. 131–140 (2009)

12. Maria, V.A., Francesca, B., Di Gregorio, S., Valeria, L., Paolo, M.: Simulation of submarine landslides by cellular automata methodology. In: Margottini, C., Canuti, P., Sassa, K. (eds.) Landslide Science and Practice, pp. 65–72. Springer, Heidelberg (2013). https://doi.org/10.1007/978-3-642-31427-8_8

13. Avolio, M.V., Di Gregorio, S., Lupiano, V., Mazzanti, P.: SCIDDICA-SS₃: a new version of cellular automata model for simulating fast moving landslides. J. Supercomput. **65**(2), 682–696 (2013). ISSN 0920-8542

14. Salles, T., Lopez, S., Cacas, M.C., Mulder, T.: Cellular automata model of density currents. Geomorphology **88**(1–2), 1–20 (2007)

15. D'Ambrosio, D., Di Gregorio, S., Gabriele, S., Gaudio, R.: A cellular automata model for soil erosion by water. Phys. Chem. Earth Part B **26**(1), 33–39 (2001)

16. Calidonna, C.R., De Pino, M., Di Gregorio, S., Gullace, F., Gullì, D., Lupiano, D.: A CA model for beach morphodynamics. In: AIP Conference Proceedings. Numerical Computations: Theory and Algorithms (NUMTA 2016), vol. 1776 (2016)

17. Arai, K., Basuki, A.: Simulation of hot mudflow disaster with cell automaton and verification with satellite imagery data. Int. Arch. Sci. Photogramm. Remote Sens. Spat. Inf. **38 Part 8**, 237–242 (2010)

18. Vanwalleghem, T., Jiménez-Hornero, F., Giráldez, J.V., Laguna, A.M.: Simulation of long-term soil redistribution by tillage using a cellular automata model. Earth Surf. Process. Landf. **35**, 761–770 (2010)

19. Valette, G., Prévost, S., Laurent, L., Léonard, J.: SoDA project: a simulation of soil surface degradation by rainfall. Comput. Graph. **30**, 494–506 (2006)

20. Punzo, M., et al.: Remocean X-band wave radar for wave field analysis: case study of Bagnara Calabra (South Tyrrhenian Sea, Italy). J. Sens. **1**, 131–140. Application of X-BandWave Radar for Coastal Dynamic Analysis: Case Test of Bagnara Calabra (South Tyrrhenian Sea, IT) **501**, 9 (2016). Article ID: 6236925

21. Gullace, F.: Un tentativo di Modellizzazione della Morfodinamica degli Arenili con Automi Cellulari. Master theses in Physics, Department Physics, University of Calabria, aa 2015/16 (2016)

22. Dominici R.: Personal communication (2019)

23. Avolio, E., Federico, S.: WRF simulations for a heavy rainfall event in southern Italy: verification and sensitivity tests. Atmos. Res. **209**, 14–35 (2018)
24. Auddino, M., Dominici, R., Viscomi, A.: Evaluation of yield sediment in the Sfalassà Fiumara (south western, Calabria) by using Gavrilovi method in GIS enviroment. Rendiconti online della Soc. Geol. Ita. **33**, 3–7 (2015)

Parallel Algorithms for Multifractal Analysis of River Networks

Leonardo Primavera[1]([⊠])[iD] and Emilia Florio[2][iD]

[1] Dipartimento di Fisica, Università della Calabria,
Cubo 31/C, Ponte P. Bucci, 87036 Rende, CS, Italy
`leonardo.primavera@unical.it`
[2] Dipartimento di Matematica e Informatica, Università della Calabria,
Cubo 30/B, Ponte P. Bucci, 87036 Rende, CS, Italy
`emilia.florio@unical.it`

Abstract. The dynamical properties of many natural phenomena can be related to their support fractal dimension. A relevant example is the connection between flood peaks produced in a river basin, as observed in flood hydrographs, and the multi-fractal spectrum of the river itself, according to the Multifractal Instantaneous Unit Hydrograph (MIUH) theory. Typically, the multifractal analysis of river networks is carried out by sampling large collections of points belonging to the river basin and analyzing the fractal dimensions and the Lipschitz-Hölder exponents of singularities through numerical procedures which involve different degrees of accuracy in the assessment of such quantities through different methods (box-counting techniques, the generalized correlation integral method by Pawelzik and Schuster (1987), the fixed-mass algorithms by Badii and Politi (1985), being some relevant examples). However, the higher accuracy in the determination of the fractal dimensions requires considerably higher computational times. For this reason, we recently developed a parallel version of some of the cited multifractal methods described above by using the MPI parallel library, by reaching almost optimal speed-ups in the computations. This will supply a tool for the assessment of the fractal dimensions of river networks (as well as of several other natural phenomena whose embedding dimension is 2 or 3) on massively parallel clusters or multi-core workstations.

Keywords: Multifractal dimension · River networks · Parallel algorithms

1 Introduction

The multifractal analysis is a powerful tool in many fields of science to relate the geometric characteristics of objects with their physical properties. Since the pioneering works by Mandelbrot [10], Grassberger [8], Grassberger and Procaccia [9], many contributions of the multifractal analysis have been given in physics, chemistry, biology and engineering.

© Springer Nature Switzerland AG 2020
Y. D. Sergeyev and D. E. Kvasov (Eds.): NUMTA 2019, LNCS 11973, pp. 307–317, 2020.
https://doi.org/10.1007/978-3-030-39081-5_27

In particular, in hydrology, the multifractal analysis has given considerable contributions. De Bartolo et al. [4], showed that river networks are multifractal objects and [3] that they are non plane-filling structures (their fractal support dimension is lesser than two). Moreover, it was shown that there is a strict correlation between the hydrological response of a basin and its multifractal spectrum through the Multifractal Instantaneous Unit Hydrograph (MIUH) [2][1].

The assessment of the multifractal spectrum can be realized through the following procedure. First, a 2D projection of the river basin on the ground plane is realized and the different channels belonging to the river basin are represented through the so-called *"blue-lines"*. Second, a regular sample of the *blue-lines* is extracted by picking up a set of points, the so-called *"net-points"*, which are representative of the river basin. Finally, a multifractal analysis is performed on this set of points through several possible techniques. The most used methods are: fixed-size algorithms (FSA), the correlation integral (CI) by Pawelzik and Schuster [11], and fixed-mass algorithms (FMA) by Badii and Politi [1].

However, some methods can be more relevant than others for hydrological studies. The FSA is by far the easiest to be implemented. It is based on the idea of partitioning the multifractal set in squares of decreasing size and counting the number of points (the so called *"mass"*) falling into each element of the partition. The fractal dimension of the object can then be obtained through a suitable scaling procedure of the mass as a function of the size of the partition. The FMA uses the opposite approach, namely one fixes the mass in each partition and then computes the size of the circle containing the given mass inside it.

As pointed out by De Bartolo et al. [7], the important parameter for the hydrological response of the basin, directly used in the MIUH, is the $D_{-\infty}$ fractal dimension, which corresponds to the right part of the multifractal spectrum. However, FSAs are not well suitable to give a correct assessment of this parameter, since they present strong oscillations in the scalings of τ_q for negative values of q, that corresponds just to the right part of the multifractal spectrum. The correlation integral method by Pawelzik and Schuster, as suggested by De Bartolo et al. [7], seems to be able to improve this situation, although the most suitable method seems to be the FMA. This happens because the oscillations in the scalings observed in the FSAs are mainly due to the poor statistics of net-points in the partitions where there is a less dense distribution of the points. The FMA, on the converse, does not suffer for this problem, since the number of net-points in each subset of the partition is fixed *a priori* and the size of the subsets is computed accordingly, which is the opposite of what happens in FSAs.

However, FMA can be very expensive, from the computational point of view, due to the procedure needed to compute the size of the partition once the "mass" is chosen. This problem was solved in the past by extracting randomly chosen net-points and applying the algorithm only on such selected sub-sets. On the one hand, this allows one to obtain the results of the multifractal analysis in reasonable CPU times, on the other hand, this sampling of the net-points lowers

---

[1] The problem of the relation between the floods in rivers and meteorologic forecasting has interested scientists in different epochs. See, for instance [6].

the precision of the calculation, as well as making useless to select large sets of net-points to improve the representation of the river network.

However, nowadays, with the development of multi-cores CPUs, which can be found even on moderately expensive personal computers, workstations or even small clusters, one can make an effort to parallelize the code in order to exploit the power of parallel CPUs or clusters to improve the computational times. At the best of our knowledge, this attempt has never been done for this algorithm. This contribution will try to fill this gap.

The organization of the contribution is the following: in next section, we briefly recall the fixed-size and fixed-mass techniques; then, we explain the parallelization strategy we adopted to speed-up the numerical computation; after that, to be sure of the results of our numerical procedure, we make a comparison between the numerical and theoretical results obtained by analyzing a random multifractal for which the fractal dimensions and the multifractal spectra are known; finally we show the results of the parallelization in terms of the scaling of the computational time with the number of cores, computed on a relatively small multi-node cluster. Finally, in the last section, we draw some conclusions and future outlooks.

2   Fixed-Size and Fixed-Mass Multifractal Analysis

Both the FSA and FMA multifractal rely on the study of the scaling properties of the set of points representing the multifractal object (let us call N the number of such points). The assessment of the fractal dimension is then carried out by partitioning the set in a number n_c of non-overlapping cells, with a size ϵ_i, $i = 1, \ldots, n_c$. Let us call *"measure"* or *"mass"* $p_i(\epsilon_i)$, the number of points falling inside each cell. For a fractal set, it is possible to show that:

$$\lim_{\epsilon_i \to 0} \frac{\sum_{i=1}^{n_c} p_i^{q-1} \epsilon_i^\tau}{n_c} = k \tag{1}$$

where k is a constant for a suitable choice of the real numbers q and τ, which are related to the fractal dimension of the object.

For a multifractal, the values of q and τ are not unique, but a ensemble of solutions q, τ_q exists, that satisfies the relation (1). These two quantities are related to the generalized fractal dimensions D_q of the set through the relation:

$$D_q = \frac{\tau_q}{q-1}, \qquad \forall q \neq 1 \tag{2}$$

Finally, the Lipschitz-Hölder exponents α_q, are given by the relation:

$$\alpha_q = \frac{d\tau_q}{dq} \tag{3}$$

while the multifractal spectrum $f(\alpha_q)$ is defined as:

$$f(\alpha_q) = q\alpha_q - \tau_q \tag{4}$$

In practical terms, the determination of the solutions $(q,\ \tau_q)$ is performed numerically in the following way. For FSA, for instance the famous "*box-counting*" algorithm, the dimension $\epsilon_i = \epsilon$ is chosen the same for all the subsets of the partition, in such a way that Eq. (1) gives:

$$\lim_{\epsilon \to 0} \log M(\epsilon, q) = \tau_q \lim_{\epsilon \to 0} \log \epsilon + \log k \tag{5}$$

where:

$$M(\epsilon, q) = \frac{\sum_{i=1}^{n_c} p_i^{q-1}}{n_c}$$

is the moment of order $q - 1$ of the "mass". Numerically, it is impossible to take the limit for vanishing values of ϵ, therefore the assessment of the solution τ_q of Eq. (5) is carried out by fixing a value of q, by choosing a sequence of exponentially decreasing values of the size ϵ of the partitioning cells and "counting" the "mass" p_i falling into each of the intervals and finally computing $\log M(\epsilon, q)$ for each value of $\log \epsilon$. Then τ_q will result as the slope of the linear best-fit approximation of this curve, for each fixed value of q. The drawback of this method is easily understood: for negative values of q, in the subsets of the partition where few points are present (namely for small values of p_i), the function $M(\epsilon, q)$ exhibits strong, nonphysical, oscillations for decreasing values of ϵ, due to the poor statistics in those cells. Negative values of q correspond to the right part of the multifractal spectrum $f(\alpha_q)$, whose fractal dimension $D_{-\infty}$ enters in the MIUH forecasting model. Therefore, the assessment of $D_{-\infty}$ accomplished through FSA can be affected by strong errors and should be adopted with particular care.

The approach followed in the FMA is opposite. In this case, the "mass" $p_i = p$ is chosen the same for each subset of the partition and the corresponding "size" of the partitioning cells is evaluated by finding, for each point, the p nearest neighbors. The radius of the circle containing that "mass" is then taken as the size ϵ_i of the partitioning cell. For a fixed value of τ_q, then, the mass p can be taken out from the summation, by yielding, for Eq. (1) a relation of the form:

$$\lim_{\epsilon_i \to 0} \log M(\epsilon_i, \tau_q) = (1 - q) \lim_{p \to 0} \log p + \log k \tag{6}$$

where we used the fact that $p \to 0$ for $\epsilon_i \to 0$, and:

$$M(\epsilon_i, \tau_q) = \frac{\sum_{i=1}^{n_c} \epsilon_i^{-\tau_q}}{n_c} \tag{7}$$

is the moment of order $-\tau_q$ of the size ϵ_i of the partitioning cells. Finally, the solution $1 - q$, and therefore q, will result from computing the scaling of $\log M(\epsilon_i, \tau_q)$ with $\log p$, for exponentially decreasing values of p, as in Eq. (6).

In the latter case, since the "mass" content in each cell of the partition is fixed *a-priori*, the problems present with the FSA are completely overcome. However, the computational times required to complete the analysis can be really very long, even for a total number of net-points N relatively small (the analysis of

10^4 net-points may require up to three days of computation on a workstation equipped with a Xeon E3-1225 V2 CPU, at 3.20 GHz, on a single core). The true bottleneck of the computation lies in the search of the nearest neighbors points for each single net-point, which is a N^2 algorithm and, therefore, the CPU time increases very quickly with the number of points.

A way to mitigate this problem and make the FMAs attractive for realistic cases found in hydraulic science (where the number of net-points can easily exceed $N = 10^5 \div 10^6$) is the parallelization of the algorithm. In the past, another approach widely adopted was to limit the search of the nearest neighbors points to a randomly extracted sub-set of the net-points, so decreasing the complexity of the computation, but also the precision of the assessment of the fractal dimensions. Although the parallelization of the algorithm is absolutely feasible, at the best of our knowledge this was never attempted before. In next section we describe how we parallelized the numerical code and, afterwards, we show how the scaling with the number of processors is not far away from an almost linear speed-up, which indicates a promising way to obtain the results of the analysis in reasonable times in realistic cases.

3   Parallelization of the Code

In order to parallelize the code in such a way that it can be run on the largest possible configurations (either for single- and multi-processor machines, or on massively parallel clusters) we used the Message-Passing-Interface (MPI) parallel library.

The serial FMA works in the following way:

1. the coordinates of the points of the fractal object are read from a file and put into a vector, each element being a structure containing two records, one for the x and the other for the y coordinates;
2. according to the input data, the values for τ_q are selected and an exponentially increasing partitioning scheme for the values of nearest neighbors points is set-up, that corresponds to a partitioning in the mass p_i;
3. in a main loop, for each point, a vector is built with all the distances between the point under consideration and the other points of the fractal set;
4. from this vector, all the distances including the first p_i nearest neighbors points are selected, which corresponds to consider the radius of the circle containing p_i nearest neighbors (or, in other words, the sizes ϵ_i of the partition);
5. the logarithms of the quantities $M(\epsilon_i, \tau_q)$ for all values selected for p are computed;
6. these quantities are finally written in a file, along with the logarithm of the mass p;

At the end of the simulation, a file containing the scalings of $\log M(\epsilon_i, \tau_q)$ as a function of $\log p_i$, for each fixed value of τ_q, is available. The corresponding values of q are obtained by plotting these quantities for the different values of

τ_q, so that the corresponding values of $1 - q$ (and, therefore, of q) are obtained through a best-fit procedure, as the angular coefficient of the linear relation (6). This gives us the curve τ_q as a function of q. The Lipschitz-Hölder exponents α_q can then be computed from the values of τ_q through Eq. (3). This can be accomplished in several ways: in some cases (when the number of values of (q, τ_q) is not very high), it may be convenient to interpolate the curve τ_q with a polynomial function and then compute the values of α_q as the derivatives of this interpolating function. Another way to do this, more suitable when a higher number of values for (q, τ_q) is available, is to numerically compute the derivative by using, for instance, a finite difference formula. A little problem with this approach comes from the fact that the values of τ_q are equally spaced, while the values of q, being obtained as the results of the linear interpolation procedure of Eq. (6), are not. This can be overcome, for instance, by choosing a centered scheme for the first derivative with a different step-size on the left and right side of the value of q under consideration.

We chose this second approach and computed the derivatives of τ_q by using the formula:

$$\left.\frac{d\tau_q}{dq}\right|_{q=q_i} = -\frac{h_2}{h_1(h_1 + h_2)}\tau_q(q_{i-1}) + \frac{h_2 - h_1}{h_1 h_2}\tau_q(q_i) + \frac{h_1}{h_2(h_1 + h_2)}\tau_q(q_{i+1}) \quad (8)$$

where: $h_1 = q_i - q_{i-1}$, $h_2 = q_{i+1} - q_i$, and the numerical error is proportional to the product $h_1 h_2$. In the case when $h_1 = h_2$ this scheme reduces to the simple second order central difference scheme.

It is immediately understandable that the most computationally expensive point in the algorithm is the third one, namely the evaluation of the distances of each point from the others, in order to obtain the nearest neighbors. This is a N^2 algorithm. Our parallelization strategy has focused mainly on this point. We start all MPI processes at the beginning of the code and let all the computational cores read the points of the fractal set and store them in a vector. We adopted this approach in order to avoid subsequent communications among the processes in order to exchange parts of this vector, which would result in higher latency times and a worse scaling of the computational time. This may appear odd, but it is actually not a serious problem in the majority of situations. For instance, the memory requirement per core to store this vector would be $16 \times N$ bytes for a double precision input file made of N net-points. For instance, for $N = 10^6$ this would require a memory allocation of less than $16\,\text{MB}$ per core, which is fairly easy to find on the majority of modern machines. This is also the main contribution to memory allocation, all the other vectors being of very small sizes.

Once each core has its own private copy of the net-points, it is assigned a specific interval of points on which it has to operate. However, thanks to the fact that the coordinates of the points are known to each core, the distances between each point in this interval from all the other ones can be easily calculated. Therefore, also the evaluation of the nearest neighbors for each of such points can be done in parallel (point 4) and the evaluation of the moments (point 5) can be performed "locally" on each core. However, this last point requires a final average of the moments for all the points, according to Eq. (7). This is accomplished with an MPI_REDUCE call, which adds up all the contributions to the moments on each core and sends the average to the master process. The file with the results (point 6 above) is then written only on the master process. This introduces a serial part in the code, whose weight is anyway considerably small, compared to the rest of the computation.

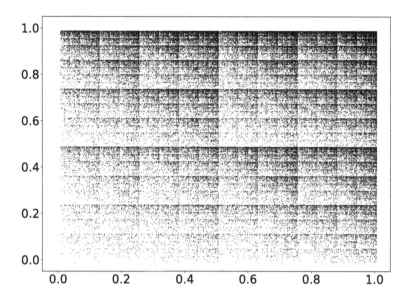

Fig. 1. A random multifractal set obtained according to the multiplicative process described by (9) with $N = 10^6$ points.

4   Numerical Results

4.1   Results of the Analysis for a Random Multifractal

In order to check the validity of the parallel algorithm, we ran several analyses either with the serial or the parallel version of the code on a deterministic multifractal. The multifractal set is obtained according to a recursive procedure by starting from a random seed (see, for instance, Falconer [5]), in the following way: let us start from the unit square $Q = [0, 1] \times [0, 1]$ and let us consider the following transformations:

$$S_1 : (x, y) \rightarrow (x', y') = (\frac{1}{2}x, \frac{1}{2}y)$$

$$S_2 : (x, y) \rightarrow (x', y') = (\frac{1}{2}(x+1), \frac{1}{2}y)$$

$$S_3 : (x, y) \rightarrow (x', y') = (\frac{1}{2}x, \frac{1}{2}(y+1))$$

$$S_4 : (x, y) \rightarrow (x', y') = (\frac{1}{2}(x+1), \frac{1}{2}(y+1))$$

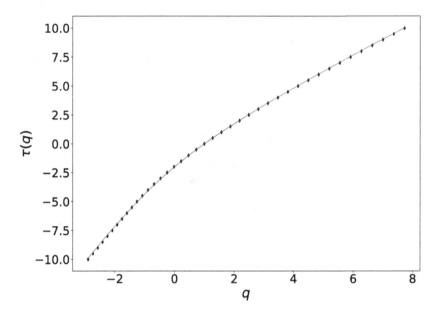

Fig. 2. Comparison between theoretical and numerical results for τ_q. Theoretical curves, obtained through Eq. (9) are plotted as red lines, numerical values are plotted as diamond marks. (Color figure online)

These transformations map a point of the original unit square Q into one of the four squares Q_i ($i = 1, \ldots, 4$) of side $1/2$ in which Q can be partitioned. We assign to each of the four squares Q_i, the following probabilities: $p_1 = 1/10$; $p_2 = 2/10$; $p_3 = 3/10$; $p_4 = 4/10$. We then generate a set of N random numbers in the interval $[0, 1[$. If the number falls in the interval $[0, 0.1[$ (that happens with a probability p_1), or $[0.1, 0.3[$ (happens with a probability p_2), or $[0.3, 0.6[$ (with a probability p_3) or, finally, $[0.6, 1[$ (with a probability p_4), the transformation corresponding to that probability is applied and by iterating this procedure, one obtains the multifractal set visualized in Fig. 1.

The theoretical prediction for τ_q is given by the relation:

$$r_1^{-\tau_q} p_1^q + r_2^{-\tau_q} p_2^q + r_3^{-\tau_q} p_3^q + r_4^{-\tau_q} p_4^q = 1$$

where r_i are the "contraction rates" used to construct the multifractal set. In our case: $r_i = 1/2$, therefore, the value of τ_q is given by:

$$\tau_q = -\frac{\log(\sum_{i=1}^4 p_i^q)}{\log 2} \tag{9}$$

Finally, the theoretical values for α_q and $f(\alpha_q)$ can be computed through Eqs. (3) and (4).

In Figs. 2 and 3, the theoretical (red lines) and numerical curves (black diamonds) are shown for τ_q and $f(\alpha_q)$, respectively. As visible, the agreement of the theoretical and numerical curves for τ_q is almost perfect. A similar situation holds for the $f(\alpha_q)$ plot, although some small differences are present in the right part of the spectrum, close to the maximum. However, this is probably due to the approximations introduced by the finite difference formula (8) we used for the evaluation of α_q.

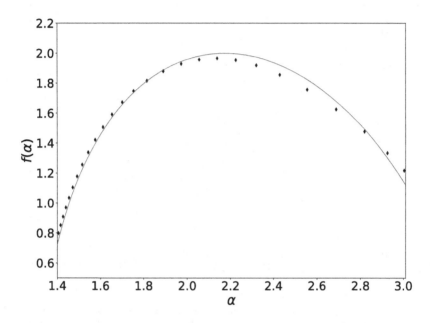

Fig. 3. Comparison between theoretical and numerical results for $f(\alpha_q)$. Theoretical curves are plotted as red lines, numerical values are plotted as diamond marks. (Color figure online)

4.2   Results of the Parallelization

In order to evaluate the effectiveness of the parallelization procedure, we computed the speed-up of the code. We ran the FMA analysis on the random theoretical multifractal shown in Fig. 1, with $N = 10^6$ points.

It is worth mentioning that the numerical results cannot be exactly the same for all the runs. This is due to the fact that the computation of the moments $M(\epsilon_i, \tau_q)$ requires an `MPI_REDUCE` operation among the processors and the sums are done in a different order, therefore the results can be slightly different in the various cases. We checked anyway that the maximum relative difference in all the cases was lesser than 10^{-8}.

As usual, we define: T_n as the CPU time necessary to run the code on n cores, and the speed-up as:

$$S(n) = \frac{T_1}{T_n}$$

that is, the ratio of the serial over the parallel CPU times. The "ideal" value, $S(n) = n$, represents the case of perfect parallelization (although it is not unusual, in some special cases, to find even superlinear speed-ups because of insufficient memory storage or excessive cache occupation in the serial case).

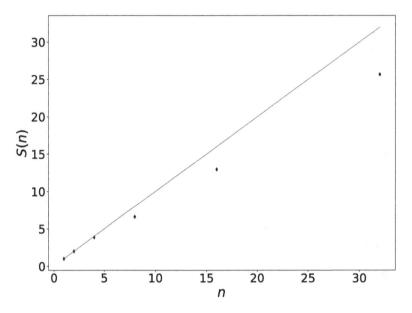

Fig. 4. Speed-up curve for the execution times vs. the number of processors (black-diamonds marks) along with the theoretical scaling $S(n) = n$ (red line). (Color figure online)

In Fig. 4, we show the speed-up $S(n)$ for the tests we did on a relatively small cluster, made of 32 nodes, 2 CPUs/node, 10 cores/CPU, for a total of 640 computational cores. The CPUs are Intel Xeon processors ES-2680 with a clock frequency of 2.8 GHz and the node interconnection is realized through an Infiniband switch with a bandwidth of 40 Gb/s. The results show a fairly good speed-up of the code (black-diamonds marks) with respect to the theoretical curve $S(n) = n$ (red line). The difference for increasing values of n are likely due to the latency of the communications among different nodes.

5   Conclusions

We used the MPI library to parallelize a code to perform the fixed-mass multi-fractal analysis. Such a code was widely used in the past, in its serial version, for applications to the study of river networks to get useful parameters to be used to study the hydrological response of a basin through the Multifractal Instantaneous Unit Hydrograph (MIUH).

For large numbers of net-points extracted from the blue-lines of the river network, the numerical complexity of the calculation requires very long computational times that are drastically reduced in the parallel version. This will allow the code to run on multi-core workstations and/or multi-node clusters by exploiting the whole potential of the CPUs.

Future possible improvements could consist in: (1) realizing an hybrid parallelization with the Open-Message-Passing (OPEN/MP) paradigm, which would avoid the inter-node communications, and (2) the porting of the code on GPUs, that would allow a very efficient massively parallel execution of the code without the need to buy expensive clusters or extremely powerful workstations.

References

1. Badii, R., Politi, A.: Hausdorff dimension and uniformity factor of strange attractors. Phys. Rev. Lett. **52**, 1661–1664 (1984)
2. De Bartolo, S.G., Ambrosio, L., Primavera, L., Veltri, M.: Descrittori frattali e caratteri morfometrici nella risposta idrologica. In: Caroni, E., Fiorotto, V., Mancinelli, A., Salandin, P. (eds.) La Difesa Idraulica del Territorio, pp. 47–60. Tergeste, Trieste (2003)
3. De Bartolo, S.G., Gabriele, S., Gaudio, R.: Multifractal behaviour of river networks. Hydrol. Earth Syst. Sci. **4**, 105–112 (2000)
4. De Bartolo, S.G., Veltri, M., Primavera, L.: Estimated generalized dimensions of river networks. J. Hydrol. **322**, 181–191 (2006)
5. Falconer, K.: Fractal Geometry: Mathematical Foundations and Applications, 2nd edn. Wiley, Chichester (2003)
6. Florio, E., Maierù, L.: The scientific knowledge of book XVI De subtilitate by G. Cardano used in the Trattato sulla divinatione naturale cosmologica by P.A. Foscarini. Open J. Human. **1**, 385–422 (2019)
7. Gaudio, R., De Bartolo, S.G., Primavera, L., Veltri, M., Gabriele, S.: Procedures in multifractal analysis of river networks, vol. 286, pp. 228–237. IAHS Publication (2004)
8. Grassberger, P.: Generalized dimensions of strange attractors. Phys. Lett. A **97**, 227–230 (1983)
9. Grassberger, P., Procaccia, I.: Characterization of strange attractors. Phys. Rev. Lett. **50**, 346–349 (1983)
10. Mandelbrot, B.B.: Possible refinement of the lognormal hypothesis concerning the distribution of energy dissipation in intermittent turbulence. In: Rosenblatt, M., Van Atta, C. (eds.) Statistical Models and Turbulence. LNP, vol. 12, pp. 333–351. Springer, Berlin (1972). https://doi.org/10.1007/3-540-05716-1_20
11. Pawelzik, K., Schuster, H.G.: Generalized dimensions and entropies from a measured time series. Phys. Rev. A **35**, 481–484 (1987)

A Methodology Approach to Compare Performance of Parallel Programming Models for Shared-Memory Architectures

Gladys Utrera[✉][iD], Marisa Gil[iD], and Xavier Martorell[iD]

Computer Architecture Department, Universitat Politècnica de Catalunya,
c/ Jordi Girona, 1-3, 08034 Barcelona, Catalonia, Spain
{gutrera,marisa,xavim}@ac.upc.edu

Abstract. The majority of current HPC applications are composed of complex and irregular data structures that involve techniques such as linear algebra, graph algorithms, and resource management, for which new platforms with varying computation-unit capacity and features are required. Platforms using several cores with different performance characteristics make a challenge the selection of the best programming model, based on the corresponding executing algorithm. To make this study, there are approaches in the literature, that go from comparing in isolation the corresponding programming models' primitives to the evaluation of a complete set of benchmarks. Our study shows that none of them may provide enough information for a HPC application to make a programming model selection. In addition, modern platforms are modifying the memory hierarchy, evolving to larger shared and private caches or NUMA regions making the memory wall an issue to consider depending on the memory access patterns of applications. In this work, we propose a methodology based on Parallel Programming Patterns to consider intra and inter socket communication. In this sense, we analyze MPI, OpenMP and the hybrid solution MPI/OpenMP in shared-memory environments. We demonstrate that the proposed comparison methodology may give more accurate predictions in performance for given HPC applications and consequently a useful tool to select the appropriate parallel programming model.

Keywords: MPI · OpenMP · NUMA · HPC · Parallel programming patterns

1 Introduction

Current HPC platforms are composed of varying computation units capacities and features connected by diverse, increasingly powerful and complex networks to provide better performance not only for large size messages but also for massive receive/send from multiple nodes. These are characteristics foreseeable for the Exascale era.

© Springer Nature Switzerland AG 2020
Y. D. Sergeyev and D. E. Kvasov (Eds.): NUMTA 2019, LNCS 11973, pp. 318–325, 2020.
https://doi.org/10.1007/978-3-030-39081-5_28

From the point of view of the software, applications also tend to be composed of different data structures and the corresponding algorithms to read and modify this data. In addition, based on the data size and the order in which the data is processed, the performance in terms of scalability or reliability, can be affected depending on the programming model in use.

This scheme has led to several approaches considering the use of pure Message Passing library Interface (MPI) [7] versus OpenMP [9] primitives inside a node or exploring several levels of hybrid message-passing and shared-memory proposals to take advantage of the different cores' characteristics. Furthermore, modern platforms are also modifying the memory hierarchy differences, evolving to larger shared and private caches or NUMA (Non-Uniform Access Memory) regions.

UMA (Uniform Memory Access) architectures, commonly referred to as SMP (Symmetric Multiprocessing), have equal memory access latencies from any processor. On the contrary, NUMA architectures are organized as interconnected SMPs and the memory access latencies may differ between different SMPs. In this situation, the memory wall is an issue to consider depending on the memory access patterns the executing application exhibits: data message size, varied size of synchronization or mutex areas; and an inter-socket evaluation is necessary.

This increasing complexity at both low- and high-level makes a challenge the selection of the best programming model, to achieve the best performance on a specific platform. In this work, we take a pattern-based approach to analyze application performance, based on the scalability achieved, including data locality.

Contributions of this work:

- Review a methodology currently used to enhance parallel programming languages with the objective of comparing them.
- Performance comparison between MPI and OpenMP under the stencil and reduce parallel patterns.

The rest of the paper is organized in the following way: Sect. 2 introduces background, related work and our motivation; Sect. 3 presents the experimental results. Finally, in Sect. 4 are the conclusions and future work.

2   Background and Motivation

The choice of the parallel programming model can be determinant in performance for a given application.

Standard programming models may ensure portability. However, when combined with the platform architecture there is not a general best approach.

In this work we focus on two standards: Message Passing Model (MPI) and OpenMP. OpenMP [9] is the de facto standard model for shared memory systems and MPI [7] is the de facto standard for distributed memory systems.

In the following subsections we introduce briefly MPI and OpenMP programming models. After that we make an overview of existing performance comparison studies, which conduct us to the motivation of our proposal.

2.1 OpenMP

OpenMP is a shared-memory multiprocessing Application Program Inference (API) for easy development of shared memory parallel programs. It provides a set of compiler directives to create and synchronize threads, and easily parallelize commonly used parallel-patterns. To that end, it uses a block-structured approach to switch between sequential and parallel regions, which follows the fork/join model. When entering a parallel region, a single thread splits into some number of threads, then when finishing that region, only a sequential thread continuous execution.

2.2 MPI

MPI is a message passing library specification for parallel programming on a distributed environment. In a message passing model, the application is composed of a set of tasks which exchange the data, local or distributed among a certain number of machines, by message passing. There exist several implementations like Intel MPI, and also open source like OpenMPI, MPICH.

Each task in the MPI model has its own address space and the access to others' tasks address space has to be done explicitly with message passing. Data partitioning and distribution to the tasks of the application is required to be programmed explicitly.

MPI provides point-to-point operations, which enable communication between two tasks, and collective operations like broadcast or reduction, which implement communication among several tasks. In addition, communication can be synchronous where tasks are blocked until the message passing operation is completed, or asynchronous where tasks can defer the waiting for the completion of the operation until some predefined point in the program. The size of the data exchanged can be from bytes to gigabytes.

2.3 Related Work and Motivation

There is an interesting study by Krawezik et al. [5], which involved the comparison of some communication primitives from OpenMP and MPI, as well work-sharing constructs from OpenMP and evaluations of the NAS Benchmarks in shared-memory platforms. They recommend OpenMP for computing loops and MPI for the communication part. On the other hand, Kang et al. [4] suggest that for problems with small data sizes OpenMP can be a good choice, while if the data is moderate and the problem computation-intensive then MPI can be considered a framework. Another work by Piotrowski [10] which evaluates square Jacobi relaxation under pure MPI the hybrid version with OpenMP, showed that in a cluster of shared-memory nodes, none of the hybrid versions outperformed pure MPI due to longer MPI point-to-point communication times. In the work by Qi et al. [11] they also use the NAS benchmarks for the evaluation and show that with simple OpenMP directives the performance obtained is as good a with shared-memory MPI on IA64 SMP. They also claim that even OpenMP has easy programming, improper programming is likely to happen leading to inferior performance. In this sense, in the work by [2] they conclude that parallelization with OpenMP can be obtained with little effort but the programmer has to be aware of data management issues in order to obtain a reasonable performance. The performance comparison by Yan et al. in their work [12] on the 3D Discrete Element Method of ellipsoid-like complex-shaped particles, examine memory aspects, task allocation as well as variations in the combined approach MPI/OpenMP. They conclude that Pure MPI achieves both lower computational granularity (thus higher spatial locality) and lower communication granularity (thus faster MPI transmission) than hybrid MPI-OpenMP in 3D DEM, where computation dominates communication, and it works more efficiently than hybrid MPI-OpenMP in 3D DEM simulations of ellipsoidal and poly-ellipsoidal particles.

The reviewed works above perform interesting analysis on selected applications and/or algorithms but none of them provide enough information for a given HPC application other than the ones specifically analyzed, to make a proper programming model selection. Even more, there are contradictory recommendations derived from the fact that they were evaluated with different platform characteristics which is a relevant factor.

2.4   Selected Patterns

Pattern-based parallel programming consists on a set of customizable parallel patterns used to instantiate blocks of applications. This approach is being empowered in the last years with the objective to provide parallel programming languages with additional features that bring more flexibility and reduce the effort of parallelization. In the work by De Sensi et al. [3], the authors demonstrate the feasibility of the approach.

In this work, we exploit the idea of pattern-based parallel programming languages to compare them in terms of performance.

We selected the *loop-of-stencil-reduce* pattern for being representative of many HPC applications. This parallel pattern is general enough to subsume other patterns like *map, map-reduce, stencil, stencil-reduction* and their usage in a loop [1].

Below we can see a pseudocode of our MPI and OpenMP implementation versions:

```
{ MPI version }                      { OpenMP version }
  data-distribution
  loop                                 loop
    start-border-recv                    #pragma omp parallel for
    stencil-reduce (in, out)               stencil-reduce (in, out)
    start-border-send                      swap-matrix-in-out
    complete-border-exchange           end loop
    swap-matrix-in-out
  end loop
```

In order to make a fair comparison, we do not take into account any converge condition; the main loop has a fixed number of iterations. There are no dependencies within the blocks inside the stencil algorithm. At every loop there is an input matrix and an output matrix. Before next iteration, we perform a swap between both matrix. The parallelization is block-based. This means that a task perform the stencil algorithm on a block matrix like the one shown in Fig. 1.

The memory access pattern for our implementation is a 5-point Stencil shown in Fig. 1. In dark grey is shown the data shared in the border of each block. Each task share data with top, bottom, left and right tasks.

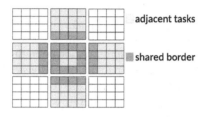

Fig. 1. Memory access pattern for a 5-point stencil

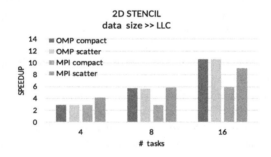

Fig. 2. OpenMP and MPI loop-stencil-reduce parallel pattern performance comparison when datasize does not fit in LLC. Bigger is better.

3 Experimental Results

In this section we show the performance results from the evaluation of the implementation in MPI, OpenMP and MPI/OpenMP of the selected parallel pattern.

The executions were performed on NordIII [8], a supercomputer based on Intel SandyBridge processors, with Linux Operating System. It has 3.056 homogeneous compute nodes (2x Intel SandyBridge-EP E5-2670/1600 20M 8-core at 2.6 GHz) with 2 GB per core. We use the Intel MPI library 4.1.3.049 and C compiler Intel/2017.4.

The evaluations are performed varying the number of tasks (1–16), task allocation (within a NUMA-node or inter NUMA-nodes) and data size. For the data size we consider two cases: (1) fits in the last-level cache (LLC); (2) does not fit in the LLC but fits in main memory.

The work distribution among tasks in our test is well-balanced, so load balancing issues are not tackle in this analysis. There are no dependencies between tasks during a given step, so tasks are embarrasingly parallel. The communication is performed between adjacent tasks (exchange of borders), but source data is from the previous step. Notice that, the input matrix is only read and the output matrix is only written (see the pseudocode at Sect. 2.4).

The initialization of data structures is done in parallel taking care of the first touch Operating system data allocation policy to minimize remote accesses during calculation. Tasks are binded to cores in order to ensure allocation policies: (1) *compact*, that is in the same NUMA node; (2) *scatter*, that is equally distributed across NUMA nodes. The results are showed in Figs. 2 and 3.

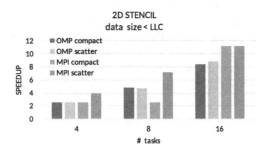

Fig. 3. OpenMP and MPI loop-stencil-reduce parallel pattern performance comparison when datasize does not fit in LLC. Bigger is better.

We can observe in Fig. 2 that the data size does not fit in the shared NUMA-node LLC, when the NUMA node is full (8 tasks) or both NUMA nodes are full (16 tasks) MPI performance degrades dramatically with respect to OpenMP. However, if data fits in LLC, as shown in Fig. 3, then MPI has better performance when having 8 tasks allocated in different NUMA nodes or when the NUMA node is full (16 tasks).

The *stencil* parallel-pattern is characterized for having memory accesses which are updated by other tasks in previous steps. This means that such data has to be exchanged before doing the current calculation. There are memory accesses not only within the block of data processed by a given task, but also for data managed by neighbouring tasks (adjacent blocks). For shared-memory programming models, collaborating tasks allocated to different NUMA-nodes have a well-documented effect on memory access performance (e.g. OpenMP) [6]). This is not the case for distributed memory programming models (e.g. MPI). In parallel programming languages where memory is not shared among tasks (i.e. MPI), this is exchanged explicitly between steps (i.e. MPI_Isend and MPI_Irecv for our current implementation shown in Sect. 2.4. However, once the data is brought becomes local (task allocation is transparent).

Taking all this into account we can appreciate the NUMA effect when data fits in LLC. As memory is not an issue for this data size, MPI obtains better performance than OpenMP. The data managed by each MPI task is local. On the other hand, when data does not fit in LLC, then as MPI duplicates data faces memory problems with the consequent increment in cache misses, degrading performance (Fig. 4). Notice that if allocating tasks in different NUMA nodes, this effect can be alleviated. Despite LLC misses in MPI for small data sizes are larger than LLC misses in OpenMP, the remote accesses generated by adjacent tasks penalize bringing better performance for MPI.

The hybrid approach MPI/OpenMPI performed worse than MPI and OpenMP in isolation when allocated one MPI task per NUMA-node, which means in our experimental platform, 2 task per node. The added memory overhead plus the thread creation and other overheads (e.g. remote memory access) do not compensate in performance. We believe that for larger NUMA-nodes this results may be different. We are currently undergoing this study.

In conclusion, for small data sizes and small number of tasks, both parallel programming languages can achieve the same performance no matter where tasks are allocated (lesser tasks, lesser interaction between them). When incrementing the number of tasks, there is more interaction between them penalizing remote accesses for OpenMP, but duplicating data at the same time for MPI.

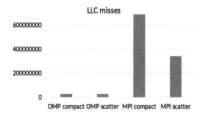

Fig. 4. Last Level Cache misses when datasize does not fit

4   Conclusions and Future Work

Based on the idea of pattern-based parallel programming we compared two standard like OpenMP, MPI and the hybrid approach MPI/OpenMP on a multicore platform with different memory accesses characteristics. We selected a scheme pattern representative of many HPC applications like a loop of stencil and reduce. We showed that both parallel programming languages can show different performance characteristics in applications depending on the data size being processed and the data locality which requires extra and conscious programming effort. Considering the hybrid approach to solve gaps in the context of a shared-platform showed to not work as good as separately.

In this work we only focused on data-parallel algorithms. We are planning to extend our study to other parallel patterns like pipeline and unstructured. In addition, the platform as already shown in terms of memory accesses has a significant role so we plan to enrich the work by studying heterogeneous platforms.

Acknowledgements. This research was supported by the following grants Spanish Ministry of Science and Innovation (contract TIN2015-65316), the Generalitat de Catalunya (2014-SGR-1051) and the European Commission through the HiPEAC-3 Network of Excellence (FP7/ICT-217068).

References

1. Aldinucci, M., et al.: A parallel pattern for iterative stencil + reduce. J. Supercomput. **74**(11), 5690–5705 (2018). https://doi.org/10.1007/s11227-016-1871-z
2. Bane, M.K., Keller, R., Pettipher, M., Computing, M., Smith, I.M.D.: A comparison of MPI and OpenMP implementations of a finite element analysis code (2000)
3. Danelutto, M., De Matteis, T., De Sensi, D., Mencagli, G., Torquati, M.: P 3 ARSEC: towards parallel patterns benchmarking. In: Proceedings of the Symposium on Applied Computing, SAC 2017, pp. 1582–1589. ACM, New York (2017). https://doi.org/10.1145/3019612.3019745
4. Kang, S.J., Lee, S.Y., Lee, K.M.: Performance comparison of OpenMP, MPI, and MapReduce in practical problems. Adv. Multimed. **2015**, 361–763 (2015). https://doi.org/10.1155/2015/575687
5. Krawezik, G.: Performance comparison of MPI and three OpenMP programming styles on shared memory multiprocessors. In: Proceedings of the Fifteenth Annual ACM Symposium on Parallel Algorithms and Architectures, SPAA 2003, pp. 118–127. ACM, New York (2003). https://doi.org/10.1145/777412.777433
6. Metzger, P., Cole, M., Fensch, C.: NUMA optimizations for algorithmic skeletons. In: Aldinucci, M., Padovani, L., Torquati, M. (eds.) Euro-Par 2018. LNCS, vol. 11014, pp. 590–602. Springer, Cham (2018). https://doi.org/10.1007/978-3-319-96983-1_42
7. MPI: Message Passing Interface (MPI) Forum. http://www.mpi-forum.org/
8. NORD3: Nord3 machine. https://www.bsc.es/support/Nord3-ug.pdf
9. OpenMP: The OpenMP API specification for parallel programming. http://www.openmp.org/
10. Piotrowski, M.: Mixed mode programming on HPCx. In: Piotrowski, M. (ed.) Parallel Scientific Computing and Optimization. SOIA, vol. 27, pp. 133–143. Springer, New York (2009). https://doi.org/10.1007/978-0-387-09707-7_12
11. Qi, L., Shen, M., Chen, Y., Li, J.: Performance comparison between OpenMP and MPI on IA64 architecture. In: Bubak, M., van Albada, G.D., Sloot, P.M.A., Dongarra, J. (eds.) ICCS 2004. LNCS, vol. 3038, pp. 388–397. Springer, Heidelberg (2004). https://doi.org/10.1007/978-3-540-24688-6_52
12. Yan, B., Regueiro, R.A.: Comparison between pure MPI and hybrid MPI-OpenMP parallelism for discrete element method (DEM) of ellipsoidal and poly-ellipsoidal particles. Comput. Part. Mech. **6**(2), 271–295 (2019). https://doi.org/10.1007/s40571-018-0213-8

Numbers, Algorithms, and Applications

New Approaches to Basic Calculus: An Experimentation via Numerical Computation

Luigi Antoniotti, Fabio Caldarola$^{(\boxtimes)}$ ⓘ, Gianfranco d'Atri,
and Marco Pellegrini ⓘ

Deparment of Mathematics and Computer Science,
Università della Calabria, Cubo 31/A, 87036 Arcavacata di Rende, CS, Italy
`antoniottiluigi@tiscali.it`, {`caldarola,datri`}`@mat.unical.it`,
`m19700530@hotmail.com`

Abstract. The introduction of the first elements of calculus both in the first university year and in the last class of high schools, presents many problems both in Italy and abroad. Emblematic are the (numerous) cases in which students decide to change their course of study or give it up completely cause the difficulties with the first exam of mathematics, which usually deals with basic calculus. This work concerns an educational experimentation involving (with differentiated methods) about 170 students, part at the IPS "F. Besta" in Treviso (IT) with main focus on two 5th classes where the students' age is about 19 years old, and part at the Liceo Classico Scientifico "XXV Aprile" in Pontedera, prov. of Pisa (IT). The experimental project aims to explore the teaching potential offered by non-classical approaches to calculus jointly with the so-called "unimaginable numbers". In particular, we employed the computational method recently proposed by Y.D. Sergeyev and widely used both in mathematics, in applied sciences and, recently, also for educational purposes. In the paper will be illustrated tools, investigation methodologies, collected data (before and after the teaching unit), and the results of various class tests.

Keywords: Mathematical education · Learning/teaching models and strategies · Interactive learning environments · Infinite · Infinity Computer · Grossone · Unimaginable numbers

1 Introduction

The difficulties that high school students encounter in their approach to university mathematics courses are fairly well known, both because of the direct experience of those who teach them, and because of the extensive literature that deals with various aspects of the problem. For example, see [19] for the intrinsic difficulties related to the transition to the third level of education, or [24,26–29] for detailed analyses of students' problems and approaches to calculus courses.

© Springer Nature Switzerland AG 2020
Y. D. Sergeyev and D. E. Kvasov (Eds.): NUMTA 2019, LNCS 11973, pp. 329–342, 2020.
https://doi.org/10.1007/978-3-030-39081-5_29

A research topic, very lively as well, which often overlaps with the previous ones, deals with the role assumed by software, simulations and computer technologies in general, as a support to academic teaching, and to the learning and personal elaboration of concepts and methods by the students (see, e.g., [23,43]). In this view, very interesting results were obtained in the paper [4], which in part served as a model to build the working environments described here.[1]

A whole line of research, within the theme concerning the students' approach to calculus, focuses on methods and problems related to teaching and learning the concept of limit and the processes with infinitely many steps. For example, [14,15,18,29,42] investigate student's difficulties, common misconceptions and obstacles for the learning, [46–48] elaborate on conceptual images and students' models of limit, and the very interesting paper [45] deals with the mathematical intuition in limiting processes.[2]

We are interested, instead, in the interactions of the students, in particular of the last year of high schools, with the first rudiments of non-classical analysis systems and similar mathematics. The teaching of non-classical mathematics, in many countries of the world, has become a consolidated fact from many years; for example, the broader phenomenon is the teaching of Robinson's *non-standard analysis* both in university courses[3] and in high schools. In Italy, in general, traditional methods are taught almost everywhere, but the focus on non-classical mathematics is strongly growing and many people are convinced of the usefulness of teaching alternative theories and methods also in high schools.[4]

The research exposed in this paper wants to investigates the response of students with respect the easy computational system proposed by Y. Sergeyev and, in smaller measure, the notational system introduced by D.E. Knuth to write the so-called *unimaginable numbers* (see Sect. 2 for a brief theoretical overview). The experimentations described here involve 8 high school classes in two different

---

[1] The paper [4] concerns a rather complex two-year experimental project conducted at the University of Calabria within the Master's Degree Program in Electronic Engineering. There, once two groups were formed each year, an experimental one equipped with computer-based tools and a control group associated with more traditional teaching methods, the aim was to analyze a series of student performances. We inform the reader that in the experimentations described in the present paper, we will find some slight traces of part of the methods used in [4]; the most visible is the employ of the computational system called *Infinity Computer* which will be used by a (very small) part of the sample, creating a hint of parallelism with the role of the software used in [4].

[2] The mathematical intuition will be important also for us, when our students will work with the infinite and, in particular, when they will approach the new concept of *grossone* (see also [44]).

[3] See, for instance, the manual [22] adopted in many academic courses and now in its third edition. See also [17] for an interesting comparison between Leibniz' and Robinson's systems and "Sarah"s conceptions".

[4] See, for instance, [16, page 2] and the proceedings [7] of the national Italian conference "Analisi nonstandard per le scuole superiori. VII Giornata di studio". This conference takes place every year and has now reached its ninth edition, Verona, October 5, 2019.

Italian institutes in Treviso and Pontedera (Pi), see Sect. 3 for details. The aim of the research are multiple, and can be summarized in three main groups/points to investigate:

(a) The students' mathematical intuition and first approaches with respect the grossone system without preliminary class lectures;
(b) The students' responses, empathy and performances during and after a brief cycle of class lectures;
(c) The students' individual home working with also the support of the Infinity Computer.

In Sect. 4 we will give the results and some brief conclusions.

One last important piece of information for the reader: the paper [21] concerns a twin but independent experimentation, concerning the same themes and carried out approximately simultaneously.

2   A Brief Overview on the Grossone-Based System and on Unimaginable Numbers

Sergeyev began to develop the grossone-based methodology in the early 2000's and, in brief, we can say that his numerical system is based on two fundamental units: the familiar 1 that generates natural numbers and a new infinite unit ①, called *grossone*, which allows to write infinite numbers and to execute computations with them in a *intuitive* and easy way, similar at all to what people daily do with finite numbers and ordinary integers. Moreover, infinitesimal numbers then appear by taking the inverses of infinite, grossone-based numbers. We also recall that, in Sergeyev's system, ① is the number of elements of $\mathbb{N}$ and $n \leq ①$ for all $n \in \mathbb{N}$.

The *extended set of natural numbers*, denoted by $\widehat{\mathbb{N}}$, can be written as

$$\widehat{\mathbb{N}} = \Big\{ \underbrace{1, 2, \ldots, ① - 1, ①}_{\mathbb{N}}, ① + 1, \ldots, 2①, \ldots, 3①, \ldots,$$
$$①^2, ①^2 + 1, \ldots, ①^2 + ①, \ldots, 2①^2, \ldots, ①^3, \ldots, 2^①, \ldots, \quad (1)$$
$$2^① + ①, \ldots \ldots, ①^①, \ldots \ldots \Big\},$$

while the *extended set of integers* as $\widehat{\mathbb{Z}} = \widehat{\mathbb{N}} \cup \{0\} \cup \{-N : N \in \widehat{\mathbb{N}}\}$. Hence, the quotients of elements of $\widehat{\mathbb{Z}}$ obviously yield the *extended set of rational numbers* denoted with $\widehat{\mathbb{Q}}$.

Introductory books for the grossone-based numerical system are [30,32], written in popular or didactic way, while the interested reader can see [31,36,38,40] for more detailed surveys and [1,8–10,13,20,33,34,37–41] (and the references therein) for some applications. Sergeyev's *Infinity Computer* [35] is, moreover, an on-line computing software developed for the grossone-based system which gives an extra level to our experimentation and connects it to previous works like [2,4] and to other similar ones in progress using the frameworks of [3,5].

"Unimaginable numbers" are instead finite but very large natural numbers with a completely different and ancient origin: the first unimaginable number comes back in fact to Archimedes of Syracuse (see [6,11]) Usually an integer $n \in \mathbb{N}$ is said *unimaginable* if it is greater than 1 *googol* which is equal to 10^{100}. Writing unimaginable numbers in common scientific notation is almost always impossible and we need notations developed *ad hoc* like Knuth's up-arrow notation that at its lower kevels gives the usual addition, multiplication and exponentiation, then tetration, pentation, hexation, and so on (see [6,11,12,25] and the references therein).

3   Description of the Samples and the Methods of the Research

The participants to this research are divided in more groups:

(P) A group P of students at the *Liceo Classico Scientifico "XXV Aprile"* in Pontedera, prov. of Pisa, Italy. This group consists of 45 students coming from a second and a third class named, by convenience, $P2$ and $P3$, respectively. More details on this sample will be given in Table 1.

(T) A group T of students at the *Istituto Superiore Commerciale Professionale "F. Besta"* in Treviso, Italy. This group comes from 2 third and 2 fourth classes conventionally indicated by $T3$, $T3'$, $T4$ and $T4'$, respectively.

($\mathcal{T}$) A second group of students at the same institute in Treviso as group T, is now indicated by the uppercase calligraphic letter $\mathcal{T}$: such a group comes from 2 fifth classes conventionally indicated by $\mathcal{T}5$ and $\mathcal{T}5'$. This group is the more interesting both for being the last class before the university career (for those who will choose to continue their studies) and because the organization, started before for this group, allowed to carry out a more extensive and articulated research.

A more accurate description of the groups listed above can be found in Table 1 where some data are shown for each class of the samples P, T and $\mathcal{T}$. The second column of Table 1 enumerates the students in each class and the third one gives the subdivision male-female. The 4th column lists the "mean age" computed at April 1, 2019 (in some case there is a certain margin of approximation caused by incomplete data). The last three columns regard the final votes obtained in mathematics at the end of the first semester (January 31, 2019): the 5th column computes the "mean vote" of all the students, the 6th one gives separated means male-female, and the last provides the maximum final vote in the class, separately again male-female.

3.1   The Group P: Experimental Activities and Methods

For the group P the experimental activity was limited to a class test at the end of May 2019, in agreement with the aim (a) described in the Introduction. The class test was administered without prior knowledge of the grossone-based

Table 1. The composition of the 8 classes constituting the samples, and some other data.

Class	Students	M - F	Mean age	Mean vote	Mean vote (M - F)	Max vote (M - F)
$P2$	25	11 - 14	15.2	6.7/10	6.5/10 - 6.8/10	9/10 - 9/10
$P3$	20	10 - 10	16.2	7.6/10	7.9/10 - 7.3/10	9/10 - 8/10
$T3$	26	8 - 18	16.6	5.6/10	5.7/10 - 5.6/10	8/10 - 9/10
$T3'$	17	6 - 11	16.6	5.8/10	5.9/10 - 5.7/10	9/10 - 9/10
$T4$	15	4 - 11	17.5	5.8/10	5.9/10 - 5.7/10	8/10 - 9/10
$T4'$	27	8 - 19	17.7	5.4/10	5.6/10 - 5.0/10	8/10 - 9/10
$T5$	23	7 - 16	18.7	5.5/10	6.1/10 - 5.2/10	9/10 - 9/10
$T5'$	15	4 - 11	19.1	5.6/10	5.8/10 - 5.5/10	9/10 - 8/10

system and without students have seen before the symbol ①, (in brief, a "zero-knowledge test"). The only information given to them was ① $\in$ ℕ and ① $\geq n$ for all $n \in$ ℕ. The test had 15 multiple-choice questions and other 6 open ones. The contents were elementary operations with ①, and about the order relation in $\widehat{\mathbb{Q}}$, i.e., the *extended set of rational numbers* in Sergeyev's framework (see [31,32,36,38,40]). Below we report in English some examples of the proposed questions.

Question 1.[5] Consider the writing ①+① and make a mark on the option that seems most correct to you among the following:

(a) ① + ① has no sense;
(b) ① + ① = ①;
(c) ① + ① is impossible to execute;
(d) ① + ① = 2①;
(e) ① + ① = 0.

Question 2. Consider the writing ① − ① and make a mark on the option that seems most correct to you among the following:

(a) ① − ① = −①;
(b) ① − ① is indeterminate;
(c) ① − ① = 0;
(d) ① − ① = ①;
(e) ① − ① has no sense.

Question 3. Consider the expression −3①+① and make a mark on the option that seems most correct to you among the following:

---

[5] The numbers used here to enumerate questions are different from those in the students' test (cf. [21, Sect. 3]). We moreover precise that in some classes in Trento we prepared two or four test versions changing for the order of questions and answers, to prevent, together with other appropriate measures, any kind of influence among students.

(a) $-3① + ① = ①$;
(b) $-3① + ①$ is a writing without sense;
(c) $-3① + ① = -3①$;
(d) $-3① + ① = -2①$;
(e) $-3① + ① = ①$;
(f) $-3① + ①$ is an indeterminate expression;
(g) $-3① + ① = 0$.

Question 4. Consider $①$, $① + ①$ and $① \cdot ①$: mark the option (or the options) that seem correct to you among those below.

(a) $① < ① + ①$, but $① \cdot ① = ①$;
(b) $① < ① + ① < ① \cdot ①$;
(c) $① + ①$ and $① \cdot ①$ are both equal to $①$;
(d) $① \leq ① + ① \leq ① \cdot ①$;
(e) It is not possible to establish any order relation between $①$, $① + ①$ and $① \cdot ①$;
(f) $① < ① + ①$ and $① + ① \leq ① \cdot ①$;
(g) $① \leq ① \cdot ① < ① + ①$;
(h) $① \leq ① \cdot ① \leq ① + ①$;
(i) The writings $① \cdot ①$ and $① + ①$ have no sense;
(j) None of the previous is correct;
(k) Other: _____

Question 5. Consider the expression $-\frac{5}{3}① + \frac{1}{2}①$ and choose the option that seems most correct to you among the following:

(a) Both $-\frac{5}{3}①$ and $\frac{1}{2}①$ are writings without sense;
(b) $-\frac{5}{3}① + \frac{1}{2}①$ is an indeterminate expression;
(c) $-\frac{5}{3}① + \frac{1}{2}① = -①$;
(d) $-\frac{5}{3}① + \frac{1}{2}① = -\frac{7}{6}①$;
(e) $-\frac{5}{3}① + \frac{1}{2}① = 0$;
(f) $-\frac{5}{3}① + \frac{1}{2}① = ①$.

Question 6. Consider the expression $\left(-\frac{2}{3}① + 2\right) \cdot (4① - 3) + 4$ and choose the option that seems most correct to you among the following:

(a) The first factor is equal to $-①$, the second to $+①$ and the addition of 4 is irrelevant, hence $\left(-\frac{2}{3}① + 2\right) \cdot (4① - 3) + 4 = -①$;
(b) $\left(-\frac{2}{3}① + 2\right) \cdot (4① - 3) + 4 = -①^2$;
(c) $\left(-\frac{2}{3}① + 2\right) \cdot (4① - 3) + 4$ is an indeterminate expression;
(d) $\left(-\frac{2}{3}① + 2\right) \cdot (4① - 3) + 4 = -\frac{8}{3}①^2 + 10① - 2$;
(e) It is not possible to sum $\frac{2}{3}①$ with 2 and $4①$ with -3, hence the expression $\left(-\frac{2}{3}① + 2\right) \cdot (4① - 3) + 4 = -①$ has no sense.

Question 7. Consider $\frac{12}{8}①$ and $\frac{5}{3}①$: mark the option (or the options) that seem correct to you among those below.

(a) The writings $\frac{12}{8}①$ and $\frac{5}{3}①$ have no sense;

(b) $\frac{12}{8}① = \frac{5}{3}①$;

(c) $\frac{12}{8}①$ and $\frac{5}{3}①$ are both equal to $①$;

(d) $\frac{12}{8}① > \frac{5}{3}①$;

(e) $\frac{12}{8}① < \frac{5}{3}①$;

(f) $\frac{12}{8}① \geq \frac{5}{3}①$;

(g) $\frac{12}{8}① \leq \frac{5}{3}①$;

(h) It is not possible to establish an order relation between $\frac{12}{8}①$ and $\frac{5}{3}①$.

It is important to notice that the questions in the test (and also for the ones of the samples T and $\mathcal{T}$) were grouped in small groups on separate sheets, and the students were asked to read and answer the questions in the order of presentation, without the possibility of changing an answer already given.

After the day of the test, and therefore outside our experimentation, there were some discussions in classroom, often to answer the questions and curiosities of the students themselves, about the meaning and use of the symbol $①$. We specify, however, that the experimentation in question did not in any way affect the regular progress of the established school program.

The results of the test will be given and analyzed in Sect. 4.

3.2   The Group T: Experimental Activities and Methods

The experimental activities and the research methodologies used for the 4 classes of group T were almost identical to those described in Subsect. 3.1. The only exceptions concern a greater ease and a simplification of some test questions, especially of some more complex ones not reported in the previous subsection, dictated by the different type of school and, unlike the group P, to classes $T3$ and $T4'$ were administered, in the days after the test, an informative questionnaire that aimed to capture the (eventual) interest and any curiosity aroused in the students. We also precise that the test questions and their number were not the same for all the classes that made up the group T, for didactic and organizational reasons.

3.3   The Group $\mathcal{T}$: Experimental Activities and Methods

The most complex and structured experimentation concerned the group $\mathcal{T}$. In addition to the zero-knowledge test as for groups P and T, but carried out about one or two months in advance, a series of short lessons and class discussions was proposed to both classes $T5$ and $T5'$. The organization relative to the class $T5$ allowed also the possibility of a greater number of lessons, which in any case covered a very limited amount of time (about 4 or 5 h split into packages of half an hour each time, over two or three months). After the cycle of lectures, a test very similar, especially for the class $T5'$, to the one administered at the beginning was proposed for both classes: for convenience and future references we call it the *final test*.

Our experimental activity with group T included also some rudiments of calculus. In particular, it has been discussed the concept of limit for $x \to \pm\infty$ in classical analysis compared with the evaluation of a function at a grossone-based infinity in Sergeyev's framework (mainly we took $x = \pm①$ than other infinite numbers). Moreover, we showed the relation between some asymptotic behaviours of a function and its derivative both in the traditional and in the new context, and we also talked about the meaning and the way to perform computations with infinitesimal quantities written in the new system. Examples of closed and open questions proposed to the students of group T in the final test are reported below translated in English.

Question 8. Consider the writing $\frac{3}{2①} - \frac{1}{①}$ and mark the right option/options among the following:

(a) $\frac{3}{2①} - \frac{1}{①} = \frac{1}{2}①$;

(b) $\frac{3}{2①} - \frac{1}{①}$ is indeterminate;

(c) The writings $\frac{3}{2①}$ and $\frac{1}{①}$ have no sense because it is not possible to divide by infinity;

(d) $\frac{3}{2①} - \frac{1}{①}$ is zero because both $\frac{3}{2①}$ and $\frac{1}{①}$ are equal to zero;

(e) $\frac{3}{2①} - \frac{1}{①} = \frac{1}{2①}$;

(f) $\frac{3}{2①} - \frac{1}{①} = \frac{2}{①}$;

(g) $\frac{3}{2①} - \frac{1}{①} = \frac{3-1}{2①} = \frac{1}{①}$;

(h) $\frac{3}{2①} - \frac{1}{①} = 0.5①^{-1}$.

Question 9. Consider the writings $\frac{1}{2①}$, $-\frac{1}{①}$ and $-3\frac{1}{2①} + \frac{1}{2}$. Indicate the true expressions among the following:

(a) $-\frac{3}{2①} + \frac{1}{2} < -\frac{1}{①} < \frac{1}{2①}$;

(b) $-\frac{1}{①} < -\frac{3}{2①} + \frac{1}{2} < \frac{1}{2①}$;

(c) $-\frac{1}{①} < \frac{1}{2①} < -\frac{3}{2①} + \frac{1}{2}$;

(d) There are no order relations between $\frac{1}{2①}$, $-\frac{1}{①}$ and $-3\frac{1}{2①} + \frac{1}{2}$ because they are not numbers;

(e) $-\frac{1}{①} = \frac{1}{2①} = 0 < -\frac{3}{2①} + \frac{1}{2} = \frac{1}{2}$;

(f) The expressions $\frac{1}{2①}$, $-\frac{1}{①}$ and $-3\frac{1}{2①} + \frac{1}{2}$ have no sense because it is not possible to divide by infinity.

Other questions of the test had the aim to solicit a comparison with the symbol ∞ as used in traditional mathematics.

Question 10. Consider the writing $\infty + \infty$ and $\infty \cdot \infty$. Choose the true options among the following:

(a) $\infty + \infty$ and $\infty \cdot \infty$ have no sense in mathematics;
(b) $\infty + \infty = \infty$ and $\infty \cdot \infty = \infty$;
(c) $\infty + \infty = \infty$ and $\infty + \infty = \infty$;
(d) $\infty + \infty = 2\infty < \infty \cdot \infty = \infty^2$;
(e) $\infty + \infty$ and $\infty \cdot \infty$ are not comparable;
(f) $\infty < \infty + \infty < \infty \cdot \infty$.

Question 11. Consider the writing $\infty - \infty$ and mark the right options among the following:

(a) $\infty - \infty = \infty$;
(b) $-\infty < \infty - \infty < \infty$;
(c) $\infty - \infty$ is an indeterminate expression;
(d) $\infty - \infty = 0$;
(e) $-\infty < \infty - \infty < \infty$;
(f) $\infty - \infty < \infty + \infty$;
(g) $\infty - \infty$ and $-\infty$ are not comparable.

Question 12. For each of the following items make a mark on "T" or "F" if you believe the corresponding statement to be true or false, respectively.

(a) $① < +\infty$ T - F

(b) $① = +\infty$ T - F

(c) $①$ and ∞ are not comparable   T - F

(d) $① \leq +\infty$ T - F

(e) $① \geq +\infty$ T - F

(f) $①$ and ∞ cannot be used together because the belong to
    different settings   T - F

(g) $①^2 > +\infty$ T - F

(h) $① + 1 = ①$ T - F

(i) $\infty + 1 = \infty$ T - F

(j) $\infty + 1 > \infty$ T - F

Question 13. In the classical setting, consider the function given by the analytical expression $f(x) = \frac{x^2}{x+1}$.

(a) Compute the domain of the function and the limits at the extremal points of the domain.
(b) Compute the asymptotes of f (vertical, horizontal and oblique).

Question 14. In the grossone setting, consider the function $f(x) = \frac{x^2}{x+1}$.

(a) Compute the values of f at $-①$ and $①$.
(b) Compute the values of f at $-① + 2$, $-① - 2$ and $① + 1$.
(c) Let $y = a(x)$ be the right oblique asymptote of f (if it exists).
  • Compute the value $a(①)$ and $a(① + 1)$.
  • Compute the difference $f(①) - a(①)$ and $f(① + 1) - a(① + 1)$.

Question 15 (excluded from the evaluation). In your opinion there are some advantages in using ① in the place of $+\infty$ (i.e., the grossone setting in the place of the classical one)? Justify the answer and, if yes, list some of them.

The last extra question, although proposed together the final test, was excluded from any attribution of a score and this was clearly written. In any case, most of the students did not answer this question, or gave hasty and little significant answers: a different outcome would probably have been recorded if it had been proposed on a different day rather than at the end of a test with many questions.

The students of class $T5$ have also been spoken in classroom of the Infinity Computer, trying to motivate them in a deepening and individual work at home on it. From the compilation of the informative questionnaire and from a continuous dialogue with the students, however, it emerged that only 4 of them actually used the Infinity Computer at least once at home, independently. For convenience, we will denote this group of students by $T5.1$.

As regards unimaginable numbers we inform the reader that, for the class $T5$, a soft approach to Knuth's notation and very large numbers was also planned, but just in part developed with the students. In particular we presented tetrations and pentations to them, and in a first moment it seemed very successful, in particular the way to iterate the ordinary exponentiation and to write it compactly under the form of a tetration. Many problems and much confusion emerged instead in a successive lesson two weeks later, and we decided to ask no questions about these topics in the final test (cf. the conclusions in Sect. 5).

4   Results and Their Analysis

In this section we will give the results of the tests, will discuss the situation picture emerging from the questionnaires and the dialogues with the students, and will finally draw up a balance of the experimentations giving some brief conclusions jointly with the next section.

The zero-knowledge test has been proposed to all the 8 classes, with major differences between the groups P and $T \cup T$, but with several minor ones inside the sample $T \cup T$. In Table 2, the columns 2–5 are devoted to this test: the 2nd and 3rd columns give the mean student score [Mean] obtained in each class and the maximum score [Max], respectively, both over the total score of the test. The 4th column gives the normalized mean (students') score [NMS] and the 5th column the standard deviation [SD]. The column from 6 to 9 are the twin ones of 2–5, but relative to the final test (notations are hence obtained by adding "$\mathcal{F}$" to the previous ones).

The results of the zero-knowledge test are, in general, positive or very positive, in dependence of the cases, as Table 2 shows. In particular, the maximum scores in the 3rd column unequivocally give the measure of how intuitive it is to perform basic calculations with the grossone system.

The scores of the final test are higher and with more gap from the corresponding one of the initial test, in particular for the classes with less teaching

Table 2. The results of the tests for each class.

Class	Mean	Max	NMS	SD	Mean $\mathcal{F}$	Max $\mathcal{F}$	NMS $\mathcal{F}$	SD $\mathcal{F}$
$P2$	15.9/21	21/21	0.76	2.7/21	–	–	–	–
$P3$	16.7/21	20/21	0.79	2.1/21	–	–	–	–
$T3$	10.5/24	18/24	0.43	2.9/24	15.1/24	21/24	0.63	2.7/24
$T3'$	8.1/24	15/24	0.34	3.9/24	–	–	–	–
$T4$	10.8/24	15/24	0.45	4.1/24	–	–	–	–
$T4'$	8.4/24	15/24	0.35	3.7/24	13.2/24	20/24	0.55	3.1/24
$T5$	16.2/24	24/24	0.67	5.5/24	20.7/24	24/24	0.86	2.7/24
$T5'$	8.1/24	18/24	0.34	3.9/24	14.2/24	22/24	0.59	3.6/24
$T5.1$	21.8/24	24/24	0.91	1.3/24	23.3/24	24/24	0.97	1.3/24

time. This result is rather unexpected and seems to be due to the fact that the class $T5$, the one with the highest number of lessons, starts from a very high score in the initial test. But it could also mean that a few quick explanations are enough to significantly improve some good initial performances (considering that it is a text with zero knowledge) especially if not very high.

As regards the small $T5.1$ group (i.e., the one consisting of the 4 students of the class $T5$ that used the Infinity Computer at least once at home), we can observe very high performances in both columns 2 and 6, with a small increase. Probably, a more difficult extra test for this group would have been interesting both at the beginning and at the end of the experimentation. Finally, from the questionnaire and, in particular, from a continuous conversation with the students, we think that their approach with the Infinity Computer has been fruitful to arouse attraction and to give further motivation and interest, probably because it is seen as a form of concrete application (recall that we are dealing with a technical-commercial school) of the grossone-based system.

5 Conlusion

The good results obtained in the various levels of the experimentation have shown a remarkable usability for the students of the new concept of infinity represented by grossone. It should in fact be emphasized that in all the 8 classes that took part in the experimentation, most of the students succeeded in assimilating, or better, effectively understanding, the distinctive properties of ①, and the correct way of performing calculations in the associate numerical-computational system in a few minutes, already at the initial zero-knowledge test. Very interesting and useful to the students, it was also the comparison, made several times during the lessons given to the classes $T5$ and $T5'$, between the classical conception of infinity (dating back to Cantor and Weierstrass) and that related to grossone: in fact the students, stimulated by the possibility of working computationally with ①, in an "unconsciously familiar" way, showed a marked interest, very difficult to be aroused in general, also for the more theoretical aspects concerning the two models of infinity. The possibility of carrying out a wider experimentation and proposing Sergeyev's model on a larger scale could therefore have relevant educational implications.

We also believe that the Infinity Computer can also have a good educational value, which however we have not had the opportunity to investigate or test in depth in a very short experimentation, and with many new features for students like ours.

Similarly, with regard to unimaginable numbers, we conclude that they could have very interesting didactic applications (especially from the point of view of generalizing the usual operations of addition, multiplication and exponentiation via tetration, pentation, hexation, etc.), but they require experimental activities completely dedicated to them because a not so easy assimilation of such topics has emerged, at least among the students of a technical school, more used to calculations than to algebraic formalisms.

Aknowledgments. This work is partially supported by the research projects *"IoT&B, Internet of Things and Blockchain"*, CUP J48C17000230006, POR Calabria FESR-FSE 2014-2020.

References

1. Antoniotti, L., Caldarola, F., Maiolo, M.: Infinite numerical computing applied to Peano's, Hilbert's and Moore's curves. Mediterr. J. Math. to appear
2. Bertacchini, F., Bilotta, E., Pantano, P., Tavernise, A.: Motivating the learning of science topics in secondary schools: a constructivistic edutainment setting for studing Chaos. Comput. Educ. **59**(4), 1377–1386 (2012)
3. Bertacchini, F., Bilotta, E., Caldarola, F., Pantano, P.: Complex interactions in one-dimensional cellular automata and linguistic constructions. Appl. Math. Sci. **12**(15), 691–721 (2018). https://doi.org/10.12988/ams.2018.8353
4. Bertacchini, F., Bilotta, E., Caldarola, F., Pantano, P.: The role of computer simulations in learning analytic mechanics towards chaos theory: a course experimentation. Int. J. Math. Educ. Sci. Technol. **50**(1), 100–120 (2019). https://doi.org/10.1080/0020739X.2018.1478134
5. Bertacchini, F., Bilotta, E., Caldarola, F., Pantano, P., Renteria Bustamante, L.: Emergence of linguistic-like structures in one-dimensional cellular automata. In: Sergeyev, Ya.D., Kvasov, D.E., Dell'Accio, F., Mukhametzhanov, M.S. (eds.) 2nd International Conference on NUMTA 2016 - Numerical Computations: Theory and Algorithms, AIP Conference Proceedings, vol. 1776, p. 090044. AIP Publ., New York (2016). https://doi.org/10.1063/1.4965408
6. Blakley, G.R., Borosh, I.: Knuth's iterated powers. Adv. Math. **34**(2), 109–136 (1979). https://doi.org/10.1016/0001-8708(79)90052-5
7. Bonavoglia, P. (ed.): Atti del convegno della VII Giornata Nazionale di Analisi non Standard per le scuole superiori, Venezia, 30 settembre 2017. Edizioni Aracne, Roma (2017). (in Italian)
8. Caldarola, F.: The Sierpiński curve viewed by numerical computations with infinities and infinitesimals. Appl. Math. Comput. **318**, 321–328 (2018). https://doi.org/10.1016/j.amc.2017.06.024

9. Caldarola, F.: The exact measures of the Sierpiński d-dimensional tetrahedron in connection with a Diophantine nonlinear system. Commun. Nonlinear Sci. Numer. Simul. **63**, 228–238 (2018). https://doi.org/10.1016/j.cnsns.2018.02.026

10. Caldarola, F., Cortese, D., d'Atri, G., Maiolo, M.: Paradoxes of the infinite and ontological dilemmas between ancient philosophy and modern mathematical solutions. In: Sergeyev, Y.D., Kvasov, D.E. (eds.) NUMTA 2019. LNCS, vol. 11973, pp. 358–372. Springer, Cham (2020)

11. Caldarola, F., d'Atri, G., Maiolo, M.: What are the unimaginable numbers? Submitted for publication

12. Caldarola, F., d'Atri, G., Mercuri, P., Talamanca, V.: On the arithmetic of Knuth's powers and some computational results about their density. In: Sergeyev, Y.D., Kvasov, D.E. (eds.) NUMTA 2019. LNCS, vol. 11973, pp. 381–388. Springer, Cham (2020)

13. Caldarola, F., Maiolo, M., Solferino, V.: A new approach to the Z-transform through infinite computation. Commun. Nonlinear Sci. Numer. Simul. **82**, 105019 (2020). https://doi.org/10.1016/j.cnsns.2019.105019

14. Cornu, B.: Quelques obstacles à l'apprentissage de la notion de limite. Recherches en Didactique des Mathematiques **4**, 236–268 (1983)

15. Davis, R.B., Vinner, S.: The notion of limit: some seemingly unavoidable misconception stages. J. Math. Behav. **5**, 281–303 (1986)

16. Di Nasso, M.: Un'introduzione all'analisi con infinitesimi. In: Analisi nonstandard per le scuole superiori - V Giornata di studio, Atti del convegno, Verona, 10 Ottobre 2015, Matematicamente.it (2016). (in Italian)

17. Ely, R.: Nonstandard student conceptions about infinitesimals. J. Res. Math. Educ. **41**(2), 117–146 (2010)

18. Ervynck, G.: Conceptual difficulties for first year university students in the acquisition of the notion of limit of a function. In: Proceedings of the Fifth Conference of the International Group for the Psychology of Mathematics Education, pp. 330–333 (1981)

19. Guzman, M., Hodgson, B., Robert, A., Villani, V.: Difficulties in the passage from secondary to tertiary education. In: Fischer, G., Rehmann, U. (eds.) Proceedings of the International Congress of Mathematicians, Berlin 1998, vol. 3, pp. 747–762 (1998)

20. Iannone, P., Rizza, D., Thoma, A.: Investigating secondary school students' epistemologies through a class activity concerning infinity. In: Bergqvist, E., et al. (eds.) Proceedings of the 42nd Conference of the International Group for the Psychology of Mathematics Education, Umeå(Sweden), vol. 3, pp. 131–138. PME (2018)

21. Ingarozza, F., Adamo, M.T., Martino, M., Piscitelli, A.: A Grossone-Based Numerical Model for Computations with Infinity: A Case Study in an Italian High School. In: Sergeyev, Y.D., Kvasov, D.E. (eds.) NUMTA 2019. LNCS, vol. 11973, pp. 451–462. Springer, Cham (2020). https://doi.org/10.1007/978-3-030-39081-5_39

22. Keisler, H.J.: Elementary Calculus: An Approach Using Innitesimals, 3rd edn. Dover Publications (2012)

23. Khan, S.: New pedagogies on teaching science with computer simulations. J. Sci. Educ. Technol. **20**, 215–232 (2011)

24. Klymchuk, S., Zverkova, T., Gruenwald, N., Sauerbier, G.: University students' difficulties in solving application problems in calculus: student perspectives. Math. Educ. Res. J. **22**(2), 81–91 (2010)

25. Knuth, D.E.: Mathematics and computer science: coping with finiteness. Science **194**(4271), 1235–1242 (1976). https://doi.org/10.1126/science.194.4271.1235

26. Lithner, J.: Students' mathematical reasoning in university textbook exercises. Educ. Stud. Math. **52**, 29–55 (2003)
27. Lithner, J.: Mathematical reasoning in calculus textbook exercises. J. Math. Behav. **23**, 405–427 (2004)
28. Lithner, J.: University students' learning difficulties. Educ. Inquiry **2**(2), 289–303 (2011)
29. Muzangwa, J., Chifamba, P.: Analysis of errors and misconceptions in the learning of calculus by undergraduate students. Acta Didactica Napocensia **5**(2) (2012)
30. Rizza, D.: Primi passi nell'aritmetica dell'infinito. Preprint (2019)
31. Sergeyev, Y.D.: A new applied approach for executing computations with infinite and infinitesimal quantities. Informatica **19**(4), 567–96 (2008)
32. Sergeyev, Y.D.: Arithmetic of Infinity. Edizioni Orizzonti Meridionali, Cosenza (2003)
33. Sergeyev, Y.D.: Evaluating the exact infinitesimal values of area of Sierpinskinski's carpet and volume of Menger's sponge. Chaos Solitons Fractals **42**(5), 3042–6 (2009)
34. Sergeyev, Y.D.: Higher order numerical differentiation on the Infinity Computer. Optim. Lett. **5**(4), 575–85 (2011)
35. Sergeyev, Y.D.: (2004). http://www.theinfinitycomputer.com
36. Sergeyev, Y.D.: Lagrange Lecture: methodology of numerical computations with infinities and infinitesimals. Rendiconti del Seminario Matematico Università e Politecnico di Torino **68**(2), 95–113 (2010)
37. Sergeyev, Y.D.: Solving ordinary differential equations by working with infinitesimals numerically on the Infinity Computer. Appl. Math. Comput. **219**(22), 10668–81 (2013)
38. Sergeyev, Y.D.: Un semplice modo per trattare le grandezze infinite ed infinitesime. Matematica, Cultura e Società: Rivista dell'Unione Matematica Italiana **8**(1), 111–47 (2015). (in Italian)
39. Sergeyev, Y.D.: Using blinking fractals for mathematical modelling of processes of growth in biological systems. Informatica **22**(4), 559–76 (2011)
40. Sergeyev, Y.D.: Numerical infinities and infinitesimals: methodology, applications, and repercussions on two Hilbert problems. EMS Surv. Math. Sci. **4**(2), 219–320 (2017)
41. Sergeyev, Y.D., Mukhametzhanov, M.S., Mazzia, F., Iavernaro, F., Amodio, P.: Numerical methods for solving initial value problems on the Infinity Computer. Int. J. Unconv. Comput. **12**(1), 55–66 (2016)
42. Sierpinska, A.: Humanities students and epistemological obstacles related to limits. Educ. Stud. Math. **18**, 371–397 (1987)
43. Steinberg, R.N.: Computer in teaching science: to simulate or not simulate? Phys. Educ. Res.: Am. J. Phys. Suppl. **68**(7), S37 (2000)
44. Tall, D.: A child thinking about infinity. J. Math. Behav. **20**, 7–19 (2001)
45. Tall, D.: Mathematical intuition, with special reference to limiting processes. In: Proceedings of the Fourth International Conference for the Psychology of Mathematics Education, pp. 170–176 (1980)
46. Tall, D.: The transition to advanced mathematical thinking: functions, limits, infinity and proof. In: Grouws, D. (ed.) Handbook for Research on Mathematics Teaching and Learning, pp. 495–511. Macmillan, New York (1992)
47. Tall, D., Vinner, S.: Concept image and concept definition in mathematics with particular reference to limits and continuity. Educ. Stud. Math. **12**, 151–169 (1981)
48. Williams, S.R.: Models of limit held by college calculus students. J. Res. Math. Educ. **22**(3), 219–236 (1991)

A Multi-factor RSA-like Scheme with Fast Decryption Based on Rédei Rational Functions over the Pell Hyperbola

Emanuele Bellini[1] and Nadir Murru[2](✉)

[1] Technology Innovation Institute, Abu Dhabi, UAE
eemanuele.bellini@gmail.com
[2] Department of Mathematics, University of Turin, Turin, Italy
nadir.murru@unito.it

Abstract. We propose a generalization of an RSA-like scheme based on Rédei rational functions over the Pell hyperbola. Instead of a modulus which is a product of two primes, we define the scheme on a multi-factor modulus, i.e. on a product of more than two primes. This results in a scheme with a decryption which is quadratically faster, in the number of primes factoring the modulus, than the original RSA, while preserving a better security. The scheme reaches its best efficiency advantage over RSA for high security levels, since in these cases the modulus can contain more primes. Compared to the analog schemes based on elliptic curves, as the KMOV cryptosystem, the proposed scheme is more efficient. Furthermore a variation of the scheme with larger ciphertext size does not suffer of impossible group operation attacks, as it happens for schemes based on elliptic curves.

Keywords: Cryptography · Pell conic · Rédei rational functions · RSA

1 Introduction

RSA is the most widespread asymmetric encryption scheme. Its security is based on the fact that the trapdoor function $\tau_{N,e}(x) = x^e \mod N$, where $N = pq$ is the product of two large prime integers, and e an invertible element in $\mathbb{Z}_{\phi(N)}(\phi(N)$ being the Euler totient function), cannot be inverted by a polynomial-time in N algorithm without knowing either the integers p, q, $\phi(N)$ or the inverse d of e modulo $\phi(N)$. Thus the pair (N, e), called the public key, is known to everyone, while the triple (p, q, d), called the secret key, is only known to the receiver of an encrypted message. Both encryption and decryption are performed through an exponentiation modulo N. Precisely, the ciphertext C is obtained as $C = M^e$ (mod N), and the original message M is obtained with the exponentiation $M = C^d$ (mod N). While usually the encryption exponent is chosen to be small, the decryption exponent is about the size of N, implying much slower performances

© Springer Nature Switzerland AG 2020
Y. D. Sergeyev and D. E. Kvasov (Eds.): NUMTA 2019, LNCS 11973, pp. 343–357, 2020.
https://doi.org/10.1007/978-3-030-39081-5_30

during decryption with respect to encryption. Through the years many proposal have been presented trying to speed up the decryption process.

In this work we present the fastest, to the authors knowledge, of such decryption algorithms whose security is based on the factorization problem. The presented scheme exploits different properties of Rédei rational functions, which are classical functions in number theory. The proposed decryption algorithm is quadratically, on the number of primes composing the modulus N, faster than RSA.

The work is divided as follows. In Sect. 2 an overview of the main schemes based on the factorization problem which successfully improved RSA decryption step is presented. In Sect. 3 the main theoretical results underlying our scheme are described. Section 4 is devoted to the presentation of the cryptographic scheme, and in Sects. 5 and 6 its security and efficiency are discussed, respectively. Section 7 concludes the work.

2 Related Work

In this section we briefly overview the main cryptographic schemes based on the factorization problem that have been introduced in order to improve RSA decryption step.

Usually, the general technique to speed up the RSA decryption step $C = M^e$ (mod N) is to compute the exponentiation modulo each factor of N and then obtain N using the Chinese Remainder Theorem.

2.1 Multifactor RSA

There exists variants of RSA scheme which exploit a modulus with more than 2 factors to achieve a faster decryption algorithm. This variants are sometimes called Multifactor RSA [6], or Multiprime RSA [8,10]. The first proposal exploiting a modulus of the form $N = p_1 p_2 p_3$ has been patented by Compaq [9,10] in 1997. About at the same time Takagi [30] proposed an even faster solution using the modulus $N = p^r q$, for which the exponentiation modulo p^r is computed using the Hensel lifting method [11, p. 137]. Later, this solution has been generalized to the modulus $N = p^r q^s$ [28]. According to [10], the appropriate number of primes to be chosen in order to resist state-of-the-art factorization algorithms depends from the modulus size, and, precisely, it can be: up to 3 primes for 1024, 1536, 2048, 2560, 3072, and 3584 bit modulus, up to 4 for 4096, and up to 5 for 8192.

2.2 RSA-like Schemes

Another solution which allows to obtain even faster decryption is to use RSA-like schemes based on isomorphism as [3,16,17,26]. As an additional property, these schemes owns better security properties with respect to RSA, avoiding small exponent attacks to either d [31] or e [12,13], and vulnerabilities which appear when switching from one-to-one communication scenario to broadcast scenario (e.g., see [14]). The aforementioned schemes are based on isomorphism between two groups, one of which is the set of points over a curve, usually a cubic or a conic. A complete overview on RSA-like schemes based on conics can be found in [3]. In general, schemes based on cubic curves have a computationally more expensive addition operation compared to schemes based on conic equations.

2.3   Generalizing RSA-like Scheme with Multifactor Modulus

As done when generalizing from RSA to Multiprime RSA, in [7] a generalization of [16,17] has been proposed, thus generalizing a RSA-like scheme based on elliptic curves and a modulus $N = pq$ to a similar scheme based on the generic modulus $N = p^r q^s$.

In this paper we present a similar generalization of the scheme [3], which is based on the Pell's equation, to the modulus $N = p_1^{e_1} \cdot \ldots \cdot p_r^{e_r}$ for $r > 2$, obtaining the fastest decryption of all schemes discussed in this section.

3   Product of Points over the Pell Hyperbola

In [3], we introduced a novel RSA–like scheme based on an isomorphism between certain conics (whose the Pell hyperbola is a special case) and a set of parameters equipped with a non–standard product. In Sect. 4, we generalize this scheme considering a prime power modulus $N = p_1^{e_1} \cdots p_r^{e_r}$. In this section, we recall some definitions and properties given in [3] in order to improve the readability of the paper. Then, we study properties of the involved products and sets in $\mathbb{Z}_{p^r}$ and $\mathbb{Z}_N$.

3.1   A Group Structure over the Pell Hyperbola over a Field

Let $\mathbb{K}$ be a field and $x^2 - D$ an irreducible polynomial over $\mathbb{K}[x]$. Considering the quotient field $\mathbb{A}[x] = \mathbb{K}[x]/(x^2 - D)$, the induced product over $\mathbb{A}[x]$ is

$$(p + qx)(r + sx) = (pr + qsD) + (qr + ps)x.$$

The group of unitary elements of $\mathbb{A}^*[x] = \mathbb{A}[x] - \{0_{\mathbb{A}[x]}\}^1$ is $\{p + qx \in \mathbb{A}^*[x] : p^2 - Dq^2 = 1\}$. Thus, we can introduce the commutative group $(\mathcal{H}_{D,\mathbb{K}}, \otimes)$, where

$$\mathcal{H}_{D,\mathbb{K}} = \{(x, y) \in \mathbb{K} \times \mathbb{K} : x^2 - Dy^2 = 1\}$$

and

$$(x, y) \otimes (w, z) = (xw + yzD, yw + xz), \quad \forall (x, y), (w, z) \in \mathcal{H}_{D,\mathbb{K}}. \tag{1}$$

It is worth noting that $(1, 0)$ is the identity and the inverse of an element (x, y) is $(x, -y)$.

Remark 1. When $\mathbb{K} = \mathbb{R}$, the conic $\mathcal{H}_{D,\mathbb{K}}$, for D a non–square integer, is called the Pell hyperbola since it contains all the solutions of the Pell equation and $\otimes$ is the classical Brahamagupta product, see, e.g., [15].

---

[1] The element $0_{\mathbb{A}[x]}$ is the zero polynomial.

3.2   A Parametrization of the Pell Hyperbola

From now on let $\mathbb{A} = \mathbb{A}[x]$.

Starting from $\mathbb{A}^*$, we can derive a parametrization for $\mathcal{H}_{D,\mathbb{K}}$. In particular, let us consider the group $\mathbb{A}^*/\mathbb{K}^*$, whose elements are the equivalence classes of $\mathbb{A}^*$ and can be written as

$$\{[a + x] : a \in \mathbb{K}\} \cup \{[1_{\mathbb{K}^*}]\}.$$

The induced product over $\mathbb{A}^*/\mathbb{K}^*$ is given by

$$[a + x][b + x] = [ab + ax + bx + x^2] = [D + ab + (a + b)x]$$

and, if $a + b \neq 0$, we have

$$[a + x][b + x] = [\frac{D + ab}{a + b} + x]$$

else

$$[a + x][b + x] = [D + ab] = [1_{\mathbb{K}^*}].$$

This construction allows us to define the set of parameters $\mathcal{P}_{\mathbb{K}} = \mathbb{K} \cup \{\alpha\}$, with α not in $\mathbb{K}$, equipped with the following product:

$$\begin{cases} a \odot b = \dfrac{D + ab}{a + b}, & a + b \neq 0 \\ a \odot b = \alpha, & a + b = 0 \end{cases} . \tag{2}$$

We have that $(\mathcal{P}_{\mathbb{K}}, \odot)$ is a commutative group with identity α and the inverse of an element a is the element b such that $a + b = 0$. Now, consider the following parametrization for the conic $\mathcal{H}_{D,\mathbb{K}}$:

$$y = \frac{1}{m}(x + 1).$$

It can be proved that the following isomorphism between $(\mathcal{H}_{D,\mathbb{K}}, \otimes)$ and $(\mathcal{P}_{\mathbb{K}}, \odot)$ holds:

$$\Phi_D : \begin{cases} \mathcal{H}_{D,\mathbb{K}} & \to \mathcal{P}_{\mathbb{K}} \\ (x, y) & \mapsto \dfrac{1 + x}{y} \quad \forall (x, y) \in \mathcal{H}_{D,\mathbb{K}}, \quad y \neq 0 \\ (1, 0) & \mapsto \alpha \\ (-1, 0) & \mapsto 0 , \end{cases} \tag{3}$$

and

$$\Phi_D^{-1} : \begin{cases} \mathcal{P}_{\mathbb{K}} & \to \mathcal{H}_{D,\mathbb{K}} \\ m & \mapsto \left(\dfrac{m^2 + D}{m^2 - D}, \dfrac{2m}{m^2 - D}\right) \quad \forall m \in \mathbb{K} , \\ \alpha & \mapsto (1, 0) , \end{cases} \tag{4}$$

see [1] and [3].

Proposition 1. *When* $\mathbb{K} = \mathbb{Z}_p$, *p prime,* $(\mathcal{P}_{\mathbb{K}}, \odot)$ *and* $(\mathcal{H}_{D,\mathbb{K}}, \otimes)$ *are cyclic groups of order* $p + 1$ *and*

$$m^{\odot(p+2)} = m \quad (\bmod\ p), \quad \forall m \in \mathcal{P}_{\mathbb{Z}_p}$$

or, equivalently

$$(x, y)^{\otimes(p+2)} = (x, y) \quad (\bmod\ p), \quad \forall (x, y) \in \mathcal{H}_{D, \mathbb{Z}_p},$$

where powers are performed using products $\odot$ *and* $\otimes$, *respectively. See* [3].

The powers in $\mathcal{P}_{\mathbb{K}}$ can be efficiently computed by means of the Rédei rational functions [27], which are classical functions in number theory. They are defined by considering the development of

$$(z + \sqrt{D})^n = A_n(D, z) + B_n(D, z)\sqrt{D},$$

for z integer and D non–square positive integer. The polynomials $A_n(D, z)$ and $B_n(D, z)$ defined by the previous expansion are called Rédei polynomials and can be evaluated by

$$M^n = \begin{pmatrix} A_n(D, z) & DB_n(D, z) \\ B_n(D, z) & A_n(D, z) \end{pmatrix}$$

where

$$M = \begin{pmatrix} z & D \\ 1 & z \end{pmatrix}.$$

From this property, it follows that the Rédei polynomials are linear recurrent sequences with characteristic polynomial $t^2 - 2zt + (z^2 - D)$. The Rédei rational functions are defined by

$$Q_n(D, z) = \frac{A_n(D, z)}{B_n(D, z)}, \quad \forall n \geq 1.$$

Proposition 2. *Let* $m^{\odot n}$ *be the n–th power of* $m \in \mathcal{P}_{\mathbb{K}}$ *with respect to* $\odot$, *then*

$$m^{\odot n} = Q_n(D, m).$$

See [2].

Remark 2. The Rédei rational functions can be evaluated by means of an algorithm of complexity $O(\log_2(n))$ with respect to addition, subtraction and multiplication over rings [24].

3.3 Properties of the Pell Hyperbola over a Ring

In this section, we study the case $\mathbb{K} = \mathbb{Z}_{p^r}$ that we will exploit in the next section for the construction of a cryptographic scheme. In what follows, we will omit from $\mathcal{H}_{D,\mathbb{K}}$ the dependence on D when it will be clear from the context.

First, we need to determine the order of $\mathcal{H}_{\mathbb{Z}_{p^r}}$ in order to have a result similar to Proposition 1 also in this situation.

Theorem 1. *The order of the cyclic group $\mathcal{H}_{\mathbb{Z}_{p^r}}$ is $p^{r-1}(p+1)$, i.e., the Pell equation $x^2 - Dy^2 = 1$ has $p^{r-1}(p+1)$ solutions in $\mathbb{Z}_{p^r}$ for $D \in \mathbb{Z}_{p^r}^*$ quadratic non–residue in $\mathbb{Z}_p$.*

Proof. Since, by Proposition 1, the Pell equation in $\mathbb{Z}_p$ has $p+1$ solutions, then we need to prove the following

1. any solution of the Pell equation in $\mathbb{Z}_p$, generates p^{r-1} solutions of the same equation in $\mathbb{Z}_{p^r}$;
2. all the solutions of the Pell equation in $\mathbb{Z}_{p^r}$ are generated as in the previous step.

(1) Let (x_0, y_0) be a solution of $x^2 - Dy^2 \equiv 1 \pmod{p}$. We want to prove that for any integer $0 \le k < p^{r-1}$, there exists one and only one integer h such that $(x_0 + kp, y_0 + hp)$ is solution of $x^2 - Dy^2 \equiv 1 \pmod{p^r}$.
Indeed, we have

$$(x_0 + kp)^2 - D(y_0 + hp)^2 = 1 + vp + 2x_0kp + k^2p^2 - 2Dy_0hp - Dh^2p^2,$$

since $x_0^2 - Dy_0^2 = 1 + vp$ for a certain integer v. Thus, we have that $(x_0 + kp, y_0 + hp)$ is solution of $x^2 - Dy^2 \equiv 1 \pmod{p^r}$ if and only if

$$Dph^2 + 2Dy_0h - v - 2x_0k - k^2p \equiv 0 \pmod{p^{r-1}}.$$

Hence, we have to prove that there is one and only one integer h that satisfies the above identity. The above equation can be solved in h by completing the square and reduced to

$$(2Dph + 2Dy_0)^2 \equiv s \pmod{p^{r-1}}, \tag{5}$$

where $s = (2Dy_0)^2 + 4(v + 2x_0k + k^2p)Dp$. Let us prove that s is a quadratic residue in $\mathbb{Z}_{p^{r-1}}$. Indeed,

$$s = 4D((x_0 + kp)^2 - 1)$$

and surely the Jacobi symbol $\left(\dfrac{s}{p^{r-1}}\right) = \left(\dfrac{s}{p}\right)^{r-1} = 1$ if r is odd. If r is even we have that

$$\left(\frac{s}{p^{r-1}}\right) = \left(\frac{4}{p^{r-1}}\right)\left(\frac{D}{p^{r-1}}\right)\left(\frac{(x_0 + kp)^2 - 1}{p^{r-1}}\right) = 1$$

since $\left(\dfrac{4}{p^{r-1}}\right) = 1$, $\left(\dfrac{D}{p^{r-1}}\right) = \left(\dfrac{D}{p}\right)^{r-1} = -1$ by hypothesis on D,

$\left(\dfrac{(x_0 + kp)^2 - 1}{p^{r-1}}\right) = -1$, since $(x_0 + kp)^2 - 1 \equiv Dy_0^2 \pmod{p}$. Now, let $\pm t$ be the square roots of s. It is easy to note that

$$t \equiv 2Dy_0 \pmod{p}, \quad -t \equiv -2Dy_0 \pmod{p}$$

or

$$-t \equiv 2Dy_0 \pmod{p}, \quad t \equiv -2Dy_0 \pmod{p}.$$

Let us call $\bar{t}$ the only one between t and $-t$ that is equal to $2Dy_0$ in $\mathbb{Z}_p$. Hence, Eq. (5) is equivalent to the linear equation

$$ph \equiv (\bar{t} - 2Dy_0)(2D)^{-1} \pmod{p^{r-1}},$$

which has one and only one solution, since $\bar{t} - 2Dy_0 \equiv 0 \pmod{p}$. Note that, if $\bar{t}$ is not equal to $2Dy_0$ in $\mathbb{Z}_p$ the above equation has no solutions. Thus, we have proved that any solution of the Pell equation in $\mathbb{Z}_p$ generates p^{r-1} solutions of the Pell equation in $\mathbb{Z}_{p^r}$.

(2) Now, we prove that all the solutions of the Pell equation in $\mathbb{Z}_{p^r}$ are generated as in step 1.
Let $(\bar{x}, \bar{y})$ be a solution of $x^2 - Dy^2 \equiv 1 \pmod{p^r}$, i.e., $\bar{x}^2 - D\bar{y}^2 = 1 + wp^r$, for a certain integer w. Then $x_0 = \bar{x} - kp$ and $y_0 = \bar{y} - hp$, for h, k integers, are solutions of $x^2 - Dy^2 \equiv 1 \pmod{p}$. Indeed,

$$(\bar{x} - kp)^2 - D(\bar{y} - hp)^2 = 1 + wp^r - 2\bar{x}kp + k^2p^2 + 2D\bar{y}hp - Dh^2p^2.$$

As a consequence of the previous theorem, an analogous of the Euler theorem holds for the product $\otimes$.

Theorem 2. *Let p, q be prime numbers and $N = p^r q^s$, then for all $(x, y) \in \mathcal{H}_{\mathbb{Z}_N}$ we have*

$$(x, y)^{\otimes p^{r-1}(p+1)q^{s-1}(s+1)} \equiv (1, 0) \pmod{N}$$

for $D \in \mathbb{Z}_N^*$ quadratic non–residue in $\mathbb{Z}_p$ and $\mathbb{Z}_q$.

Proof. By Theorem 1, we know that

$$(x, y)^{\otimes p^{r-1}(p+1)} \equiv (1, 0) \pmod{p^r}$$

and

$$(x, y)^{\otimes q^{s-1}(s+1)} \equiv (1, 0) \pmod{q^s}.$$

Thus, said $(a, b) = (x, y)^{\otimes p^{r-1}(p+1)q^{s-1}(s+1)}$, we have

$$(a, b) \equiv (1, 0) \pmod{p^r},$$

i.e., $a = 1 + kp^r$ and $b = hp^r$ for some integers h, k. On the other hand, we have

$$(a, b) \equiv (1, 0) \pmod{q^s} \Leftrightarrow (1 + kp^r, hp^r) \equiv (1, 0) \pmod{q^s}.$$

We can observe that $1 + kp^r \equiv 1 \pmod{q^s}$ if and only if $k = k'q^s$ for a certain integer k'. Similarly, it must be $h = h'q^s$, for an integer h'. Hence, we have that $(a, b) = (1 + k'p^r q^s, h'p^r q^s) \equiv (1, 0) \pmod{N}$.

Corollary 1. *Let* $p_1, ..., p_r$ *be primes and* $N = p_1^{e_1} \cdot ... \cdot p_r^{e_r}$, *then for all* $(x, y) \in \mathcal{H}_{\mathbb{Z}_N}$ *we have*

$$(x, y)^{\otimes \Psi(N)} = (1, 0) \pmod{N},$$

where

$$\Psi(N) = p_1^{e_1 - 1}(p_1 + 1) \cdot ... \cdot p_r^{e_r - 1}(p_r + 1),$$

for $D \in \mathbb{Z}_N^*$ *quadratic non–residue in* $\mathbb{Z}_{p_i}$, *for* $i = 1, ..., r$.

Now, we can observe that when we work on $\mathbb{Z}_N$, the map Φ_D is not an isomorphism. Indeed, the orders of $\mathcal{H}_{D,\mathbb{Z}_N}$ and $\mathcal{P}_{\mathbb{Z}_N}$ do not coincide. However, it is still a morphism and we also have $|\mathbb{Z}_N^*| = |\mathcal{H}_{\mathbb{Z}_N}^*|$, because of the following proposition.

Proposition 3. *With the above notation, we have that*

1. $\forall (x_1, y_1), (x_2, y_2) \in \mathcal{H}_{\mathbb{Z}_N}^*$, $\Phi_D(x_1, y_1) = \Phi_D(x_2, y_2) \Leftrightarrow (x_1, y_1) = (x_2, y_2)$;
2. $\forall m_1, m_2 \in \mathbb{Z}_N^*$, $\Phi_D^{-1}(m_1) = \Phi_D^{-1}(m_2) \Leftrightarrow m_1 = m_2$;
3. $\forall m \in \mathbb{Z}_N^*$, *we have* $\Phi^{-1}(m) \in \mathcal{H}_{\mathbb{Z}_N}^*$ *and* $\forall (x, y) \in \mathcal{H}_{\mathbb{Z}_N}^*$, *we have* $\Phi_D(x, y) \in \mathbb{Z}_N^*$.

See [3].

As a consequence, we have an analogous of the Euler theorem also for the product $\odot$, i.e., for all $m \in \mathbb{Z}_N^*$ the following holds

$$m^{\odot \Psi(N)} = \alpha \pmod{N},$$

where $\odot$ is the special product in $\mathcal{P}_{\mathbb{Z}_N}$ defined in Eq. 2.

4   The Cryptographic Scheme

In this section, we describe our public–key cryptosystem based on the properties studied in the previous section.

4.1   Key Generation

The key generation is performed by the following steps:

- choose r prime numbers $p_1, \ldots, p_r$, r odd integers $e_1, \ldots, e_r$ and compute $N = \prod_{i=1}^r p_i^{e_i}$;
- choose an integer e such that $\gcd(e, \operatorname{lcm} \prod_{i=1}^r p_i^{e_i - 1}(p_i + 1)) = 1$;
- evaluate $d = e^{-1} \pmod{\operatorname{lcm} \prod_{i=1}^r p_i^{e_i - 1}(p_i + 1)}$.

The public or encryption key is given by (N, e) and the secret or decryption key is given by $(p_1, \ldots, p_r, d)$.

4.2   Encryption

We can encrypt pair of messages $(M_x, M_y) \in \mathbb{Z}_N^* \times \mathbb{Z}_N^*$, such that $\left(\dfrac{M_x^2 - 1}{N} \right) =$
-1. This condition will ensure that we can perform all the operations. The encryption of the messages is performed by the following steps:

- compute $D = \dfrac{M_x^2 - 1}{M_y^2} \pmod{N}$, so that $(M_x, M_y) \in \mathcal{H}_{D, \mathbb{Z}_N}^*$;
- compute $M = \Phi(M_x, M_y) = \dfrac{M_x + 1}{M_y} \pmod{N}$;
- compute the ciphertext $C = M^{\odot e} \pmod{N} = Q_e(D, M) \pmod{N}$

Notice that not only C, but the pair (C, D) must be sent through the insecure channel.

4.3   Decryption

The decryption is performed by the following steps:

- compute $C^{\odot d} \pmod{N} = Q_d(D, C) \pmod{N} = M$;
- compute $\Phi^{-1}(M) = \left(\dfrac{M^2 + D}{M^2 - D}, \dfrac{2M}{M^2 - D} \right) \pmod{N}$ for retrieving the messages (M_x, M_y).

5   Security of the Encryption Scheme

The proposed scheme can be attacked by solving one of the following problems:

1. factorizing the modulus $N = p_1^{e_1} \cdot \ldots \cdot p_r^{e_r}$;
2. computing $\Psi(N) = p_1^{e_1 - 1}(p_1 + 1) \cdot \ldots \cdot p_r^{e_r - 1}(p_r + 1)$, or finding the number of solutions of the equation $x^2 - Dy^2 \equiv 1 \mod N$, i.e. the curve order, which divides $\Psi(N)$;
3. computing Discrete Logarithm problem either in $(\mathcal{H}_{\mathbb{Z}_N}^*, \otimes)$ or in $(\mathcal{P}_{\mathbb{Z}_N}^*, \odot)$;
4. finding the unknown d in the equation $ed \equiv 1 \mod \Psi(N)$;
5. finding an impossible group operation in $\mathcal{P}_{\mathbb{Z}_N}$;
6. computing M_x, M_y from D.

5.1   Factorizing N or Computing the Curve Order

It is well known that the problem of factorizing $N = p_1^{e_1} \cdot \ldots \cdot p_r^{e_r}$ is equivalent to that of computing the Euler totient function $\phi(N) = p_1^{e_1 - 1}(p_1 - 1) \cdot \ldots \cdot p_r^{e_r - 1}(p_r - 1)$, e.g. see [23] or [29, Section 10.4].

In our case we need to show the following

Proposition 4. *The problem of factorizing N is equivalent to the problem of computing $\Psi(N) = p_1^{e_1-1}(p_1 + 1) \cdot \ldots \cdot p_r^{e_r-1}(p_r + 1)$ or the order of the group $\mathcal{P}_{\mathbb{Z}_N}^*$ (or equivalently of $\mathcal{H}_{\mathbb{Z}_N}^*$), which is a divisor of $\Psi(N)$.*

Proof. Clearly, knowing the factorization of N yields $\Psi(N)$. Conversely, suppose N and $\Psi(N)$ are known. A factorization of N can be found by applying Algorithm 1 recursively.

Remark 3. Algorithm 1 is an adaptation of the general algorithm in [29, Section 10.4], used to factorize N by only knowing $\phi(N)$ (Euler totient function) and N itself. The main idea of the Algorithm 1 comes from the fact that $x^{\odot\Psi(N)} = 1 \pmod{N}$ for all $x \in \mathbb{Z}_N^*$, which is the analog of the Euler theorem in $\mathcal{P}_{\mathbb{Z}_N}$. Notice that, because of Step 7, Algorithm 1 is a probabilistic algorithm. Thus, to find a non-trivial factor, it might be necessary to run the algorithm more than once. We expect that a deeper analysis of the algorithm will lead to a similar probabilistic behaviour than the algorithm in [29], which returns a non-trivial factor with probability $1/2$.

Algorithm 1. Find a factor of N by knowing N and $\Psi(N)$

```
 1: function FIND FACTOR(N,Ψ(N))
 2:      h = 0
 3:      t = Ψ(N)
 4:      while IsEven(t) do
 5:          h = h + 1
 6:          t = t / 2
 7:      a = Random(N − 1)
 8:      d = gcd(a, N)
 9:      if d ≠ 1 then
10:          return d
11:      b = a^⊙t  mod N
12:      for j = 0, . . . , h − 1 do
13:          d = gcd(b + 1, N)
14:          if d ≠ 1 or d ≠ N then
15:              return d
16:          b = b² mod N
17:      return 0
```

Since we proved that the problems 1 and 2 are equivalent, we can only focus on the factorization problem.

According to [10], state-of-the-art factorization methods as the Elliptic Curve Method [18] or the Number Field Sieve [4,19] are not effective if in the following practical cases

- $|N| = 1024, 1536, 2048, 2560, 3072, 3584$ and $N = p_1^{e_1}p_2^{e_2}p_3^{e_3}$ with $e_1+e_2+e_3 \leq 3$ and $p_i, i = 1, 2, 3$ greater than approximately the size of $\sqrt[3]{N}$.

- $|N| = 4096$ and $N = p_1^{e_1} p_2^{e_2} p_3^{e_3} p_4^{e_4}$ with $e_1 + e_2 + e_3 + e_4 \leq 4$ and $p_i, i = 1, \ldots, 4$ greater than approximately the size of $\sqrt[4]{N}$.
- $|N| = 8192$ and $N = p_1^{e_1} p_2^{e_2} p_3^{e_3} p_4^{e_4} p_5^{e_5}$ with $e_1 + e_2 + e_3 + e_4 + e_5 \leq 5$ and $p_i, i = 1, \ldots, 5$ greater than approximately the size of $\sqrt[5]{N}$.

Notice that currently, the largest prime factor found by the Elliptic Curve Method is a 274 bit digit integer [32]. Note also that the Lattice Factoring Method (LFM) of Boneh, Durfee, and Howgrave-Graham [5] is designed to factor integers of the form $N = p^u q$ only for large u.

5.2 Computing the Discrete Logarithm

Solving the discrete logarithm problem in a conic curve can be reduced to the discrete logarithm problem in the underlying finite field [22]. In our case the curve is defined over the ring $\mathbb{Z}_N$. Solving the DLP over $\mathbb{Z}_N$ without knowing the factorization of N is as hard as solving the DLP over a prime finite field of approximately the same size. As for the factorization problem, the best known algorithm to solve DLP on a prime finite field is the Number Field Sieve. When the size of N is greater than 1024 then the NFS can not be effective.

5.3 Solving the Private Key Equation

In the case of RSA, small exponent attacks [12,13,31] can be performed to find the unknown d in the equation $ed \equiv 1 \mod \Psi(N)$. Generalization of these attacks can be performed on RSA variants where the modulus is of the form $N = p_1^{e_1} p_2^{e_2}$ [20]. It has already been argued in [3,16] and [16] that this kind of attacks fails when the trapdoor function is not a simple monomial power as in RSA, as it is in the proposed scheme.

5.4 Finding an Impossible Group Operation

In the case of elliptic curves over $\mathbb{Z}_N$, as in the generalized KMOV cryptosystem [7], it could happen that an impossible addition between two curve points occurs, yielding the factorization of N. This is due to the fact that the addition formula requires to perform an inversion in the underlying ring $\mathbb{Z}_N$. However, as shown by the same authors of [7], the occurrence of an impossible addition is very unlikely for N with few and large prime factors.

In our case an impossible group operation may occur if $a + b$ is not invertible in $\mathbb{Z}_N$, i.e. if $\gcd(a + b, N) \neq 1$, yielding in fact a factor of N. However, also in our case, if N contains a few large prime factors, impossible group operations occur with negligible probability, as shown by the following proposition.

Proposition 5. *The probability to find an invertible element in $\mathcal{P}_{\mathbb{Z}_N}$ is approximately*

$$1 - \left(1 - \frac{1}{p_1}\right) \cdot \ldots \cdot \left(1 - \frac{1}{p_r}\right)$$

Proof. The probability to find an invertible element in $\mathcal{P}_{\mathbb{Z}_N}$ is given by dividing the number of non-invertible elements in $\mathcal{P}_{\mathbb{Z}_N}$ by the total number of elements of this set, as follows:

$$\frac{|\mathcal{P}_{\mathbb{Z}_N}| - \#\{\text{invertible elements in } \mathcal{P}_{\mathbb{Z}_N}\}}{|\mathcal{P}_{\mathbb{Z}_N}|} \tag{6}$$

$$= \frac{|\mathbb{Z}_N| + 1 - (\#\{\text{invertible elements in } \mathbb{Z}_N\} + 1)}{|\mathbb{Z}_N| + 1} \tag{7}$$

$$= \frac{N - \phi(N)}{N + 1} \tag{8}$$

$$\sim 1 - \left(1 - \frac{1}{p_1}\right) \cdot \ldots \cdot \left(1 - \frac{1}{p_r}\right) \tag{9}$$

where we used $N \sim N + 1$ and $\phi(N) = N \left(1 - \frac{1}{p_1}\right) \cdot \ldots \cdot \left(1 - \frac{1}{p_r}\right)$.

This probability tends to zero for large prime factors.

Let us notice that, in the Pell curve case, it is possible to avoid such situation, by performing encryption and decryption in $\mathcal{H}^*_{\mathbb{Z}_N}$, without exploiting the isomorphism operation. Here the group operation $\otimes$ is defined between two points on the Pell curve, as in Eq. 1, and does not contain the inverse operation. In the resulting scheme the ciphertext is obtained as $(C_x, C_y) = (M_x, M_y)^{\otimes e}$, where the operation $\otimes$ depends on D. Thus the triple (C_x, C_y, D) must be transmitted, resulting in a non-compressed ciphertext.

5.5   Recovering the Message from D

To recover the message pair (M_x, M_y) from $D = \frac{M_x^2 - 1}{M_y^2} \pmod{N}$, the attacker must solve the quadratic congruence $M_x^2 - DM_y^2 - 1 = 0 \pmod{N}$ with respect to the two unknowns M_x and M_y. Even if one of the two coordinates is known (partially known plaintext attack), it is well known that computing square roots modulo a composite integer N, when the square root exists, is equivalent to factoring N itself.

5.6   Further Comments

As a conclusion to this section, we only mention that as shown in [3], RSA-like schemes based on isomorphism own the following properties: they are more secure than RSA in the broadcast scenario, they can be transformed to semantically secure schemes using standard techniques which introduce randomness in the process of generating the ciphertext.

6   Efficiency of the Encryption Scheme

Recall that our scheme encrypts and decrypts messages of size $2 \log N$. To decrypt a ciphertext of size $2 \log N$ using CRT, standard RSA requires four full exponentiation modulo $N/2$-bit primes. Basic algorithms to compute $x^d \mod p$ requires $O(\log d \log^2 p)$, which is equal to $O(\log^3 p)$ if $d \sim p$.

Using CRT, if $N = p_1^{e_1} \cdot \ldots \cdot p_r^{e_r}$, our scheme requires at most r exponentiation modulo N/r-bit primes.

This means that the final speed up of our scheme with respect to RSA is

$$\frac{4 \cdot (N/2)^3}{r \cdot (N/r)^3} = r^2/2 \tag{10}$$

When $r = 2$ our scheme is two times faster than RSA, as it has already been shown in [3]. If $r = 3$ our scheme is 4.5 time faster, with $r = 4$ is 8 times faster, and with $r = 5$ is 12.5 times faster.

7   Conclusions

We generalized an RSA-like scheme based on the Pell hyperbola from a modulus that was a product of two primes to a generic modulus. We showed that this generalization leads to a very fast decryption step, up to 12 times faster than original RSA for the security level of a modulus of 8192 bits. The scheme preserves all security properties of RSA-like schemes, which are in general more secure than RSA, especially in a broadcast scenario. Compared to similar schemes based on elliptic curves it is more efficient. We also pointed that a variation of the scheme with non-compressed ciphertext does not suffer of impossible group operation attacks.

References

1. Barbero, S., Cerruti, U., Murru, N.: Generalized Rédei rational functions and rational approximations over conics. Int. J. Pure Appl. Math **64**(2), 305–317 (2010)
2. Barbero, S., Cerruti, U., Murru, N.: Solving the Pell equation via Rédei rational functions. Fibonacci Q. **48**(4), 348–357 (2010)
3. Bellini, E., Murru, N.: An efficient and secure RSA-like cryptosystem exploiting Rédei rational functions over conics. Finite Fields Appl. **39**, 179–194 (2016)
4. Bernstein, D.J., Lenstra, A.K.: A general number field sieve implementation. In: Lenstra, A.K., Lenstra, H.W. (eds.) The development of the number field sieve. LNM, vol. 1554, pp. 103–126. Springer, Heidelberg (1993). https://doi.org/10.1007/BFb0091541
5. Boneh, D., Durfee, G., Howgrave-Graham, N.: Factoring $N = p^r q$ for large r. Crypto **1666**, 326–337 (1999)
6. Boneh, D., Shacham, H.: Fast variants of RSA. CryptoBytes **5**(1), 1–9 (2002)
7. Boudabra, M., Nitaj, A.: A new generalization of the KMOV cryptosystem. J. Appl. Math. Comput. **57**, 1–17 (2017)
8. Ciet, M., Koeune, F., Laguillaumie, F., Quisquater, J.: Short private exponent attacks on fast variants of RSA. UCL Crypto Group Technical Report Series CG-2003/4, Université Catholique de Louvain (2002)
9. Collins, T., Hopkins, D., Langford, S., Sabin, M.: Public key cryptographic apparatus and method. Google Patents, US Patent 5,848,159 (1998)

10. Compaq: Cryptography using Compaq multiprime technology in a parallel process-
    ing environment. ftp://15.217.49.193/pub/solutions/CompaqMultiPrimeWP.pdf.
    Accessed 2019
11. Cohen, H.: A Course in Computational Algebraic Number Theory. Springer, Hei-
    delberg (2013)
12. Coppersmith, D., Franklin, M., Patarin, J., Reiter, M.: Low-exponent RSA with
    related messages. In: Maurer, U. (ed.) EUROCRYPT 1996. LNCS, vol. 1070, pp.
    1–9. Springer, Heidelberg (1996). https://doi.org/10.1007/3-540-68339-9_1
13. Coppersmith, D.: Small solutions to polynomial equations, and low exponent RSA
    vulnerabilities. J. Cryptol. **10**(4), 233–260 (1997)
14. Hastad, J.: N using RSA with low exponent in a public key network. In: Williams,
    H.C. (ed.) CRYPTO 1985. LNCS, vol. 218, pp. 403–408. Springer, Heidelberg
    (1986). https://doi.org/10.1007/3-540-39799-X_29
15. Jacobson, M.J., Williams, H.C., Taylor, K., Dilcher, K.: Solving the Pell Equation.
    Springer, New York (2009). https://doi.org/10.1007/978-0-387-84923-2
16. Koyama, K.: Fast RSA-type schemes based on singular cubic curves $y^2 + axy \equiv x^3$
    (mod n). In: Guillou, L.C., Quisquater, J.-J. (eds.) EUROCRYPT 1995. LNCS,
    vol. 921, pp. 329–340. Springer, Heidelberg (1995). https://doi.org/10.1007/3-540-
    49264-X_27
17. Koyama, K., Maurer, U.M., Okamoto, T., Vanstone, S.A.: New public-key schemes
    based on elliptic curves over the Ring Z_n. In: Feigenbaum, J. (ed.) CRYPTO 1991.
    LNCS, vol. 576, pp. 252–266. Springer, Heidelberg (1992). https://doi.org/10.1007/
    3-540-46766-1_20
18. Lenstra Jr., H.W.: Factoring integers with elliptic curves. Ann. Math. **126**, 649–673
    (1987)
19. Lenstra, A.K., Lenstra, H.W., Manasse, M.S., Pollard, J.M.: The number field
    sieve. In: Lenstra, A.K., Lenstra, H.W. (eds.) The development of the number field
    sieve. LNM, vol. 1554, pp. 11–42. Springer, Heidelberg (1993). https://doi.org/10.
    1007/BFb0091537
20. Lu, Y., Peng, L., Sarkar, S.: Cryptanalysis of an RSA variant with moduli $N = p^r q^l$.
    J. Math. Cryptol. **11**(2), 117–130 (2017)
21. McEliece, R.J.: A public-key cryptosystem based on algebraic coding theory. Deep
    Space Netw. Prog. Rep. **44**, 114–116 (1978)
22. Menezes, A.J., Vanstone, S.A.: A note on cyclic groups, finite fields, and the discrete
    logarithm problem. Appl. Algebr. Eng. Commun. Comput. **3**(1), 67–74 (1992)
23. Miller, G.L.: Riemann's hypothesis and tests for primality. In: Proceedings of Sev-
    enth Annual ACM Symposium on Theory of Computing, pp. 234–239 (1975)
24. More, W.: Fast evaluation of Rédei functions. Appl. Algebr. Eng. Commun. Com-
    put. **6**(3), 171–173 (1995)
25. NIST: Round 1 Submissions. https://csrc.nist.gov/Projects/Post-Quantum-
    Cryptography/Round-1-Submissions. Accessed 2019
26. Padhye, S.: A public key cryptosystem based on pell equation. IACR Cryptology
    ePrint Archive, p. 191 (2006)
27. Rédei, L.: Über eindeutig umkehrbare polynome in endlichen körpern redei. Acta
    Sci. Math. **11**, 85–92 (1946)
28. Lim, S., Kim, S., Yie, I., Lee, H.: A generalized takagi-cryptosystem with a modulus
    of the form $p^r q^s$. In: Roy, B., Okamoto, E. (eds.) INDOCRYPT 2000. LNCS,
    vol. 1977, pp. 283–294. Springer, Heidelberg (2000). https://doi.org/10.1007/3-
    540-44495-5_25
29. Shoup, V.: A Computational Introduction to Number Theory and Algebra. Cam-
    bridge University Press, Cambridge (2009)

30. Takagi, T.: Fast RSA-type cryptosystem modulo $p^k q$. In: Krawczyk, H. (ed.) CRYPTO 1998. LNCS, vol. 1462, pp. 318–326. Springer, Heidelberg (1998). https://doi.org/10.1007/BFb0055738
31. Wiener, M.J.: Cryptanalysis of short RSA secret exponents. IEEE Trans. Inf. Theory **36**(3), 553–558 (1990)
32. Zimmermann, S.: 50 largest factors found by ECM. https://members.loria.fr/PZimmermann/records/top50.html. Accessed 2017

Paradoxes of the Infinite and Ontological Dilemmas Between Ancient Philosophy and Modern Mathematical Solutions

Fabio Caldarola[1]([⊠]) [iD], Domenico Cortese[2], Gianfranco d'Atri[1],
and Mario Maiolo[3] [iD]

[1] Department of Mathematics and Computer Science, Università della Calabria,
Cubo 31/A, 87036 Arcavacata di Rende, CS, Italy
{caldarola,datri}@mat.unical.it
[2] Via degli Orti, 12, 89861 Tropea, VV, Italy
domcor87@hotmail.it
[3] Department of Environmental and Chemical Engineering (DIATIC),
Università della Calabria, Cubo 42/B, 87036 Arcavacata di Rende, CS, Italy
mario.maiolo@unical.it

Abstract. The concept of infinity had, in ancient times, an indistinguishable development between mathematics and philosophy. We could also say that his real birth and development was in Magna Graecia, the ancient South of Italy, and it is surprising that we find, in that time, a notable convergence not only of the mathematical and philosophical point of view, but also of what resembles the first "computational approach" to "infinitely" or very large numbers by Archimedes. On the other hand, since the birth of philosophy in ancient Greece, the concept of infinite has been closely linked with that of contradiction and, more precisely, with the intellectual effort to overcome contradictions present in an account of Totality as fully grounded. The present work illustrates the ontological and epistemological nature of the paradoxes of the infinite, focusing on the theoretical framework of Aristotle, Kant and Hegel, and connecting the epistemological issues about the infinite to concepts such as the continuum in mathematics.

Keywords: Infinite · Pythagorean school · Greek philosophy · Grossone · Unimaginable numbers

1 Introduction

"Mathematics is the science of infinity" says the very famous sentence of Hermann Weyl, 1930 (see [53, p. 17]). And a quite surprising fact is that the South of Italy, in particular Calabria and Sicily, played a historic role of the highest importance in the development of the idea of infinity in mathematics and philosophy, disciplines that at the time were often not distinguishable in many respects. One could almost say that the infinite from the mathematical-philosophical point of

© Springer Nature Switzerland AG 2020
Y. D. Sergeyev and D. E. Kvasov (Eds.): NUMTA 2019, LNCS 11973, pp. 358–372, 2020.
https://doi.org/10.1007/978-3-030-39081-5_31

view was born in Magna Graecia as well as from the computational point of view, in fact, "unimaginable numbers" were born in the Greek colonies of Sicily by the greatest mathematician of antiquity, Archimedes of Syracuse. It is therefore very interesting to investigate how the concept of infinity was born, developed and evolved on the border between mathematics and philosophy and this is what the present paper wants to do, at least in part. A further coincidence is that recent speculations on the idea and use of infinity, in mathematics but not only, once again see Calabria as the protagonist, as we will see later.

From Sect. 2, we will analyze and explain why the idea of the infinite has always been perceived as a contradiction in relation with the philosophical and scientific necessity of systematizing the entire reality within a complete and univocal set of measures and forces. Such a contradiction took two forms:

1. The idea of infinite as a being which pushes itself beyond any given limit in size or time. Once one assumes an infinite being, there is no longer a way to hypothesize a "center" in order to univocally identify the position of the parts in the existent, nor there is a way to understand a foundation which explains being in its entirety, since entirety can never be comprehended and, therefore, led back to one picture or principle.
2. The idea of infinite as a process which pushes itself beyond any given limit in size or time. This is the case of the infinite divisibility of matter or of time. Similarly to the preceding case, there is no way to identify a unity of measure (such as integer numbers) through which to construct the entire reality: there will always be a possible incommensurability.

In the last sections we will see that modern philosophy and mathematics - because of the necessity of supposing both discrete unities of measure and a continuum matter and irrational numbers - cannot overcome this conceptual short circuit which leads to the paradox of the infinite (in which, because of the recalled incommensurability, the infinite processes imply that the "part" is as large as the "whole"). Idealistic philosophy, instead, disarms such a concept of infinity by considering it as an intellectual trap and as an unimportant process from a dialectical or "pragmatic" point of view. The issue becomes to create a convention which puts in harmony all human experiences indifferently to a possible infinite process.

2   The Infinite as an Ontological Problem and the Limitless Size

The need to resolve *contradictions* in ancient philosophy is the one which is most linked to the typically philosophical necessity of outlining the Totality, that is to say of finding out the unique, original reason or ground (*arche*) which can explain the existence and the behavior of "everything". The occurrence of a contradiction stays in the fact that if something remains unexplained and unconnected to the rest of totality the very notion of knowledge of totality and *episteme* as pursued by philosophy collapses (see, for instance, [14,17,50,54]). This can happen if not everything can be reduced to one principle and we need at least two disconnected ones to make sense of phenomena, or if not all becoming can be explained with one principle. Such an unexplained differences would be arbitrary gaps which would deprive philosophy (or epistemology) of its ultimate goal: to trace the rational origin of reality in order to inscribe all its aspects within "the sense of the Whole". This goal is not at all extraneous to the intrinsic nature of modern science and physics - for instance, in its effort to locate elements which are more and more fundamental and to ultimately unify all physical laws into one great theory (theory of everything, see [21,23]).

Pythagoras was maybe the first philosopher mathematician who really had to deal with the concept of infinite and with the disruption of the idea of a "structured whole" which its existence entails. Pythagorean mathematics is based on the idea of "discontinuity", as it is exclusively anchored in integer numbers and, therefore, the increase of a magnitude proceeds by "discontinuous leaps". In such a worldview all objects were constituted by a finite number of monads, particles similar to atoms. Two magnitudes could be expressed by an integer number and were mutually commensurable, they admitted a common denominator. Pythagorean thought will be put in crisis by the discovery of incommensurable magnitudes (that is to say, which do not admit a common denominator), developed within the school itself as the relationship between diagonal and side of a square resulted to be irrational, and safeguarded as an unspeakable secret. This entailed that diagonal and side are composed not by a finite amount of points, but by infinite points: for the first time actual infinite and not only potential infinite was discussed.

Within this conceptual framework, the idea of infinite becomes an issue in several senses, which are strongly interrelated. The simplest of these senses is infinite as what pushes itself beyond any given limit in size or time. Can an "infinite" size (or time) be considered consistent with the - metaphysical or even physical - necessity of postulating a Totality which must be, at least in principle, wholly grounded? Is it more consistent to suppose a finite universe, with the danger of a possible call for "something which would always be beyond its Whole", or to envisage an infinite one, with the risk of the impossibility to make sense of a Being which is never complete and, thus, never totally established by one rational foundation? This last is the objection put forward by the Eleatic Parmenides, for whom Being is eternal in the sense of being without past and future, because it cannot come from not-being, but it is hypothesized as a finite sphere - as opposed to the idea of his disciple Melissus. For Parmenides Being is also immutable: becoming can never be rationally justified from the existence of the necessary one original founding ground, because the alteration of this "one" should be explained by another principle otherwise it would be arbitrary and contradictory, and such a double principle should be explained by another original principle and so on. Becoming must only be explained as an illusion. The ultimate paradox of the infinite stays here. If you suppose its existence - in the extension of time and space or in the divisibility of time and space - you **cannot** hypothesize an original element which explains why reality organizes itself in a certain way or in another. In fact, if you suppose an infinity divisibility of time and space you cannot locate or imagine such an original element, and if you suppose an infinite extension of time or space you can never **ensure** the entire consistency of time and space with the structure dictated by such an element. On the other hand, if you suppose the non-existence of the infinite, it would be impossible to explain why a Being stops being, or why a certain force stops acting (in its capacity to divide matter or time). The latter case can only be explained by the existence of two counterposed original elements, but their contraposition would remain unexplained, betraying the very aim of episteme and science.

3   Infinite as Infinite Divisibility of a Certain Size or Time and the Problem of Continuum

While Parmenides, maybe the first philosopher to clearly recognize the question above, disarms the paradox of the infinite by excluding - as an "illusion" - the possibility of time, becoming, alteration, infinity but even of something outside the sphere of being (not preventing some logical inconsistency), Aristotle philosophy tries to circumvents the paradox by accepting the duplicity of the original element. In his case, they are "form" and "matter". The "ground" which for the Stagirite is at the basis of ontology, in fact, does not concern so much the mere physical and material constituents of being as, rather, the formal principles which imprint the underline indefinite matter and which build individuals and characters of reality. First of all, in the Aristotelian ontology this formal principle can stand "on its own" in the sense that a form, an identity of a primary substance is complete and it does not draw its characters or existence from other structures. Also, primary substances, which correspond to the individuals, are not mere aggregates of their parts. To stand on its own, in this case, does not mean that substance cannot change. Substance can be generated, altered and corrupted by means of a "substantial transformation" (that is to say through the process in which an individual assumes or loses its form on or from its matter - which is, in turn, an union of another form and another matter until the indefinite primary matter in reached).

The transformations of substances are explained by the four interpretative causes, material, efficient, formal and final one, with the final cause being the crucially prevalent one in explaining the existence and change of things, as clarified in *Physics II* and in *Parts of Animals I*. The causes, in fact, are ultimately explained as an irresistible tension which the worldly matter has towards the supreme immobile mover, a *de facto* divinity which has, for Aristotle, the same qualities of the Eleatic Being and act as final cause of everything. Aristotle, like Plato before him, solves the paradox of movement by assuming a necessary "axiomatic" duplicity of first principle, in his case the supreme first mover and indeterminate matter which continuously needs to be "informed"; a duplicity which explains the relentless finalistic tension which corresponds to becoming. Indeed, Aristotle's metaphysics presents the interesting characteristic of interpreting all kinds of movements, included division and enlargement, as expression of a finalistic push which is itself the tension between the two "fundamental *archai*" of Being: in this way the contradiction of the existence of movement, division or alteration which the necessity of a unique ground brings is sterilized. Further, one of these two "principles" seems to be, for the Stagirite, only a principle in a negative sense, interpretable as an absence of a definite form (see Physics, 204a8–204a16, and Metaphysics, IX.6, 1048a–b, in [4, Vol. I]).

Such a theoretical scenario entails one important thing: Aristotle cannot be interested in infinite division of a magnitude (and in the infinite backward reduction it implies) as an action corresponding *per se* to an attempt to locate the Origin. The prospect of an infinite reduction is theoretically unimportant because the transformations in which its steps consist (as every kind of transformation) would already be a reflection and a function of what is for him the arche, the tension towards immobile mover. The ontological gaps (in the sense explained in the previous section) which a similar series of actions manifests are already anyway "endless"; in the sense that the inevitable existence of matter - the "second principle" - makes the tension towards perfection endless. What really counts from a scientific and an ontological point of view is, therefore, the comprehension of the nature - that is to say of the causes - of the daily phenomena and their ultimate direction. The existence of infinite division is therefore ontologically relevant in the possible daily progression of its stages (infinite in potentiality), not in its concept as "already actually present" (infinite in actuality), which stops being idealized.

Such a rational solution is no longer possible in the philosophical era which Severino characterizes for the dissolution of the cohesion of "certainty" and "truth" (see [50, vol. II]). Modern philosophy, from Descartes to Kant included, introduced the opposition between "certainty" and "truth": the majority of philosophers had to deal with the fact that what appears rational to human thought may be at odds with what is outside it. Kant fulfills this attitude by claim in that the chain of conditions, and also the world, are a subjective representations and their epistemological value stays in how we manage - by means of sensibility and understanding, the subjective categories - to build a consistent picture of experience. Reason is still an arbiter of empirical truths, useful to assess consistency and harmony of the set of concepts which our understanding produces. But outside experience it does not make sense to rationally argue for one solution or the other of the aporia earlier recalled, for finitude or for infinity of original conditions and constituents: it is outside human reaching, for instance, to verify whether a series of causes is actually infinite, such a totality is never to be met with in experience, it's only the "natural attitude" of human reason to relentlessly chase this task (see [28, p. 460–461]). As a consequence, the first two antinomies of pure reason involve the very issue of infinity in relation to entirety: *"Thesis 1: The world has a beginning in time, and is also limited as regards space. Antithesis 1: The world has no beginning, and no limits in space; it is infinite as regards both time and space. Thesis 2: Every composite substance in the world is made up of simple parts, and nothing anywhere exists save the simple or what is composed of the simple. Antithesis 2: No composite thing in the world is made up of simple parts, and there nowhere exists in the world anything simple."* (See [34, p. 48] and also [5, 6]).

This coincides to the fact that Kant's philosophy cannot outflank the paradoxes of the infinite or, better to say, he cannot propose a theory of knowledge which metaphysically or even existentially systematizes the world. Once the faculties of reason are limited in their scrutiny of radical backward causes and ontological elements by the inevitable gap between certainty and truth, one cannot claim any definitive solution to potential contradictions. Theoretical reason and categories cannot embrace the knowledge of the first cause and of the totality - and, consequently, of the solutions to the paradoxes of the infinite.

Modern mathematics and modern philosophy seem to share the impossibility to rationally unarm the tension caused by the existence of something which is incommensurable to the notion of entireness and complete systematicity. As hinted before, if one supposes an infinity divisibility of time and space one cannot locate or imagine an original element which structures the fabric of reality with its regularities, and if one supposes an infinite extension of time or space it can never been ensured the entire consistency of time and space with the structure dictated by such an element. In mathematics such a problem corresponds with that of the continuum. If empirical reality is rationally perceived as a "continuum" both in its temporal or spatial extension and in its subdivision, in other words, you can always pick and isolate a further fragment of magnitude which disrupts what has been systematized so far in term of relations among different magnitudes to picture the structure of reality, without rationality being legitimate in restricting such a process with a justifications which has not empirical bases. All this makes reciprocal commensurability among unities of measures and imagines of completeness impossible. Hence the *paradox of the infinite* whereby the part is as "large" as the entire. Mathematics and modern philosophy issues may be synthesized in this way. The character of the incommensurable to entireness and complete systematicity is brought, in mathematics, by the very necessity of systematicity to compare and measure reciprocally incommensurable magnitudes and, in modern philosophy, by the necessity not to resort to ideal rational hypotheses in the origin of Being to contain the logical infinite ontological processes explained in this article.

Because of the existence of the continuum in mathematics and physics, you cannot get rid of the "actual" infinite because you cannot get rid of the incommensurable. Therefore you cannot do without supposing an "already happened infinite approach" which converges to a measure, in order to "perfectly" measure things - as opposed to Aristotle. Limits, or better series, are a way to explicit that entire, structurally finite magnitude to which the sum of some infinite fragments is convergent.

4   Infinite as a "Convention" and the Conventional Resolutions to the Contradiction

Infinite has also been thought as resolution to the intellectual and existential discomfort which derives from the finitude of things within becoming. This position has been adopted, in the history of philosophy, first by pre-Socractic thinkers who did not have yet a clear account of the paradox of the infinite as described above and, secondly, by idealistic philosophers such as Hegel who decided that the only way to circumvent such a paradox is to resort to the acceptation of "conventions" as ontologically valid. In other words, infinite is the character of the totality of being once we have given a *reason* to the chaotic turning up of differences, to the apparent contradictions and all kinds of obstacles which make our experience of the world appear full of uncomfortable discrepancies. A similar rationalization aims at finding the intrinsic harmony of things, so to show the ground, the origin (arche) which accounts for their coming into beings and, therefore, describes the Totality as the very Logic of their opposition. A Logic which, taken "in itself", is not limited to one contingent thing, it does not find arbitrary discrepancies: it is *un*limited, like a circle. The philosopher who manages to be aware of that is able to experience the non-contradiction of existence. The rational unity of all things is infinite, therefore, in the sense that it never presents a particular which is impossible to make sense of, and therefore it cannot contemplate a feeling of limit or restriction due to a contradiction: it is always already "full of Being". Such a conception of infinite and arche first appears in Greek philosophy in Anaximander and his idea of apeiron (see [51]), the original identity of all things which is without peculiar determinations and, thus, "without boundaries". It was famously elaborated by Heraclitus and his emphasis on the strive of oppositions as the logic principle of reality (see [27, 29]) and it will be, as we will see, a crucial feature of Hegel's idealism.

German idealism - and its major representative, Hegel - introduces an innovative insight to modern philosophy: "construction" is real, what seems to be the fruit of human mind or human culture is the expression, the reflection or, even, an active part of the rational structure of reality and of its logic. From this point of view it does not make any sense to distinguish between "empirical" and "metaphysical" construction as everything draws its "epistemo-logical dignity" from its being a logical or "practical" passage which contributes to harmonize and rationalize experience. With Hegel we meet the conceptions of the infinite as "rationalization" which aims at finding the intrinsic harmony of things, so to show the origin which accounts for their coming into beings and, therefore, describes the Totality as the very Logic of their opposition. Being infinite, in this sense, is the character of the idealistic Absolute and its ontological sense is the very opposite of the infinite as a limitless addition of *finite* aspects of reality. To continue adding an element to a concept or to a magnitude without elaborating a rational synthesis of the sense of its coming into being or of its *telos* - and the other parts of reality means to maintain a certain "arbitrary" discrepancy. The infinite series of this concept is what Hegel calls "bad infinity" (see [24]).

The contradiction between the necessity to find a first condition to make our conception of the world "complete" on the one hand and the necessity not to "arbitrarily stop" without reason to a certain element (proper of a world which is continuum) on the other hand should be synthesized by the idea of the universe as eternal (without time) and "circular" as rationally explained in itself, as the existence of a feature is always explained as the synthesis of a previous apparent discrepancy, without the need to resort to a condition which is out of the already "existent" dialectic of beings. If a thought manages to invent an instrument which makes sense of a process, harmonize its meaning with the surroundings without resorting to virtually infinite actions, such a thought is real and rational. Even if the cosmological problem is not central to Hegel's reflection, his dialectics brings to the non-necessity of an infinite research of spatial and temporal limits or conditions. As he says in the Science of Logic, *"the image of true infinity, bent back onto itself, becomes the circle"* (see [24, p. 149, §302]).

The most original and pregnant insight of Hegel's rationalization of reality is that it does not matter, for dialectical speculation, how many "finite" and irrational characters are present in a certain context, or how deep a contradiction is structured in a concept. Since the aim of philosophy - and, therefore, of human existence - is to make thought and Being coincide (see [15]) and since thought and Being are not two separate entities, any practical convention which fulfills the task of making such irrational characters unimportant and of making our acting and experience "satisfactory" has the dignity of a real character of reality. One way to "synthesize" the necessity of modern mathematics calculus of resorting to the concept of incommensurability between magnitudes - and, thus, of limits and "actual" infinite - and the necessity of logic and common sense to avoid the relative paradoxes is the concept of grossone, proposed by Sergeyev. The grossone appears to be what has just be named as a conventional systematization which interprets the infinite as an Entire in order to outflank its contradictions, and which consciously overlooks the fact of its being constituted by "infinite finitudes" because practically unimportant - and for this reason "irrational" in a dialectical sense - in the context of its applications.

Sergeyev starts from a literary interpretation of the principle "the part is less than the whole" and he applies it both to finite and, especially, infinite quantities, in strong contrast with the conception of Cantor's infinite, for which the infinite processes imply that the "part" is as large as the "whole". As we have seen, Idealistic philosophy too, agrees with such a principle that leads to a "multiple concept of infinity", that results also in accord with the dialectical or "pragmatic" point of view. At the University of Calabria, in the early 2000, Sergeyev began to develop his idea of an infinite with good computational properties and, at the same time, easy to use at every level, from schools to academic research. In a few words, Sergeyev developed a new computational system based on two fundamental units or "atoms": the ordinary unit 1 to generate finite quantities as the familiar natural or integer numbers, and a new infinite unit, named *"grossone"*, to generate infinite and infinitesimal numbers. For example, [38, 40, 43, 46, 48] are introductory surveys to this new system and the book [39] is also written in a popular way and well understandable for high school students. This new computational methodology has even been applied in a number of areas both in mathematics, applied sciences, philosophy, and recently also in mathematical education. For instance, optimization and numerical differentiation [1, 16, 20, 33, 42, 44, 49], calculus of probabilities and geometry

[13,31,32], fractals and complex systems [3,8,9,12,18,19,41,45,47], paradoxes and supertasks [35,37], didactic experimentations [2,26,36], and others.

The peculiar main characteristic of the grossone, that leads it to a growing diffusion, is probably his double nature of infinite number, but with the behavior of an ordinary natural one. This intrinsic characteristic is in fact the basis of several of its properties among which we recall, since this article deals with paradoxes from different points of view, the ability to solve some of them focused on the infinite. For example, consider what is probably the most famous of them, the Hilbert's paradox of the Grand Hotel, which was designed to highlight the strange properties of the infinite countable cardinality $\aleph_0$ in the Cantorian sense. Even when the hotel is full, it is always possible to host a finite or infinite countable number of new guests by moving the present guests in the hotel in a suitable manner.[1] This leads to many semantic problems at various levels, as well as questions of a logical nature. To take a simple example, it is sufficient just to consider the semantics of the word "full": commonly, if a tank or container is full it is not possible to add more. Adopting the grossone system instead, such kinds of paradoxes are avoided because a hotel with ① rooms cannot accept other guests when it is full (see for instance [39, Section 3.4]).

Another example, of different nature, that yields a paradox is the following, known as the *Thompson lamp paradox*.

Example 1 (The Thompson lamp paradox). Given a lamp, assume that we turn it on at time zero and turn it off after $1/2$ minute, then we turn it on after $1/4$ minute and turn it back off again after $1/8$ minute, and so on, by acting on the switch to each successive power of $1/2$. At the end of a minute, will the lamp be on or off? This puzzle was originally proposed by the philosopher J.F. Thompson in 1954 to analyze the possibility of completing a supertask, i.e. a larger task made up of an infinite number of simple tasks (see [52]).

Note that the classical geometric series of common ratio $1/2$ and starting from $1/2$ converges to 1, in symbols

$$\sum_{t=1}^{+\infty} \left(\frac{1}{2}\right)^t = 1, \tag{1}$$

hence in a minute the switch will be moved infinitely many times (admitting that this is possible) and the question above has no answer. Using the grossone-based system, it is offered in [39, Section 3.4] the solution that the lamp is "off" after one minute, and it is essentially due to the parity of grossone.

We propose here a more detailed interpretation: the switch will have the first motion at the time zero, the second at the time $\frac{1}{2}$, the third at the time $\frac{1}{2} + \left(\frac{1}{2}\right)^2$, and the n-th at the time $\frac{1}{2} + \left(\frac{1}{2}\right)^2 + \ldots + \left(\frac{1}{2}\right)^{n-1}$. We recall that a sequential

---

[1] If there are n new guests the simplest choice is to use the function $\mathbb{N} \to \mathbb{N}$, $m \mapsto m + n$, instead, in case of an infinite countable number, the function $\mathbb{N} \to \mathbb{N}$, $m \mapsto 2m$ (see [22,39]).

process, in Sergeyev's theory, cannot have more than ① steps, hence the last action on the switch, i.e. the ①-th action, will be made at the time

$$\sum_{t=1}^{①-1} \left(\frac{1}{2}\right)^t = -1 + \sum_{t=0}^{①-1} \left(\frac{1}{2}\right)^t$$

$$= -1 + \frac{1 - \left(\frac{1}{2}\right)^{①}}{\frac{1}{2}} = 1 - \left(\frac{1}{2}\right)^{①-1}$$

(cf. Eq. (1)). In other words this means that the last action on the switch will be made at an infinitesimal time before the end of the minute, more precisely $\left(\frac{1}{2}\right)^{①-1}$ minute before, and from this moment on, the switch will remain off (hence it will be off as well at the time 1 min).

Sergeyev, by means of an "infinite" which behaves as an ordinary natural finite number, disarms the anarchy of the continuum. Such an "anarchy" springs from the absence of a limit in a process of division – of space or time - or in the growth of a size, absence which is the reason why you cannot locate an original element which can justify a univocal system of references capable of ultimately describing the features of reality or – which is the same thing – of "creating a stable totality". This infinite process of addition of "fragments of finitude" is what create the inconvenience whereby the "part is as big as the entire" and the consequent impossibility to systematize the entire. To sterilize such a process means to conventionally set aside the traditional and problematic sense of the infinite to embrace what idealistic philosophy and Hegel refers to as "true infinite" (see [24]), that is to say a practical convention which makes thought and experience more fluid and comforting (without ontological "finitudes"). "Infinite" is here used, of course, in a completely unusual acceptation if compared to its common mathematical meaning. The function of grossone is therefore the same as a dialectical (in a philosophical sense) device which ensures that our approach to a certain concept is deprived of discrepancies: we perceive it without ontological "limits" in the sense that we do not find compromising alteration or obstacle in its logic.

Unimaginable numbers[2] seem to respond to the same pragmatic and, therefore, ontological necessity: even if they does not contemplate actually "infinite" quantities, they give "form" to numbers which are as large as to question their commensurability to a rational systematization of our picture of reality, generating the same discomfort of "relentless process" as an infinite process does. But from a closer point of view, these notions can assume a more specific ontological

---

[2] The *unimaginable numbers* are numbers extremely large so that they cannot be written through the common scientific notation (also using towers of exponents) and are behind every power of imagination. To write them some special notations have been developed, the most known of them is *Knuth's up-arrow notation* (see [30]). A brief introduction to these numbers can be found in [10], while more information is contained in [7,11,25].

connotation, which communicate with Aristotle's approach as we have already hinted.

We have seen how Aristotle is not interested in infinite division of a magnitude (and in the infinite backward reduction it implies) as an action corresponding *per se* to an attempt to locate the Origin. The prospect of an infinite reduction is theoretically unimportant for him because the transformations in which its steps consist (as every kind of transformation for Aristotle's philosophy) would already be a function of what is the arche, the tension towards immobile mover, the *de facto* divinity external to the empirical perimeter, at the origin of all causes.

Aristotle's philosophy can afford to outflank the issues about the infinite because to conceive it would be impossible and redundant since this very idea is incommensurable to the idea of first cause and first principle, whose action will always be "in progress". The infinite is always only potential, never actual.

The rejection of an ontological hypostatization of the concept of the infinite is also confirmed in its relation to the notion of substance. The infinite is not a formal principle which "is" in the primary sense of the term, it is not a substance whose structure cannot be divided because if so it would lose its unique and original identity: it is an accident which can be "divided" and may potentially happen and be altered by applying to completely different circumstances: *"it is impossible that the infinite should be a thing which is in itself infinite, separable from sensible objects. If the infinite is neither a magnitude nor an aggregate, but is itself a substance and not an accident, it will be indivisible; for the divisible must be either a magnitude or an aggregate. But if indivisible, then not infinite, except in the way in which the voice is invisible. But this is not the way in which it is used by those who say that the infinite exists, nor that in which we are investigating it, namely as that which cannot be gone through."*[3] The infinite is reduced to what we would call today "an idea in a Kantian sense" - with the difference that, as said, its paradoxes does not even have a crucial role in the determination of ultimate metaphysical truths: *"the infinite does not exist potentially in the sense that it will ever actually have separate existence; it exists potentially only for knowledge. For the fact that the process of dividing never comes to an end ensures that this activity exists potentially, but not that the infinite exists separately."*[4] Similarly, the unimaginable numbers can be interpreted as an idea in "a Kantian sense", having only a regulatory function in the sense that they own a linguistic and rational denotation and can give a sensible, "reasonable" context to the regularities which we ordinarily experiences, but that will never actually exist because no thoughts or no computing machine will never be able to "see" (or even think) them. Obviously, in the logical context of modern mathematics, the idea of original immobile mover is note relevant as the concept of "natural number", which share with the first one a similar regulatory theoretical (and, therefore, pragmatic) "action".

---

[3] See Physics, 204a8–204a16, in [4, Vol. I].

[4] See Metaphysics, IX.6, 1048a-b, in [4, Vol. I].

5   Conclusions

The sense of the introduction of grossone into a mathematical formula cannot be reduce to a "technical stratagem", and to understand it is necessary to understand the ontological relevance of the problem of the infinite in the context of a consistent picture of the world which any discipline try to achieve. In this paper we have seen how the original philosophical effort to comprehend the entireness of reality, both in the sense of its fundamental elements which justify its structure and in the sense of the "total amount" of its features, inevitably clashes with the problem that the process of locating these elements or this totality can be thought as without an end. This problem produces the paradox whereby a "part" of reality can contain as many amounts of hypothetical "fundamental elements" or features as the entire. This is an issue which belongs to ontology in a broad sense as to mathematics and its elements in a more specific acceptation, since mathematics - in order to work and to be applied to "experience" - has to create its specific reality consistent in its entireness and in its axioms. One of the solutions to this philosophical paradox elaborated by modern philosophy is to take this problem as a very "problem of thought", making reality coincide with our very conception of reality and with our practical and cultural instruments. In this sense, for Hegel's philosophy, reality is dialectical not "for our thought": reality coincides with our thought and *vice versa*, and any ontological problem is reduced to a problem of internal consistency and contradiction. But if thought and reality are two sides of the same coin, then to create a logic which allows our thought to work with the "reality" of a discipline without any logical arrest or short circuit which would generate practical issues is to create a new dialectical advance in reality - or, at least, in the reality internal to our discipline. The innovation of grossone is that it seems to engage with such a task in the context of the paradoxes of the infinite, by proposing an explicitly "conventional" notion which challenges the very necessity of the ontological mathematical turmoil experienced so far.

Aknowledgments. This work is partially supported by the research projects *"IoT&B, Internet of Things and Blockchain"*, CUP J48C17000230006, POR Calabria FESR-FSE 2014–2020.

References

1. Amodio, P., Iavernaro, F., Mazzia, F., Mukhametzhanov, M.S., Sergeyev, Y.D.: A generalized Taylor method of order three for the solution of initial value problems in standard and infinity floating-point arithmetic. Math. Comput. Simul. **141**, 24–39 (2017)
2. Antoniotti, A., Caldarola, F., d'Atri, G., Pellegrini, M.: New approaches to basic calculus: an experimentation via numerical computation. In: Sergeyev, Y.D., Kvasov, D.E. (eds.) NUMTA 2019. LNCS 11973, pp. 329–342. Springer, Cham (2020)
3. Antoniotti, L., Caldarola, F., Maiolo, M.: Infinite numerical computing applied to Hilbert's, Peano's, and Moore's curves. Mediterr. J. Math. (to appear)

4. Barnes, J. (ed.): The Complete Works of Aristotle. The Revised Oxford Translation, vol. I and II. Princeton University Press, Princeton (1991). 4th printing
5. Bennet, J.: Kant's Dialectic. Cambridge University Press, Cambridge (1974)
6. Bird, G.: Kant's Theory of Knowledge: An Outline of One Central Argument in the Critique of Pure Reason. Routledge & Kegan Paul, London (1962)
7. Blakley, G.R., Borosh, I.: Knuth's iterated powers. Adv. Math. **34**(2), 109–136 (1979). https://doi.org/10.1016/0001-8708(79)90052-5
8. Caldarola, F.: The Sierpiński curve viewed by numerical computations with infinities and infinitesimals. Appl. Math. Comput. **318**, 321–328 (2018). https://doi.org/10.1016/j.amc.2017.06.024
9. Caldarola, F.: The exact measures of the Sierpiński d-dimensional tetrahedron in connection with a Diophantine nonlinear system. Commun. Nonlinear Sci. Numer. Simul. **63**, 228–238 (2018). https://doi.org/10.1016/j.cnsns.2018.02.026
10. Caldarola, F., d'Atri, G., Maiolo, M.: What are the unimaginable numbers? Submitted for publication
11. Caldarola, F., d'Atri, G., Mercuri, P., Talamanca, V.: On the arithmetic of Knuth's powers and some computational results about their density. In: Sergeyev, Y.D., Kvasov, D.E. (eds.) NUMTA 2019. LNCS 11973, pp. 381–388. Springer, Cham (2020)
12. Caldarola, F., Maiolo, M., Solferino, V.: A new approach to the Z-transform through infinite computation. Commun. Nonlinear Sci. Numer. Simul. **82**, 105019 (2020). https://doi.org/10.1016/j.cnsns.2019.105019
13. Calude, C.S., Dumitrescu, M.: Infinitesimal probabilities based on grossone. Spinger Nat. Comput. Sci. **1**, 36 (2019). https://doi.org/10.1007/s42979-019-0042-8
14. Cavini, W.: Ancient epistemology naturalized. In: Gerson, L.P. (ed.) Ancient Epistemology. Cambridge University Press, Cambridge (2009)
15. Cesa, C.: Guida a Hegel. Laterza, Bari (1997)
16. Cococcioni, M., Pappalardo, M., Sergeyev, Y.D.: Lexicographic multi-objective linear programming using grossone methodology: theory and algorithm. Appl. Math. Comput. **318**, 298–311 (2018)
17. Crivelli, P.: Aristotle on Truth. Cambridge University Press, Cambridge (2004)
18. D'Alotto, L.: A classification of one-dimensional cellular automata using infinite computations. Appl. Math. Comput. **255**, 15–24 (2015)
19. D'Alotto, L.: Cellular automata using infinite computations. Appl. Math. Comput. **218**(16), 8077–8082 (2012)
20. De Leone, R.: The use of grossone in mathematical programming and operations research. Appl. Math. Comput. **218**(16), 8029–8038 (2012)
21. David Peat, F.: Superstrings and the Search for the Theory of Everything. Contemporary Books, Chicago (1988)
22. Faticoni, T.G.: The Mathematics of Infinity: A Guide to Great Ideas. Wiley, Hoboken (2006)
23. Gribbin, J.: The Search for Superstrings, Symmetry, and the Theory of Everything. Back Bay Books, New York (2000)
24. Hegel, G.W.F.: The Science of Logic. Cambridge University Press, Cambridge (2010 [1817]). Ed. by G. Di Giovanni
25. Hooshmand, M.H.: Ultra power and ultra exponential functions. Integral Transforms Spec. Funct. **17**(8), 549–558 (2006). https://doi.org/10.1080/10652460500422247

26. Ingarozza, F., Adamo, M.T., Martino, M., Piscitelli, A.: A grossone-based numerical model for computations with infinity: a case study in an Italian high school. In: Sergeyev, Y.D., Kvasov, D.E. (eds.) NUMTA 2019. LNCS 11973, pp. 451–462. Springer, Cham (2020)

27. Kahn, C.H.: The Art and Thought of Heraclitus. Cambridge University Press, Cambridge (1979)

28. Kant, E.: Critique of Pure Reason. Cambridge University Press, Cambridge (1998 [1791]). Transl. by P. Guyer, A.W. Wood (eds.)

29. Kirk, G.S.: Heraclitus: The Cosmic Fragments. Cambridge University Press, Cambridge (1978)

30. Knuth, D.E.: Mathematics and computer science: coping with finiteness. Science **194**(4271), 1235–1242 (1976). https://doi.org/10.1126/science.194.4271.1235

31. Margenstern, M.: An application of grossone to the study of a family of tilings of the hyperbolic plane. Appl. Math. Comput. **218**(16), 8005–8018 (2012)

32. Margenstern, M.: Fibonacci words, hyperbolic tilings and grossone. Commun. Nonlinear Sci. Numer. Simul. **21**(1–3), 3–11 (2015)

33. Mazzia, F., Sergeyev, Y.D., Iavernaro, F., Amodio, P., Mukhametzhanov, M.S.: Numerical methods for solving ODEs on the Infinity Compute. In: Sergeyev, Y.D., Kvasov, D.E., Dell'Accio, F., Mukhametzhanov, M.S. (eds.) 2nd International Conference "NUMTA 2016 - Numerical Computations: Theory and Algorithms", AIP Conference Proceedings, vol. 1776, p. 090033. AIP Publishing, New York (2016). https://doi.org/10.1063/1.4965397

34. Nicolau, M.F.A., Filho, J.E.L.: The Hegelian critique of Kantian antinomies: an analysis based on the Wissenchaft der Logik. Int. J. Philos. **1**(3), 47–50 (2013)

35. Rizza, D.: A study of mathematical determination through Bertrand's Paradox. Philos. Math. **26**(3), 375–395 (2018)

36. Rizza, D.: Primi passi nell'aritmetica dell'infinito (2019, preprint)

37. Rizza, D.: Supertasks and numeral system. In: Sergeyev, Y.D., Kvasov, D.E., Dell'Accio, F., Mukhametzhanov, M.S. (eds.) 2nd International Conference "NUMTA 2016 - Numerical Computations: Theory and Algorithms", AIP Conference Proceedings, vol. 1776, p. 090005. AIP Publishing, New York (2016). https://doi.org/10.1063/1.4965369

38. Sergeyev, Y.D.: A new applied approach for executing computations with infinite and infinitesimal quantities. Informatica **19**(4), 567–596 (2008)

39. Sergeyev, Y.D.: Arithmetic of infinity. Edizioni Orizzonti Meridionali, Cosenza (2003)

40. Sergeyev, Y.D.: Computations with grossone-based infinities. In: Calude, C.S., Dinneen, M.J. (eds.) UCNC 2015. LNCS, vol. 9252, pp. 89–106. Springer, Cham (2015). https://doi.org/10.1007/978-3-319-21819-9_6

41. Sergeyev, Y.D.: Evaluating the exact infinitesimal values of area of Sierpinskinski's carpet and volume of Menger's sponge. Chaos Solitons Fractals **42**(5), 3042–3046 (2009)

42. Sergeyev, Y.D.: Higher order numerical differentiation on the Infinity Computer. Optim. Lett. **5**(4), 575–585 (2011)

43. Sergeyev, Y.D.: Lagrange lecture: methodology of numerical computations with infinities and infinitesimals. Rend Semin Matematico Univ Polit Torino **68**(2), 95–113 (2010)

44. Sergeyev, Y.D.: Solving ordinary differential equations by working with infinitesimals numerically on the infinity computer. Appl. Math. Comput. **219**(22), 10668–10681 (2013)

45. Sergeyev, Y.D.: The exact (up to infinitesimals) infinite perimeter of the Koch snowflake and its finite area. Commun. Nonlinear Sci. Numer. Simul. **31**, 21–29 (2016)

46. Sergeyev, Y.D.: Un semplice modo per trattare le grandezze infinite ed infinitesime. Mat. Soc. Cult. Riv. Unione Mat. Ital. **8**(1), 111–147 (2015)

47. Sergeyev, Y.D.: Using blinking fractals for mathematical modelling of processes of growth in biological systems. Informatica **22**(4), 559–576 (2011)

48. Sergeyev, Y.D.: Numerical infinities and infinitesimals: methodology, applications, and repercussions on two Hilbert problems. EMS Surv. Math. Sci. **4**(2), 219–320 (2017)

49. Sergeyev, Y.D., Mukhametzhanov, M.S., Mazzia, F., Iavernaro, F., Amodio, P.: Numerical methods for solving initial value problems on the Infinity Computer. Int. J. Unconv. Comput. **12**(1), 55–66 (2016)

50. Severino, E.: La Filosofia dai Greci al nostro Tempo, vol. I, II, III. RCS Libri, Milano (2004)

51. Theodossiou, E., Mantarakis, P., Dimitrijevic, M.S., Manimanis, V.N., Danezis, E.: From the infinity (apeiron) of Anaximander in ancient Greece to the theory of infinite universes in modern cosmology. Astron. Astrophys. Trans. **27**(1), 162–176 (2011)

52. Thompson, J.F.: Tasks and super-tasks. Analysis **15**(1), 1–13 (1954)

53. Weyl, H.: Levels of Infinity/Selected Writings on Mathematics and Philosophy. Dover (2012). Ed. by P. Pesic

54. Zanatta, M.: Profilo Storico della Filosofia Antica. Rubettino, Catanzaro (1997)

The Sequence of Carboncettus Octagons

Fabio Caldarola[1]($\boxtimes$)(iD), Gianfranco d'Atri[1], Mario Maiolo[2](iD),
and Giuseppe Pirillo[3]

[1] Department of Mathematics and Computer Science, Università della Calabria,
Cubo 31/A, 87036 Arcavacata di Rende, CS, Italy
{caldarola,datri}@mat.unical.it
[2] Department of Environmental and Chemical Engineering (DIATIC),
Università della Calabria, Cubo 42/B, 87036 Arcavacata di Rende, CS, Italy
mario.maiolo@unical.it
[3] Department of Mathematics and Computer Science U. Dini, University of Florence,
viale Morgagni 67/A, 50134 Firenze, Italy
pirillo@math.unifi.it

Abstract. Considering the classic Fibonacci sequence, we present in
this paper a geometric sequence attached to it, where the word "geometric" must be understood in a literal sense: for every Fibonacci number F_n we will in fact construct an octagon C_n that we will call the
n-th Carboncettus octagon, and in this way we obtain a new sequence
$\{C_n\}_n$ consisting not of numbers but of geometric objects. The idea of
this sequence draws inspiration from far away, and in particular from a
portal visible today in the Cathedral of Prato, supposed work of *Carboncettus marmorarius*, and even dating back to the century before that of
the writing of the *Liber Abaci* by *Leonardo Pisano* called Fibonacci (AD
1202). It is also very important to note that, if other future evidences will
be found in support to the historical effectiveness of a Carboncettus-like
construction, this would mean that Fibonacci numbers were known and
used well before 1202. After the presentation of the sequence $\{C_n\}_n$, we
will give some numerical examples about the metric characteristics of the
first few Carboncettus octagons, and we will also begin to discuss some
general and peculiar properties of the new sequence.

Keywords: Fibonacci numbers · Golden ratio · Irrational numbers ·
Isogonal polygons · Plane geometric constructions

1 Introduction

The names here proposed of "*n*-th Carboncettus octagon" and "Carboncettus
sequence/family of octagons", or better, the inspiration for these names, comes
from far away, sinking its roots in the early centuries of the late Middle
Ages. They are in fact connected to the cathedral of Prato, a jewel of Italian Romanesque architecture, which underwent a profound restructuring in the
11th century, followed by many others afterwards. The side portal shown in

© Springer Nature Switzerland AG 2020
Y. D. Sergeyev and D. E. Kvasov (Eds.): NUMTA 2019, LNCS 11973, pp. 373–380, 2020.
https://doi.org/10.1007/978-3-030-39081-5_32

Fig. 1 (which we will later call simply the *portal*) at the time of its construction seems to have been the main portal of the cathedral. The marble inlays on its sides and the figures represented have aroused many discussions among scholars for many years and in particular have always aroused the attention and interest of G. Pirillo, an interest that he recently transmitted also to the other authors. Pirillo studied the figures of the portal for a long time and traced a fascinating symbolism, typical of medieval culture (see for example [11]). According to these studies, the right part of the portal, for instance, through a series of very regular and symmetrical figures, would recall the divine perfection, while the left part, through figures that approximate the regular ones but are not themselves regular, the imperfection and the limits of human nature. The very interesting fact is that the artist/architect who created the work (which is thought to be a certain *Carboncettus Marmoriarius*, very active at that time and in those places, [11]) seems to have been in part used the mathematical language to express these concepts and ideas, and this thing, if confirmed, would assume enormous importance, because before the 12th century we (and many experts of the field) have no knowledge of similar examples. The construction of the Carboncettus octagon (or better, of the Carboncettus octagons, since they are infinitely many) originates from Fibonacci numbers and yields a sequence not of numbers but of geometrical figures: we will explain the details starting from Sect. 2.

From the historical point of view we cannot avoid to note an interesting, particular coincidence: probably, the most known and most important octagonal monument existing in Calabria dates back to the same period as the construction of the portal of the *Duomo* of Prato, and it is the octagonal tower of the Norman-Swabian Castle in Cosenza. But it is important to specify, for the benefit of the reader, that, in Cosenza, on the site of the actual Norman-Swabian Castle, a fortification had existed from immemorial time, which underwent considerable changes over the years: first a Bruttuan fortress, then Roman, Norman and Swabian, when it had the most important restructuring due to Frederick II of Swabia. In particular, it is Frederick who wanted the octagonal tower visible today, his preferred geometric shape: remember, for example, the octagonal plan of the famous *Castel del Monte* near Bari, in Apulia.

With regard to Fibonacci numbers, we would like to point out to the reader for completeness of information, a recent thesis by G. Pirillo often and many times discussed within this group of authors. In [10,12–14] Pirillo presented the audacious thesis that the first mathematicians who discovered Fibonacci numbers were some members of the Pythagorean School, well documented and active in Crotone in the 6th, 5th and 4th centuries B.C., hence about 1,700 years before that *Leonardo Pisano*, known as *"Fibonacci"*, wrote his famous *Liber Abaci* in 1202. Such a thesis is mainly supported by computational evidences arising from pentagon and pentagram about the well-known Pythagorean discovery of the existence of incommensurable numbers. The interested reader can find further information and references on the Pythagorean School, incommensurable lengths, Fibonacci numbers and some recent developments in [6,8,10,14–17].

Fig. 1. The side portal of the cathedral of Prato. The two topmost figures have octagonal shape: the one on the right is based on a regular octagon, while the one on the left seems to allude to a very particular construction that inspires thus paper and the now called *Carboncettus octagons*.

Similarly to the above thesis note that, since the portal in Prato is dating back to the 12th century, if other future evidences will support the employ of Fibonacci numbers in its geometries, this would mean that they were known before 1202 as well, even if only a few decades.

A final remark on notations: we denote by $\mathbb{N}$ the set of positive integers and by $\mathbb{N}_0$ the set $\mathbb{N} \cup \{0\}$. A sequence of numbers or other mathematical objects is denoted by $\{a_n\}_{n\in\mathbb{N}}$, $\{a_n\}_n$, or simply $\{a_n\}$. If, moreover, A, B, C are three points of the plane, AB denotes the line segment with endpoints A and B, $|AB|$ its length, and $\angle ABC$ the measure of the angle with vertex in B.

2   The Carboncettus Family of Octagons

If r is any positive real number, we denote by Γ_r the circumference of radius r centered in the origin. As usual, let F_n be the n-th Fibonacci number for all $n \in \mathbb{N}_0$, i.e.,

$$F_0 = 0, \quad F_1 = 1, \quad F_2 = 1, \quad F_3 = 2, \quad F_4 = 3, \quad F_5 = 5, \quad \text{etc.}$$

If $n \in \mathbb{N}$ we consider a couple of concentric circumferences having radii of length F_n and F_{n+2}, respectively. If $n = 1$ they are represented in green in Fig. 2, were the radius of the inner circumference is 1 and that of the outer one is 2, i.e. F_3. Then we draw two couples of parallel tangents, orthogonal between them, to the inner circumference and we consider the eight intersection points A, B, C, D, E, F, G, L with the outer circumference $\Gamma_{F_{n+2}}$, as in Fig. 2. The octagon obtained by drawing the polygonal through the points $A, B, C, D, E, F, G, L, A$, in red in Fig. 2, is called the n-th Carboncettus octagon and is denoted by C_n. Therefore, the red octagon in Fig. 2, is the first Carboncettus octagon C_1.

From a geometrical point of view, the Carboncettus octagon C_n is more than a cyclic polygon; it is in fact an isogonal octagon for all $n \in \mathbb{N}$, that is, an equiangular octagon with two alternating edge lengths.[1] More recently it is also used to say a vertex-transitive octagon: all the vertices are equivalent under the symmetry group of the figure and, in the case of C_n, for every couple of vertices, the symmetry which send the first in the second is unique. The symmetry group of C_n is in fact isomorphic to the one of the square, the dihedral group D_4.[2]

An interesting property of the Carboncettus sequence $\{C_n\}_{n\in\mathbb{N}}$ is the fact that, with the exception of the first three elements C_1, C_2, C_3 (or, at most, also C_4), all the subsequent ones are completely indistinguishable from a regular octagon (see, for example, Fig. 3 representing C_2: it is yet relatively close to a regular octagon). Due to the lack of space, we will deepen these and other important aspects mentioned in the following, in a subsequent paper in preparation.

---

[1] In this view, a recent result established that a cyclic polygon is equiangular if and only if is isogonal (see [7]). Of course, an equiangular octagon is not cyclic in general, while it is true for 3- and 4-gons (see [2]).

[2] Note, for didactic purposes, how the multiplication table of D_4 emerges much more clearly to the mind of a student thinking to C_1 than thinking to a square.

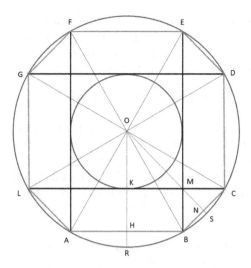

Fig. 2. The construction of the Carboncettus octagon. In the picture, in particular, it is shown in red the octagon C_1. (Color figure online)

3   The First Four Octagons of the Carboncettus Sequence: Geometric Properties and Metric Data

In this section we will give some numerical examples looking closely at the first elements of the sequence $\{C_n\}_{n \in \mathbb{N}}$.

Example 1 (The octagon C_1). The first Carboncettus octagon C_1 is built starting from the circumferences Γ_1 and Γ_2, as said in Sect. 2. In this case we obtain a very particular isogonal octagon: drawing the eight radii

$$OA,\ OB,\ OC,\ OD,\ OE,\ OF,\ OG,\ OL \tag{1}$$

of the circumference Γ_2 as in Fig. 2, the resulting shape has commensurable angle measures, in fact all them are integer multiples of $\pi/12 = 15°$. Not only; in this way C_1 results formed by 4 equilateral triangles (congruent to ABO, see Fig. 2) and 4 isosceles triangles (congruent to BCO). The lengths of their sides and heights are

$$\begin{aligned}
|AB| = |OA| = 2, && |OH| = |KC| = \sqrt{3}, \\
|BC| = \sqrt{6} - \sqrt{2}, && |ON| = \frac{\sqrt{6}+\sqrt{2}}{2},
\end{aligned} \tag{2}$$

which, for example, are all incommensurable in pairs. Instead, for the widths of the angles we trivially have

$$\angle AOB = \angle OBA = \pi/3 = 60°, \qquad \angle BOC = \angle HOB = \pi/6 = 30°, \\ \angle OBC = 5\pi/12 = 75°. \tag{3}$$

Discussing the commensurability of the angles for all the sequence $\{C_n\}_n$ is interesting, but we are forced to postpone this elsewhere. The same, as well, considering the commensurability, along all the sequence, of some of the side lengths made explicit in (2). Note lastly that perimeter and area are

$$\mathrm{Per}(C_1) = 8 + 4\sqrt{6} - 4\sqrt{2}, \qquad \mathrm{Area}(C_1) = 4 + 4\sqrt{3}.$$

The second Carboncettus octagon C_2 originates from the circumferences Γ_1 and Γ_3, with radii $F_2 = 1$ and $F_4 = 3$, respectively, and the result is the black octagon in Fig. 3, compared with a red regular one inscribed in the circumference Γ_3 itself. Using the letters disposition of Fig. 2, the lengths of the correspondent sides and heights considered in (2), the angle widths, perimeter and area, are those listed in the second column of Table 1.

Table 1. Some metric data relative to the first three elements of the Carboncettus sequence, after C_1. The letters are displayed in the construction as in Fig. 2.

	C_2	C_3	C_4
$|OK|$	1	2	3
$|OA|$	3	5	8
$|AB|$	2	4	6
$|BC|$	$4 - \sqrt{2}$	$\sqrt{42} - 2\sqrt{2}$	$\sqrt{110} - 3\sqrt{2}$
$|OH|$	$2\sqrt{2}$	$\sqrt{21}$	$\sqrt{55}$
$|ON|$	$2 + \sqrt{2}/2$	$\sqrt{2} + \sqrt{42}/2$	$(3\sqrt{2} + \sqrt{110})/2$
$\angle AOB$	$\approx 38.942°$	$\approx 47.156°$	$\approx 44.049°$
$\angle BOC$	$\approx 51.058°$	$\approx 42.844°$	$\approx 45.951°$
$\angle OAB$	$\approx 70.529°$	$\approx 66.421°$	$\approx 67.976°$
Perim.	$24 - 4\sqrt{2}$	$16 + 4\sqrt{42} - 8\sqrt{2}$	$24 + 4\sqrt{110} - 12\sqrt{2}$
Area	$14 + 8\sqrt{2}$	$34 + 8\sqrt{21}$	$92 + 12\sqrt{55}$

4   The "limit octagon" and Future Researches

Many aspects of the new sequence $\{C_n\}_n$ are interesting to investigate. For example, scaling the octagon C_n by a factor equal to the n-th Fibonacci number F_n, the sequence will converge to a limit octagon C_∞^N (where the top N stands for "normalized") that can be drawn through the "Carboncettus construction" described at the beginning of Sect. 2, by starting from the circumferences with radii given by the following limit ratios

$$\lim_{n\to\infty} \frac{F_n}{F_n} = 1 \qquad \text{and} \qquad \lim_{n\to\infty} \frac{F_{n+2}}{F_n}, \tag{4}$$

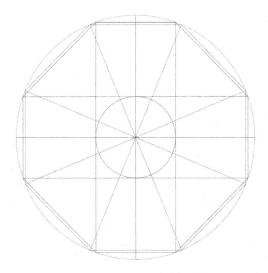

Fig. 3. The second element of the Carboncettus sequence, the octagon C_2, is drawn in black. A regular octagon inscribed in the same circumference Γ_3, is also represented in red. (Color figure online)

respectively. It is well known that the limit of the ratio of two consecutive Fibonacci numbers F_{n+1}/F_n converges to the *golden ratio*

$$\phi := (1 + \sqrt{5})/2 \approx 1.618033987, \tag{5}$$

hence, the second limit in (4) is simple to compute as follows[3]

$$\lim_{n \to \infty} \frac{F_{n+2}}{F_n} = \lim_{n \to \infty} \frac{F_{n+2}}{F_{n+1}} \cdot \frac{F_{n+1}}{F_n} = \phi^2 \approx 2.618033987, \tag{6}$$

and we conclude that C_∞^N can be constructed using the circumferences Γ_1 and Γ_{ϕ^2}.

Another approach to directly study the "limit octagon" C_∞ instead of the "limit normalized octagon" C_∞^N, could come by using the computational system introduced for example in [18–20] and applied as well to limit curves, limit polytopes, fractals and similar geometric shapes in [1,3–5,20] (or even to Fibonacci numbers in [9]).

Aknowledgments. This work is partially supported by the research projects *"IoT&B, Internet of Things and Blockchain"*, CUP J48C17000230006, POR Calabria FESR-FSE 2014–2020.

---

[3] The reader certainly remembers the well know property $\phi^2 = 1 + \phi$ of the golden ratio that causes the coincidence of the fractional parts of (5) and (6).

References

1. Antoniotti, L, Caldarola, F., Maiolo, M.: Infinite numerical computing applied to Hilbert's, Peano's, and Moore's curves. Mediterr. J. Math. (in press)
2. Ball, D.: Equiangular polygons. Math. Gaz. **86**(507), 396–407 (2002)
3. Caldarola, F.: The Sierpiński curve viewed by numerical computations with infinities and infinitesimals. Appl. Math. Comput. **318**, 321–328 (2018). https://doi.org/10.1016/j.amc.2017.06.024
4. Caldarola, F.: The exact measures of the Sierpiński d-dimensional tetrahedron in connection with a Diophantine nonlinear system. Commun. Nonlin. Sci. Numer. Simul. **63**, 228–238 (2018). https://doi.org/10.1016/j.cnsns.2018.02.026
5. Caldarola, F., Maiolo, M., Solferino, V.: A new approach to the Z-transform through infinite computation. Commun. Nonlin. Sci. Numer. Simul. **82**, 105019 (2020). https://doi.org/10.1016/j.cnsns.2019.105019
6. Caldarola, F., Cortese, D., d'Atri, G., Maiolo, M.: Paradoxes of the infinite and ontological dilemmas between ancient philosophy and modern mathematical solutions. In: Sergeyev, Y., Kvasov, D. (eds.) NUMTA 2019. LNCS, vol. 11973, pp. 358–372. Springer, New York (2020)
7. De Villiers, M.: Equiangular cyclic and equilateral circumscribed polygons. Math. Gaz. **95**, 102–107 (2011)
8. Koshy, T.: Fibonacci and Lucas Numbers with Applications. Wiley, New York (2001)
9. Margenstern, M.: Fibonacci words, hyperbolic tilings and grossone. Commun. Nonlin. Sci. Numer. Simul. **21**(1–3), 3–11 (2015)
10. Pirillo, G.: A characterization of Fibonacci numbers. Chebyshevskii Sbornik **19**(2), 259–271 (2018)
11. Pirillo, G.: Figure geometriche su un portale del Duomo di Prato. Prato Storia e Arte **121**, 7–16 (2017). (in Italian)
12. Pirillo, G.: La scuola pitagorica ed i numeri di Fibonacci. Archimede **2**, 66–71 (2017). (in Italian)
13. Pirillo, G.: L'origine pitagorica dei numeri di Fibonacci. Periodico di Matematiche **9**(2), 99–103 (2017). (in Italian)
14. Pirillo, G.: Some recent results of Fibonacci numbers, Fibonacci words and Sturmian words. Southeast Asian Bull. Math. **43**(2), 273–286 (2019)
15. Pirillo, G.: Fibonacci numbers and words. Discret. Math. **173**(1–3), 197–207 (1997)
16. Pirillo, G.: Inequalities characterizing standard Sturmian and episturmian words. Theoret. Comput. Sci. **341**(1–3), 276–292 (2005)
17. Pirillo, G.: Numeri irrazionali e segmenti incommensurabili. Nuova Secondaria **7**, 87–91 (2005). (in Italian)
18. Sergeyev, Y.D.: Arithmetic of Infinity. Edizioni Orizzonti Meridionali, Cosenza (2003)
19. Sergeyev, Y.D.: Lagrange lecture: methodology of numerical computations with infinities and infinitesimals. Rend. Semin. Matematico Univ. Polit. Torino **68**(2), 95–113 (2010)
20. Sergeyev, Y.D.: Numerical infinities and infinitesimals: methodology, applications, and repercussions on two Hilbert problems. EMS Surv. Math. Sci. **4**(2), 219–320 (2017)

On the Arithmetic of Knuth's Powers and Some Computational Results About Their Density

Fabio Caldarola[1] , Gianfranco d'Atri[1], Pietro Mercuri[2] ,
and Valerio Talamanca[3]([✉])

[1] Department of Mathematics and Computer Science, Università della Calabria,
Cubo 31/A, 87036 Arcavacata di Rende, Italy
{caldarola,datri}@mat.unical.it
[2] Università "Tor Vergata", 00133 Rome, Italy
mercuri.ptr@gmail.com
[3] Università Roma Tre, 00146 Rome, Italy
valerio@mat.uniroma3.it

Abstract. The object of the paper are the so-called "unimaginable numbers". In particular, we deal with some arithmetic and computational aspects of the Knuth's powers notation and move some first steps into the investigation of their density. Many authors adopt the convention that unimaginable numbers start immediately after 1 *googol* which is equal to 10^{100}, and G.R. Blakley and I. Borosh have calculated that there are exactly 58 integers between 1 and 1 googol having a nontrivial *"kratic representation"*, i.e., are expressible nontrivially as Knuth's powers. In this paper we extend their computations obtaining, for example, that there are exactly $2\,893$ numbers smaller than $10^{10\,000}$ with a nontrivial kratic representation, and we, moreover, investigate the behavior of some functions, called *krata*, obtained by fixing at most two arguments in the Knuth's power $a \uparrow^b c$.

Keywords: Unimaginable numbers · Knuth up-arrow notation ·
Algebraic recurrences · Computational number theory

1 Introduction: The Unimaginable Numbers

An unimaginable number, intuitively and suggestively, is a number that go beyond the human imagination. There is not a completely accepted standard formal definition of unimaginable numbers, but one of the most used is the following: a number is called *unimaginable* if it is greater than 1 googol, where *1 googol* is equal to 10^{100}. To better understand the size of numbers like these, consider that it is estimated that in the observable universe there are at most 10^{82} atoms; this justifies the term *unimaginable*. The first appearance of the unimaginable numbers was, to our knowledge, in Magna Graecia in the work of Archimedes of Syracuse. Archimedes in his work called *"Arenarius"* in Latin, or

© Springer Nature Switzerland AG 2020
Y. D. Sergeyev and D. E. Kvasov (Eds.): NUMTA 2019, LNCS 11973, pp. 381–388, 2020.
https://doi.org/10.1007/978-3-030-39081-5_33

the *"Sand Reckoner"* in English, describes, using words of the natural language, an extremely large number that, in exponential notation, is equal to

$$10^{8 \cdot 10^{16}} = 10^{80\,000\,000\,000\,000\,000}. \tag{1}$$

Obviously, writing this number without any kind of modern mathematical notation, as Archimedes did, is very very difficult. Let us jump to modern times and introduce the most used notation that allows to write numbers so large that are definitely beyond the common experience of a human being.

Definition 1 (Knuth's up-arrow notation). *For all non-negative integers a, b, n, we set*

$$a \uparrow^n b := \begin{cases} a \cdot b & \text{if } n = 0; \\ 1 & \text{if } n \geq 1 \text{ and } b = 0; \\ a \uparrow^{n-1} (a \uparrow^n (b-1)) & \text{if } n \geq 1 \text{ and } b \geq 1. \end{cases} \tag{2}$$

For $n = 1$ we obtain the ordinary exponentiation, e.g., $3 \uparrow 4 = 3^4$; for $n = 2$ we obtain *tetration*, for $n = 3$ *pentation*, and so on. Hence (2) represents the so called *n-hyperoperation*. Now, using Knuth's notation, the Archimedes' number (1) can be easily written as follows

$$((10 \uparrow 8) \uparrow^2 2) \uparrow (10 \uparrow 8).$$

In this paper we use an alternative notation to that introduced in Definition 1. Denoting by $\mathbb{N}$ the set of natural numbers (i.e., non-negative integers) we define the *Knuth's function* k as follows

$$\begin{aligned} k : \mathbb{N} \times \mathbb{N} \times \mathbb{N} &\longrightarrow \mathbb{N} \\ (B, d, T) &\mapsto k(B, d, T) := B \uparrow^d T \end{aligned} \tag{3}$$

and we call the first argument of k (i.e., B) the *base*, the second (i.e., d) the *depth* and the third (i.e., T) the *tag* (see [3]).

The paper is organized as follows: in Sect. 2 we introduce some general computational problems, while in Sect. 3, which is the core of this work, we deal with density and representational problems related to Knuth's powers. In particular, Proposition 2 and Corollary 1 give some simple results which characterize the difference of digits in base 10 between two "consecutive" Knuth's powers of the simplest "non-trivial" type, i.e., $a \uparrow^2 2$. Proposition 3 extends, instead, a computation by Blakley and Borosh (see [3, Proposition 1.1]): they found that there are exactly 58 numbers smaller than 1 *googol* ($= 10^{100}$) nontrivially expressible through the Knuth's function k. We obtained that such number increases to 2893 if we consider integers lesser than $10^{10\,000}$. Among these 2893 numbers, 2888 are expressible through the aforementioned form $a \uparrow^2 2$.

We conclude the introductory section by giving the reader some brief information on some useful references to deepen the issues related to unimaginable numbers. In addition to article [3] which, for our purposes, represents the main reference, the same authors investigate the modular arithmetic of Knuth's powers in [4]. Knuth himself had instead introduced the notation (2) a few years earlier in [16] (1976), but these ideas actually date from the beginning of the century (see [1, 2, 13, 19]). More recent works that start from "extremely large" or "infinite" numbers are [7, 8, 12, 14, 15, 17, 18, 20, 21, 23]. There are also the online resources [5, 6, 22]. Finally [10] provides the reader with a brief general introduction with some further reference.

2   Representational and Computational Problems

It is obvious that almost all the numbers are unimaginable, hence a first natural question is: can we write every unimaginable number using Knuth's up-arrow notation? The answer is trivial: we cannot. Actually there are just very few numbers that are expressible using this notation. More precisely let $\mathcal{K}_0$ denote the image of the map k, i.e., the set of those natural numbers that are expressible via Knuth's notation. As customary one can consider the ratio

$$\rho_0(x) = \frac{\#\big(\mathcal{K}_0 \cap \{m \in \mathbb{N} : m < x\}\big)}{x}. \tag{4}$$

The computed values of $\rho_0(x)$ are very close to zero and $\rho_0(x)$ appears to be quickly converging to zero as $x \to +\infty$. In the next section we compute some values of a ratio related to this.

In recent years dozens of systems and notations have been developed to write unimaginable numbers (for example see [1, 6, 7, 14, 15]), most of them can reach bigger numbers in a more compact notation than Knuth's notation can, but the difference between two consecutive numbers with a compact representation in a specific notation often increases quicker than in Knuth's notation. Hence, almost all unimaginable numbers remain inaccessible to write (and to think about?) and the problem of writing an unimaginable number in a convenient way is open.

A strictly related open problem is to find a good way to represent an unimaginable number on a computer. It is not possible to represent with usual decimal notation numbers like $3 \uparrow^3 3$ on a computer at the present time. Hence, to do explicit computations involving these numbers is quite hard. Therefore finding a way, compatible with classical operations, to represent these kind of numbers on a computer not only would make many computations faster but it would also help to deeper develop the mathematical properties and the applications related to unimaginable numbers.

We recall, for the convenience of the reader, some basic properties of Knuth's up-arrow notation that will be used in the next section.

Proposition 1. *For all positive integers a, b, n, with $b > 1$, we have:*

(i) $a \uparrow^n b < (a + 1) \uparrow^n b$;
(ii) $a \uparrow^n b < a \uparrow^n (b + 1)$;
(iii) $a \uparrow^n b \leq a \uparrow^{n+1} b$, where the equality holds if and only if $a = 1$ or $a = b = 2$.

Proof. See [3, Theorem 1.1].

3   About the Density of Numbers with Kratic Representation

We follow the nomenclature used in [3] and we say that a positive integer x has a *non-trivial kratic representation* if there are integers a, b, n all greater than 1 such that $x = a \uparrow^n b$. Note that a kratic representation should not be confused with a kratos: a *kratos* (pl. *krata*)[1] is a function h that comes from the Knuth's function k by fixing at most two arguments (see [3]). It is then a natural question to ask "how many" numbers have a non-trivial kratic representation.

Example 1. The least positive integer with non-trivial kratic representation is 4, in fact $2 \uparrow^n 2 = 4$ for all positive integers n.

It is easy to see that numbers with kratic representation of the form $a \uparrow^2 2 = a^a$ are more frequent than those with other types of kratic representation. The following proposition states how often they appear with respect to the number of digits, i.e., it calculates the increment of the number of digits between two "consecutive" numbers with kratic representation of that form. We need a further piece of notation: as usual Log denotes the logarithm with base 10, $\lfloor \alpha \rfloor$ the floor of a real number α and $\nu(a)$ the number of digits of a positive integer a (in base 10). Using these notation we have

$$\nu(a) = \lfloor \text{Log } a \rfloor + 1 \tag{5}$$

for all positive integers a.

Proposition 2. *For every integer $a \geq 1$ we have*

$$\nu\left((a+1)^{a+1}\right) - \nu\left(a^a\right) = \left\lfloor \text{Log}(a+1) + a \text{Log}\left(1 + \frac{1}{a}\right) \right\rfloor \tag{6}$$

or

$$\nu\left((a+1)^{a+1}\right) - \nu\left(a^a\right) = \left\lfloor \text{Log}(a+1) + a \text{Log}\left(1 + \frac{1}{a}\right) \right\rfloor + 1. \tag{7}$$

Proof. The proposition states that the difference between the number of digits of $(a+1)^{a+1}$ and a^a is given by Formula (6) or (7). For any integer $a \geq 1$ we have

$$\begin{aligned}
\nu\left((a+1)^{a+1}\right) - \nu\left(a^a\right) &= \lfloor \text{Log}(a+1)^{a+1} \rfloor - \lfloor \text{Log } a^a \rfloor \\
&= \lfloor (1+a)\text{Log}(a+1) \rfloor - \lfloor a \text{Log } a \rfloor \\
&= \lfloor \text{Log}(a+1) + a \text{Log}(a+1) - a \text{Log } a \\
&\quad + a \text{Log } a \rfloor - \lfloor a \text{Log } a \rfloor \\
&= \left\lfloor \text{Log}(a+1) + a \text{Log}\left(1 + \frac{1}{a}\right) + a \text{Log } a \right\rfloor \tag{8} \\
&\quad - \lfloor a \text{Log } a \rfloor .
\end{aligned}$$

---

[1] Kratos, written in ancient Greek κράτος, indicated the "personification of power".

Since for all real numbers α and β the following inequalities hold

$$\lfloor\alpha\rfloor + \lfloor\beta\rfloor \leq \lfloor\alpha+\beta\rfloor \leq \lfloor\alpha\rfloor + \lfloor\beta\rfloor + 1, \tag{9}$$

then, combining (9) with (8) we obtain

$$\left\lfloor \mathrm{Log}(a+1) + a\,\mathrm{Log}\left(1+\frac{1}{a}\right)\right\rfloor \leq \nu\left((a+1)^{a+1}\right) - \nu\left(a^a\right)$$
$$\leq \left\lfloor \mathrm{Log}(a+1) + a\,\mathrm{Log}\left(1+\frac{1}{a}\right)\right\rfloor + 1,$$

proving the proposition.

Corollary 1. *For every integer $a \geq 1$ the following inequalities hold*

$$\lfloor\mathrm{Log}(a+1)\rfloor \leq \nu\left((a+1)^{a+1}\right) - \nu\left(a^a\right) \leq \lfloor\mathrm{Log}(a+1)\rfloor + 2. \tag{10}$$

Proof. The first inequality in (10) is an immediate consequence of (6). For the second one note that, using the previous proposition, the second inequality in (9) and the well-known bound

$$\left(1+\frac{1}{a}\right)^a < e, \qquad \text{for all } a \geq 1, \tag{11}$$

we obtain

$$\nu\left((a+1)^{a+1}\right) - \nu\left(a^a\right) \leq \left\lfloor \mathrm{Log}(a+1) + \mathrm{Log}\left(1+\frac{1}{a}\right)^a\right\rfloor + 1$$
$$\leq \lfloor\mathrm{Log}(a+1)\rfloor + \left\lfloor \mathrm{Log}\left(1+\frac{1}{a}\right)^a\right\rfloor + 2$$
$$= \lfloor\mathrm{Log}(a+1)\rfloor + 2.$$

The two possibilities given by (6) and (7) in Proposition 2 and the three given by Corollary 1 (that is, $\nu\left((a+1)^{a+1}\right) - \nu\left(a^a\right) - \lfloor\mathrm{Log}(a+1)\rfloor = 0,\ 1,\ 2$) are all effectively realized: it is sufficient to look at the values $a = 1, 2, 7$ in Table 1.

Table 1. The first 10 values of a^a.

a	1	2	3	4	5	6	7	8	9	10
a^a	1	4	27	256	3 125	46 656	823 543	16 777 216	387 420 489	10 000 000 000

Remark 1. Note that using (11) and the lower bound $2 \leq (1+1/a)^a$, for $a \geq 1$, we obtain

$$2(a+1) \leq \frac{(a+1)\uparrow^2 2}{a\uparrow^2 2} < e(a+1) \tag{12}$$

for all integers $a \geq 1$. It is also interesting that the ratio of two consecutive numbers of that form can be approximated by a linear function in the base a.

The previous remark implies that given a number with kratic representation of the form $a \uparrow^2 2$, the subsequent one, $(a + 1) \uparrow^2 2$, is rather close to it. Instead, numbers with kratic representation of other forms are much more sporadic: the following proposition gives a more precise idea of this phenomenon.

Proposition 3. *There are exactly 2893 numbers smaller than $10^{10\,000}$ that admit a non-trivial kratic representation. Among them, 2888 have a representation of the form $a \uparrow^2 2$, and only 5 do not have such a representation.*

Proof. By [3, Proposition 1.1] there are exactly 58 numbers with less than 10^2 digits in decimal notation that have a non-trivial kratic representation; we collect them in the following set

$$E_2 = \{a \uparrow^2 2 : 2 \leq a \leq 56\} \sqcup \{2 \uparrow^2 3,\ 3 \uparrow^2 3,\ 2 \uparrow^2 4\}.$$

Note also that some of them have more than one representation:

$$2 \uparrow^2 2 = 4 = 2 \uparrow^d 2 \ \forall d \geq 2, \quad 3 \uparrow^2 3 = 3^{27} = 3 \uparrow^3 2, \quad 2 \uparrow^2 4 = 2^{16} = 2 \uparrow^3 3.$$

We look for the numbers we need to add to E_2 to obtain the desired set

$$E := \left\{ n \in \mathbb{N} : n < 10^{10\,000} \text{ and } n \text{ has a non-trivial kratic representation} \right\}.$$

We consider different cases depending on the depth d.

(i) "$d = 2$". Since

$$\mathrm{Log}(2889 \uparrow^2 2) \approx 9998.1 \quad \text{and} \quad \mathrm{Log}(2890 \uparrow^2 2) \approx 10001.99,$$

we have to add to E_2 the numbers from $57 \uparrow^2 2$ to $2889 \uparrow^2 2$. Then, since

$$\mathrm{Log}(5 \uparrow^2 3) \approx 2184.28 \quad \text{and} \quad \mathrm{Log}(6 \uparrow^2 3) \approx 36305.4,$$

the numbers $4 \uparrow^2 3$ and $5 \uparrow^2 3$ belong to E as well. Instead,

$$\mathrm{Log}(3 \uparrow^2 4) \approx 3638334640024.1 \quad \text{and} \quad \mathrm{Log}(2 \uparrow^2 5) \approx 19728.3 \qquad (13)$$

guarantee, by using Proposition 1, that there are no other elements with $d = 2$ in E.

(ii) "$d = 3$". Note that $4 \uparrow^3 2 = 4 \uparrow^2 4 > 3 \uparrow^2 4$, and $3 \uparrow^3 3 = 3 \uparrow^2 3 \uparrow^2 3 > 3 \uparrow^2 4$, and $2 \uparrow^3 4 = 2 \uparrow^2 2^{16}$, hence, by using (13), we have that E does not contain any new element with $d = 3$.

(iii) "$d = 4$". Since $3 \uparrow^4 2 = 3 \uparrow^3 3$ and $2 \uparrow^4 3 = 2 \uparrow^3 4$, part (ii) yields that they do not belong to E. Therefore, it has no new elements with $d = 4$.

Now, only the (trivial) case "$d \geq 5$" remains, but since $3 \uparrow^d 2 > 3 \uparrow^4 2$, (iii) yields that there are no new elements with $d \geq 5$ in E. In conclusion, we have proved that

$$E = E_2 \sqcup \{a \uparrow^2 2 : 57 \leq a \leq 2889\} \sqcup \{4 \uparrow^2 3,\ 5 \uparrow^2 3\}$$

and its cardinality is 2893. From the proof, it is also clear that the only elements of E having no representation of the type $a \uparrow^2 2$ are $2 \uparrow^2 3,\ 3 \uparrow^2 3,\ 2 \uparrow^2 4,\ 4 \uparrow^2 3$ and $5 \uparrow^2 3$.

We conclude the paper with some observations about the frequency of numbers with non-trivial kratic representation. Let $\mathcal{K}$ denote the set of integers that admit a non-trivial kratic representation. Define the *kratic representation ratio* $\rho(x)$ as:

$$\rho(x) := \frac{\#(\mathcal{K} \cap \{m \in \mathbb{N} : m < x\})}{x}.$$

(Note the differences with respect to (4).) We find the following values:

$$\rho(10) = \frac{1}{10} = 0.1 \cdot 10^{-1},$$

$$\rho\left(10^2\right) = \frac{3}{10^2} = 0.3 \cdot 10^{-1},$$

$$\rho\left(10^4\right) = \frac{5}{10^4} = 0.5 \cdot 10^{-3},$$

$$\rho\left(10^{10}\right) = \frac{9}{10^{10}} = 0.9 \cdot 10^{-9},$$

$$\rho\left(10^{10^2}\right) = \frac{58}{10^{10^2}} = 0.58 \cdot 10^{-98},$$

$$\rho\left(10^{10^4}\right) = \frac{2893}{10^{10^4}} = 0.2893 \cdot 10^{-9996}.$$

These data seems to indicate that $\rho(x)$ tends rapidly to zero for $x \to +\infty$. However, it is not known, to our knowledge, an explicit formula for $\rho(x)$ or, equivalently, for the cardinality of the set $\mathcal{K}(x) = \mathcal{K} \cap \{m \in \mathbb{N} : m < x\}$ itself.

Aknowledgments. This work is partially supported by the research projects *"IoT&B, Internet of Things and Blockchain"*, CUP J48C17000230006, POR Calabria FESR-FSE 2014–2020. The third author was also supported by the research grant "Ing. Giorgio Schirillo" of the Istituto Nazionale di Alta Matematica "F. Severi", Rome.

References

1. Ackermann, W.: Zum hilbertschen aufbau der reellen zahlen. Math. Ann. **99**, 118–133 (1928). https://doi.org/10.1007/BF01459088
2. Bennett, A.A.: Note on an operation of the third grade. Ann. Math. Second Series **17**(2), 74–75 (1915). https://doi.org/10.2307/2007124
3. Blakley, G.R., Borosh, I.: Knuth's iterated powers. Adv. Math. **34**(2), 109–136 (1979). https://doi.org/10.1016/0001-8708(79)90052-5
4. Blakley, G.R., Borosh, I.: Modular arithmetic of iterated powers. Comp. Math. Appl. **9**(4), 567–581 (1983)
5. Bowers, J.: Exploding Array Function. Accessed 25 Apr 2019. http://www.polytope.net/hedrondude/array.htm
6. Bowers, J.: Extended operator notation. Accessed 21 Apr 2019. https://sites.google.com/site/largenumbers/home/4-1/extended_operators
7. Bromer, N.: Superexponentiation. Mathematics Magazine **60**(3), 169–174 (1987). JSTOR 2689566

8. Caldarola, F.: The exact measures of the Sierpiński d-dimensional tetrahedron in connection with a Diophantine nonlinear system. Commun. Nonlinear Sci. Numer. Simul. **63**, 228–238 (2018). https://doi.org/10.10.16/j.cnsns.2018.02.026

9. Caldarola, F., Cortese, D., d'Atri, G., Maiolo, M.: Paradoxes of the infinite and ontological dilemmas between ancient philosophy and modern mathematical solutions. In: Sergeyev, Y.D., Kvasov, D.E. (eds.) NUMTA 2019. LNCS 11973, pp. 358–372. Springer, Cham (2020). https://doi.org/10.1007/978-3-030-39081-5_31

10. Caldarola, F., d'Atri, G., Maiolo, M.: What are the "unimaginable numbers"? (submitted)

11. Donner, J., Tarski, A.: An extended arithmetic of ordinal numbers. Fundamenta Mathematicae **65**, 95–127 (1969)

12. Friedman, H.M.: Long finite sequences. J. Comb. Theor. Series A **95**(1), 102–144 (2001). https://doi.org/10.1006/jcta.2000.3154

13. Goodstein, R.L.: Transfinite ordinals in recursive number theory. J. Symbolic Logic **12**(4), 123–129 (1947). https://doi.org/10.2307/2266486

14. Hooshmand, M.H.: Ultra power and ultra exponential functions. Integr. Transforms Spec. Functions **17**(8), 549–558 (2006). https://doi.org/10.1080/1065246050042224

15. Knobel, R.A.: Exponentials reiterated. American Mathematical Monthly **88**(4), 235–252 (1981). https://doi.org/10.2307/2320546

16. Knuth, D.E.: Mathematics and computer science: coping with finiteness. Science **194**(4271), 1235–1242 (1976). https://doi.org/10.1126/science.194.4271.1235

17. Lameiras Campagnola, M., Moore, C., Costa, J.F.: Transfinite ordinals in recursive number theory. J. Complex. **18**(4), 977–1000 (2002). https://doi.org/10.1006/jcom.2002.0655

18. Leonardis, A., d'Atri, G., Caldarola, F.: Beyond Knuth's notation for "Unimaginable Numbers" within computational number theory. arXiv:1901.05372v2 [cs.LO] (2019)

19. Littlewood, J.E.: Large numbers. Math. Gaz. **32**(300), 163–171 (1948). https://doi.org/10.2307/3609933

20. MacDonnell, J.F.: Some critical points of the hyperpower function $x^{x^{\cdots}} x^{x^{\cdots}}$. Int. J. Math. Educ. **20**(2), 297–305 (1989). https://doi.org/10.1080/0020739890200210

21. Marshall, A.J., Tan, Y.: A rational number of the form $^a a$ with a irrational. Math. Gaz. **96**, 106–109 (2012)

22. Munafo, R.: Large Numbers at MROB. Accessed 19 May 2019

23. Nambiar, K.K.: Ackermann functions and transfinite ordinals. Appl. Math. Lett. **8**(6), 51–53 (1995). https://doi.org/10.1016/0893-9659(95)00084-4

Combinatorics on n-sets: Arithmetic Properties and Numerical Results

Fabio Caldarola(ORCID), Gianfranco d'Atri, and Marco Pellegrini$^{(\boxtimes)}$(ORCID)

Department of Mathematics and Computer Science, Università della Calabria,
Cubo 31/A, 87036 Arcavacata di Rende, CS, Italy
{caldarola,datri}@mat.unical.it, m19700530@hotmail.com

Abstract. The following claim was one of the favorite "initiation question" to mathematics of Paul Erdős: for every non-zero natural number n, each subset of $I(2n) = \{1, 2, \ldots, 2n\}$, having size $n + 1$, contains at least two distinct elements of which the smallest divides the largest. This can be proved using the pigeonhole principle. On the other side, it is easy to see that there are subsets of $I(2n)$ of size n without *divisor-multiple pairs*; we call them *n-sets*, and we study some of their combinatorial properties giving also some numerical results. In particular, we give a precise description of the elements that, for a fixed n, do not belong to every n-set, as well as the elements that do belong to all the n-sets. Furthermore, we give an algorithm to count the n-sets for a given n and, in this way, we can see the behavior of the sequence $a(n)$ of the number of n-sets. We will present some different versions of the algorithm, along with their performances, and we finally show our numerical results, that is, the first 200 values of the sequence $a(n)$ and of the sequence $q(n) := a(n + 1)/a(n)$.

Keywords: n-sets · Natural numbers · Divisibility relations · Combinatorics on finite sets · n-tuples

1 Introduction

During a dinner Paul Erdős posed a question to the young Lajos Pósa: is it true that, for every integer $n \geq 1$, each subset of $I(2n) = \{1, 2, \ldots, 2n\}$ having size $n + 1$ contains at least two distinct elements of which the smallest divides the largest? Before the dinner ended, he proved this fact using the pigeonhole principle and equivalence classes of different cardinalities (see [1]).

On the other side, it is easy to see that there are subsets of $I(2n)$ of size n without *divisor-multiple pairs*; we call them *n-sets* (see Definition 1) and we asked ourselves some questions about them: How many are they? What can we say about their elements? If we fix n, are there elements included in every n-set? Are there elements not included in any n-set? There is already some research about n-sets, see for instance [2–4].

This paper is structured as follows. In Sect. 2 we introduce the problem and we describe an equivalence relation that is the key to answer to many of these

© Springer Nature Switzerland AG 2020
Y. D. Sergeyev and D. E. Kvasov (Eds.): NUMTA 2019, LNCS 11973, pp. 389–401, 2020.
https://doi.org/10.1007/978-3-030-39081-5_34

questions, as presented in [5]. In Sect. 3 we find some characterizations of the elements that are not included in any n-set, as well as the elements that are included in any n-set; some results are already present in [6]. Finally, in Sect. 4, we describe an algorithm to count the n-sets for a given integer $n > 0$. In this way we can see the behavior of the sequence $a(n)$ of the number of n-sets and we will also show some different versions of the algorithm.

2   Notations and Preliminaries

2.1   Erdős Problem

Throughout the paper, lowercase letters indicate natural numbers, that is non-negative integers. Given two integers $y > x > 0$, we say that they form a *divisor-multiple pair* if x divides y, in symbols $x|y$. Moreover, the cardinality of a set A is denoted by $|A|$ and, for short, we indicate the set $\{1, 2, \ldots, n\}$ of the first n positive natural numbers by $I(n)$.

For the following proposition see [1, Chap. 25.1] or [5].

Proposition 1. *For any integer $n \geq 1$, every subset X of $I(2n)$ such that $|X| \geq n + 1$ contains at least a divisor-multiple pair.*

For instance, if $n = 3$ the set $\{3, 4, 5, 6\} \subset I(6)$ contains only one divisor-multiple pair, while the set $\{1, 2, 4, 6\} \subset I(6)$ contains exactly 5 divisor-multiple pairs. Instead, what can happen if $X \subset I(2n)$ and $|X| = n$? For instance, $\{2, 3, 5\} \subset I(6)$ and $\{6, 7, 8, 9, 10\} \subset I(10)$ do not contain any divisor-multiple pair, and these examples lead us to the following definition.

Definition 1. *If X is a subset of $I(2n)$ such that $|X| = n$ and X does not contain any divisor-multiple pair, we say that X is a n-set.*

Remark 1. We observe that for every positive integer n, n-sets do exist, that is, the value $n + 1$ in Proposition 1 is optimal: indeed, for every $n \geq 1$, the sets $Y_n = \{n + 1, n + 2, \ldots, 2n - 1, 2n\}$ and $Z_n = \{n, n + 1, \ldots, 2n - 2, 2n - 1\}$ are n-sets.

2.2   An Interesting Equivalence Relation

The proof of Proposition 1 uses some of the following facts, that will be also useful to study the structure of n-sets. First, we observe that any positive integer x can be written in a unique way as $x = 2^k d$, with $k \geq 0$ and d odd. Then, we define a function ε such that $\varepsilon(x) = d$.

Definition 2. *If $x, y \in I(2n)$, we say that $x \sim y$ if $\varepsilon(x) = \varepsilon(y)$; we denote $[x]_n$ the set $\{y \in I(2n) : y \sim x\}$, that is, the equivalence class of x.*

It is easy to see that $\sim$ is an equivalence relation and each equivalence class $[x]_n$ is a set of the form $\{d, 2d, \ldots, 2^t d\}$ with d odd, $2^t d \leq 2n$ and $2^{t+1}d > 2n$; thus, every equivalence class $[x]_n$ contains exactly one odd number. This means that here are n equivalence classes in $I(2n)$ because there are exactly n odd numbers in $I(2n)$. We moreover observe that if $x \sim y$ and $x < y$ then $x|y$, since $x = 2^k d$ and $y = 2^{k'} d$, with $k < k'$.

Example 1. For $n = 10$ we have $[1]_{10} = \{1, 2, 4, 8, 16\}, [3]_{10} = \{3, 6, 12\}$, $[5]_{10} = \{5, 10, 20\}, [7]_{10} = \{7, 14\}, [9]_{10} = \{9, 18\}, [11]_{10} = \{11\}, [13]_{10} = \{13\}$, $[15]_{10} = \{15\}, [17]_{10} = \{17\}, [19]_{10} = \{19\}$.

Definition 3. *Given $n \geq 1$ we denote the family of all the n-sets by $\mathcal{N}(n)$ and its cardinality, $|\mathcal{N}(n)|$, by $a(n)$, i.e., $a(n)$ represents the number of (unordered) n-tuples contained in $I(2n)$ without divisor-multiple pairs.*

For example, if $n = 1$ we have $\mathcal{N}(1) = \{\{1\}, \{2\}\}$ and $a(1) = 2$. Instead, if $n = 3$ we have $\mathcal{N}(3) = \{\{2, 3, 5\}, \{3, 4, 5\}, \{4, 5, 6\}\}$, hence $a(3) = 3$.

For future references we state the following

Remark 2. Fixed $n \geq 1$, every n-set contains exactly one element of any equivalence class $[x]_n$.

This gives a (rather unpractical) method to construct an n-set: we choose exactly one element from every equivalence class and we check if there are divisor-multiple pairs. If not, we have an n-set, otherwise we change our choices.

If we study what happens when increasing n, we see that, for any n and for any x, we have $[x]_n \subseteq [x]_{n+1}$, that is, equivalence classes are "nested". Moreover, we can easily observe that if d is odd with $n < d < 2n$, we have $[d]_n = \{d\}$: so $|[d]_n| = 1$ and d belongs to every n-set. This motivates the following

Definition 4. *Let $n \geq 1$ be fixed. The set $S(n) := \bigcap_{X \in \mathcal{N}(n)} X$ is called the n-kernel and the set $E(n) := I(2n) - \bigcup_{X \in \mathcal{N}(n)} X$ is called the set of the n-excluded elements.*

For instance, $S(1) = \varnothing$, $S(3) = \{5\}$, while $E(1) = \varnothing$ and $E(3) = \{1\}$ (see the example after Definition 3). Some simple results about $S(n)$ and $E(n)$ are hence the following (see [5])

Proposition 2. *(a) If d is odd with $n < d < 2n$, we have $d \in S(n)$.*
(b) If $1 \leq n < m$, then $E(n) \subseteq E(m)$.

3   Excluded Integers and Kernel

Proposition 2 (b) implies that if x does not belong to any n-set and $1 \leq n < m$, then x does not belong to any m-set. Then, what is the minimum n that "excludes" a given x? We start to consider odd numbers and, from now on, d will be always an odd natural number.

Definition 5. *Given d, we call d*-threshold value *the number $n_d := (3d+1)/2$.*

The reason of this definition comes from the following (see [5,6]).

Proposition 3. *For every $n \geq 1$, for every d, we have that $d \in E(n)$ if and only if $n \geq n_d$, i.e. n is equal or greater than the d-threshold value $n_d = (3d+1)/2$.*

The previous proposition represents an effective criterion to evaluate if an odd number d is n-excluded for a certain n or, *vice versa*, to calculate the minimum n that makes a certain odd number d an n-excluded. In order to see whether any integer is n-excluded or not, we need further considerations. From now on it could be useful to label odd numbers, so we will introduce the following notation: $d_i := 2i - 1, \forall i \geq 1$. Obviously, $d_i < d_j$ if and only if $i < j$.

Definition 6. *Fixed $n \geq 1$, we set $M_i = M_{i,n} := \max[d_i]_n$ and $m_i = m_{i,n} := \min([d_i]_n \setminus E(n))$ for all integers i with $1 \leq i \leq n$.*

Example 2. For $n = 5$ we have $[1]_5 = \{1, 2, 4, 8\}$, $[3]_5 = \{3, 6\}$, $[5]_5 = \{5, 10\}$, $[7]_5 = \{7\}$ and $[9]_5 = \{9\}$. Thus $M_1 = 8$, $M_2 = 6$, $M_3 = 10$ $M_4 = 7$, $M_5 = 9$ and $Y_5 = \{6, 7, 8, 9, 10\} = \{M_i : 1 \leq i \leq 5\}$.

Observe that the M_i are never n-excluded because they are exactly all the elements of the n-set Y_n described in Remark 1. On the other hand, enumerating the set $\{m_{i,n} : 1 \leq i \leq n\}$ completely is not as simple as enumerating the set $\{M_{i,n} : 1 \leq i \leq n\}$ because we previously need to determine $E(n)$. Nevertheless these numbers will play an important role in obtaining the following (see [6]).

Proposition 4. *Let $n \geq 1$ be fixed, $x \in I(2n)$ and $\varepsilon(x) = d_i$. Then, the following are equivalent:*

(a) $x \in E(n)$;
(b) x is a divisor of a certain m_j, the minimum non-excluded element of the class $[d_j]_n$, with $j > i$.

Corollary 1. *If $x = 2^k d \in E(n)$ and $h < k$, then $2^h d \in E(n)$.*

This means that, in every class $[d_i]_n$, the set of non-excluded elements is of the form $\{m_i = 2^k d_i, 2^{k+1} d_i, \ldots, 2^{k+t} d_i = M_i\}$, for some $k, t \geq 0$. In particular, if $k = 0$ all the elements of $[d_i]_n$ are not excluded, instead if $t = 0$ we have $m_i = M_i \in S(n)$. In any case, the maximum element is never excluded.

We now need a generalized definition of the *threshold value*, connected to any even or odd integer, as follows:

Definition 7. *Given $x = 2^k d$, $k \geq 0$, we call x*-threshold value *the number $n_x := (3^{k+1} d + 1)/2$.*

Note that if $k = 0$ then $x = d$ and the x-threshold value in Definition 7 coincides with the d-threshold value in Definition 5. Hence n_x is a true generalization of n_d, extended to even integers x.

Remark 3. It is interesting to note that different integers can generate the same threshold value. In fact, for any integer h with $0 \leq h \leq k$, we have that all the numbers of the form $y_h = 2^{k-h}3^h d$ generate the same *threshold value* $n_{y_h} = (3^{(k-h)+1}3^h d + 1)/2 = (3^{k+1}d + 1)/2$.

Finally, we have a criterion to decide whether an integer is n-excluded or not (see [6]).

Proposition 5. *For every* $x = 2^k d$, *we have that* $x \in E(n)$ *if and only if* $n \geq n_x$, *where* n_x *is the* x-*threshold value defined as* $n_x = (3^{k+1}d + 1)/2$.

Example 3. Now we take $n = 35$ and we want to determine the set $E(35) \subset I(70)$. Note that the condition $n \geq n_x$ is equivalent to $d \leq (2n - 1)/3^{k+1}$.

- Excluded odd numbers: $1, 3, 5, \ldots, 21, 23$ (because $23 \leq 69/3$ and $25 > 69/3$).
- Excluded numbers of the form $2d$ ($k = 1$): $(2 \cdot 1), (2 \cdot 3), (2 \cdot 5), (2 \cdot 7)$ (because $7 \leq 69/3^2$ and $9 > 69/3^2$).
- Excluded numbers of the form $4d$ ($k = 2$): $(4 \cdot 1)$ (because $1 \leq 69/3^3$ and $3 > 69/3^3$).
- Excluded numbers of the form $2^k d$ with $k \geq 3$: none (because $1 > 69/3^{k+1}$).

In conclusion we have $E(35) = \{1, 2, 3, 4, 5, 6, 7, 9, 10, 11, 13, 14, 15, 17, 19, 21, 23\}$.

The previous proposition, about excluded integers, can also be helpful to give a precise description of the n-kernel $S(n)$.

Proposition 6. *For every* $n \geq 1$ *we have*

$$S(n) = \{d : n + 1 \leq d \leq 2n - 1\} \cup \left\{2d : \frac{n+1}{2} \leq d \leq \frac{2n-1}{3}\right\}.$$

Proof. Observe that $x \in S(n)$ if and only if $|[x]_n \setminus E(n)| = 1$, then we distinguish between odd and even numbers.

For odd numbers, by Proposition 2 (a), we already know that $\{d : n+1 \leq d \leq 2n - 1\} \subset S(n)$; instead, if $d \leq n$ we know that $|[d]_n| \geq 2$, then the maximum of $[d]_n$ is even and $d \notin S(n)$.

For even numbers: if $x = 2^k d$ with $k \geq 1$, we have that $2^k d \in S(n)$ if and only if $2^{k-1}d \in E(n)$ and $2^{k+1}d \geq 2n + 2$. By Proposition 5, we know that $n \geq (3^k d + 1)/2$ and, in addition, $2^k d \geq n + 1$. Putting all together, we obtain

$$\frac{n+1}{2^k} \leq d \leq \frac{2n-1}{3^k} \qquad \Longrightarrow \qquad \left(\frac{3}{2}\right)^k \leq \frac{2n-1}{n+1} < 2.$$

This forces $k = 1$, because if $k \geq 2$ then $(3/2)^k > 2$; thus, $x = 2d$ and

$$\frac{n+1}{2} \leq d \leq \frac{2n-1}{3}.$$

Example 4. As in Example 3 we take $n = 35$ and we want to determine the set $S(35) \subset I(70)$.

- Odd numbers: $37, 39, 41, \ldots, 67, 69$.
- Even numbers: with $36/2 \leq d \leq 69/3$, we get $38, 42, 46$.

In conclusion we have:

$$S(35) = \{37, 38, 39, 41, 42, 43, 45, 46, 47, 49, 51, 53, 55, 57, 59, 61, 63, 65, 67, 69\}.$$

4   Counting n-sets

In this section we come to the problem of counting n-sets. By Remark 2 we must choose exactly one element from every equivalence class, in order to get an n-set. Now, if an element belongs to the n-kernel its choice is forced. We can ignore all the n-excluded numbers, then we can define the *restricted* equivalence classes $C_i := [d_i]_n \setminus E(n)$. These C_i are actually equivalence classes in $I(2n) \setminus E(n)$. Working "by hand" with small integers $n \geq 1$, we observe a new phenomenon: some classes contain exactly two non-excluded numbers, say y and $2y$, and none of them is a divisor or a multiple of every other non-excluded numbers.

Definition 8. *We say that two classes C_i and C_j are* related *if there are two non-excluded integers x, y such that $x \in C_i$, $y \in C_j$, and x, y form a divisor-multiple pair. If a class C_i is not related to any other class, we say that C_i is* unrelated.

Proposition 7. *If C is an unrelated class, then it contains exactly one or two non-excluded elements.*

Proof. Suppose, by *reductio ad absurdum*, that C is an unrelated class with at least 3 non-excluded elements. Such 3 elements are then of the form $2^k d$, $2^{k+1}d$, $2^{k+2}d$ and, by Remark 3, we know that $2^{k+1}d$ has the same threshold value of $2^k 3d$. Now this gives $2^k 3d < 2^k 4d = 2^{k+2}d \leq 2n$ and, consequently, $2^k 3d \in I(2n)$, but since $2^k 3d$ is a multiple of $2^k d$, then we obtain $2^k 3d \in [3d]_n = C'$, while $2^k d \in [d]_n = C$. In conclusion this means that C and C' are related, hence a contradiction.

For completeness, we notice that unrelated classes with exactly one non-excluded element actually exist (i.e., the classes that contain the elements of the kernel) as well as unrelated classes with exactly two non-excluded elements (see, for instance, Example 5).

Definition 9. *An unrelated class with exactly two non-excluded elements is called an* independent pair.

The reason for this name is the following: for an independent pair $\{y, 2y\}$, whether we choose y or we choose $2y$, the choices of elements of other classes are not affected. Then, if we count the total number of choices of elements from other classes and we multiply it by 2 we get $a(n)$. Obviously, this can be done for every independent pair.

Proposition 8. *Fixed $n \geq 1$, if we have r independent pairs and x choices of the elements that belong to the other classes, then*

$$a(n) = 2^r x.$$

The proof is trivial from the previous discussion. We now describe a first version (version 1.0) of our algorithm to count n-sets:

(1) Divide $I(2n)$ in n equivalence classes, according to $\sim$.
(2) If the minimum element of a class is a divisor of the minimum element of a following class, remove it (by Proposition 4 it is in fact excluded) and repeat. Alternatively, remove excluded elements using Proposition 5.
(3) Remove the classes with only one element (these elements form the kernel, then their choice is forced).
(4) Find and count the independent pairs. If r is their number, remove then the corresponding classes.
(5) Using all remaining elements, count the choices for them, one element for each class so that no element is a divisor or a multiple of anyone else (for this purpose we can use a simple backtracking algorithm). Let x the number of such choices.
(6) Finally, write out $a(n) = 2^r x$.

In order to execute step (5), we consider a graph $\mathcal{G}(n)$ whose vertices are the numbers of $I(2n)$ that are both non-excluded and in no unrelated class. In $\mathcal{G}(n)$ there is an edge between the vertices y and z if and only if y, z belong to different classes and form a divisor-multiple pair. Sometimes we can have two classes C_i and C_j with $i < j$, an edge between m_i and M_j, and no edges going "outside" these classes. In this case C_i is related only to C_j and *vice versa*. This means that the choices involving the elements of these two classes do not affect the choices of the elements in other classes. Therefore it is easy to count these choices: since there is only one forbidden combination, namely (m_i, M_j), they are $|C_i| \cdot |C_j| - 1$ (we recall that we already removed excluded elements).

Definition 10. *If C_i, C_j are two classes as above, we say that they are strictly related.*

By our software computations, in this case we see that $|C_j|$ is always 2, while $|C_i|$ can be 2 or 3; then, $|C_i| \cdot |C_j| - 1$ can only take the value 3 or 5. Hence we have a number (possibly zero) of factors 3 and 5. After removing these classes, experimentally we see that all the other classes seem to be "connected", and this means that there is not a simple way to calculate the number of choices... In any case, we have that $a(n) = 2^r 3^{e_3} 5^{e_5} x$, where r is the number of independent pairs, e_p is the number of factors p that come from the strictly related classes and x is the number of choices in the remaining part of the graph $\mathcal{G}(n)$. We remark that the number x can still have some (or all) of the prime factors $2, 3, 5$.

Thus, the previous exposed algorithm can be improved by adding the following step (4.2) after the old step (4) called here (4.1), and modifying the last step (6) as shown below:

(4.1) Find and count the independent pairs.If r is their number, remove then the correspondent classes.

(4.2) Find if there are strictly related classes C_i, C_j. If $|C_i| \cdot |C_j| - 1 = 3$, add 1 to e_3, instead, if $|C_i| \cdot |C_j| - 1 = 5$, add 1 to e_5. Remove then C_i and C_j.

(6) Write out $a(n) = 2^r 3^{e_3} 5^{e_5} x$.

We refer to such a modified algorithm as the version 1.1 of the algorithm to compute $a(n)$. We now give a complete and elucidating example of computing for $n = 44$.

Example 5. Take $n = 44$.

– Excluded elements (see Proposition 5):

- odd numbers: $1, 3, 5, \ldots, 29$;
- numbers of the form $2d$: $2, 6, 10, 14, 18$;
- numbers of the form $4d$: $4, 12$;
- numbers of the form $8d$: 8.

– Kernel: $\{45, 47, 49, \ldots, 87\} \cup \{46, 50, 54, 58\}$ (see Proposition 6).

– Independent pairs: $\{34, 68\}$, $\{38, 76\}$, $\{31, 62\}$, $\{35, 70\}$, $\{37, 74\}$, $\{41, 82\}$, $\{43, 86\}$. Hence $r = 7$.

– Strictly related classes: $\{28, 56\}$ and $\{42, 84\}$, $\{26, 52\}$ and $\{39, 78\}$, $\{22, 44, 88\}$ and $\{33, 66\}$. Thus $e_3 = 2$ and $e_5 = 1$.

– Remaining numbers: $16, 20, 24, 30, 32, 36, 40, 48, 60, 64, 72, 80$. See also the graph $\mathcal{G}(44)$ in Fig. 1.

– By using a backtracking algorithm, as in step (5), we find that for the graph $\mathcal{G}(44)$ in Fig. 1 there are 33 choices, from $(16, 24, 20, 36, 30)$ up to $(64, 48, 80, 72, 60)$, then we have $x = 33$.

In conclusion, counting the 44-sets we obtain $a(44) = 2^7 \cdot 3^2 \cdot 5 \cdot 33 = 190080$.

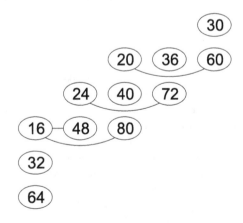

Fig. 1. Remaining numbers and their divisibility relations: the graph $\mathcal{G}(44)$.

The version 1.0 of our algorithm was able to calculate $a(156)$ in $12''$ c.a. by using a common desk pc, while version 1.1 can calculate $a(156)$ in about $0.4''$, $a(200)$ in $1'72''$ c.a, $a(229)$ in $1^h 13' 20''$ c.a. To go further, we developed a version 2.0 of the algorithm, in order to get a better performance of step (5), and to do it we use a recursion process. More precisely, we replace step (5) in version 1.1 by the following three new steps:

(5.1) Consider the first column of the graph.
(5.2) For each element of the first column, remove all of its multiples and count the choices for the subgraph starting from the second column (the subgraph can have unrelated classes, of any size, and strictly related classes; we consider these facts in order to improve the performance).
(5.3) Sum the choices for every subgraph.

Further improvements were obtained considering that, for $n \geq 122$, the main graph is not anymore connected, then we can count separately the choices for any connected component and multiply them. This is also true for subgraphs considered in Step (5.2), even for lower values of n. Moreover, we can store data about any connected component, because a graph can contain more components with the same structure, especially small components, so they are studied only once. In fact, once we considered the graph, one can ignore the numbers written inside any vertex and focus only on its structure. The version 2.0 can calculate $a(2000)$ in less than $1''$, $a(3000)$ in about $6''$ and $a(4000)$ in about $59''$.

Now we show a table with the values of $a(n)$ for n between 141 and 200. We notice that a table for $n \leq 140$ can be found in [5].

Table 1. The values of r, e_3, e_5, x and $a(n)$, for $141 \leq n \leq 200$.

n	r	e_3	e_5	x	$a(n)$
141	23	6	1	2874768	87900275217530880
142	24	6	1	2874768	175800550435061760
143	24	6	1	2874768	175800550435061760
144	24	6	1	3535116	216182780193669120
145	25	6	1	3535116	432365560387338240
146	25	6	1	3535116	432365560387338240
147	24	7	1	3535116	648548340581007360
148	24	6	2	3535116	1080913900968345600
149	26	7	2	207948	762998047742361600
150	26	7	2	289731	1063074361717555200
151	27	7	2	289731	2126148723435110400
152	26	6	3	289731	1771790602862592000
153	26	6	2	2028117	2480506844007628800
154	27	6	2	2028117	4961013688015257600
155	27	6	2	2028117	4961013688015257600
156	27	6	2	2652153	6487479438173798400

(*continued*)

Table 1. (*continued*)

n	r	e_3	e_5	x	$a(n)$
157	28	6	2	2652153	12974958876347596800
158	29	6	2	1041250	10188081856512000000
159	28	7	2	1041250	15282122784768000000
160	28	7	2	1570324	23047187687748403200
161	28	7	2	1570324	23047187687748403200
162	28	7	2	2097494	30784308137636659200
163	29	7	2	2097494	61568616275273318400
164	28	6	3	2097494	51307180229394432000
165	27	7	3	2097494	76960770344091648000
166	28	7	3	2097494	153921540688183296000
167	30	7	2	2097494	123137232550546636800
168	30	7	2	2696778	158319298993559961600
169	31	7	2	2696778	316638597987119923200
170	30	7	2	5008302	294021555273754214400
171	30	7	1	35058114	411630177383255900160
172	30	6	2	35058114	686050295638759833600
173	30	6	2	35058114	686050295638759833600
174	29	7	2	35058114	1029075443458139750400
175	30	7	2	35058114	2058150886916279500800
176	31	7	2	10296468	1208946080003339059200
177	30	8	2	10296468	1813419120005008588800
178	31	8	2	10296468	3626838240010017177600
179	31	8	2	10296468	3626838240010017177600
180	31	8	2	12106458	4264390937307355545600
181	32	8	2	12106458	8528781874614711091200
182	32	8	2	12106458	8528781874614711091200
183	31	9	2	12106458	12793172811922066636800
184	31	8	3	12106458	21321954686536777728000
185	33	8	2	12106458	17057563749229422182400
186	32	9	2	12106458	25586345623844133273600
187	33	9	2	12106458	51172691247688266547200
188	32	8	3	12106458	42643909373073555456000
189	32	8	3	16351517	57596747872934559744000
190	32	8	3	30367103	106965388906878468096000
191	32	8	3	30367103	106965388906878468096000
192	32	8	3	38775698	136583908471798235136000
193	33	8	3	38775698	273167816943596470272000
194	35	8	2	38775698	218534253554877176217600
195	34	9	2	38775698	327801380332315764326400
196	34	8	2	190430552	536619592706793091891200
197	34	8	2	190430552	536619592706793091891200
198	34	8	2	266586008	751220187789456624844800
199	35	8	2	266586008	1502440375578913249689600
200	34	8	2	405038413	1141369102446680132812800

We see clearly from Table 1 that the sequence $a(n)$ is not monotonic, in fact, sometimes we have $a(n) < a(n+1)$, sometimes $a(n) = a(n+1)$ and other times $a(n) > a(n+1)$. Moreover, the ratio $a(n+1)/a(n)$ seems to be bounded. After a first look at the table, one can state the following conjecture.

Conjecture 1. Let $q(n) := a(n+1)/a(n)$, then:

(a) $n \equiv 1 \pmod 3$ yields $1/2 < q(n) \le 1$;
(b) $n \equiv 2 \pmod 3$ yields $1 < q(n) \le 3/2$;
(c) $n \equiv 0 \pmod 3$ yields $7/5 < q(n) \le 2$.

Now we show a table with some values of $q(n)$ (Table 2).

Table 2. Some rounded values of $q(n)$, with $1 \le n \le 200$, divided according to the congruence class of n mod. 3

n	$q(n)$	n	$q(n)$	n	$q(n)$
1	1	2	1.5	3	1.6667
4	0.8	5	1.5	6	2
7	0.8333	8	1.4	9	1.8571
10	1	11	1.3077	12	2
13	0.7059	14	1.5	15	1.6667
16	1	17	1.4	18	2
19	0.7857	20	1.5	21	2
22	0.7879	23	1.3077	24	2
25	1	26	1.3529	27	1.6377
28	1	29	1.3982	30	2
31	0.6519	32	1.5	33	2
34	1	35	1.233	36	2
37	1	38	1.5	39	1.5354
40	0.6769	41	1.5	42	2
43	0.8333	44	1.3939	45	2
46	1	47	1.3043	48	2
49	0.74	50	1.5	51	1.6667
52	1	53	1.3514	54	2
55	0.8178	56	1.5	57	2
58	0.8	59	1.2446	60	2
61	1	62	1.3985	63	1.6284
64	1	65	1.5	66	2
67	0.5769	68	1.5	69	1.856
70	1	71	1.2328	72	2
73	1	74	1.5	75	1.6667
76	0.8	77	1.5	78	2
79	0.7676	80	1.3383	81	2
82	1	83	1.2876	84	2
85	0.8	86	1.5	87	1.6645

(continued)

Table 2. (*continued*)

n	$q(n)$	n	$q(n)$	n	$q(n)$
88	1	89	1.3038	90	2
91	0.8333	92	1.5	93	2
94	0.7024	95	1.2822	96	2
97	1	98	1.3999	99	1.5568
100	1	101	1.5	102	2
103	0.6665	104	1.5	105	2
106	1	107	1.1986	108	2
109	0.9286	110	1.5	111	1.6237
112	0.7866	113	1.5	114	2
115	0.8333	116	1.4	117	2
118	1	119	1.2228	120	2
121	0.6544	122	1.5	123	1.6667
124	1	125	1.3985	126	2
127	0.8142	128	1.5	129	1.8571
130	0.8	131	1.3077	132	2
133	1	134	1.3461	135	1.6667
136	1	137	1.5	138	2
139	0.6177	140	1.5	141	2
142	1	143	1.2297	144	2
145	1	146	1.5	147	1.6667
148	0.7059	149	1.3933	150	2
151	0.8333	152	1.4	153	2
154	1	155	1.3077	156	2
157	0.7852	158	1.5	159	1.5081
160	1	161	1.3357	162	2
163	0.8333	164	1.5	165	2
166	0.8	167	1.2857	168	2
169	0.9286	170	1.4	171	1.6667
172	1	173	1.5	174	2
175	0.5874	176	1.5	177	2
178	1	179	1.1758	180	2
181	1	182	1.5	183	1.6667
184	0.8	185	1.5	186	2
187	0.8333	188	1.3506	189	1.8571
190	1	191	1.2769	192	2
193	0.8	194	1.5	195	1.637
196	1	197	1.3999	198	2
199	0.7597	200	1.5		

We can see that the upper inequalities in Conjecture 1 (a)–(c) become equalities very often, while we can not say anything about lower bounds in each congruence class mod. 3. In fact, by way of example, a previous version of this conjecture stated that "if $n \equiv 0 \,(\mathrm{mod}\ 3)$ then $3/2 < q(n) \leq 2$", but this is false because, for instance, $q(639) \approx 1.4945$ and $q(1119) \approx 1.4669$.

Acknowledgments. This work is partially supported by the research projects *"IoT&B, Internet of Things and Blockchain"*, CUP J48C17000230006, POR Calabria FESR-FSE 2014-2020.

References

1. Aigner, M., Ziegler, G.M.: Proofs from the BOOK, 4th edn. Springer, Heidelberg (2010). https://doi.org/10.1007/978-3-642-00856-6
2. The On-Line Encyclopedia of Integer Sequences. http://oeis.org/A174094. Accessed 3 June 2019
3. Liu, H., Pach, P.P., Palincza, R.: The number of maximum primitive sets of integers (2018). arXiv:1805.06341
4. Vijay, S.: On large primitive subsets of 1,2,...,2n (2018). arXiv:1804.01740
5. Bindi, C., et al.: Su un risultato di uno studente di Erdős. Periodico di Matematiche, Organo della Mathesis (Società italiana di scienze matematiche e fisiche fondata nel 1895), Serie 12, Anno CXXVI, N. 1 Gen-Apr 2016, vol. 8, pp. 79–88 (2016)
6. Bindi, C., Pellegrini, M., Pirillo, G.: On a result of a student of Erdős (submitted for publication)

Numerical Problems in XBRL Reports and the Use of Blockchain as Trust Enabler

Gianfranco d'Atri[1,4], Van Thanh Le[2], Dino Garrì[3], and Stella d'Atri[4(✉)]

[1] Department of Mathematics and Computer Science, University of Calabria,
Rende, Italy
datri@mat.unical.it
[2] Faculty of Computer Science, Free University of Bolzano, Bolzano, Italy
vanle@unibz.it
[3] BlockchainLab, Milan, Italy
garridino@gmail.com
[4] Blockchain Governance, Cosenza, Italy
s.datri@dcgovernance.it

Abstract. Financial statements are formal records of the financial activities that companies use to provide an accurate picture of their financial history. Their main purpose is to offer all the necessary data for an accurate assessment of the economic situation of a company and its ability to attract stakeholders. Our goal is to investigate how Benford's law can be used to detect fraud in a financial report with the support of trustworthiness by blockchain.

Keywords: Financial models · Blockchain · Laws of Bradford and Zipf

1 Introduction

The eXtensible Business Reporting Language (XBRL) [9] is the world-leading standard for business reporting, which opens a new era for report digitalization that enables machines to read reports as humans. By scanning reports, we can investigate financial aspects as well as doing forecasts for the organization, in the paper, we will examine a well-known and proven Benford law, apply into an XBRL report to review its consistency, also give recommendations for analyzers.

Traditionally, a company or an organization release their financial reports based on their own standard and only for the internal usages, that results in statistic issues for authorities, especially when governments would like to investigate their gross national metrics as GDP or GDI. Widely, exchanging information in different reporting languages ruins business professionals over the world. Therefore a new standard of XBRL is needed and could be spread for any kind of report in any language. We will explore XBRL concept in details in next subsections.

The work is the extension of the two conference papers [3,4] with the apply of DLV. The DLV system is already used in a number of real-world applications

© Springer Nature Switzerland AG 2020
Y. D. Sergeyev and D. E. Kvasov (Eds.): NUMTA 2019, LNCS 11973, pp. 402–409, 2020.
https://doi.org/10.1007/978-3-030-39081-5_35

including agent-based systems, information extraction, and text classification. DLV is a rather solid candidate to constitute the basis of a system supporting automatic validation of XBRL data in concrete application domains. An introduction to DLV can be found in [5,8] and an in-depth description of OntoDLP is contained in [14,15].

1.1 XBRL

Financial reports contain sensitive data that might have a huge impact on the future of an organization in terms of investments and collaborations. This mandates careful management and control mechanisms able to capture any inconsistencies or manipulation of the published reports. The first step towards this goal started with the introduction of the eXtensible Business Reporting Language [9], which is the world's leading standard for financial reporting. It facilitates inter-organization communication and enables automatic reports processing and analysis. XBRL was started from an XML standard from Charles Hoffman in 1998, the version was developed and then the first international meeting for XBRL was held in London, 2001 until now the recent release version is XBRL 2.0.

XBRL relies on XML and XML based schema to define all its constructs. Its structure consists of two main parts:

1. XBRL instance, containing primarily the business facts being reported (see Fig. 1).

```
<rp:RevenueTotal unitRef="EUR">5000</rp:RevenueTotal>
<rp:CostOfSales unitRef="EUR">3000</rp:CostOfSales>
<rp:GrossProfit unitRef="EUR">2000</rp:GrossProfit>
```

Fig. 1. Facts

2. XBRL taxonomy, a collection of arcs which define metadata about these facts and their relationship with other facts (see Fig. 2).

```
<loc xlink:type="locator"
  xlink:href="taxonomy#rp_RevenueTotal" />
<loc xlink:type="locator"
  xlink:href="taxonomy#rp_CostOfSales" />
<loc xlink:type="locator"
  xlink:href="taxonomy#rp_GrossProfit" />
<calculationArc xlink:type="arc"
  xlink:from="rp_GrossProfit" xlink:to="rp_RevenueTotal"
  weight="1" />
<calculationArc xlink:type="arc"
  xlink:from="rp_GrossProfit" xlink:to="rp_CostOfSales"
  weight="-1" />
```

Fig. 2. XBRL Linkbase example

1.2   Blockchain

Blockchain is a distributed, decentralized ledger that stores transactions and it addresses the double-spending problem in a trust-less peer-to-peer network, without the need for a trusted third party or an administrator. Blockchain is maintained by a network of computers called nodes. Whenever a new transaction arrives, the nodes verify its validity and broadcast it to the rest of the network. The main building elements of a Blockchain are [11]:

- Transactions, which are signed pieces of information created by the participating nodes in the network then broadcast to the rest of the network.
- Blocks, that are collections of transactions that are appended to the blockchain after being validated.
- A blockchain is a ledger of all the created blocks that make up the network.
- The blockchain relies on Public keys to connect the different blocks together (similar to a linked list).
- A consensus mechanism is used to decide which blocks are added to the blockchain.

Blockchain contains mathematical aspects that make it become trust and stable over the network:

- Cryptographic Hash Functions: is a function H if and only if it is infeasible to find two value x, y that $x \neq y$ and $H(x) = H(y)$
- Hiding: the function H is hiding if $y = H(r||x)$ with r is chosen from a probability distribution, it is infeasible to find x with known value of y.
- Puzzle friendliness: with known value of y in $y = H(r||x)$, the H is called puzzle friendly if we have n bit of y, we can not find x in 2^n times.

In order to build a blockchain-based solution for financial report, we need a blockchain model supporting smart contract, there are several possible models, such as Hyperledger Fabric [1], Bitcoin [10] or NEO [12], but we choose Ethereum [6] as the best candidate, since it is an open-source blockchain platform that enables developers to build and deploy decentralized applications, interaction with web-based application is also much more easier than other models. The platform runs smart contracts, a computer protocol running on top of a blockchain, performing as a contract. Smart contracts can include data structures and function calls that are executed in a centralized fashion. This guarantees the fact that the contract execution will persist on the chain.

1.3   Benford's Law

Benford's law was discovered by Simon Newcomb in 1881 and was published in the American Journal of Mathematics [2], he observed that in logarithm tables the earlier pages of library copies were much more worn than the other pages. It is similar with the situation for scientists preferred using the table to look up numbers which started with number one more than others.

It has been shown that this result applies to a wide variety of data sets and applications [13], including electricity bills, street addresses, stock prices... It tends to be most accurate when values are distributed across multiple orders of magnitude.

Using Benford's law to detect fraud was investigated in [2]. Following the law, a set of numbers is said to satisfy Benford's law if the leading first digit d $(d \in 1, ..., 9)$ occurs with probability:

$$P(d) = log_{10}(d+1) - log_{10}(d) = log_{10}(\frac{d+1}{d}) = log_{10}(1 + \frac{1}{d})$$

That presents for the distribution (Fig. 3):

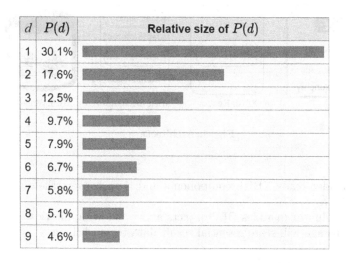

d	P(d)	Relative size of P(d)
1	30.1%	
2	17.6%	
3	12.5%	
4	9.7%	
5	7.9%	
6	6.7%	
7	5.8%	
8	5.1%	
9	4.6%	

Fig. 3. Benford's law for the first digit

Besides this, the distribution for the second and following digits also are demonstrated but the most well-known one is the first. The law has a brief explanation in [7] where R. Fewster showed that: any number X could be written by this way: $X = r * 10^n$. Where r is a real number with $1 \leq r < 10$ and n is an integer. The leading digit of X is the same as r, the value will be 1 if and only if $1 \leq r < 2$. We can isolate r by taking logs to base 10:

$log_{10}X = log_{10}(r * 10^n) = log_{10}(r) + n$

The leading digit of X is 1 when $1 \leq r < 2$:

$0 \leq log_{10}(r) < log_{10}2 = 0.301$ Thus $n \leq log_{10}(X) < n + 0.301$

Benford laws are widely discussed in the academic area with its application.

Our contribution is threefold: (i) providing a methodology to automatically evaluate and validate the consistency of the generated reports, (ii) to use Blockchain to store information as an immutable and uninterruptible worldwide database, (iii) to apply numerical tools to detect possible errors or even frauds in the reports.

For (i) we use ASP (answer set programming) to automatize reasoning on the XBRL data. For (ii) we analyze the implementation of smart contracts on the Ethereum Blockchain. For (iii) we explore the so-called Bradford's and Zipf's laws as tools for checking numerical values in the specific domain of company balance sheets.

2 Architecture Based Blockchain

As discussed in our previous work [4], the architecture composes three main parts: XBRL Reader, XBRL Evaluator and XBRL Storage as Fig. 4.

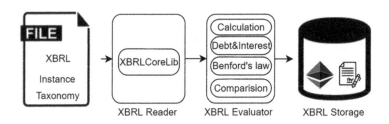

Fig. 4. Architecture

- XBRLReader reads XBRL components and then convert them to DLVFact and DLVArc
- XBRL Evaluator qualifies DLV objects above following metrics
- XBRL Storage will store essential result data to blockchain.

3 Numerical Methodology

In our scope, we use Benford's law as a metric to make a recommendation for the system management in case their dataset does not fit the law. The law will be applied in different size of dataset to have a comprehensive view of analyzing. Our queries are presented in Fig. 5.

```
 1  firstDigit(V,CY, CP,CT1,X,U,GR) :- fact(CY, CP,CT1,X,U,GR), &firstDigitInCY("Global",X;V).
 2  benfordLaw("1","30.1").
 3  benfordLaw("2","17.6").
 4  benfordLaw("3","12.5").
 5  benfordLaw("4","9.7").
 6  benfordLaw("5","7.9").
 7  benfordLaw("6","6.7").
 8  benfordLaw("7","5.8").
 9  benfordLaw("8","5.1").
10  benfordLaw("9","4.6").
11
12  countNeg("Global",C,PC, DBL) :- firstDigit(_,_, _,_,_,_,_), &percentCountInCY("Global","-","0";C,PC,DBL).
13  countZero("Global",C,PC, DBL) :- firstDigit(_,_, _,_,_,_,_), &percentCountInCY("Global","0","0";C,PC,DBL).
14
15  percentCount("Global",V, C,PC, DBL) :- benfordLaw(V,BL), firstDigit(_,_, _,_,_,_,_), &percentCountInCY("Global",V,BL;C,PC,DBL).
16  validBenfordLaw("Global",TH) :- percentCount("Global",V,C,PC,DBL) , &isValidBenfordLawInCY("Global","1.0";"True",TH).
17  nonValidBenfordLaw("Global",TH) :- percentCount("Global",V,C,PC,DBL) , &isValidBenfordLawInCY("Global","2.0";"False",TH).
```

Fig. 5. iDLV queries for Benford's law

3.1   Experiment Environment

These queries are executed in our dataset with 122 Financial reports of Italian companies from 2000 to 2015 and provided by a top provider of financial information in Italy.

With evaluation aspects related to the law:

- All first digits in all reports
- All first digits from conto_economico group
- All first digits from stato_patrimoniale group
- All first digits for individual reports

For the group of *all report* and *individual reports*, it facilitate us to have a view from general to detail, the other two groups presented for two major and indispensable parts in an Italian financial report.

3.2   Numerical Results

We set *scarto* value as the mean of the differences between Benford's law and our actual results (Table 1 for the group of all financial reports)

$$Scarto = \sqrt{\frac{\sum_{i=1}^{9}(P_i - B_i)^2}{9}}$$

The scarto we got is 0.33 for all documents in our dataset, this value is acceptable for an accountancy company. In reality, the auditors could set a baseline for them to follow the value of *scarto*.

Table 1. Costs of smart contract functions execution

Number	Count	Percentage (P_i)	Benford's law(B_j)	Differences
9	750	4.72	4.6	0.12
8	806	5.08	5.1	−0.02
7	873	5.5	5.8	−0.3
6	985	6.21	6.7	−0.49
5	1303	8.21	7.9	0.31
4	1469	9.25	9.7	−0.45
3	2001	12.61	12.5	0.11
2	2820	17.76	17.6	0.16
1	4867	30.66	30.1	0.56
Negative	1731			
Zero	4945			
			Scarto	0.33

Considering other evaluation aspects: conto_economico group and stato_patrimoniale group (Fig. 6). The result is affected by the dataset, the smaller dataset the higher scarto value we get.

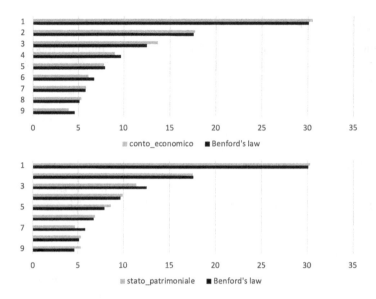

Fig. 6. conto_economico and stato_patrimoniale

- conto_economico: Number of negative is 505, Number of Zero is 483, Scarto is 0.96
- stato_patrimoniale: Number of negative is 135, Number of Zero is 195, Scarto is 1.33

The scarto value becomes even worse when we scale down the dataset for each report. With each report, the value of *scarto* is commonly in the range of 5.0 to 9.0 with around 55 reports having the nearest values.

4   Conclusion

In the paper, we demonstrated an application of Benford's law for financial reports in XBRL format. The tests showed that the i-DLV tool is adaptable for big data with a combination of complex calculations. For Benford's law, it is more sufficient with a big database and could become a useful technique for accountants to do investigation for the whole dataset of a specific company. The result from the law enables them to have a big picture of the data status and could become a baseline for further analysis. However, with a small dataset, we need other tools or other standards to qualify. Blockchain performs a trust enabler role in the system that prevents insecurity behaviors and maintains system consistency.

Acknowledgements. This research is partially supported by POR Calabria FESR-FSE 2014-2020, research project IoT&B, CUP J48C17000230006.

References

1. Cachin, C.: Architecture of the hyperledger blockchain fabric. In: Workshop on Distributed Cryptocurrencies and Consensus Ledgers (2016)
2. Durtschi, C., Hillison, W., Pacini, C.: The effective use of benford's law to assist in detecting fraud in accounting data. J. Forensic Account. **5**(1), 17–34 (2004)
3. D'Atri, G.: Logic-based consistency checking of XRBL instances. IJACT **3–6**, 126–131 (2014)
4. D'Atri, G., Le, V.T., Ioini, N.E., Pahl, C.: Towards trustworthy financial reports using blockchain. Cloud Comput. **2019**, 47 (2019)
5. Eiter, T., Faber, W., Leone, N., Pfeifer, G.: Declarative problem-solving using the DLV system. In: Minker, J. (ed.) Logic-Based Artificial Intelligence. The Springer International Series in Engineering and Computer Science, vol. 597, pp. 79–103. Springer, Boston (2000). https://doi.org/10.1007/978-1-4615-1567-8_4
6. Ethereum Foundation: Ethereum (2013). www.ethereum.org. Accessed 10 Apr 2019
7. Fewster, R.: A simple explanation of Benford's law. Am. Stat. **63**, 26–32 (2009)
8. Gelfond, M., Lifschitz, V.: Classical negation in logic programs and disjunctive databases. New Gener. Comput. **9**(3), 365–385 (1991). https://doi.org/10.1007/BF03037169
9. Malhotra, R., Garritt, F.: Extensible business reporting language: the future of e-commerce-driven accounting. Int. J. Bus. **9**(1) (2004)
10. Nakamoto, S., et al.: Bitcoin: a peer-to-peer electronic cash system (2008)
11. Narayanan, A., Bonneau, J., Felten, E., Miller, A., Goldfeder, S.: Bitcoin and Cryptocurrency Technologies: A Comprehensive Introduction. Princeton University Press, Princeton (2016)
12. NEO: NEO Smart Economy (2013). https://docs.neo.org/docs/en-us/index.html. Accessed 06 Oct 2019
13. Nigrini, M.J.: Benford's Law: Applications for Forensic Accounting, Auditing, and Fraud Detection, vol. 586. Wiley, Hoboken (2012)
14. Ricca, F., Gallucci, L., Schindlauer, R., Dell'Armi, T., Grasso, G., Leone, N.: Ontodlv: an ASP-based system for enterprise ontologies. J. Logic Comput. **19**, 643–670 (2009)
15. Ricca, F., Leone, N.: Disjunctive logic programming with types and objects: the DLV+ system. J. Appl. Logic **5**, 545–573 (2007)

Modelling on Human Intelligence a Machine Learning System

Michela De Pietro$^{(\boxtimes)}$ ⓘ, Francesca Bertacchini ⓘ, Pietro Pantano ⓘ,
and Eleonora Bilotta ⓘ

University of Calabria, via Pietro Bucci, 17b, Arcavacata di Rende, Italy
{michela.depietro,francesca.bertacchini,pietro.pantano,
eleonora.bilotta}@unical.it
http://www.unical.it

Abstract. Recently, a huge set of systems, devoted to emotions recognition has been built, especially due to its application in many work domains, with the aims to understand human behaviour and to embody this knowledge into human-computer interaction or human-robot interaction. The recognition of human expressions is a very complex problem for artificial systems, caused by the extreme elusiveness of the phenomenon that, starting from six basic emotions, creates a series of intermediate variations, difficult to recognize by an artificial system. To overcome these difficulties, and expand artificial knowledge, a Machine Learning (ML) system has been designed with the specific aim to develop a recognition system modelled on human cognitive functions. Cohn-Kanade database images was used as data set. After training the ML, it was tested on a representative sample of unstructured data. The aim is to make computational algorithms more and more efficient in recognizing emotional expressions in the faces of human subjects.

Keywords: Machine learning · Artificial intelligence · Emotion recognition

1 Introduction

The face is the first instrument of non-verbal communication and is the first interaction between human beings. The face changes based on what a person feels at a given moment. Starting from the slightest change in the facial muscles and continuing to change until emotion is expressed. This change provides information about the emotional state of a person [1].

When we talk about the analysis of facial expressions, we refer to the recognition of face, the different facial movements and the changes in the face. Emotion is often expressed through subtle changes in facial features, such as in stiffening the lips when a person is angry or in lowering the corners of the lips when a person is sad [2] or again in the different change of eyebrows or of eyelids [3].

Facial expressions are an important instrument in non-verbal communication. The role of the facial expressions' classification could be a helpful used in

© Springer Nature Switzerland AG 2020
Y. D. Sergeyev and D. E. Kvasov (Eds.): NUMTA 2019, LNCS 11973, pp. 410–424, 2020.
https://doi.org/10.1007/978-3-030-39081-5_36

behavioural analysis [1]. For effective human-computer interaction, automated analysis of facial expression is important.

Ekman & Friesen devised the coding of actions of the human face system, called FACS [4]. FACS allows different facial movements to be coded in Action Units (AU) based on the muscular activity of the face that generates temporary changes in facial expression. They identified 44 FACS AUs, of which 30 AUs are related to the contractions of specific facial muscles: 12 for the upper face and 18 for the lower face. AUs can happen singly or in combination [5]. The face movement takes place in a space and takes into consideration landmarks points of a face. The dynamics of change in this space can reveal emotions, pain and cognitive states and regulate social interaction. The movement of landmark points of a face, such as the corners of the mouth and eyes, constitutes a 'reference point space' [6].

A lot of researchers [7,8] have used different neural networks for facial expression analysis. The performance of a neural network depends on several factors including the initial random weights, the training data, the activation function used, and the structure of the network including the number of hidden layer neurons, etc.

The main purpose of this study is to create a system of recognition of facial emotions in some images. The work is carried out using a computational algorithm capable of performing an automatic recognition job pre-training the system using a set of data and a set of tests. In particular, the system has been trained in trying to recognise emotions already from the first change in the facial muscles in the subjects, that is at the onset of expression. In this way the computational algorithms become more and more efficient in recognizing emotional expressions in the faces of human subjects and they are being trained to the complexity of the dynamics of emotional expressions.

2   Background

Since Darwin's revolutionary book [9] on the expression of the emotions, neurophysiological and cognitive theories advanced the processes of recognition of emotions in human subjects. The peripheral theory of James [10] and the central theory of Cannon [11], with different visions, belong to the first strand. According to James, an emotional response occurs when there is an emotional stimulus that can be internal or external to our body. Both the emotional stimulus and the information that it brings with it are perceived by the sense organs, thus causing a neurogenerative modification and, only later, in the central nervous system, the processing at the cortical level is activated, becoming an emotionally felt event.

According to Cannon [11] emotions originate from the central nervous system and, later, they reach the periphery. Then they are displayed through the expression of emotions, revealed by the face and the whole body. The coding of information takes place in the central nervous system, precisely in the thalamus, the centre for gathering, processing and sorting information from the centre to the periphery and from the periphery to the centre and the amygdala, a structure located in the brain, considered as an "emotional computer", which allows the sorting and recognition of emotions. Patients with amygdala damage present difficulties in recognising emotions (whether pleasant or not) and abnormalities in emotional behaviour [12]. The amygdala is therefore recognised as the centre for the exchange of emotional messages [13]. Starting from the seventies, cognitive theories introduce the psychological element, considered the most relevant aspect in emotional research. Both the physiological and psychological components, the latter defined arousal, interact with each other, causing the emotional event.

According to the cognitive-activation theory of Schachter and Singer [8], by assigning to the emotional state a specific emotional event, our brain succeeds in contextualising its own emotional experience [14]. A further contribution is provided by Lazarus [15], with the introduction of the concept of appraisal, the process that allows the cognitive processing of the event. Indeed, as Frijda [16] will extend later, emotions originate from the causal evaluation of the events, related to the meanings that people attach to the event itself. They do not appear by chance, but from an in-depth analysis of the situations that arise, determining the influence that emotions can ultimately have on a person's well-being [17]. Events that meet expectations and desires enable positive emotions; on the contrary, events that can harm, activate negative emotions. For this reason, each individual, based on his/her experience, may experience the same situations or events expressing different and sometimes divergent emotions. Emotions are therefore placed in a subjective dimension and are very flexible and variable.

Some studies talk about the 3D reconstruction [18], they tell about the expression of a face can be reconstructed with 3D techniques that allow to simulate and help to understand the emotions present on a face [19,20]. Often, the psychological conflict is represented by the division and multiplication of the inner characters of a subject and the images of "strange attractors" [21,22], which represent chaos [23–25]. Each cerebral hemisphere maintains a network without scales that generates and maintains a global state of chaos [26]. Sometimes chaos can be approached in cases of people suffering from neurological and psychological disorders, often in these people it is difficult to recognise the emotions they express [27]. Numerous studies show that 3D virtual learning environments and information communication technologies can be excellent tools to support people, for example, suffering from autism spectrum disorders [28,29]. In particular, they can be very effective in helping to understand social behaviours and emotions [30].

Very significant advances in the study of emotions and subsequent models of interpretation in computer vision were obtained thanks to the FACS (Facial Action Coding System) coding system, devised by Ekman & Friesen [4]. According to this system, emotional expressions are expressed from the face muscular contraction according to specific organisations. The FACS system makes it possible to identify the individual muscles (called Action Units) (and their combination in emotional units), involved in all the emotional expressions that humans produce. Thus, the movements of one or more facial muscles determine the six expressions recognised as universal, that are neutral, happy, sad, angry, disgusting and surprise facial expressions. These movements play a very important role in conveying the individual's emotional states to the observer, especially in face-to-face social interaction. The research has developed different approaches and methods for the analysis of fully automatic facial expressions, useful in human-computer interaction or computer-robotic systems [31–33]. In automatic recognition, the facial expression of the image is processed to extract this information from it, which can help to recognise the six basic expressions. The steps of the process are as follows: image acquisition, extraction of basic features, and finally classification of expressions.

3   Materials and Methods

3.1   Facial Image Database

In this research we used is the Cohn-Kanade database which contains 486 sequences of 97 subjects extended in the 2010 with the name CK+ database that added 107 sequences of other 26 subjects, with a total of 593 sequences from 123 subjects [34]. Each sequence contains images ranging from onset (neutral frame) to peak expression (last frame). The peak frame belongs to the FACS coded for the facial action unit (AUs). In particular, the FACS system, that was designed by Ekman and Friesen [4], devised the coding of the actions of the human face system. FACS allows different facial movements to be coded in Action Units (AU) based on the muscular activity of the face that generates temporary changes in facial expression. They identified 44 FACS AUs, of which 30 AUs are related to the contractions of specific facial muscles: 12 for the upper face and 18 for the lower face. AUs can happen singly or in combination [35].

At all subjects in CK database were asked to perform a series of expressions, with a neutral background behind them and only one person was present in the diagram [36]. Image sequences have been digitised into 640 arrays of 480 or 490 pixels with 8-bit precision for gray-scale values. In order for the internal validity of the data to be guaranteed, they have been manually coded and the reliability of the coding has subsequently been verified [34]. Only 118 subjects are annotated with the principal emotions (anger, disgust, fear, happy, sad and surprise) [35].

The database was composed by subjects were between 18 and 50 years, 69% women, 81%, Euro-American, 13% African Americans and 6% other groups. Their facial behaviour was recorded using two Panasonic AG-7500 cameras synchronised to the hardware.

Images that were not labelled by the original CK database system and CK+ were deleted and were not used in this study.

The Dataset of this study is composed by 1544 photos (Fig. 1).

Fig. 1. Part of dataset.

3.2 Mathematica Wolfram Machine Learning

To analyse data, we applied *Wolfram Mathematica* software that use its own computer language, called *Wolfram Language.*

Wolfram Mathematica is able to learn what to do only by looking at examples, using the idea of machine learning. In particular, it is possible to construct algorithms to analyse data or images automatically [36]. There are two types of machine learning: supervised and unsupervised learning. Supervised learning is defined as the process of an algorithm that learns from the training data set. The algorithm formulates forecasts on the control set based on what was described by the supervisor. The unsupervised learning algorithms, instead, model the structure of the input data or their distribution to obtain other information on the data. *Wolfram Mathematica* includes several types of integrated automatic learning. In particular, functions like Predict and Classify are automated and work on many types of data, including numerical, categorical, time series, text, images and audio [36]. In this study, Classify supervised automatic learning was used.

In our work, we processed image using automatic machine learning feature provided by the software. The neural network to which *Wolfram Mathematica* refers is the *LeNet Network.* The software allows you to immediately use the available pre-trained networks or manipulate and reassemble them to train new data. Indeed, through the use of Wolfram language, it is possible to implement the automatic learning system of the neural network contained in it.

3.3   Image Processing

A research proposal is to ensure that the software analyse the images and automatically recognises the emotions expressed in the image.

To understand how the software takes out the emotion in the images, we must refer to the slight changes of face or, in particular, to the changes of landmarks. Indeed, subjects modify their facial expression by moving the fundamental characteristics of their faces, that is, by modifying the position of the reference points.

The fundamental facial points are used to locate and to represent the salient regions of face, such as:

- Eyes (right and left);
- Eyebrows (right and left);
- Nose;
- Mouth (internal and external);
- Outline of the face.

These points are starting points for the software's automatic recognition of emotions. *Wolfram Mathematica* extracts all parameters of the reference points (Fig. 2).

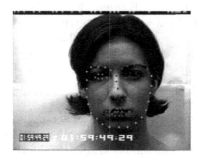

Fig. 2. Different colours of different landmarks. (Color figure online)

The first step was to manipulate and reassemble the pre-trained networks available in the *Wolfram Mathematica* software. This in order to create a more solid and functional network by training the machine with database described above.

To implement the training of the machine, the images corresponded a particular emotion have been assigned to each emotion, taking the images of each subject from the onset to the peak of emotion (excluding neutral emotion). In this way the software learned to understand the corresponding emotion already from the first variation of the landmarks in the subjects' faces (Fig. 3).

To start recognition process we applied *FindFaces* module in *Wolfram Mathematica* that allows to find people's faces in the images proposed and returns a list of bounding boxes. In addition, we use Classify command was assigned

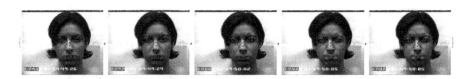

Fig. 3. Change of landmarks with variation of emotion.

to each emotion the corresponding package of photos (Fig. 1), manipulating and reassembling the pre-trained *Mathematica's networks*. *Classify* generates *ClassifierFunction* based on an association of classes with their examples. From here it is possible to extract information about the method used by the calculator to classify emotions. The same procedure was carried out for all six emotions.

3.4 Image Analysis

The pre-trained *Wolfram Mathematica* network has been increased using a total of 1544 samples adopted for machine training. The machine took about 1 min to do the learning, having 6 classes to learn from (Fig. 4). In this case *Wolfram Mathematica* Machine Learning used the "Forecasting method" for its learning. This method, as defined by *Wolfram* [36], predicts the value or class of an example using a set of decision trees. It is a type of machine learning algorithm named *Bootstrap Aggregation* or *Bagging* (*Bootstrap Aggregation Algorithm*). The *Bootstrap* method is used for estimating statistical quantities from samples and creates models from a single set of training data. In particular, it is a learning method for classification and regression that operates by building a multitude of decision trees. The forest forecast is obtained by taking the most common class or the tree's predictions of the average value. Each decision tree is then trained on a random subset of the training set [36].

The total accuracy of the method is 91.3% (Fig. 5); the accuracy of individual emotions is high for some emotions (89% happiness, 81% disgust), while low due to negative emotions (0.02% anger, 27% fear) (Fig. 6).

The last step was to test the trained machine to verify its validity. Having chosen a group of unused photos in the initial database, named Test Dataset. The Test Dataset is composed by 37 images for each emotion. The *FacialExpression* module was used, which allows to recognise the expression of a face displayed in an image. *FacialExpression* has an integrated classifier that automatically chooses the best algorithm to adopt based on the available models built and trained through a large database called Wolfram's Net Neural Repository, an archive consisting of 70 neural network models [36]. In particular, the output is a prediction of what the neural network interprets or returns the most probable alternative to the results, using the trained network in step 2. In this way using the network trained in step 2 the *Mathematica Machine Learning* identifies the emotions included in each image and the relative probability of correct identification of the expressed emotion.

Classifier information

Input type	Image
Number of classes	6
Method	RandomForest
Accuracy	91.3% ± 2.9%
Loss	0.426 ± 0.050
Single evaluation time	21.8 ms/example
Batch evaluation speed	33.6 examples/s
Classifier memory	496. kB
Training examples used	1544 examples
Training time	1 min 14 s

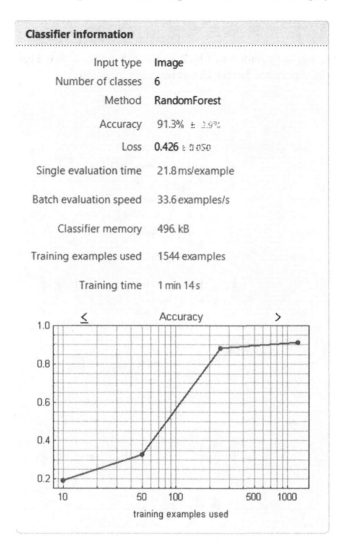

Fig. 4. Information about training.

```
ClassifierInformation[emozioni, "Accuracy"]
0.912584 ± 0.0294156
```

Fig. 5. Method accuracy.

```
Table[cm[[b]]["Accuracy"], {b, 6}]
{0.0810811, 0.810811, 0.27027, 0.864865, 0.594595, 0.891892}
```

Fig. 6. Classes accuracy.

4   Results

Emotion detection is provided in Fig. 7 and it can be seen that Disgust, Happiness and Surprise perform better than the other emotions.

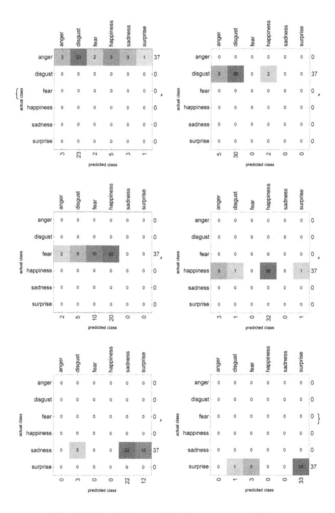

Fig. 7. Confusion matrix for every emotion.

This result is common and intuitive as these 3 are particular emotions that cause great deformation in the facial features. Indeed, the movement of the facial area for these emotions was easily detected by the trained system. On the contrary, other emotions (i.e. Anger, Sadness and Fear) do not behave as well, as the movements of the face are less than those of positive emotions. An explanation of this arises from the fact that emotions like Rage and Fear have subtle movements and are easily confused with other stronger emotions. This is also confirmed by other studies such as Lucey, Cohn, Kanade [5].

The machine has now been tested on a single subject to further verify its validity. The results obtained are discrete; for each subject the right emotion is recognised. From here it can be stated that the method works well on small image tests. In particular, in Fig. 8 the recognition of the fear emotion is fulfilled, in Fig. 9 the recognition of the disgust emotion, in Fig. 10 the recognition of the fear emotion.

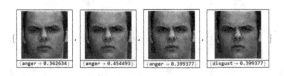

Fig. 8. Recognition of anger emotion in one subject.

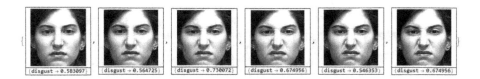

Fig. 9. Recognition of disgust emotion in one subject.

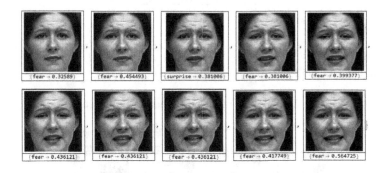

Fig. 10. Recognition of fear emotion in one subject.

Figure 11 shows the recognition of the happiness emotion, Fig. 12 the recognition of the sad emotion, Fig. 13 the recognition of the surprised emotion. For each subject the emotion is already recognised at the onset of emotion.

Now, *Mathematica Machine Learning* has been asked to provide information about a single emotion, taken happiness as an example, it displays the progress

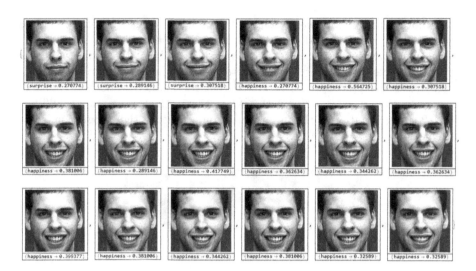

Fig. 11. Recognition of happiness emotion in one subject.

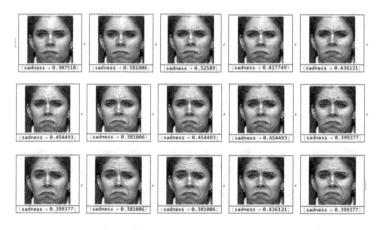

Fig. 12. Recognition of sadness emotion in one subject.

Fig. 13. Recognition of surprise emotion in one subject.

of the percentage of recognition of the happiness. As shown in Fig. 14, the recognition rate is low for the first images, i.e. the onset of emotion, this is because at first a few facial muscles move slowly and the emotion can be confused with another emotion (such as example surprise), the percentage of recognition however goes back up to about 40% when the emotion is actually recognised.

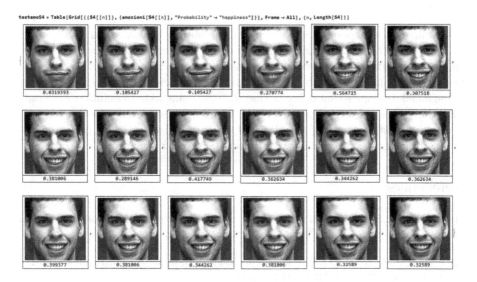

Fig. 14. Recognition of surprise emotion in one subject.

5   Conclusion

In this work, the recognition of emotions using *Wolfram Mathematica* software has been realised.

The reasons that led to this work concern:

a. making computational algorithms more and more efficient in recognising emotional expressions in the faces of human subjects;
b. train algorithms to the complexity of the dynamics of emotional expressions.

In fact, to overcome the traditional stillness of the data, usually used in this sector, we have used the different articulations of the onset of emotional behaviour, until its complete appearance in the face of the subject, up to the disappearance to allow for other types of expressions. This allowed us to train a computational machine first on static faces and then to train it precisely on the expressive dynamics, using a specially organised experimental set.

The results show that:

– The system is able to recognise the six main facial expressions used in an extremely efficient way;

- Recognises with reliable percentages almost all the emotions expressed in dynamic form, even if the recognition percentages are fluctuating and do not progress linearly from the onset of the expression until the return of the facial muscles to the neutral expression;
- It does not properly recognise the facial expressions related to the anger emulation: in this case the performance of the computational system is very low.

Evolutionary explanations of why this emotion that was not adequately recognised by the system may reside in the fact that, according to some authors, anger is not functional to social relations, so it may not be adaptive from the Darwinian point of view or because it is very insignificant from the point of view of the facial dynamics that this emotion realises, being able to manifest even only with the look and leaving the rest of the facial muscles completely immobile.

Further system training and data set choices focused on this emotion could give us different results, which we will pursue in other experiments.

References

1. Kulkarni, S.S., Reddy, N.P., Hariharan, S.I.: Facial expression (mood) recognition from facial images using committee neural networks. Biomed. Eng. Online **8**(1), 16 (2009)
2. Carroll, J.M., Russell, J.A.: Facial expressions in Hollywood's protrayal of emotion. J. Pers. Soc. Psychol. **72**(1), 164 (1997)
3. Eibl-Eibesfeldt, I.: Human Ethology. Aldine de Gruvter, New York (1989)
4. Ekman, P., Friesen, W.V.: Facial Action Coding System: Investigator's Guide. Consulting Psychologists Press, Palo Alto (1978)
5. Lucey, P., Cohn, J.F., Kanade, T., Saragih, J., Ambadar, Z., Matthews, I.: The extended Cohn-Kanade dataset (CK+): a complete dataset for action unit and emotion-specified expression. In: Computer Society Conference on Computer Vision and Pattern Recognition-Workshops, pp. 94–101. IEEE (2010)
6. Sebe, N., Lew, M.S., Sun, Y., Cohen, I., Gevers, T., Huang, T.S.: Authentic facial expression analysis. Image Vis. Comput. **25**(12), 1856–1863 (2007)
7. Kobayashi, H., Hara, F.: Facial interaction between animated 3D face robot and human beings. In: International Conference on Systems, Man, and Cybernetics. Computational Cybernetics and Simulation, vol. 4, pp. 3732–3737. IEEE, Orlando (1997)
8. Schachter, S., Singer, J.: Cognitive, social, and physiological determinants of emotional state. Psychol. Rev. **69**(5), 379 (1962)
9. Darwin, C.: On the Origin of Species. D. Appleton & Company, New York (1860)
10. James, W.: Essays in Psychology, vol. 13. Harvard University Press, Cambridge (1983)
11. Cannon, W.B.: The James-Lange theory of emotions: a critical examination and an alternative theory. Am. J. Psychol. **39**(1/4), 106–124 (1927)
12. Walker, D.L., Davis, M.: The role of amygdala glutamate receptors in fear learning, fear-potentiated startle, and extinction. Pharmacol. Biochem. Behav. **71**(3), 379–392 (2002)

13. Doré, B.P., Tompson, S.H., O'Donnell, M.B., An, L.C., Strecher, V., Falk, E.B.: Neural mechanisms of emotion regulation moderate the predictive value of affective and value-related brain responses to persuasive messages. J. Neurosci. **39**(7), 1293–1300 (2019)
14. LeDoux, J.E., Hofmann, S.G.: The subjective experience of emotion: a fearful view. Curr. Opin. Behav. Sci. **19**, 67–72 (2018)
15. Lazarus, R.S.: Thoughts on the relations between emotion and cognition. Am. Psychol. **37**(9), 1019 (1982)
16. Frijda, N.H.: Appraisal and beyond. Cogn. Emot. **7**(3–4), 225–231 (1993)
17. Bourgais, M., Taillandier, P., Vercouter, L., Adam, C.: Emotion modeling in social simulation: a survey. J. Artif. Soc. Soc. Simul. **21**(2), 1–5 (2018)
18. Wibowo, H., Firdausi, F., Suharso, W., Kusuma, W.A., Harmanto, D.: Facial expression recognition of 3D image using facial action coding system (FACS). Telkomnika **17**(2), 628–636 (2019)
19. Gabriele, L., Bertacchini, F., Tavernise, A., Vaca-Cárdenas, L., Pantano, P., Bilotta, E.: Lesson planning by computational thinking skills in Italian pre-service teachers. Inform. Educ. **18**(1), 69–104 (2019)
20. Gabriele, L., Tavernise, A., Bertacchini, F.: Active learning in a robotics laboratory with university students. In: Increasing Student Engagement and Retention Using Immersive Interfaces: Virtual Worlds, Gaming, and Simulation, pp. 315–339. Emerald Group Publishing Limited (2012)
21. Abdechiri, M., Faez, K., Amindavar, H., Bilotta, E.: The chaotic dynamics of high-dimensional systems. Nonlinear Dyn. **87**(4), 2597–2610 (2017)
22. Freeman, W.J.: Emotion is from preparatory brain chaos; irrational action is from premature closure. Behav. Brain Sci. **28**(2), 204–205 (2005)
23. Bilotta, E., Bossio, E., Pantano, P.: Chaos at school: Chua's circuit for students in junior and senior high school. Int. J. Bifurcat. Chaos **20**(01), 1–28 (2010)
24. Bilotta, E., Pantano, P., Vena, S.: Artificial micro-worlds part I: a new approach for studying life-like phenomena. Int. J. Bifurcat. Chaos **21**(2), 373–398 (2011)
25. Bilotta, E., Pantano, P.: Artificial micro-worlds part II: cellular automata growth dynamics. Int. J. Bifurcat. Chaos **21**(03), 619–645 (2011)
26. Bilotta, E., Stranges, F., Pantano, P.: A gallery of Chua attractors: part III. Int. J. Bifurcat. Chaos **17**(3), 657–734 (2007)
27. Lorincz, A., Jeni, L., Szabo, Z., Cohn, J., Kanade, T.: Emotional expression classification using time-series kernels. In: Proceedings of the IEEE Conference on Computer Vision and Pattern Recognition Workshops, pp. 889–895. IEEE (2013)
28. Bertacchini, F., et al.: An emotional learning environment for subjects with Autism Spectrum Disorder. In: International Conference on Interactive Collaborative Learning (ICL), pp. 653–659. IEEE (2013)
29. Cárdenas, L.A.V., et al.: An educational coding laboratory for elementary pre-service teachers: a qualitative approach. Int. J. Eng. Pedagogy (iJEP) **6**(1), 11–17 (2016)
30. Bertacchini, F., Bilotta, E., Gabriele, L., Pantano, P., Servidio, R.: Using Lego MindStorms in higher education: cognitive strategies in programming a quadruped robot. In: Workshop Proceedings of the 18th International Conference on Computers in Education, pp. 366–371 (2010)
31. Bertacchini, F., Tavernise, A.: Knowledge sharing for cultural heritage 2.0: prosumers in a digital agora. Int. J. Virtual Communities Soc. Netw. **6**(2), 24–36 (2014)
32. Bertacchini, F., Bilotta, E., Pantano, P.: Shopping with a robotic companion. Comput. Hum. Behav. **77**, 382–395 (2017)

33. Mane, S., Shah, G.: Facial recognition, expression recognition, and gender identification. In: Balas, V.E., Sharma, N., Chakrabarti, A. (eds.) Data Management,
Analytics and Innovation. AISC, vol. 808, pp. 275–290. Springer, Singapore (2019).
https://doi.org/10.1007/978-981-13-1402-5_21
34. Kanade, T., Cohn, J.F., Tian, Y.: Comprehensive database for facial expression
analysis. In: Proceedings Fourth IEEE International Conference on Automatic Face
and Gesture Recognition, pp. 46–53. IEEE (2000)
35. Tian, Y.I., Kanade, T., Cohn, J.F.: Recognizing action units for facial expression
analysis. IEEE Trans. Pattern Anal. Mach. Intell. **23**(2), 97–115 (2001)
36. Wolfram, S.: An Elementary Introduction to the Wolfram Language. Wolfram
Media, Incorporated, Champaign (2017)

Algorithms for Jewelry Industry 4.0

Francesco Demarco[1]([✉])[iD], Francesca Bertacchini[2][iD], Carmelo Scuro[1][iD], Eleonora Bilotta[1][iD], and Pietro Pantano[1][iD]

[1] Department of Physics, University of Calabria, Rende, Italy
{francesco.demarco,carmelo.scuro,pietro.pantano}@unical.it
[2] Department of Mechanical, Energy and Management Engineering, University of Calabria, Rende, Italy
francesca.bertacchini@unical.it

Abstract. The industrial and technological revolution and the use of innovative software allowed to build a virtual world from which we can control the physical one. In particular, this development provided relevant benefits in the field of jewelry manufacturing industry using parametric modeling systems. This paper proposes a parametric design method to improve smart manufacturing in 4.0 jewelry industry. By using constrained collection of schemata, the so called Direct Acyclic Graphs (DAGs) and additive manufacturing technologies, we created a process by which customers are able to modify 3D virtual models and to visualize them, according to their preferences. In fact, by using the software packages Mathematica and Grasshopper, we exploited both the huge quantity of mathematical patterns (such as curves and knots), and the parametric space of these structures. A generic DAG, grouped into a unit called User Object, is a design tools shifting the focus from final shape to digital process. For this reason, it is capable to returns a huge number of unique combinations of the starting configurations, according to the customers preferences. The configurations chosen by the designer or by the customers, are 3D printed in wax-based resins and, later, ready to be merged, according to artisan jewelry handcraft. Two cases studio are proposed to show empirical evidences of the designed process to transform abstract mathematical equations into real physical forms.

Keywords: Parametric jewelry · Algorithm for jewelry · Smart manufacturing

1 Introduction

Since their introduction CAD (Computer Aided Design) software has expanded to all fields of design including jewelry one. CAD system allows designers to generate and display extremely complex objects and they are generally employed in detailed design stage rather than in conceptual design stage. It's well established that most of these systems cannot allow design exploration because shape transformation and reinterpretations aren't supported. So, the conceptual design stage, in which designers generates ideas and explore possibilities, is

© Springer Nature Switzerland AG 2020
Y. D. Sergeyev and D. E. Kvasov (Eds.): NUMTA 2019, LNCS 11973, pp. 425–436, 2020.
https://doi.org/10.1007/978-3-030-39081-5_37

often assigned to hand sketches. However, these systems Parametric Design (PD) are emerging as a distinctive method of design. PD is a tool involving an explicit visual dataflow in the form of a graph and it supports the designers by offering a wide design exploration. This dataflow model a design as a constrained collection of schemata [1] in the form of a graph so called Direct Acyclic Graph (DAG). Parametric design aid human creativity by providing the opportunity to explore a larger range of design possibilities than in manual production [4], design variations can be made simply by adjusting the input parameters. In the recent years, parametric modeling has become a widespread tool in computational designer community, basically, two main approaches have been developed for this tool. First approach, adopted in the present case, is based on a visual explicit dataflow program called Directed Acyclic Graph (DAG). While the second approach is an implicit bottom-up method based on cognitive natural system in the form of heuristic algorithm. In the first phase of the development, parametric design found its application in the design of large-scale object and therefore in architecture design. Now, the technological development with the consequent introduction of the numerical control machine and the 3D printer strictly linked to the development of tools and software applications allowing the physicalizations of small-scale objects. These innovations lend designers to research more complex ideas, parametric design has had an impact not only on formal characteristic but also influences the logic of approach to design, indeed, we can talk about a new paradigm of design thinking [5]. Nowadays, the parametric modelling is present in every design area and its use increases day after day. Therefore, we can assist at a continuous metamorphosis and redefinition of parametric design theories and, simultaneously, at the development of parametric tools and machineries.

This paper introduces an approach to parametric design thinking oriented to the creation of jewels inspired by mathematical shapes, Mathematics can offer a huge library of aesthetic patterns, that integrated with algorithmic logic, adapt well to applications in art and design. The advantages of this approach are demonstrated using two example of real parametric modeling that maintains the benefit of design exploration until the detailed design stage, this allows final customer to explore different combination and personalizes the model. The paper is organized in five main section. Section 2 provide a brief review on the related works. Section 3 described the methodology adopted for this research. Section 4 provide the experimental results and their discussion. The conclusions of this work are presented in Sect. 5.

2   Literature Review

Parametric design has found its maximum applications in the industry 4.0 paradigm, it represents the systemic transformation of production thanks to the combination of physical-digital system. This industrial revolution is linked to the technological phenomenon of digitalization that allows us to build a virtual world from which we can control the physical one. In this paradigm virtual simulations, IOT strategies [34,35] and additive manufacturing technologies [44,45] guide product development processes allowing the study and realization of complex objects, a complexity that can only be achieved through parametric modeling processes. "Parametric design is about to change"[6] this is the simplest and most effective description given to parametric design, it allows to overcome the static design system. In literature, we can find several researches and different new approaches and theories, however, an approach considered more appropriate and efficacy than other does not exist. The scientific community is still engaged in research for the definition of a general theory of parametric design thinking [5,6,9], theoreticians are defining and characterizing the formalizations of parametric design processes and their related key concepts [7,41]. Research developed towards both parametric design approaches generative design (GD) and graph based [1,31]. GD is a method that uses computational capabilities to support design. GD, mainly diffused among the scientific community, uses Five techniques that are spread in all fields of design. Those techniques have been described by Singh et al. [8,9] and those are: cellular automata, shape grammar, L-systems, swarm intelligence and genetic algorithms. Several researchers have studied various issues of generative design. Bilotta et al. propose a design based on cellular automata, to represent chaos and hyperchaos object [10–29]. There are few publications on parametric design system related to jewelry applications. An example is the work of Kielarova et al. who proposed an approach on generative design based on a shape grammar applied to jewelry ring design [2]. It also provides an effective description of the designer's main aim who is the development of a generative design system that applies affine transformations to original elements and create new 3D shapes based on a finite set of shape rules. Another example is the work of Sansri et al. [30] they studied the applications of a genetic algorithm to an Art Deco double clip broach.

2.1 Lack of Literature on Design and Industry Interaction

If it is true that there is a broad scientific discussion on parametric modeling software, such as Generative Components (Bentley Systems), Dynamo (Autodesk) and Grasshopper (McNeel and Associates) [30], the analysis of the sources revealed a complete lack of paper dealing with the design theme in the paradigm of industry 4.0 [44]. The presence of this connection is demonstrated by the development of tools and software applications designed to directly connect the virtual world to the productive field. These processes, which have been kept as industrial secrets, are beginning to impact not only the productive system but also the society on a large scale. An important example of this phenomenon is Shapewa is a platform designed for direct physicalization of virtual model, anyone using this platform can become a designer and physicalize a unique object. Another example is Shapediver a tool that automatically turns parametric CAD file into interactive 3D model accessible through any web browser. This tool allows companies to provide a product with a high rate of customization by keeping the DAG hidden and eliminating high application development costs. The cause of the lack of scientific production on this theme is probably to be found on its multidisciplinary nature, indeed, the two topics are part of two distinct fields of research. Parametric design research is mainly carried out by researchers from architecture and civil engineering departments [36–40], while industry 4.0 is mainly treated by completely different classes of researchers [42–46]. Therefore, this article introduces itself in a multidisciplinary sector with the main purpose of creating a scientific discussion on new production processes starting from design up to the finished object.

3 Methodology

Mathematics leads man's thinking to the limit and becomes a source of inspiration for design research, there is an infinite catalog of mathematical shapes and models that can be adopted in jewelry design. Jewelry design, as any other industrial design process, need to consider the balance of aesthetic and functionality, therefore, for the correct design of an object it is not enough to create an aesthetically attractive shape, but it is necessary to take into account many ergonomic parameters. This last process implies not only the generation of inferences and judgements, but also the planning of actions in order to act on shapes and transforming them. The aim of a designer is the individuating of users' needs, thinking about the functions that fit well with these needs, for creating objects that are capable of embodying these functions by using some formal properties [10]. Explicit Direct Acyclic Graph generated in parametric design acts as a cognitive artifact shifting the focus from the final form to the digital process, [3,30–33] or rather it acts as a generative algorithm. Building the structure of the DAG designers define the geometric constrain and the relationship between the parts, this involves the creation of a shape grammar. Most of the resulting rules in this grammar are based on affine transformations, such as, translation, scaling, rotation, and repetition [1]. The resulting constrained collection of schemata allows to explore the design space when input parameters are changed in real-time [38]. In this case study, we started from analyzing some mathematical shapes and

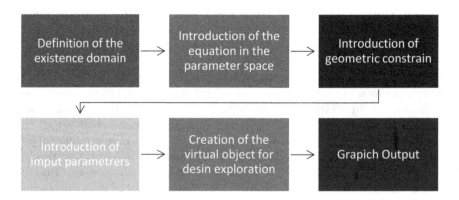

Fig. 1. A schematic representation of the adopted methodology: DAG for jewelry design. (Color figure online)

we have chosen a couple with completely different characteristics. Then, shape transformations in jewelry design process were studied to identify how shapes are transformed from one state to another with the aim of guaranteeing aesthetics and functionality, this involves creativity, analysis and development. In the two examples, presented in the fourth section, has been developed the methodology described in Fig. 1. This methodology corresponds to the creation of DAG developed in Grasshopper with the aid of a series of extensions such as peacock and lunchbox. Each of the graph-based models can be divided into six parts, since the parametric model is well-structured it has the potential to be understood by others due to their explicit representation. We can define the methodology adopted as a digital workflow that can always be replicated by the following six steps:

- Step 1. Creation of the conditions of existence necessary to define any mathematical function;
- Step 2. Entry of parametric equations;
- Step 3. Introduction of the shape modifiers;
- Step 4. Introduction of sliders and panels for transformations management;
- Step 5. Definition of the generative system;
- Step 6. Graphic output interface.

The main advantage in this graph-based application is to keep the possibility of exploring the design space even when the detailed model has been completed. This allows final users to move sliders, type new numbers or press buttons and their eyes see the design morph and evolve, so user can find his unique solution in the parameter space. In this process a .STL file is generated and transmitted through a data network and the resulting object can be physicalized in polymeric resin using additive manufacturing technologies. The prototype, so generated, is used in classic jewelry manufacturing process, raising the technological threshold of manufacturing industry without making any change in the productive process.

4   Results

In our analysis we considered the development of two generative DAGs, each of them consists of a generative algorithm structured the six main parts. These structuring parts of the algorithm are those introduced in Fig. 1. The most important part of this generative structure is the Step 3: the geometric constrains are introduced. The constraints constitute of a set of rules that allow to transform a pure shape into an object capable to satisfy a specific function.

The role of the designer is not just to choose the shape of the material object, but he has to be able to identify the suitable transformations that can be applied to the original shape. For this aim, the shapes and the operations used in jewelry manufacturing have been deeply analyzed in order to define certain rules and constraints that could guide the designer in the parametric tanking process. From the analysis conducted it emerged that the most commonly used transformations are:

- Translation: moving of a shape;
- Rotation; turning a shape around a center or an axis;
- Scaling: changing size of a shape;
- Mirror: reflecting a shape;
- Repetition: create a pattern of shape;
- Combination: combining different shape.

These shape alterations are based on the affine transformations and very often they are combined between them for the creation of the so-called ornamental groups. These transformations are controlled by a group of input parameters. These parameters constitute a semi-automatic user communication interface. Although the basic forms and transformation rules were previously defined by the designer the user is free to make explorations in the design space by manipulating the input data. The system is capable to calculate all the solutions obtained as combinations of the different input slider and provide, almost, in real time the shape transformation, in fact the maximum computational time is less than 5 second. In the logic diagram shown in Fig. 1 the group that allows the interaction with the shape is the one in step 4 represented in blue. The two case studies proposed in the paper are analyzed below. Of both cases studied the generative systems, using graph based method is presented. The generative systems were implemented following the logic set out in Sect. 3 of the paper, it is possible to see the correspondence between the logical scheme and the corresponding functions in the DAG by the use of color groups. In order to create an easy-to-use user interface the input parameter, which are introduced in step 4, are moved to the beginning of explicit digital dataflow.

4.1   Ring with Two Stones DAG

The first shape analyzed was derived from a heuristic algorithm created by Wolfram Mathematica. This algorithm is based on chaotic development systems, unlike what was expected by these complex systems the patterns that emerged were regular and they presented alternation of symmetries and breaks of the symmetries themselves.

As you can see this DAG has eight input parameters, each modification of the input parameters corresponds to a variation, in real time, of the geometry, therefore, there are countless outputs rings that this single algorithm is capable to generate. In Fig. 2 is showed the Grasshopper3D workflow designed for modelling the Ring with stones. Figure 3 shows only a few examples of the results that can be obtained, the user can evolve the model by varying the shape and size of the stone, changing the stone's material as well as the material of the ring, its size and the number of faces that delimit the surface.

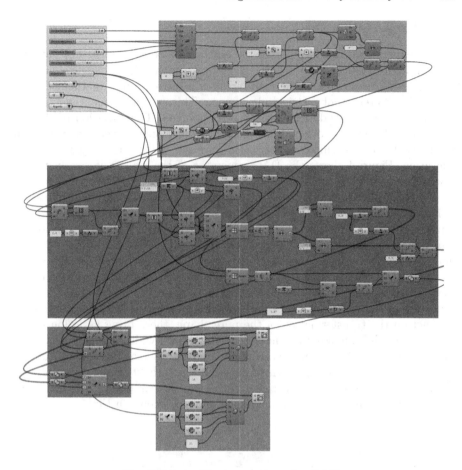

Fig. 2. Ring with stones generative DAG.

Fig. 3. Five different possible results of ring with stone generative DAG.

4.2   Lissajous Ring DAG

Further experimentation was conducted on the Lissajous pattern. This latter is representative of the graph of a curve given by a system of parametric equations. The curves represent one of the main areas of investigation of mathematics for their ability to represent physical phenomena. This system of equations is often used as a means of studying oscillatory phenomena. The variation of the frequencies of this system of equations generates harmonic patterns of particular beauty that are well suited to aesthetic purposes as the use in jewelry manufacturing. Figure 4 shows the DAG generator of the ring in its explicit form in which the different functions can be identified. This generator is capable to enlarge the exploration in the design space by providing the user multiple alternatives. Countless sequences of this figure are generated while maintaining the characteristics and functions that the ring have to satisfy. The main transformations related to the DAG are based on the frequency variations in the harmonic functions generating the curve. The variation of the frequency as in every harmonic equation the variation of the frequency leads to a drastic variation of the curve. In particular, as we can notice in Fig. 5, these types of variations cause a substantial change in the grid that contracts or expands. Many other parameters act on the transformation of the pure shape, these bring variation in the number of modules making up the object, the overall width of the ring and the thickness of the section. This gives the possibility to personalize the object and also the possibility of acting on the ergonomic characteristics and on the weight of the object. The user can then choose to generate completely different rings from extremely rarefied objects to complex plots as shown in Fig. 5.

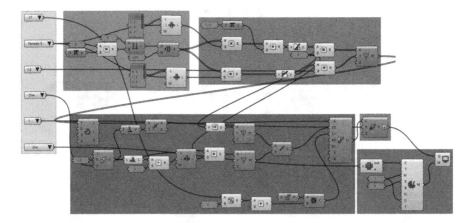

Fig. 4. Lissajous ring generative DAG.

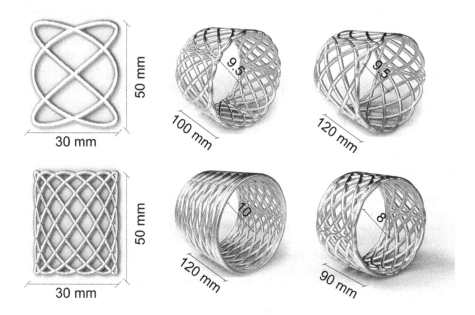

Fig. 5. Possible Lissajous pattern and their transformation through DAG.

5 Conclusion and Future Work

This paper is based on an experimental investigation to connect parametric and generative design with o field of industrial design in the new production framework of Industry 4.0 (I4.0). In our opinion, the jewelry sector is the manufacturing sector with the greatest potential for development compared to these new technologies. However, it still remains tied to traditional low-tech methodologies. This investigation helps to understand how abstract mathematical equation systems, algorithmic and genetic theories can be used to become effective tools of industrial design. This is due to the fact that such fusion allows creating unique and beautiful forms with a high scientific and mathematical background. The generative system conceived by the designer helps the user to automatically create his own configuration of a specific project allowing a high level of customization. Product customization has become an essential attribute in the industrial framework created by the recent I4.0 revolution.

The study of the transformations of shapes and the ergonomic functions used for the correct design of jewels has led to the definition of a correct grammar. A so created grammar is used precisely for the transformation of pure shapes into objects capable of satisfying specific features. The conducted research showed that it is possible to: (1) adopt any abstract mathematical models as a source to produces real objects; (2) show how new industrial designers can use the field of parametric and computational modelling to foster their creativity.

For the experimental part of this work we used the explicit parametric modeling method, since the necessity of a common language between humans and machines leaded to the creation of a clear interface that could be easily used. However, we referred also to implicit systems in order to extend and deepen the research on the parametric design by applying genetic algorithms. Indeed, further developments of this work regard the use of heuristic algorithms to optimize the solutions provided by generative systems, applying further mathematical concepts to the jewelery manufacturing field.

References

1. Aish, R., Woodbury, R.: Multi-level interaction in parametric design. In: Butz, A., Fisher, B., Krüger, A., Olivier, P. (eds.) SG 2005. LNCS, vol. 3638, pp. 151–162. Springer, Heidelberg (2005). https://doi.org/10.1007/11536482_13
2. Kielarova, S.W., et al.: An approach of generative design system: jewelry design application. In: 2013 IEEE International Conference on Industrial Engineering and Engineering Management, Piscataway, New Jersey, USA, pp. 1329–1333 (2013)
3. Harding, J.E., Shepherd, P.: Meta-parametric design. Des. Stud. **52**, 73–95 (2017)
4. Oxman, R.: Theory and design in the first digital age. Des. Stud. **27**, 229–265 (2006)
5. Oxman, R.: Thinking difference: theories and models of parametric design thinking. Des. Stud. **52**, 4–39 (2017)
6. Woodbury, R.G., Peters, B., Sheikholesami, M.: Element of Parametric Design, 1st edn. Taylor & Francis, Abingdon (2010)
7. Oxman, R.: Parametric design thinking. Des. Stud. **52**, 1–3 (2017)
8. Singh, V., Gu, N.: Towards an integrated generative design framework. Des. Stud. **33**(2), 185–207 (2012)
9. Bentley, P., Kumar, S.: Three ways to grow designs: a comparison of embryogenies for an evolutionary design problem. In: Proceedings of the 1st Annual Conference on Genetic and Evolutionary Computation, pp. 35–43. Morgan Kaufmann Publishers (1999)
10. Bilotta, E., Pantano, P., Stranges, F.: Computer graphics meets chaos and hyperchaos: some key problems. Comput. Graph. **30**(3), 359–367 (2006)
11. Gabriele, L., Tavernise, A., Bertacchini, F.: Active learning in a robotics laboratory with university students. In: Increasing Student Engagement and Retention Using Immersive Interfaces: Virtual Worlds, Gaming, and Simulation, pp. 315–339. Emerald Group Publishing Limited (2012)
12. Adamo, A., Bertacchini, P.A., Bilotta, E., Pantano, P., Tavernise, A.: Connecting art and science for education: learning through an advanced virtual theater with "talking heads". Leonardo **43**(5), 442–448 (2010)
13. Bertacchini, F., Bilotta, E., Pantano, P.: Shopping with a robotic companion. Comput. Hum. Behav. **77**, 382–395 (2017)
14. Bertacchini, F., Bilotta, E., Caldarola, F., Pantano, P., Bustamante, L.R.: Emergence of linguistic-like structures in one-dimensional cellular automata. In: AIP Conference Proceedings, vol. 1776, p. 090044. AIP Publishing (2016)
15. Bertacchini, F., Bilotta, E., Carini, M., Gabriele, L., Pantano, P., Tavernise, A.: Learning in the smart city: a virtual and augmented museum devoted to Chaos theory. In: Chiu, D.K.W., Wang, M., Popescu, E., Li, Q., Lau, R. (eds.) ICWL 2012. LNCS, vol. 7697, pp. 261–270. Springer, Heidelberg (2014). https://doi.org/10.1007/978-3-662-43454-3_27

16. Bertacchini, F., Tavernise, A.: Knowledge sharing for cultural heritage 2.0: prosumers in a digital agora. Int. J. Virtual Communities Soc. Network. (IJVCSN) **6**(2), 24–36 (2014)

17. Bertacchini, F., Bilotta, E., Gabriele, L., Pantano, P., Tavernise, A.: Toward the use of Chua's circuit in education, art and interdisciplinary research: some implementation and opportunities. Leonardo **46**(5), 456–463 (2013)

18. Bilotta, E., Bossio, E., Pantano, P.: Chaos at School: Chua's circuit for students in junior and senior High School. Int. J. Bifurcat. Chaos **20**(1), 1–28 (2010)

19. Bilotta, E., et al.: ImaginationTOOLS (TM)-A 3D environment for learning and playing music. In: Eurographics Italian Chapter Conference Proceedings, pp. 139–144, (2007)

20. Bilotta, E., Di Blasi, G., Stranges, F., Pantano, P.: A gallery of Chua attractors part VI. Int. J. Bifurcat. Chaos **17**(6), 1801–1910 (2007)

21. Bilotta, E., Stranges, F., Pantano, P.: A gallery of Chua attractors: part III. Int. J. Bifurcat. Chaos **17**(3), 657–734 (2007)

22. Bilotta, E., Di Blasi, G., Stranges, F., Pantano, P.: A gallery of Chua attractors: part IV. Int. J. Bifurcat. Chaos **17**(4), 1017–1077 (2007)

23. Bilotta, E., Pantano, P., Vena, S.: Artificial micro-worlds part I: a new approach for studying life-like phenomena. Int. J. Bifurcat. Chaos **21**(2), 373–398 (2011)

24. Bilotta, E., Pantano, P.: Artificial micro-worlds part II: cellular automata growth dynamics. Int. J. Bifurcat. Chaos **21**(3), 619–645 (2011)

25. Lombardo, M.C., Barresi, R., Bilotta, E., Gargano, F., Pantano, P., Sammartino, M.: Demyelination patterns in a mathematical model of multiple sclerosis. J. Math. Biol. **75**(2), 373–417 (2017)

26. Abdechiri, M., Faez, K., Amindavar, H., Bilotta, E.: The chaotic dynamics of high-dimensional systems. Nonlinear Dyn. **87**(4), 2597–2610 (2017)

27. Bertacchini, F., et al.: An emotional learning environment for subjects with Autism Spectrum Disorder. In: 2013 International Conference on Interactive Collaborative Learning (ICL), pp. 653–659. IEEE, September 2013

28. Vaca-Cárdenas, L.A., et al.: Coding with Scratch: the design of an educational setting for Elementary pre-service teachers. In: 2015 International Conference on Interactive Collaborative Learning (ICL), pp. 1171–1177. IEEE, September 2015

29. Bertacchini, F., Bilotta, E., Gabriele, L., Pantano, P., Servidio, R.: Using Lego MindStorms in higher education: cognitive strategies in programming a quadruped robot. In: Workshop Proceedings of the 18th International Conference on Computers in Education, ICCE, pp. 366–371 (2010)

30. Sansri, S., Kielarova, S.W.: Multi-objective shape optimization in generative design: art deco double clip brooch jewelry design. In: Kim, K.J., Kim, H., Baek, N. (eds.) ICITS 2017. LNEE, vol. 449, pp. 248–255. Springer, Singapore (2018). https://doi.org/10.1007/978-981-10-6451-7_30

31. Aish, R., Hanna, S.: Comparative evaluation of parametric design systems for teaching design computation. Des. Stud. **52**, 144–172 (2017)

32. Hudson, R., Shepherd, P., Hines, D.: Aviva Stadium: a case study in integrated parametric design. Int. J. Architect. Comput. **9**(2), 187–203 (2011)

33. Oxman, R., Gu, N.: Theories and models of parametric design thinking. In: Proceedings of the 33rd eCAADe Conference, pp. 477–482 (2015)

34. Ashton, K., et al.: That 'internet of things' thing. RFID J. **22**(7), 97–114 (2009)

35. Sundmaeker, H., et al.: Vision and challenges for realising the Internet of Things. In: Cluster of European Research Projects on the Internet of Things, vol. 3, no. 3, pp. 34–36. European Commission (2010)

36. Prats, M., et al.: Transforming shape in design: observations from studies of sketching. Des. Stud. **20**(5), 503–520 (2009)
37. Derix, C.: Mediating spatial phenomena through computational heuristics. In: Proceedings of the 30th Annual Conference of the Association for Computer Aided Design in Architecture, pp. 61–66 (2010)
38. Gero, J.S., Kumar, B.: Expanding design spaces through new design variables. Des. Stud. **14**(2), 210–221 (1993)
39. Harding, J., Joyce, S., Shepherd, P., Williams, C.: Thinking Topologically at Early Stage Parametric Design. na. (2012)
40. French, M.J., Gravdahl, J.T., French, M.J.: Conceptual Design for Engineers. Springer, Heidelberg (1985). https://doi.org/10.1007/978-3-662-11364-6
41. Cho, S.B.: Towards creative evolutionary systems with interactive genetic algorithm. Appl. Intell. **16**(2), 129–138 (2002)
42. Gibson, I., Rosen, D.W., Stucker, B.: Additive Manufacturing Technologies, 2nd edn. Springer, New York (2014). https://doi.org/10.1007/978-1-4939-2113-3
43. Kruth, J.P., Leu, M.C., Nakagawa, T.: Progress in additive manufacturing and rapid prototyping. CIRP Ann. **47**(2), 525–540 (1998)
44. Huang, S.H., Liu, P., Mokasdar, A., Hou, L.: Additive manufacturing and its societal impact: a literature review. Int. J. Adv. Manuf. Technol. **47**(2), 1191–1203 (2013)
45. Lee, J., Bagheri, B., Kao, H.A.: A cyber-physical systems architecture for industry 4.0-based manufacturing systems. Manuf. Lett. **3**, 18–23 (2015)
46. Rüßmann, M., et al.: Industry 4.0: the future of productivity and growth in manufacturing industries. Boston Consult. Group **9**(1), 54–89 (2015)

Clustering Analysis to Profile Customers' Behaviour in POWER CLOUD Energy Community

Lorella Gabriele$^{(\boxtimes)}$ ⓘ, Francesca Bertacchiniⓘ, Simona Giglioⓘ,
Daniele Mennitiⓘ, Pietro Pantanoⓘ, Anna Pinnarelliⓘ, Nicola Sorrentinoⓘ,
and Eleonora Bilottaⓘ

University of Calabria, via Pietro Bucci, 17b, Arcavacata di Rende, Italy
{lorella.gabriele,francesca.bertacchini,simona.giglio,daniele.menniti,
pietro.pantano,anna.pinnarelli,nicola.sorrentino,
eleonora.bilotta}@unical.it

Abstract. This paper presents a cluster analysis study on energy consumption dataset to profile "groups of customers" to whom address POWERCLOUD services. POWER CLOUD project (PON I& C2014–2020) aims to create an energy community where each consumer can become also energy producer (PROSUMER) and so exchange a surplus of energy produced by renewable sources with other users, or collectively purchase or sell wholesale energy. In this framework, an online questionnaire has been developed in order to collect data on consumers behaviour and their preferences. A clustering analysis was carried on the filled questionnaires using Wolfram Mathematica software, in particular FindClusters function, to automatically group related segments of data. In our work, clustering analysis allowed to better understand the energy consumption propensity according the identified demographic variables. Thus, the outcomes highlight how the availability to adopt technologies to be used in PowerCloud energy community, increases with the growth of the family unit and, a greater propensity is major present in the age groups of 18–24 and 25–34.

Keywords: Machine learning · Cluster analysis · Consumer behaviour

1 Introduction

A sustainable energy transition with the adoption of highly advanced technologies such as Smart Grid involves changes in a wide range of customers' energy behaviors, including the adoption of sustainable energy sources, energy efficient technologies, investments in energy efficiency processes in buildings, and above all, changes in direct and indirect behavior by costumers in energy consumption. Some research has shown that such measures still struggle to take off for a number of reasons, including the acceptance of technologies by users, of crucial importance for the development of a new culture of energy saving. These

Y. D. Sergeyev and D. E. Kvasov (Eds.): NUMTA 2019, LNCS 11973, pp. 437–450, 2020.
https://doi.org/10.1007/978-3-030-39081-5_38

technologies are more advanced than the previous ones. Smart Grids contain clusters of electrical engineering technologies merged with network technologies, connected with powerful instrumentation of smart sensors and meters, placed between the system that supplies electricity and the end users [1]. The advantages are noteworthy for users. In fact, the main features of Smart Grids are their ability to recover following damage, the possibility to all types of costumers of actively participating for energy saving, adaptability to different types of both wholesale and retail market, for industrial or domestic uses, resistance to attacks and natural disasters, a higher quality supply [2]. Since the interfaces of these systems are highly specialized and specific, they require a high level of knowledge, and certainly a non-trivial interaction by costumers. Therefore, communication plays a crucial role in the adoption of smart grid technologies. Only 5–6 years ago, some demographic surveys have shown that many adult users have not only never heard of such technologies, but that they did not even have an understanding of the amount of energy consumed in their home. However, they claimed to be willing to cooperate if they had the information they needed. So usually, the acceptance of new technologies for energy saving is directly linked to the amount of information a costumer has, accompanied by the perceived usefulness, in relation to how much this technology positively influences the quality of life [3, 4]. Thus, human behavior is complex and acceptance rarely is in line with centralized energy saving policies. In fact, unlike the decision-making choices of daily products, where consumers make rational choices in line with their values and intentions, it is not the case for the choice of energy products-services, on which the lack of knowledge or misinformation consider the costs and benefits of all the existing and optimal technological alternatives. Moreover, even if the choices, at the social level, have been implemented and the end users have accepted the technologies, the construction of significant relationships between energetic behavior and daily behavior is still scarce. However, an increasingly refined knowledge about all the customers' behavior allows a very important process because, with the liberalization of electricity markets, sales companies increasingly need to define depletion patterns for their electricity customers. Many studies have been carried out precisely to define consumption models, using Machine Learning and clustering systems in order to classify types of customers connected with their consumption profiles. These results have been gathered by using some interesting technological tools such as AMR (Automatic Meter Reading) [5] and the demand response management (DRM) schemes to manage energy for residential buildings in smart grids [6]. This allowed an optimization of the energy markets. Currently, however, new needs are emerging in the energy communities. Therefore, it is not only necessary to have an idea of customers' electricity consumption behavior, but also how users exchange electricity in the community, how sell and buy again, redefining energy production and marketing methods. In short, in order to optimize the energy reserves produced more and more, from the research point of view it is interesting to know which are the social dynamics of energy exchange, whether it is for sale or purchase. Therefore, in this article, a clustering analysis is proposed to profile the customers' behavior useful for the

PowerCloud project aims. The remainder of the paper is organised as follows. After this introduction, Sect. 2 informs about the related works on the topic. Section 3 introduces the PowerCloud project at a glance, highlighting the Realization Objective 3 (OR3). Section 4 argues about Machine Learning and cluster analysis; Sect. 5 presents the methodology adopted to implement questionnaire and to analyze dataset. Section 6 discusses Data and Statistical analysis and Sect. 7 summarizes the Clustering results. Finally, concluding remarks close the work.

2   Related Work

Smart grids (SGs) for energy saving are now a reality in almost all countries of the world. Although there is no a commonly accepted meaning for intelligent network, the US Department of Energy (DOE) classifies it as a two-way flow of electricity and information that can be used to monitor everything from power plants to customers' preferences. According to DOE, this definition embodies one of the main prerogatives of an intelligent network or SGs. The SGs are usually connected with the Advanced Metering Infrastructure (AMI), distributed worldwide, which is why we talk about Big Data and in general of all the connected 4.0 Technologies for the energy sector [7–9]. A large number of heterogeneous data have been produced by these infrastructures, distributed all over the world. In a simplified model of the SG, each consumer is considered as an actor who has the main aim to reduce his/her energy consumption. To evaluate the consumers' harvesting behavior, they are endowed with a Smart Meter (SM), which is a device between the local source and the grid, capable of tracking the energy consumed by load or injected into the grid. These processes are promptly transferred to a billing system capable of reporting on each energy withdrawal or injection, establishing from time to time the intervals of interest for data recording or recording all the consumption behavior of customers throughout the entire period of a day. In order to give to the end-customers a more detailed feedback on their electricity consumption, many methods have been tested, from price reduction to monetary incentives, creating elastic tariffs and dynamic pricing or by direct-load control [10], by allowing to others the control of all consuming devices and machines. The information that can be extracted from the SM are of enormous value for all the players of the SGs domain, both for consumers and for stakeholders, returning into economic benefits and improved services for both ones. However, analyzing such information flow presents any problems. As Big Data, they are huge and complex data sets, which are difficult for traditional tools to store, process and analyze [11], while the computational capabilities to analyze them should be extraordinarily powerful and time-consuming. This is why several automatic methods of analysis have been developed [12]. Measured in the field, collected for a long enough period on the costumers' load behavior, data are useful to formulate algorithms capable of clustering customers into macro-categories, based on their behavioral models. In the related literature, the used methods foresee the following steps: (1) Data collection and processing which allow to progress with the association of each customer with its representative load model, to be considered for categorization purposes. Data must be indicative of customer's consumptions behavior collected on a daily load model. The duration must be long enough to ensure that behavioral patterns emerge. Therefore, no less than two or three weeks of observation are required in the same loading condition. At the end of data collection, they are processed and those that present anomalies are eliminated; (2) In the second phase of the clustering process, the definition of the main functions used as input for allowing the analysis of the customers' grouping algorithms is carried out, making a selection of the most representative features of each user; (3) In the third phase of the clustering method, the evaluation of the clustering procedures, illustrating the types of algorithms and the validity of the chosen indicators is fulfilled in order to assess clustering effectiveness; (4) In the last phase, called post-clustering phase, there is the emergence of the customer classes, on the attributes used to establish the client classes based on the results of algorithms applied. In our research we applied Wolfram Mathematica clustering techniques, since it allows to run a number of automatically operations in several areas of applications and it can be applied in the estimating, assessing and monitoring of data processing [13]. In Zotos [13] opinion, Wolfram Mathematica is a tool developed primarily to handle several aspects of technical computing and engineering in a coherent and unified way and it is a "faster and more capable functional language approach".

3   POWER CLOUD Project and User Profile

POWER CLOUD project aims to create an energy community, where each consumer can become also energy producer (PROSUMER) and so exchange a surplus of energy produced by renewable sources with other users, or collectively purchase or sell wholesale energy. In particular, the project will implement the POWERCLOUD platform that will be capable to manage energy consumption and energy production of users in two ways: working in real-time data or on planned data. In the planning way, it will be made an estimation of the consumption profiles of each user, allowing the optimal scheduling of the energy loads according both to the availability of energy produced and the price of the electricity market. In real-time way, any surplus or energy deficit will be managed, allowing the sale or supply, according to the market modality at a local level among the users of the energy community. Both modalities, allow to obtain an economic optimization of energy supply and consumption. The platform collect a large amount of data and process them according specific needs and returning useful suggestions regarding the to the different energy players of the energy community. The idea is to interact with each PROSUMER (consumers/producers) so that he can modulate his consumption and possibly, as energy producers, exchange the surplus of energy with other users without further intermediaries; or by collectively buying the wholesale energy being able to dynamically manage their loads optimally to obtain considerable savings in terms of tariffs. As regards the Realization Objective 3 (OR3), it aims to collect and analyse data on users' habits who are part of the energy community, grouping them through advanced mathematical and statistical techniques, to elaborate an energy profile for individual and/or class of users. The Realization Objective 3 is managed by the ESG group. ESG is a research group that operates at the University of Calabria and has a long tradition in carrying out interdisciplinary research activities, conjugating different knowledges belonging to different fields, such as Mathematics [14–20], Complex systems [21–25], Machine learning [26], Artificial Intelligence [27] and Cognitive psychology [28–35], as required by the activities foreseen by the OR3. In this regard, an online questionnaire has been developed and data has been analysed in order to understand consumer's behaviour and their preferences. The ultimate purpose is to profile "groups of customers" to whom address POWERCLOUD services taking into account their daily behaviour regarding electricity consumption and other external factors (daily and seasonal rhythms on the one hand, meteorological factors etc.).

4   Machine Learning and Cluster Analysis

The advent of big data era lead some researchers to generate innovate techniques and apply new processing models named Machine Learning. Machine Learning is an area of computer science connected to Artificial Intelligence (AI) and focused on the development of computer systems able to learn and act as humans do [36]. Through Machine Learning algorithms, we can to carry out research on database of structured and unstructured data, acquiring information about different systems including multimedia systems such as images, texts, videos, building predictive models and extracting values from Big Data. Machine Learning tasks are classified into two main categories: supervised learning (supervised machine) and unsupervised learning (unsupervised machine). In supervised learning it is given examples in the form of possible inputs and the respective requested outputs, the aim is to extract a general rule that associates the input to the correct output creating a model; in unsupervised learning, the model aims to find a structure in the inputs provided, without the inputs being labelled in any way [37]. Currently, Machine Learning analysis has become one of the most important topics in several fields, such as engineering, economics, physics, and many others [26,38] due the development of advanced analytics techniques that provide to leverage Big Data as a competitive advantage. Advanced statistical methods are used to conduct Big Data analysis including Naive Bayes for classification, Logistic Regression for regression and Support Vector Machines (SVM) for regression and classification [39]. In recent years, the scientific research has increasingly considered Machine Learning algorithms a useful method to analyse electrical load consumption data [40,41], in particular using cluster analysis, a powerful method to explore patterns structures within data. Indeed, cluster analysis, based on an unsupervised learning process, is one of the most commonly applied data mining techniques and useful to form partitioning a dataset into sub-groups that are similar to each other and are quite dissimilar to the other groups [42, 43]. The most popular clustering approaches are Density-Based Clustering (DBSCAN - density-based spatial clustering of applications with noise), Hierarchical Clustering (algorithm that groups similar objects into groups named clusters) and K-Means Clustering (algorithm that allows to clustering N data points into K disjoint subsets based on their attributes) [44]. In our research, given a set of data composed by the most relevant information about energy production and consumption at home, we have applied cluster analysis, using Wolfram Mathematica [45].

5   Methodology

5.1   Scope

The OR3 POWERCLOUD aimed at investigate which are users available to get closer to the renewable energy market, to domotic and to be part of an energy community and to recognize users with different approaches towards energy consumption and use.

5.2   Procedure

To achieve the scope above descripted, an online survey was implemented and submitted to representative sample of households energy consumption. A first draft of the questionnaire has been assembled and administered to a small group of five households, through an interview. The aim of this paper is to identify energy consumption dataset to profile "groups of customers" to whom address POWERCLOUD services. In this view, Jakob Nielsen, one of the world's leading experts on usability, states that to conduct an effective usability test of any product, no more than five users are needed [46, 47]. A large sample is not necessary, but it is sufficient that the level of experience of the chosen subjects corresponds with a good approximation to that of the population for which the product is addressed. The usability test with users is an empirical evaluation methodology that foresees observation during the survey, it allows to obtain valuable information on the degree of understanding of the "product" without excessively wasting resources. Thus, users were asked to complete the questionnaire highlighting whether the questions had elements that were difficult to understand and/or were contradictory, in order to individuate strengths and weaknesses of the pilot questionnaire. This phase was followed by a brainstorming with the project group to obtain clarifications on the problematic or neglected points and lastly to elaborate the final questionnaire. The online questionnaire developed with Google module consists of 4 section and 22 items: the first section was devoted to collect demographic information, the second section to collect data relating to user preferences on propensity for energy saving, the third section concerns willingness to adopt renewable energies, the fourth section on domotics and automation systems knowledge. The questions are in mixed mode (open and closed). For closed-ended questions, subjects were asked to express their opinion using a 5-point Likert scale, where: 1 = Strongly disagree 2 = Disagree 3 =

Uncertain 4 = Agree 5 = Strongly agree. The average duration of completing the questionnaire was approximately 8 min. 261 questionnaires have been compiled on households' energy consumption useful to study users' habits. Questionnaire answers have been encoded numerically both for multiple choices and several categories. Answers with missing input has been eliminated (4 questionnaires). This is a very important and critical process for the construction of the data set. Consequently, 257 vectors of answers have been analysed with Mathematica. We followed two-step of data analysis: a descriptive analysis to examine a raft of consumers attitudes, motivations and knowledge variables, along with traditional covariates such as online service used and other demographic characteristics, and a cluster analysis to identify distinct groups with similar features in the sample. As previously done by Frank et al. [48], we carried out Descriptive analysis on the results of the clusters to understand the relationship of these groups of consumers (allocated in the different clusters) with different levels of Propensity for energy saving and opinion on current energy supplier, Willingness to adopt renewable energies and Domotics and automation systems knowledge. In Fig. 1 the macro steps that allowed to obtain the "Energy customers' behaviour profile" are schemed.

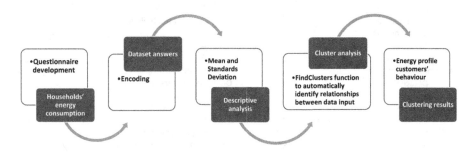

Fig. 1. A schematic representation of the adopted methodology: from the collection of the dataset to the customers' profile.

6   Findings

6.1   Descriptive Analysis

Descriptive analysis carried out on a datataset of 257 answers, shows that the sample is equally composed by women and man. Predominantly they are aged between 18 to 34. 61% and live in an apartment, while 39% in a detached house. 42% of them have energy class A house. They use the web to carry out numerous activities, from finding travel and vacation packages to using online services, even if, in the majority, they use social networks (59%). The house is mostly heated by an autonomous boiler, in fact there is a very high percentage 84% and use Gas for heat water (79%). The demographic characteristics of the sample are summarized in Table 1. The survey instrument employed a five-point Likert scale to solicit level of agreement with three factors towards Propensity for energy saving and opinion on current energy supplier, Willingness to adopt renewable energies, Domotic and automation systems knowledge. The average Likert scale scores and the standard deviation are displayed in Table 2. Most respondents "agree" or "strongly agree" (mean score is 4.32) if "in my city more energy will be produced by renewable sources".

Table 1. Characteristics of the study sample (percentages).

Characteristics	Sample survey (n = 257)	Characteristics	Sample survey (n = 257)
Gender		**Water house heating**	
Male	43%	Electric water heater	16%
Female	57%	Gas	79%
Age		Condominium district heating	5%
18 to 24	38%		
25 to 34	39%	**Heating distribution system**	
35 to 44	10%	Heating units	82%
Over 45	13%	Floor heating	5%
Family unit		−Thermoconvectors	3%
1 component	20%	Other/Nothing	10%
2 components	11%	**Energy class of the building**	
3 components	28%	A+	29%
4 components	26%	A	42%
5 components	13%	B	15%
Accommodation		Other	14%
Detached home	39%	**Online services used**	
Apartment	61%	Online purchase	26%
Heating house system		Social network	59%
Condominium boiler	8%	Utilities payment	4%
Solar panels	5%	Search for vacation packages, air flights, train travel	10%
Autonomous boiler	84%		
District heating	3%		
Summer cooling system			
Convector	7%		
Fixed air conditioner	38%		
Mobile conditioner	14%		
Other/Nothing	40%		

6.2 Clustering Analysis

A cluster analysis procedure is applied to three identified demographic variables: age, family unit and online services used, in relation with the different levels of Propensity for energy saving and opinion, Willingness to adopt renewable energies and Domotic and automation systems knowledge. Some results are displayed in Tables 3 and 4. We used a specific function "FindClusters" to select and group homogeneous elements in a set of data to reduce the search space and to find the optimal solutions [45]. By default, "FindClusters" function identifies clusters on

Table 2. Propensity for energy saving and opinion on current energy supplier, Willingness to adopt renewable energies, Domotic and automation systems knowledge.

	Descriptive Statistics	
	Mean Likert score value	Standard deviation
Propensity for energy saving and opinion on current energy supplier		
I am available to use only class A ++ household appliances	3.91	1.29
My current energy supplier is reliable	3.39	1.06
I think that is important to have transparency and clarity in the invoice by the electricity supplier	4.33	1.14
I think that energy suppliers' offers currently on the market are cheap	2.86	1.01
Willingness to adopt renewable energies	4.19	1.18
I am available to use renewable energy sources and/or products/technologies capable to energy save	4.32	1.16
I agree if in my city more energy will be produced by renewable sources	3.00	1.26
I agree to pay extra to use energy produced by renewable sources		
Domotics and automation systems knowledge	3.70	1.15
Home automation is a valid aid for elderly or disabled people	3.65	1.08
Home automation systems allow energy save and therefore reduce consumption and waste	3.73	1.10
Home automation systems are capable to make home safer		

Table 3. Cluster Analysis on Age and Propensity for energy saving and opinion on current energy supplier [a], Willingness to adopt renewable energies [b] and Domotic and automation systems knowledge [c].

Cluster Analysis on Age and Propensity for energy saving and opinion on current energy supplier

	Cluster 1 Mean (S.D.)	Cluster 2 Mean (S.D.)
I am available to use only class A ++ household appliances	1.97 (1.09)	4.26 (0.96)
My current energy supplier is reliable	2.17 (0.87)	3.60 (0.93)
I think that is important to have transparency and clarity in the invoice by the electricity supplier	2.22 (1.09)	4.72 (0.59)
I think that energy suppliers' offers currently on the market are cheap	2.47 (0.98)	2.93 (0.99)
Number of cluster members	40	217
Proportion of respondents	15.6%	84.4%

[a]

Cluster Analysis on Age and Willingness to adopt renewable energies

	Cluster 1 Mean (S.D.)	Cluster 2 Mean (S.D.)	Cluster 3 Mean (S.D.)
I am available to use renewable energy sources and/or products/technologies capable to energy save	1.82 (0.83)	4.82 (0.41)	3.67 (0.75)
I agree if in my city more energy will be produced by renewable sources	2.14 (1.32)	4.96 (0.18)	3.64 (0.65)
I agree to pay extra to use energy produced by renewable sources	2.02 (0.96)	3.2 (1.32)	3 (0.85)
Number of cluster members	34	170	53
Proportion of respondents	13.23%	66.15%	20.62%

[b]

Cluster Analysis on Age and Domotics and automation systems knowledge

	Cluster 1 Mean (S.D.)	Cluster 2 Mean (S.D.)	Cluster 3 Mean (S.D.)	Cluster 4 Mean (S.D.)	Cluster 5 Mean (S.D.)	Cluster 6 Mean (S.D.)	Cluster 7 Mean (S.D.)	Cluster 8 Mean (S.D.)	Cluster 9 Mean (S.D.)	Cluster 10 Mean (S.D.)	Cluster 11 Mean (S.D.)
Home automation is a valid aid for elderly or disabled people	2.80 (0.70)	4.2 (0.80)	4. (0.94)	4.71 (0.48)	4.62 (0.51)	2.62 (0.51)	1.33 (0.5)	4.63 (0.67)	2.5 (0.54)	1.66 (0.57)	3.5 (0.52)
Home automation systems allow energy save and therefore reduce consumption and waste	2.92 (0.46)	4.28 (0.58)	4.01 (0.70)	4.14 (0.69)	2.12 (0.64)	2.62 (0.51)	1.11 (0.33)	5. (0.)	2.16 (0.40)	2. (0.)	3.5 (0.52)
Home automation systems are capable to make home safer	2.97 (0.51)	4.45 (0.50)	4.05 (0.81)	2.57 (0.53)	3.62 (0.74)	2.37 (0.74)	1.11 (0.33)	4.90 (0.30)	2.5 (1.04)	2. (0.)	3.5 (0.52)
Number of cluster members	42	91	59	7	8	8	12	11	6	3	10
Proportion of respondents	16.3%	35.4%	23%	2.7%	3.1%	3.1%	4.7%	4.2%	2.3%	1.1%	3.9%

[c]

the basis of the shortest distance comparing data without acquiring any information about it. To generate the clusterization on the input of dataset, Wolfram Mathematica software used Gaussian Mixture modelling the probability density

Table 4. Cluster Analysis on Online services and Propensity for energy saving and opinion on current energy supplier [a], Willingness to adopt renewable energies [b], Domotic and automation systems knowledge [c].

Cluster Analysis on Online Services and Propensity for energy saving and opinion on current energy supplier

	Cluster 1 Mean (S.D.)	Cluster 2 Mean (S.D.)
I am available to use only class A ++ household appliances	2.13 (1.08)	4.29 (0.96)
My current energy supplier is reliable	2.28 (0.93)	3.62 (0.92)
I think that is important to have transparency and clarity in the invoice by the electricity supplier	2.32 (1.09)	4.77 (0.51)
I think that energy suppliers' offers currently on the market are cheap	2.47 (0.91)	2.94 (1.01)
Number of cluster members	46	211
Proportion of respondents	17.9%	82.1%

[a]

Cluster Analysis on Online Services and Willingness to adopt renewable energies

	Cluster 1 Mean (S.D.)	Cluster 2 Mean (S.D.)	Cluster 3 Mean (S.D.)
I am available to use renewable energy sources and/or products/technologies capable to energy save	2.51 (1.15)	4.75 (0.45)	4.09 (1.04)
I agree if in my city more energy will be produced by renewable sources	2.66 (1.23)	4.87 (0.33)	4.29 (1.03)
I agree to pay extra to use energy produced by renewable sources	2.42 (1.04)	3.25 (1.28)	2.64 (1.01)
Number of cluster members	56	170	31
Proportion of respondents	21.8%	66.1%	12.1%

[b]

Cluster Analysis on Online Services and Domotics and automation systems knowledge

	Cluster 1 Mean (S.D.)	Cluster 2 Mean (S.D.)	Cluster 3 Mean (S.D.)
Home automation is a valid aid for elderly or disabled people	3.61 (1.13)	3.93 (1.18)	3.61 (1.10)
Home automation systems allow energy save and therefore reduce consumption and waste	3.58 (1.09)	3.82 (1.03)	3.58 (1.07)
Home automation systems are capable to make home safer	3.69 (1.08)	3.86 (1.20)	3.55 (0.92)
Number of cluster members	149	74	34
Proportion of respondents	58%	28.8%	13.2%

[c]

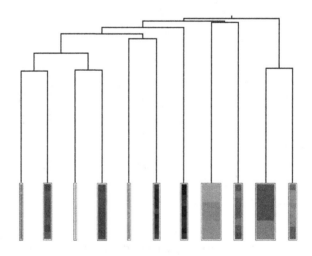

Fig. 2. Dendrogram shows the number of clusters related Age and Domotic analysis

of a numeric space using a mixture of multivariate normal distribution. Furthermore, as regarding the distance, the software clusterized data according to a similarity function. The similarity function automatically apply the Euclidean distance, which groups elements into clusters automatically according to their proximity. The different analyses carried out on dataset show similar results both on mean and standard deviation and number of clusters created. In fact, emerged 2 or 3 main clusters and only in one case "Map [Length, cluster]" is equal to 11. However, even in this latter case the greatest number of elements is grouped in the first 3 segments. A segment (133 respondents) belong to 18–24 year age range are aware that "Domotic" is not a valid aid only for the elderly or disabled people but can contribute to energy saving and therefore reduce consumption and waste and capable to make home safer. A strong awareness is noted in the 25 and 34 year age range (74) as regards Domotic and automation systems knowledge. While the population belonging to an age group over 45 does not show a clear knowledge of these sistems. On the base of the sample's demographic characteristics, we identified the following consumer profiles: A first profile with 18–24 year age range, family unit composed by 3 persons, uses both e-commerce services and social networks. This subjects are inclined to adopt technologies useful to energy saving and to employ renewable source. The second profile with 25–34 year age range, family unit composed by 3 persons, strongly agree to energy saving and the adoption of technologies for renewable energy. They believe that is important to have clarity and transparency on the invoice and are willing to use A++ household appliances. However, they consider today's energy market cheap. The third profile with 35 to 44 age, family unit composed by 3 persons, are not agree with production and use of renewable energy sources and consider their energy supplier unreliable. Furthermore, they are not willing to pay extra in the invoice for renewable energy production and are strongly discouraged by the energy market. Figure 2 shows the dendrogram of the performed hierarchical cluster analysis using age and Domotic variables. The dendrogram represents the similarities between respondents, grouped into 11 main clusters.

7   Conclusions

Cluster analysis determines a reduction of the total quantity of information to reach homogeneous groups. Each consumer analysed has x variables (demographic, behavioural or affective) that can be quantitative and qualitative. Hence, dividing into categories means considering, in a group of individuals common characteristics with the group to which he belongs. This analysis is applied in different pattern-analysis and fields and it is relevant to make assessments on data structure [49]. In this work the purpose is to detect similarities among dataset collected by the questionnaire in order to identify distinct patterns capable to describe customers behaviour profile. In particular, this is useful to define the profile for each customer class, and the calculation of the global power and energy information for the customer classes in order to adapt the billing tariffs.

As regards POWERCLOUD project, it is important to collect information about consumer's profile, on their preferences and their habits in order to promote specific marketing activities, to handle customer needs and formulate effective business strategies and identify firms strengths and weaknesses. For example, advertising could be targeted to persuade consumers that are unwilling to pay extra to adopt renewable energy. It will be useful to highlight that the initial outlay, will allow to obtain economic advantages later. Moreover, specific cookies could be showed on websites usually used by consumers for "booking holidays". Broader audience could be directly engaged through social networks in acquiring POWERLCOUD community benefits also in real time.

Acknowledgements. This research was supported by the following grants POWER-CLOUD (PON I&C2014-2020-MISE F/050159/01-03/X32).

References

1. Yu, X., Xue, Y.: Smart grids: a cyber-physical systems perspective. Proc. IEEE **104**(5), 1058–1070 (2016)
2. Dada, J.O.: Towards understanding the benefits and challenges of Smart/Micro-Grid for electricity supply system in Nigeria. Renew. Sustain. Energy Rev. **38**, 1003–1014 (2014)
3. Davis, F.D.: Perceived usefulness, perceived ease of use, and user acceptance of information technology. MIS Q. **13**(3), 319–340 (1989)
4. Park, C.K., Kim, H.J., Kim, Y.S.: A study of factors enhancing smart grid consumer engagement. Energy Policy **72**, 211–218 (2014)
5. Mutanen, A., Ruska, M., Repo, S., Jarventausta, P.: Customer classification and load profiling method for distribution systems. IEEE Trans. Power Delivery **26**(3), 1755–1763 (2011)
6. Li, W.T., et al.: Demand response management for residential smart grid: from theory to practice. IEEE Access **3**, 2431–2440 (2015)
7. Tu, C., He, X., Shuai, Z., Jiang, F.: Big data issues in smart grid-a review. Renew. Sustain. Energy Rev. **79**, 1099–1107 (2017)
8. Zhou, K., Fu, C., Yang, S.: Big data driven smart energy management: from big data to big insights. Renew. Sustain. Energy Rev. **56**, 215–225 (2016)
9. Wang, K., et al.: Wireless big data computing in smart grid. IEEE Wirel. Commun. **24**(2), 58–64 (2017)
10. Strüker, J., Dinther, C.: Demand response in smart grids: research opportunities for the IS discipline. In: Proceedings of the Eighteenth Americas Conference on Information Systems, Seattle, Washington, 9–12 August, pp. 1–10 (2012)
11. Chen, H.H., Chen, S., Lan, Y.: Attaining a sustainable competitive advantage in the smart grid industry of China using suitable open innovation intermediaries. Renew. Sustain. Energy Rev. **62**, 1083–1091 (2016)
12. Chicco, G.: Overview and performance assessment of the clustering methods for electrical load pattern grouping. Energy **42**(1), 68–80 (2012)
13. Zotos, K.: Performance comparison of Maple and Mathematica. Appl. Math. Comput. **188**(2), 1426–1429 (2007)
14. Alfano, I., Carini, M., Gabriele, L.: Building SCIENAR, a virtual community of artists and scientists: usability testing for the system improvement. In: Lazakidou,

A. (ed.) Virtual Communities, Social Networks and Collaboration, pp. 147–161. Springer, New York (2012). https://doi.org/10.1007/978-1-4614-3634-8_8

15. Bilotta, E., Di Blasi, G., Stranges, F., Pantano, P.: A gallery of Chua attractors: Part VI. Int. J. Bifurcat. Chaos **17**(06), 1801–1910 (2007)

16. Bilotta, E., Stranges, F., Pantano, P.: A gallery of Chua attractors: Part III. Int. J. Bifurcat. Chaos **17**(3), 657–734 (2007)

17. Bilotta, E., Di Blasi, G., Stranges, F., Pantano, P.: A gallery of Chua attractors: Part IV. Int. J. Bifurcat. Chaos **17**(04), 1017–1077 (2007)

18. Bilotta, E., Bossio, E., Pantano, P.: Chaos at school: Chua's circuit for students in junior and senior High School. Int. J. Bifurcat. Chaos **20**(01), 1–28 (2010)

19. Bilotta, E., Pantano, P., Vena, S.: Artificial micro-worlds part I: a new approach for studying life-like phenomena. Int. J. Bifurcat. Chaos **21**(2), 373–398 (2011)

20. Bilotta, E., Pantano, P.: Artificial micro-worlds part II: cellular automata growth dynamics. Int. J. Bifurcat. Chaos **21**(03), 619–645 (2011)

21. Bertacchini, F., Bilotta, E., Carini, M., Gabriele, L., Pantano, P., Tavernise, A.: Learning in the smart city: a virtual and augmented museum devoted to chaos theory. In: Chiu, D.K.W., Wang, M., Popescu, E., Li, Q., Lau, R. (eds.) ICWL 2012. LNCS, vol. 7697, pp. 261–270. Springer, Heidelberg (2014). https://doi.org/10.1007/978-3-662-43454-3_27

22. Bertacchini, F., Bilotta, E., Gabriele, L., Pantano, P., Tavernise, A.: Toward the use of Chua's circuit in education, art and interdisciplinary research: some implementation and opportunities. Leonardo **46**(5), 456–463 (2013)

23. Abdechiri, M., Faez, K., Amindavar, H., Bilotta, E.: The chaotic dynamics of high-dimensional systems. Nonlinear Dyn. **87**(4), 2597–2610 (2017)

24. Abdechiri, M., Faez, K., Amindavar, H., Bilotta, E.: Chaotic target representation for robust object tracking. Sig. Process. Image Commun. **54**, 23–35 (2017)

25. Bertacchini, F., Bilotta, E., Caldarola, F., Pantano, P.: The role of computer simulations in learning analytic mechanics towards chaos theory: a course experimentation. Int. J. Math. Educ. Sci. Technol. **50**(1), 100–120 (2019)

26. Giglio, S., Bertacchini, F., Bilotta, E., Pantano, P.: Using social media to identify tourism attractiveness in six Italian cities. Tour. Manag. **72**, 306–312 (2019)

27. Firouznia, M., Faez, K., Amindavar, H., Koupaei, J.A., Pantano, P., Bilotta, E.: Multi-step prediction method for robust object tracking. Digit. Signal Proc. **70**, 94–104 (2017)

28. Adamo, A., Bertacchini, P.A., Bilotta, E., Pantano, P., Tavernise, A.: Connecting art and science for education: learning through an advanced virtual theater with "talking heads". Leonardo **43**(5), 442–448 (2010)

29. Bertacchini, F., Bilotta, E., Gabriele, L., Pantano, P., Servidio, R.: Using Lego MindStorms in higher education: cognitive strategies in programming a quadruped robot. In: Workshop Proceedings of the 18th International Conference on Computers in Education, pp. 366–371 (2010)

30. Bertacchini, F., et al.: An emotional learning environment for subjects with Autism Spectrum Disorder. In: International Conference on Interactive Collaborative Learning (ICL), pp. 653–659. IEEE (2013)

31. Bertacchini, F., Tavernise, A.: Knowledge sharing for cultural heritage 2.0: prosumers in a digital agora. Int. J. Virtual Commun. Soc. Netw. **6**(2), 24–36 (2014)

32. Vaca-Cárdenas, L.A., et al.: Coding with scratch: the design of an educational setting for elementary pre-service teachers. In: International Conference on Interactive Collaborative Learning (ICL), pp. 1171–1177. IEEE (2015)

33. Gabriele, L., Marocco, D., Bertacchini, F., Pantano, P., Bilotta, E.: An educational robotics lab to investigate cognitive strategies and to foster learning in an arts and humanities course degree. Int. J. Online Eng. (iJOE) **13**(04), 7–19 (2017)
34. Bertacchini, F., Bilotta, E., Lombardo, M.C., Sammartino, M., Pantano, P.: Brain-like large-scale cognitive networks and dynamics. Eur. Phys. J. Spec. Top. **227**(7–9), 787–797 (2018)
35. Gabriele, L., Bertacchini, F., Tavernise, A., Vaca-Cárdenas, L., Pantano, P., Bilotta, E.: Lesson planning by computational thinking skills in Italian pre-service teachers. Inf. Educ. **18**(1), 69–104 (2019). https://doi.org/10.15388/infedu.2019.04
36. Zhou, L., Pan, S., Wang, J., Vasilakos, A.V.: Machine learning on big data: opportunities and challenges. Neurocomputing **237**, 350–361 (2017)
37. Love, B.C.: Comparing supervised and unsupervised category learning. Psychon. Bull. Rev. **9**(4), 829–835 (2002)
38. Bertacchini, F., Bilotta, E., Pantano, P.: Shopping with a robotic companion. Comput. Hum. Behav. **77**, 382–395 (2017)
39. Duque, J., Patino, J., Betancourt, A.: Exploring the potential of machine learning for automatic slum identification from VHR imagery. Remote Sens. **9**(9), 895 (2017)
40. Tanwar, A.K., Crisostomi, E., Ferraro, P., Raugi, M., Tucci, M., Giunta, G.: Clustering analysis of the electrical load in European countries. In: 2015 International Joint Conference on Neural Networks (IJCNN), pp. 1–8. IEEE (2015)
41. Benítez, I., Quijano, A., Díez, J.L., Delgado, I.: Dynamic clustering segmentation applied to load profiles of energy consumption from Spanish customers. Int. J. Electr. Power Energy Syst. **55**, 437–448 (2014)
42. Antoniadis, A., Brossat, X., Cugliari, J., Poggi, J.M.: Clustering functional data using wavelets. Int. J. Wavelets Multiresolut. Inf. Process. **11**(01), 1350003 (2013)
43. Raju, S., Chandrasekaran, M.: Performance analysis of efficient data distribution in P2P environment using hybrid clustering techniques. Soft Comput. **23**(19), 9253–9263 (2019)
44. Nerurkar, P., Shirke, A., Chandane, M., Bhirud, S.: Empirical analysis of data clustering algorithms. Procedia Comput. Sci. **125**, 770–779 (2018)
45. Wolfram, S.: An elementary introduction to the Wolfram Language. Wolfram Media, Incorporated (2017)
46. Nielsen, J., Tahir, M.: Homepage Usability: 50 Websites Deconstructed. New Riders Publishing, Thousand Oaks (2001)
47. Nielsen, J.: Designing Web Usability. Verlag: Markt+Technik Verlag; Adresse: München (2001)
48. Frank, A.G., Dalenogare, L.S., Ayala, N.F.: Industry 4.0 technologies: implementation patterns in manufacturing companies. Int. J. Prod. Econ. **210**, 15–26 (2019)
49. Jain, A.K., Murty, M.N., Flynn, P.J.: Data clustering: a review. ACM Comput. Surv. (CSUR) **31**(3), 264–323 (1999)

A Grossone-Based Numerical Model for Computations with Infinity: A Case Study in an Italian High School

Francesco Ingarozza$^{(\boxtimes)}$, Maria Teresa Adamo, Maria Martino, and Aldo Piscitelli

Liceo Scientifico Statale "Filolao", via Acquabona, 88821 Crotone, KR, Italy
francesco81tfa@gmail.com, mt.adamo@hotmail.it,
mariafilomena.martino@gmail.com, aldo.piscitelli@inwind.it

Abstract. The knowledge and understanding of abstract concepts systematically occur in the studies of mathematics. The epistemological approach of these concepts gradually becomes of higher importance as the level of abstraction and the risk of developing a "primitive concept" which is different from the knowledge of the topic itself increase. A typical case relates to the concepts of infinity and infinitesimal. The basic idea is to overturn the normal "concept-model" approach: no longer a concept which has to be studied and modeled in a further moment but rather a model that can be manipulated (from the calculation point of view) and that has to be associated to a concept that is compatible with the calculus properties of the selected model. In this paper the authors want to prove the usefulness of this new approach in the study of infinite quantities and of the infinitesimal calculus. To do this, they expose results of an experiment being a test proposed to students of a high school. The aim of the test is to demonstrate that this new solution could be useful in order to enforce ideas and acknowledgment about infinitesimal calculus. In order to do that, the authors propose a test to their students a first time without giving any theoretical information but only using an arithmetic/algebraic model. In a second moment, after some lectures, the students repeat the test showing that new better results come out. The reason is that after lessons, students could join new basic ideas or primitive concepts to their calculus abilities. By such doing they do not use a traditional "concept–model" but a new "model–concept" solution.

Keywords: Mathematics education · Teaching/learning methods and strategies · Grossone · Computer tools

1 Introduction

The understanding and knowledge of abstract concepts or of infrequent concepts, are that of an element which systematically occurs in the study of mathematics

© Springer Nature Switzerland AG 2020
Y. D. Sergeyev and D. E. Kvasov (Eds.): NUMTA 2019, LNCS 11973, pp. 451–462, 2020.
https://doi.org/10.1007/978-3-030-39081-5_39

(see [16,23]). Therefore the epistemological approach in regards to these concepts, becomes fundamental to give guarantees of success in terms of didactic results. The sense of this observation is very important. Besides it increases as the same level of abstraction of mathematics concepts and related models increases. The reason for that sentence is that as the level of abstraction increases and becomes considerable, it increases the risk to develop those fundamental ideas which could be different form the real being of knowledge. Those fundamental ideas are defined primitive concepts; they could be very far from the real subject of knowledge. Not only, that ideas could be misleading with the respect to the whole ideas of knowledge (see [3,5,13–15,17,18]). One of the abstraction elements that high school students encounter is related to the concepts of infinity and infinitesimal (see, for example, [12,16,22,24,27,28]). Therefore it becomes mandatory a path of knowledge, of discussion about traditional teaching. Such a strategy can be one, which makes use of a symbolic association of a "symbol-concept" model. This way it becomes possible to normalize a process based on an abstract concept, whose in-depth discussion requires a convergence of different signifiers.

This normalization process can be reached through the assumption of a single shared model, such a syllabus based on an epistemological model (see [19,21]). The basic idea is to overturn the normal "concept-model" approach: no longer a concept to be studied and modelled (only in that of a second time), but a model that can be manipulated (from the point of view of calculation) and to which it is possible to associate a concept. The associated (to the model) concept is compatible with the calculation properties of the identified model and it's also compatible with the theoretical results which can be presented (see, for instance, [3,5,21]). The described approach is the one which has been used in this didactic experiment. It represents a new idea for the presentation of the concepts of infinity and infinitesimal. The objective is therefore the ability to express a concept regardless of the intrinsic difficulties of the same representation. For the infinite quantity, the symbol ① ("grossone") is used. By using this model-based approach, the "symbol-concept" association overcomes the operational limits of a no adequate language. In the following discussion we present the results of a didactic experimentation. This experiment is aimed at demonstrating the validity and effectiveness, in terms of didactic repercussions. The demonstration is based on a real didactic case which is the study of infinitesimal calculus. Through this case we want to prove the validity and effectiveness of the "symbol-concept" approach and of the model based approach in computational arithmetic (see [4,12,19,21,23,25]). It is important to note that another similar experiment has been carried out independently during the same periods (see [1]).

It is important to remark that the content of this paper concerns about the Arithmetic of infinite (based on the use of the symbol ①). This idea was developed starting from 2003 by Sergeyev (see, e.g., [22,23,25,26]). We should notice that there are different examples in the scientific literature that can confirm the usefulness of such an approach in different fields of mathematics (see [2,6–11,19,20,22–27]). Infinite Arithmetic represents a new and different point of view on the concept of number and its related concept of counting. Starting from their own experiences, the authors can confirm the difficulties of the traditional calculus that students face, especially when they deal with counting. For example, difficulties related to the study of probabilities are originated by the difficulties of counting events. The new approach can be applied both to already known problems and to new ones. For the first class of problems this method can confirm well known results while for new ones it can represent an alternative way for finding solutions. Regarding the study of infinitesimal calculus, the grossone based model can represent a good solution which can be easily understood by high school students.

2   Description of the Experimentation

In the following two subsections we deal with the sample and the pursued objectives and purposes.

2.1   The Sample

The didactic experiment was directed to the students in their 4th year at the secondary school *Liceo Scientifico Filolao*, in Crotone, Italy. In particular, the sample was evaluated by three classes (from now we call them class A, class B and class C in order to save the privacy of the students) and involved a total of about 70 students; it lasted about 8 h and was carried out in the period of April/May 2019. The choice of the sample and of the period is not random but it has a precise reason: during this period those students started to approach infinitesimal calculus and exactly limits operations. Because of these considerations, it is correct to suppose that there are at least four reasons to consider the chosen sample and the period to be good choices:

- Confidence with the concepts exposed in the experiment;
- Students know the difficulties in regards to a symbolic-arithmetic manipulation of these subjects;
- The possibility to verify, after the experiment, that it is possible to have an algebraic and computational manipulation of the topics;
- The possibility to verify the advantages deriving from this model-based approach for infinitesimal calculus.

It should be reminded that all the classes involved in the experiment have, as planned, already operated with elements of infinitesimal calculus and with the study of indeterminate forms.

2.2   Objectives and Purposes

The experiment carried out on the selected sample was not only a didactic experiment but also an epistemological experiment: the target of the experiment was twofolded. On the one hand, the well-known importance of presenting concepts that require a good deal of abstraction as clearly as possible and far from ambiguities, doubts and perplexities, and jointly, a good mastery of arithmetical, algebraic and computational skills with regard to presented concepts. Near to this aspect, purely didactic, from an epistemological point of view, it was/is important to verify the didactic efficacy of the approach used, starting from the verification and clear comprehension, and with the mastery of the calculation properties of the concepts presented to be able to draw appropriate considerations about the usefulness of the presented approach (see [13,15,18]).

3   Results of the Experiment

The sample was submitted to a didactic experiment which was made up by three distinct phases:

- Test administration;
- Didactic lectures oriented on commenting the test;
- Test administration.

The basic idea of the experiment is to highlight some shadow areas which are related to some aspects inherited by infinitesimal calculus and especially to the formal treatment of the concept of infinity. In order to demonstrate the didactic efficiency of the model based approach, a preliminary test was submitted to the students; the aim of this test was to verify good computational capacities linked to some symbols in the face of a limited understanding of some concepts or of some obvious gaps in the complete and full understanding of those same concepts. In particular, the grossone symbol was presented within the preliminary test but a complete description of grossone and its properties was not given. Only one information was provided to the students, in the introduction of the preliminary test. The information is the clarification of the property $①> n$, $\forall$ finite $n \in \mathbb{N}$ (for obvious reasons of formal completeness, see also [19]). After doing that clarification, the students were given a test of 46 questions divided into four macro areas:

- Section 1 - Order relationships;
- Section 2 - Arithmetic;
- Section 3 - Equality;
- Section 4 - Counting of the elements of known numerical sets.

In Section 1 there are questions underlying the order relationships constructed starting from the symbol $①$. This section presents 5 open questions. It is therefore possible to answer each question by selecting one of the 5 proposed solutions. With regard to the questions contained in Section 1 it is reasonable to expect sufficiently high results. Moreover, the almost non-existent knowledge about $①$ is not a limitation for a successful result in Section 1. Indeed, the starting hypothesis ($①> n$, $\forall$ finite $n \in \mathbb{N}$) is more than sufficient to answer with a certain degree of accuracy and with a reduced limit of error, to the questions contained in Section 1 concerning trivial order relationships between arbitrary quantities containing ①. In Section 2, six algebraic questions/exercises were submitted. In particular, to pupils were given some expressions to solve. The expressions contain within them the symbol ① (which must be treated as a trivial and simple algebraic quantity, starting from the previous consideration ① > n, $\forall n \in \mathbb{N}$). It is evident that since these are exercises of a purely algebraic nature (expressions!) the submission method of Section 2 is that of closed questions: the students have to perform the required calculations and enter the results. We can confirm the same observations made for Section 1 even for Section 2; in this section of a pure arithmetic/algebraic exercises are submitted. More precisely they (the exercises) are expressions containing the quantity ①. They can be treated as expressions of literal calculation that contain precisely the quantity ① instead of a monomial with any literal part (no clarifications or further properties are required on the numerical set which contains the quantity ①).

It is reasonable to expect quite high results for the entire Section 2 and it is quite obvious to suppose that different results (far from expected ones) may be due to trivial calculation errors. In Section 3, questions were proposed in the form of open ended questions, and they are related to the "parity" of quantities obtained starting from ①. Finally, in Section 4, students have to answer to questions regarding the counting of elements of known numerical sets or parts of them. In Section 3 and Section in 4 results change considerably. In fact, the only given information about ① is not more sufficient to provide exhaustive clarifications that allow unequivocal and correct answers to the questions which are proposed in Section 3 and Section 4 (apart from rarely cases such as the first 7/8 questions of Section 4 in which it is possible to answer by using the normal notions of infinitesimal calculation and limit operation). Besides it is important to remember that the test was made in such a way to avoid the incidence on the final result of randomly right answers (especially in sections in which open ended questions are submitted). In order to avoid false positive cases (randomly correct answers) which could distort the evaluation on the didactic impact of the experimentally presented model, it was decided to assign a highly penalizing negative score for each wrong answer. In particular, wrong answers generate a score of -1, which therefore cancels the score generated by a right answer. In this way it is reasonable to expect that the students won't answer the questions which are related to unknown subjects and topics. By using this strategy is possible to suppose a strong reduction of false positive cases; in this way it is possible to immunize the experiment against this potential risk that could affect its actual validity.

Starting from the previous considerations regarding Section 3 and Section 4, including the strategy of answer evaluation, it is therefore reasonable to expect a significant lowering of the results obtained by the students in Section 3 and Section 4. Notice also that in a total of 46 total applications, Section 4 evaluates more than 54% and Section 3 evaluates more than 20% of the total; for this reason the two sections jointly account for over 75% and therefore it is reasonable to expect very low results from the preliminary test. The fundamental idea of the experiment is just this: to verify a radical change of the results of the same test proposed during the initial and final phases, on the same sample. As was already said, there is only one significant difference between these two phases of the test: the test was submitted for the second time to the same sample of students only after they had followed a cycle of lectures regarding the Grossone theory and model. In such way the students could develop a new basic idea and a new way to approach infinitesimal calculus.

The aim of the experiment in its whole is to demonstrate that this new approach could improve the development of new primitive concepts for students. The first aim of the lesson is to speak to the students about the model used for the representation of infinite quantities, that is, the symbol ① which starts as a numerical approach and extends to all its properties (see [19, 23, 25]). The target of the test is precisely to demonstrate that starting form an easy algebraic manipulation of a symbol, an association with more abstract concepts can be useful in order to understand aspects related to the concept. In particular, and in detail (Section 3 and Section 4) a few expedients are sufficient to start from the innovation associated with a mere and new definition. This definition is that of a new numerical set, made up of an infinite number of elements some of which are infinite elements related to infinite quantities. Thanks to this new and different approach, (grossone model based) students could be able to reach a very high level of knowledge, in a new way which could be a logical unexceptionable manner. This is probably the only way, or one of the few available solutions in order to reach this knowledge, in regard to these subjects. The same concepts could be precluded or not easily deducible by using a different strategy so far from that described on this paper, for example, the traditional way (as used and intended in Italian schools).

The last consideration can be confirmed by results of the test which had been submitted for the first time to students. In fact, these students were in the same conditions as the most of students of Italian high schools. They had never used the new approach which is proposed in this article. Such knowledge allows, in the final phase of the experimentation, to be able to carry out the same preliminary test (even if it's called final test) from which better results come out. It can be said that this experiment is an evaluation and measurement model for the behavior of didactic efficiency of the model based approach.

In the following, some examples of questions of the test are reported.

For Section 1:

Let us consider ① symbol as a positive quantity such that $①> n$ ($\forall$ finite $n \in \mathbb{N}$).

Choose the right order relationship among the different proposed solutions

(1) ① and -①
    (A) $①> -①$
    (B) We can't establish an order relationship
    (C) $-①\leq ①$
    (D) $-①= ①$
    (E) $①< -①$
(2) 2① and 7①
    (A) $2①> 7①$

(B) 7①= 2①
(C) We can't establish an order relationship
(D) 2①≥ 7①
(E) 2①< 7①

(3) (1/2) ① and (2/3)①
    (A) (1/2)①> (2/3)①
    (B) (1/2)①= (2/3)①
    (C) (2/3)①> (1/2)①
    (D) We can't establish an order relationship
    (E) -(2/3)①> (1/2)①

For Section 2:
Solve the following exercises:

(1) $6(① - 3) - 8(9 + 3①)$
(2) $①(3 + ①) - 4①^2 - 3①(1 - ①)$
(3) $2[①(3① - 5) + 3(① + 3) - ①(2① + 4① - 11)]$
(4) $\frac{1}{4}(① - 2)(① + 2) - \left(\frac{1}{2}① - 1\right)^2$
(5) $(①^2 + ① + 1)^2 - (①^2 + ①)^2$
(6) $(①^2 + ① + 1)(① - 1) - (① + 1)(①^2 - ① + 1) + 2$

For Section 3:
Determine whether the following quantities are even or odd numbers, by indicating the letter E (even) or O (odd).

(1) $2①$
(2) $5①$
(3) $7① + 1$
(4) $2① - 3$
(5) $\frac{1}{2}①$
(6) $\frac{3}{5}① + 3$
(7) $\frac{3}{7}①$
(8) $\frac{5}{4}① - 2$
(9) $\frac{1}{2}① - 3$
(10) $①$

For Section 4 (remind that in this Section 25 questions have been asked):

(1) Indicate the correct value of the sum $\infty + 2$, justifying your answer
(2) Indicate the correct value of the sum $\infty + \infty$, justifying your answer
(3) Indicate the correct value of the product $2 \cdot \infty$, justifying your answer
(4) Indicate the correct value of the product $\infty \cdot \infty$, justifying your answer
(5) Indicate the correct value of the sum $\infty - \infty$, justifying your answer

    ...

(9) Determine the number of elements of the set $\mathbb{N}$
(10) Determine the number of elements of the set $\mathbb{Z}$
(11) Let us consider E as the set of even numbers. Determine if the number of elements of the set $\mathbb{N} \setminus P$ is even or odd.

(12) Let us consider O as the set of odd numbers. Determine if the number of elements of the set $\mathbb{Z} \setminus O$ is even or odd.

(13) Determine the nature (even or odd) of elements of the set $\mathbb{Z} \setminus \{0\}$

(14) Determine the nature (even or odd) of elements of the set made up by the first 100 natural numbers minus the first 30 odd numbers

It should be noticed that the whole test is made up of 46 questions: 5 questions for Section 1 (order relationships), 6 questions, more exactly, 6 exercises (or expressions) for Section 2, 10 questions for Section 3 (parity problems) and 25 questions for Section 4. It is important to note that Section 3 and Section 4 are strictly related each to other. In Section 3 students have to find the nature (even or odd) of some quantities made up by the symbol ①. In Section 4 they are called to determine the nature of the number of elements of sets which have infinite elements. How can they do this? How can they count an infinite number of elements? The only way is to use the ① based model as described in the following section.

Another important thing to be noted in Section 4 is the following: students should be able to answer very easily to some questions by studying infinitesimal calculus according to the traditional approach which is used in Italian school. On the other hand, they are not prepared to overcame the second part of Section 4. According to the consideration which has been just done, it is possible to use the results of Section 4 to have an idea of the behavior of the new didactic approach for the study of infinitesimal calculus.

4 The Lessons

Lessons took place in the form of frontal teaching. They were characterized by the use of methodologies based on the model of social collaboration and by using a metacognitive approach. Due to these reasons, the model of the Socratic lesson was used. By doing this choice students were led by the teacher to build their knowledge of concepts step by step. The starting point was the commentary of the answers given (and above all not given) during the preliminary test. This comment has revealed two facts: few difficulties in performing simple complex arithmetic operations and at the same time a very big difficulty in counting (this problem is often ignored in the literature, see [3, 13, 18]). This finding was derived from questions such as:

- Is the number of elements of the set $\mathbb{N}$ even or odd?
- How many are the elements of the set $\mathbb{Z}$?

To these questions we tried to give an answer by using a logically unexceptionable path: it is possible not to know if the number of elements of the set $\mathbb{N}$ is even or odd but surely it will be the sum of two numbers which constitute a partition of $\mathbb{N}$, i.e. even numbers and odd numbers. These two numerical sets have the same number of elements and so, it is possible to deduce that whatever the number of these elements (let's call α this number), the number of the elements of $\mathbb{N}$ will always be double (and so 2α). Thus, it is easy to understand that the number of elements of $\mathbb{N}$ is even.

From this consideration we then moved to the definition of ①: the scale value at which all the elements of $\mathbb{N}$ are counted. From this point on the road has been downhill: the set $\mathbb{N}$ of natural numbers having elements from 1 to ① is defined. This set contains infinitely many elements (precisely, ① elements), some of them are finite, the others are infinite (see, e.g., [19, 23, 25, 26]). Then, the set $\widehat{\mathbb{N}}$ is defined as the set which contains positive integers larger than ①, i.e., $\mathbb{N} \subset \widehat{\mathbb{N}}$. From there it was possible with a few tricks that referred to the properties of the rest classes, to go on to discuss all the aspects which allow the students to answer to the most of the question of the test, especially to those of Section 3 and Section 4.

5 Analysis of the Results

In this section we show that the experimentation led to results which are broadly in line with the forecasts. The preliminary test has scores which never reach the level of sufficiency (related as the half of the maximum value, equal to 46 points). It is important to note that the most of scores levels are widely achieved in sections 1 and 2 concerning traditional arithmetic. As it was expected, it can be seen from Tables 1, 2 and 3 below, there is a considerable difference between the results which are carried out before the lectures and the results related to the second administration of the test (results related to the first administration of the test are always on the left for each comparison). On the one hand, it is possible to note that there are no significant differences (between the first administration of the test and the second one) related to Section 1 or Section 2. Moreover, in these sections, students have reached a score really closed to the maximum allowed value. On the other hand, it is not possible to say the same thing for Sections 3 and 4. In these cases, differences between the first administration of the test and the second one are very strong and they are in the direction to confirm the positive behavior of the proposed approach in terms of didactic efficiency.

The negative trend of the initial test results is mainly produced by the results of section 3 and section 4 which are based on questions that require a minimum level of knowledge at a conceptual level. The approach described above demonstrates, at the outcome of the test, its effectiveness since the same test, administered after only two lessons, produced completely different results. As it is possible to see, the difference is mainly due to the results of the questions posed in sections 3 and 4 which, in the preliminary test, provided the major difficulties to the students and had affected the final result.

Table 1. The mean score obtained by the students. The rows refer to the classes and the columns refer to the 4 different sections in which the tests have been divided. Columns for each section are double, "before" and "after" the cycle of lectures.

Class	Section 1 before/5	Section 1 after/5	Section 2 before/6	Section 2 after/6	Section 3 before/10	Section 3 after/10	Section 4 before/25	Section 4 after/25
A	3.14	3.52	3.86	4.56	1.1	9.71	2.71	24.33
B	1.64	4.45	3.86	5.77	1.82	9.45	0.27	22.5
C	3.08	4.83	4.8	4.63	0.32	9.75	0.68	23.96

Table 2. The standard deviations relative to the means listed in Table 1.

Class	Section 1 before/5	Section 1 after/5	Section 2 before/6	Section 2 after/6	Section 3 before/10	Section 3 after/10	Section 4 before/25	Section 4 after/25
A	2.41	1.01	2.89	1.28	1.23	0.39	3.06	0.7
B	6.23	0.79	8.66	0.36	2.6	0.43	0.74	1.25
C	1.19	0.22	2.24	1.98	6.37	0.35	1.26	1.96

Table 3. Table presents the mean vote of all the sections 1–4 with weights. The last two columns yield the relative standard deviations as in Table 2.

Class	Mean, sections 1–4 before lectures/46	Mean, sections 1–4 after lectures/46	Standard dev. before lectures/46	Standard dev after lectures/46
A	10.81	42.52	9.58	4.82
B	7.59	42.1	29.7	1.48
C	8.88	43.17	11.63	4.81

6   Conclusions

The experimentation performed in Liceo Scientifico "Filolao" using the grossone-based methodology has shown that the symbol-concept approach can be an improved solution with respect to the traditional concept-symbol approach at least in the discussion of those topics that are intrinsically difficult to treat and understand. It has been shown that the ①-based methodology allows students to understand better concepts related to infinity and infinitesimals. The authors hope that this work can be a starting point for similar educational experiments carried out on a large scale and even in different initial conditions regarding periods of the test, age of the students of the sample, etc. In this way it is possible to have a better overview of the real effectiveness of this strategy in terms of educational impact. Starting from these considerations, the authors intend to identify new didactic paths for the introduction in teaching of new computational algorithms that make use of the acquired knowledge.

Acknowledgements. The authors thank Fabio Caldarola, University of Calabria, for the supervision of the project and the Headmistress of *Liceo Scientifico "Filolao"*, Antonella Romeo, for the economic support. The authors thank the anonymous reviewers for their useful comments that have improved the presentation. Special thanks go to Irene Dattolo for her valuable support provided for the translation of the text.

References

1. Antoniotti, L., Caldarola, F., d'Atri, G., Pellegrini, M.: New approaches to basic calculus: an experimentation via numerical computation. In: Sergeyev, Ya.D., Kvasov, D.E. (eds.) NUMTA 2019. LNCS, vol. 11973, pp. 329–342. Springer, Heidelberg (2019)
2. Antoniotti, L., Caldarola, F., Maiolo, M.: Infinite numerical computing applied to Hilbert's, Peano's, and Moore's curves. Mediterranean J. Math. (to appear)
3. Asubel, D.: Educazione e processi cognitivi. Franco Angeli (1978)
4. Bertacchini, F., Bilotta, E., Caldarola, F., Pantano, P.: The role of computer simulations in learning analytic mechanics towards chaos theory: a course experimentation. Int. J. Math. Educ. Sci. Technol. **50**(1), 100–120 (2019)
5. Bonaiuti, G., Calvani, A., Ranieri, M.: Fondamenti di didattica. Teoria e prassi dei dispositivi formativi. Carrocci, Roma (2007)
6. Caldarola, F.: The exact measures of the Sierpiński d-dimensional tetrahedron in connection with a Diophantine nonlinear system. Commun. Nonlinear Sci. Numer. Simul. **63**, 228–238 (2018). https://doi.org/10.1016/j.cnsns.2018.02.026
7. Caldarola, F.: The Sierpiński curve viewed by numerical computations with infinities and infinitesimals. Appl. Math. Comput. **318**, 321–328 (2018). https://doi.org/10.1016/j.amc.2017.06.024
8. Caldarola, F., Cortese, D., d'Atri, G., Maiolo, M.: Paradoxes of the infinite and ontological dilemmas between ancient philosophy and modern mathematical solutions. In: Sergeyev, Y.D., Kvasov, D.E. (eds.) NUMTA 2019. LNCS, vol. 11973, pp. 358–372. Springer, Heidelberg (2019)
9. Caldarola, F., Maiolo, M., Solferino, V.: A new approach to the Z-transform through infinite computation. Commun. Nonlinear Sci. Numer. Simul. **82**, 105019 (2020). https://doi.org/10.1016/j.cnsns.2019.105019
10. Cococcioni, M., Pappalardo, M., Sergeyev, Y.D.: Lexicographic multi-objective linear programming using grossone methodology: theory and algorithm. Appl. Math. Comput. **318**, 298–311 (2018)
11. De Cosmis, S., De Leone, R.: The use of grossone in mathematical programming and operations research. Appl. Math. Comput. **218**(16), 8029–8038 (2012)
12. Ely, R.: Nonstandard student conceptions about infinitesimals. J. Res. Math. Educ. **41**(2), 117–146 (2010)
13. Faggiano, E.: "Integrare" le tecnologie nella didattica della Matematica: un compito complesso. Bricks **2**(4), 98–102 (2012)
14. Gastaldi, M.: Didattica generale. Mondadori, Milano (2010)
15. Gennari, M.: Didattica generale. Bompiani, Milano (2006)
16. Iannone P., Rizza D., Thoma A.: Investigating secondary school students' epistemologies through a class activity concerning infinity. In: Bergqvist E., et al. (eds.) Proceedings of the 42nd Conference of the International Group for the Psychology of Mathematics Education, Umeå, Sweden, vol. 3, pp. 131–138. PME (2018)
17. La Neve, C.: Manuale di didattica. Il sapere sull'insegnamento. La Scuola, Brescia (2011)
18. Palumbo, C., Zich, R.: Matematica ed informatica: costruire le basi di una nuova didattica **2**(4), 10–19 (2012)
19. Rizza, D.: Primi Passi nell'Aritmetica dell'Infinito. Un nuovo modo di contare e misurare (2019, Preprint)
20. Rizza, D.: A study of mathematical determination through Bertrand's Paradox. Philosophia Mathematica **26**(3), 375–395 (2018)

21. Scimone, A., Spagnolo, F.: Il caso emblematico dell'inverso del teorema di Pitagora nella storia della trasposizione didattica attraverso i manuali. La matematica e la sua didattica **2**, 217–227 (2005)

22. Sergeyev, Y.D.: A new applied approach for executing computations with infinite and infinitesimal quantities. Informatica **19**, 567–596 (2008)

23. Sergeyev, Y.D.: Arithmetic of infinity. 2nd electronic ed. 2013. Edizioni Orizzonti Meridionali, Cosenza (2003)

24. Sergeyev, Y.D.: Numerical point of view on Calculus for functions assuming finite, infinite, and infinitesimal values over finite, infinite, and infinitesimal domains. Nonlinear Anal. Ser. A: Theory Methods Appl. **1**(12), 1688–1707 (2009)

25. Sergeyev, Y.D.: Un semplice modo per trattare le grandezze infinite ed infinitesime. Matematica, Società Cultura: Rivista dell'Unione Matematica Italiana **8**(1), 111–147 (2015)

26. Sergeyev, Y.D.: Numerical infinities and infinitesimals: Methodology, applications, and repercussions on two Hilbert problems. EMS Surv. Math. Sci. **4**(2), 219–320 (2017)

27. Sergeyev, Y.D., Mukhametzhanov, M.S., Mazzia, F., Iavernaro, F., Amodio, P.: Numerical methods for solving initial value problems on the infinity computer. Int. J. Unconv. Comput. **12**, 3–23 (2016)

28. Tall, D.: A child thinking about infinity. J. Math. Behav. **20**, 7–19 (2001)

A Computational Approach with MATLAB Software for Nonlinear Equation Roots Finding in High School Maths

Annarosa Serpe$^{(\boxtimes)}$ (iD)

Department of Mathematics and Computer Science, University of Calabria,
Rende, CS, Italy
annarosa.serpe@unical.it

Abstract. The paper focuses on solving the nonlinear equation $f(x) = 0$, one of the classic topics of Numerical Analysis present in the syllabus of experimental sections of Italian high schools in secondary education. The main objective of this paper is to propose an example of constructivist teaching practice emphasizing the computational approach with the use of MATLAB software.

MATLAB is a high-performance language for technical computing, but it is also suitable for high school maths class teaching because of its powerful numeric engine, combined with interactive visualization tools. All this helps to keep teaching and learning of this part of mathematics alive and attractive.

Keywords: Iterative methods · Math education · MATLAB software

1 Introduction

Solving nonlinear equations has been a part of the Maths syllabus of the experimental sections of the Italian secondary school for three decades and is often present in the questions of the written test of the State Exams (National High School Examination).

Despite its considerable relevance, in classroom teaching practice, this topic is often overlooked or treated more from a theoretical than from a computational point of view. It is believed that this constitutes a serious gap in educational standards because the topic is understood in its essence only if various examples are resolved using a computer, and not by reducing the topic to a series of calculations, mostly done by hand, often insignificant and detached from real contexts. Nowadays, the computer can act as mediator between the concrete and the abstract thanks to the relationship between Mathematics and computer technology [14, 18]. For example, computer programming favours a highly educational mental training [12, 20, 27] at the same time sensible reality problems can effectively be solved or simulated. Programming involves the ability to generate a solution to a problem. Generating solutions means that one of the learning outcomes is the ability to solve problems and also, if the problem is a big problem, the ability to split the problem into sub-problems and create a generalizable central solution. In addition, the student achieves the ability to create usable, readable and attractive solutions. Programming with MATLAB software not only supports an

© Springer Nature Switzerland AG 2020
Y. D. Sergeyev and D. E. Kvasov (Eds.): NUMTA 2019, LNCS 11973, pp. 463–477, 2020.
https://doi.org/10.1007/978-3-030-39081-5_40

education of Applied Mathematics but at the same time provides teachers with tools and ideas conducive to engaging the students in activities-based learning of mathematics encouraging several aspects of maths - empirical, speculative, formal, and reflexive [6].

This paper is part of the debate surrounding programming in high school Maths, with special focus on implementing algorithms related to appropriate methods for solving the nonlinear equation $f(x) = 0$, with the use of MATLAB software. The aim is to favour the algorithm as a synthesis between the syntactic and semantic aspects of mathematical objects as well as to encourage computer simulation interpreted as a 'physical experiment' and a source of conjecture. The use of algorithms is a computational attempt to model and thus make problems in the real world 'effectively computable'.

The computational approach is very relevant to high school Maths education, and in this specific instance it helps students to relate the different semiotic representations of the iterative methods studied, to compare them and to analyse the pros and cons of applying them. In this way the students discover maths concepts and make generalizations, thus developing and promoting a different kind of mathematical thinking.

The paper is structured as follows. Section 2 presents the methodological framework used. In Sect. 3 the teaching design of the computational approach is implemented, followed by reflections on the comparative analysis of the algorithms, and the advantages and disadvantages of iterative methods to estimate roots of equations.

Finally, in Sect. 4 conclusions and recommendations are suggested.

2   Methodological Framework

In general, the appropriate methods for solving the nonlinear equation $f(x) = 0$ are iterative methods. They are divided into two groups: Open methods (OM) and Bracketing methods (BM).

The OMs require only a single starting value or two starting values that do not necessarily bracket a root. They may diverge as the computation progresses, but when they do converge, they usually do so much faster than BMs. Some of the best known OMs are Secant method, Newton-Raphson method, and Muller's method; whereas the BMs are Bisection method and False Position Method. In Italian high school, the teaching of iterative methods is often characterized by over-emphasizing algorithmic developments and procedural handling of the symbolic aspects of mathematical objects. In this style of teaching, the students construct a partial mathematical knowledge consisting mainly of algorithms, a situation that makes them manipulate symbols routinely, without giving significance to the basic concepts of iterative methods. The lack of articulation between the different semiotic registers [9] that should be acquired does not allow students to form an adequate comprehension of the mathematical concepts involved.

The difficulty lies in making students understand the main difference between a numerical method and a numerical algorithm. The immediate consequence of this distinction is that the same numerical method can be implemented using different algorithms; in particular, the different algorithms will not be equivalent, but some will be better than others. The support of mathematical software such as MATLAB can facilitate this learning process because it provides a means to articulate the different semiotic registers of a concept. In fact, a computational approach in the classroom helps to bring out two fundamental aspects:

1. a numerical method allows us to approximate a value otherwise impossible to determine with algebraic methods;
2. a numerical method is often characterized by the need to perform many, relatively simple, calculations based on a repetitive scheme, for this reason numerical algorithms are almost exclusively used for the realization of calculating programs performed by a computer.

The algorithm is a strong conceptual tool in that it involves all sorts of technical and intellectual inferences, interventions and filters [11]; therefore, it has always been a driving force in the history of mathematics [4]. Due to this, the choice of a particular algorithm for solving a given problem is the result of an analysis of the problem and a comparative analysis of the various algorithms, based on their cost and their accuracy, as well as intuition refined by experience. Therefore, computer programming is an unavoidable part in the design of algorithm.

The contribution of programming in the learning of school mathematics has been demonstrated in numerous projects and research settings [2, 5, 10, 17, 18, 25, 30].

Computer programming has been described by Nickerson (1982) [21] as a creative endeavour that requires planning, accuracy in language use, generation and testing of hypotheses, and ability to identify action sequences. It ought to represent a fundamental part of the literacy for twenty-first century citizens [26] as it is now a skill required for most jobs and spare time activities. While this is certainly true, one should not forget the psychological perspectives of computing in Mathematics [16].

The value of programming in educational and pedagogical terms has been recognized internationally [1, 5, 8, 19, 22–24, 28, 29]. To this end, it is important to stress the importance that algorithms are independent of the programming language used, and each algorithm can be expressed in different programming languages. The design of an algorithm is a demanding intellectual activity, significantly more difficult than expressing the algorithm as a program.

Programming with MATLAB not only helps to reinforce traditional mathematics learning, but also the teaching of numerical methods for solving some classical mathematical problems using computers. The students can create the algorithm to get the solution or they can plot the graph of the function and view it as a graphic visualization. In this way, students assimilate the notions and procedures of modelling real-world problems as mathematical problems as well as translating mathematical solutions into real-world solutions. All this motivates the students to appreciate the relevance of computational components in classical mathematics materials. In bringing students closer to the culture of analysis and numerical methods, the teacher plays a fundamental role because she/he must harmonize theory, methods and algorithms in a constructive, lively and understandable way. This requires a change of vision in the teaching which - in most cases - prefers a theoretical style which is scarcely usable for students as it lacks models and examples. A different approach, such as the computational one, makes it easier for students to make sense of what they study by maturing the awareness that the numerical methods constitute an aspect of applied mathematics. The teaching of iterative methods for solving the nonlinear equation $f(x) = 0$, effectively, is dependent upon a wide range of factors, but among the most important are those which are associated with activities and practice within the educational process. Coming from an applied mathematics perspective, simulations with the computer can be seen as a part of the modelling process. As a result of this, the teaching practice design proposed is based on the 'Numerical modelling cycle' (Fig. 1) by Sonar [31] with the use of MATLAB.

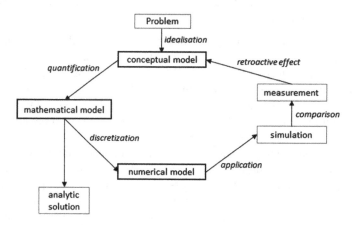

Fig. 1. Numerical modelling cycle [31].

From Fig. 1, it is clear that the conclusion of each modelling cycle is the implementation of a computer simulation program. Of course, numerical algorithms play a central role in this implementation of the model because they encode a particular type of abstraction. The algorithm bridges the gap between code and implementation, between software and experience. An implemented algorithm is, on the one hand, an intellectual gesture, and, on the other, a functioning system that incorporates in its structure the material assumptions about perception, decision and communication. In the end, MATLAB is used to translate the language of mathematics into the language of the computer [15].

3   Teaching the Iterative Methods with MATLAB

One of the basic principles of numerical analysis is iteration. In general, the idea of iteration indicates the repetition of a simple process to improve the estimation of the solution of a more complicated problem. Iterative methods are important in solving many of today's real-world problems [7], so it is important that your first approach to these methods be as positive and simple, as well as informative and educational, as possible. Based on the outlined methodological framework, the first step is to introduce a real-world problem.

3.1   Real-World Problem

The houses of Charles and David are located on two adjacent plots separated by a 2-meter-high fence. As summer approaches, Charles and David decide to paint the exterior walls of their houses overlooking the fence. To do this, Charles and Davis use two ladders respectively 8 m and 6 m long (see Fig. 2).
  Calculate the distance between the walls of the two houses.

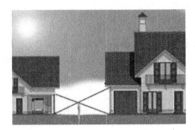

Fig. 2. Real task representation.

3.2   Mathematical Model

The position of the two ladders recalls two right-angled triangles with a common cathetus (see Fig. 3). From a closer observation it is possible to identify four right-angled triangles two by two similar to each other.

Fig. 3. Geometrical task representation.

  Let $\overline{AB} = x$, applying the Pythagorean theorem to right-angled triangles ABD and ACB we have:

$$\overline{BD} = \sqrt{\overline{AD}^2 - \overline{AB}^2} = \sqrt{64 - x^2}, \text{ and } \overline{AC} = \sqrt{\overline{CB}^2 - \overline{AB}^2} = \sqrt{36 - x^2}.$$

  Applying Thales' theorem to triangles ACB, HEB, and ABD, AHE respectively, we have the following proportions:

$$x : \overline{HB} = \sqrt{36 - x^2} : 2; \quad x : \overline{AH} = \sqrt{64 - x^2} : 2$$

By hypothesis $\overline{AH} = \overline{AB} - \overline{HB}$, and after the easy substitutions we have that:

$$2x\left(\frac{1}{\sqrt{64 - x^2}} + \frac{1}{\sqrt{36 - x^2}} - \frac{1}{2}\right) = 0$$

Now by the cancellation law for multiplication:

$$2x = 0 \text{ or } \left(\frac{1}{\sqrt{64 - x^2}} + \frac{1}{\sqrt{36 - x^2}} - \frac{1}{2}\right) = 0$$

In the first case, $2x = 0$ is not possible since $x \neq 0$ by hypothesis ($x = \overline{AB}$ is the distance between the walls of the two houses); then this implies that:

$$\left(\frac{1}{\sqrt{64 - x^2}} + \frac{1}{\sqrt{36 - x^2}} - \frac{1}{2}\right) = 0 \tag{1}$$

Quantitative Analysis. Equation (1) is not easy to solve; In fact, in this case we do not have any solving formula. Numerically, (1) can be equivalently expressed as $f(x) = 0$, where $f : [a, b] \rightarrow \mathcal{R}$ is a real function in a finite interval $[a, b] \subset \mathcal{R}$.

The domain of $f(x)$ is: $\forall x \in \,]-6, 6[\subset \mathcal{R}$. Since x represents the distance between the walls of the two houses, then the interval will be considered $]0, 6[\subset \mathcal{R}$. With the use of the MATLAB, it is possible to draw the graph of $f(x)$. To do this let us consider a grid of the domain by point $x_i \in [0, 6[, i = 1, \ldots, n$, and let us compute the corresponding values $y_i = f(x_i), i = 1, \ldots, n$. The naturalness of the method is accompanied by a theoretical support that guarantees its convergence property: the plot of the approximant is approaching the exact one by increasing n, i.e. the number of considered points[1]. Implementing MATLAB we need few instructions on the 'prompt of command':

```
>> x = [0:0.001:5.9];
>> y = 1./sqrt(64-x.^2) + 1./sqrt(36-x.^2)-1./2;
>> plot(x,y);
>> plot(x,y,'LineWidth',1.4);
>> plot(x,y,'b',x,0*x,'k','LineWidth',1.2);
```

The boundary point of the domain and the number of equispaced points are fixed, the function and vector of points are created (abscissa and ordinate of the points, respectively) and then the function is displayed through the code plot(x,y).

This creates a 2-D line plot of the data in Y versus the corresponding values in X. The last two commands are needed to set the plot's properties of the displayed curve (colour, kind of line, etc.).

---

[1] This is polynomial interpolation field.

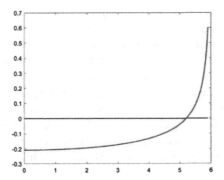

Fig. 4. Plot of function (1) in the interval [0, 5.9), for $n = 100$.

This plot (Fig. 4) help the students to visualize the proprieties of function (1), and to explicit the fact that a function typically changes sign in the vicinity of a root.

The presence of one root is suggested at about $x = 5.4$ where $f(x)$ appears to be tangent to the x axis. The graphic interpretation, as well as offering a rough estimate of the root, is an important tool for understanding the properties of the function and anticipating the pitfalls of numerical methods. Beyond its usefulness, the graphic technique used has a limited practical value because it is not precise.

3.3   The Bisection Method

Quantitative analysis returns a rough estimate of the root. This estimate can be used as an initial hypothesis to introduce iterative methods. In fact, function (1) is continuous on the interval $[0, 6[\subset \mathcal{R}$ and changes sign on opposite sides of the root. Starting from this observation it possible to more precisely identify the root by dividing the interval into a number of subintervals. Each of these subintervals is searched to locate the sign change. This process can obviously be continued until we hit the point where the interval has become so small that the root is determined with sufficient accuracy [3].

For example, see Fig. 5.

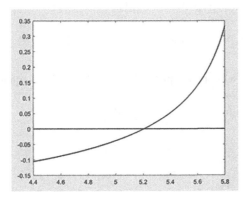

Fig. 5. Plot of function (1) in the interval $[4.5; 5.5]$ for $n = 100$.

By using this information, most numerical methods for $f(x) = 0$ compute a sequence of increasingly accurate estimates of the root.

These methods are called iterative methods. We assume that $f : \mathcal{R} \to \mathcal{R}$ i.e., $f(x) = 0$ is a function that is real valued and that x is a real variable. Suppose that $f(x) = 0$ is continuous on an interval $[a, b]$, and $f(a) \cdot f(b) < 0$.

Then $f(x)$ changes sign on $[a, b]$, and $f(x)$ has at least one root on the interval. The simplest numerical procedure for finding a root is to repeatedly halve the interval $[a, b]$, keeping the half for which $f(x)$ changes sign. This procedure is called the Bisection method, and is guaranteed to converge to a root denoted here by α.

Specifically, the procedure is as follows. Define $a_1 = a, b_1 = b$. Then for $n = 1, 2, 3, \ldots$ do

$$
\begin{vmatrix}
x_n = \frac{1}{2}(a_n + b_n) \\
\text{if } f(x_n) = 0 \quad \text{then } \alpha = x_n. \\
\text{if } f(x_n) < 0 \quad \text{then } a_{n+1} = b_{n+1} \quad \text{else} \\
a_{n+1} = a_n, \quad b_{n+1} = x_n
\end{vmatrix}
$$

Since $b_n - a_n = 2^{-(n-1)}(b - a), n = 1, 2, 3, \ldots$, and x_n is the midpoint of $[a_n, b_n]$ if α is the root eventually captured, we have

$$
|x_n - \alpha| \leq \frac{1}{2}(a_n + b_n) = \frac{b - a}{2^n}. \tag{2}
$$

Although convergence of an iterative process is certainly desirable, it takes more than just convergence to make it practical. What one wants is fast convergence. A basic concept to measure the speed of convergence is the order of convergence [13].

Therefore, it is necessary to recall the definition of linear convergence.

Definition. *Linear convergence* [13]. We can say that x_n converges to α (at least) linearly if

$$
|x_n - \alpha| \leq \varepsilon_n, \tag{3}
$$

where $\{\varepsilon_n\}$ is a positive sequence satisfying

$$
\lim_{n \to \infty} \frac{\varepsilon_n + 1}{\varepsilon_n} = c, 0 < c < 1. \tag{4}
$$

If (3) and (4) hold with the inequality in (3) replaced by an equality, then c is called the *asymptotic error constant*.

Thus, (3) holds with $\varepsilon_n = 2^{-n}(b - a)$ and

$$
\frac{\varepsilon_n + 1}{\varepsilon_n} = \frac{1}{2}, \text{all } n. \tag{5}
$$

This shows that the Bisection method converges (at least) linearly with asymptotic error constant (for the bound ε_n) equal to $\frac{1}{2}$.

Given an (absolute) error tolerance $Tol > 0$, the error in (2) will be less than or equal to Tol if

$$\frac{b-a}{2^n} \leq Tol.$$

Solved explicitly for n, this will be satisfied if

$$n = \frac{\log \frac{b-a}{Tol}}{\log 2},$$

where $\lceil x \rceil$ denotes the "ceiling" of x (i.e., the smallest integer $\geq x$).

Thus, we know *a priori* how many steps are necessary to achieve a prescribed accuracy [13]. The Bisection method is also known as dichotomy or binary chopping or the half-interval method. The procedure is called the bracketing method because two initial guesses for the root are required. As the name implies, the guesses must 'bracket', or be on either side of, the root.

Bisection Algorithm. The construction of the algorithm is an important phase to define the 'finite sequence of steps' that allows the computer to get to the solution. A simple algorithm for bisection calculation is as follows. The step are as follows:

[Initialization] $n = 1, a_n = a, b_n = b$

[Bisection Iteration] $x_n = \frac{1}{2}(a_n + b_n)$

[Convergence Test]   If $|f(x_n)| \leq \varepsilon$ then the zero is x_n. Stop.

If $f(a_n) \cdot f(x_n) < 0$ then $b_n = x_n$,   else $a_n = x_n$.

$n = n + 1$ and repeat step 2 until convergence is achieved.

3.4   Computer Simulation of Bisection Method in MATLAB Code

When implementing the procedure on a computer, it is clearly unnecessary to provide arrays to store a_n, b_n, x_n; one simply keeps overwriting. Assuming a and b have been initialized and TOL assigned, one could use the following MATLAB program code:

```
% Bisection method
  function [alfa,n]=bisection(f,aa,bb,TOL)
  n=1;
  a(n)=aa; b(n)=bb;
  error(n)=(bb-aa)/2;
  x(n)=(a(n)+b(n))/2;
  while error(n)>=TOL
        if f(x(n))==0
            alfa=x(n);
            return
        else if f(a(n))*f(x(n))<0
              a(n+1)=a(n);
              b(n+1)=x(n);
        else
              a(n+1)=x(n);
              b(n+1)=b(n);
        end
        n=n+1;
        error(n)=(b(n)-a(n))/2;
        x(n)=(a(n)+b(n))/2;
  end
        alfa=x(n-1);
    % The analysis of error
        [a',b',x',f(a)',f(b)',f(x)',abs(b-a)', error']
        subplot(1,2,1)
        plot(1:n,error,'-r.')
        title('error')
        xlabel('iterations')
        subplot(1,2,2)
        semilogy(1:n,error,'-k')
        grid on
        title('log(error)')
        xlabel('iterations');
        % Check on the number of iterations
        ceil(log((b(1)-a(1))/TOL)/log(2));
```

For the above mentioned MATLAB code an M-file used as a function was created[2].

---

[2] To write our program in MATLAB it is important to distinguish between a program (or MATLAB script) and a function. MATLAB scripts are "main programs" and functions are that which is written and can be used in them. The functions must be stored in the same directory where the script is which calls them. A MATLAB-script is stored as an M-file (a file with the suffix .m) and is executed in the command window by typing its name without the suffix.

A function has input parameters and delivers results as output parameters. So for our programme code (Bisection method), the function utilized in the 'prompt of command' is: `[alfa,n] = bisection(f,aa,bb,TOL)`.

This function has four input parameters and two output parameters. The input parameters are the boundary points a and b, the required tolerance (*TOL*) and the function (1). The output parameters are the root of the equation and the number of iterations.

In this contest it is important to highlight that in MATLAB the index number 0 (zero) is not allowed. Therefore, the iterative numbering (and that of the vector components) always starts from 1 (one).

3.5   Measurement and Output with MATLAB for Bisection Method

From the code proposed in Sect. (3.4), executing the following commands

```
>> f = @(x) 1./sqrt(64-x.^2) + 1./sqrt(36-x.^2)-1./2;
>> [alfa,n] = bisection(f,5,5.6,10^(-4))
```

you obtain the information requested. The output (Fig. 6) shows in the columns:

- 1 through to 3 the value of a_n, b_n, x_n.
- 4 through to 6 the value of $f(a_n), f(b_n), f(x_n)$.
- 7 through to 8 the absolute value $|b_n - a_n|$ and $\frac{b-a}{2^n}$.

The numerical value of the root and the iteration are shown at the end of the output. After thirteen iterations the estimation of the root searched for is:

$$\alpha = 5.206689453125000;$$

and the *Error* $= 0.000073242187500$.

The fact that the error has a linear trend on a logarithmic scale tells us that the Bisection method has a *convergence order* of 1 or simply *converges in a linear way*: it tends to 0 with the same speed as *1/n*. This translates into the fact that at each step you always gain the same number of exact significant digits. In particular, we can observe from the output (Fig. 6) that in about every 3 steps a correct decimal figure is gained. The explanation lies in the fact that the Bisection method divides the current interval into two equal parts. Translated into machine terms[3], a binary figure (BIT = BInary digiT) is exactly arrived at in each step. Since it takes more than 3 bits to do 10^4 this explains why you get a figure in base 10 roughly every 3 steps of the method. In this way the students "touch" computer arithmetic.

---

[3] Which are stored in base 2 in the systems currently in use (Standard IEEE754).

[4] Indeed, $2^3 = 8$ e $2^4 = 16$ or, more simply $log_2 10 = 3.32$.....

```
ans =
     5.000000000000000    5.600000000000000    5.300000000000000
     5.000000000000000    5.300000000000000    5.150000000000000
     5.150000000000000    5.300000000000000    5.225000000000000
     5.150000000000000    5.225000000000000    5.187500000000000
     5.187500000000000    5.225000000000000    5.206250000000000
     5.206250000000000    5.225000000000000    5.215624999999999
     5.206250000000000    5.215624999999999    5.210937500000000
     5.206250000000000    5.210937500000000    5.208593750000000
     5.206250000000000    5.208593750000000    5.207421875000000
     5.206250000000000    5.207421875000000    5.206835937499999
     5.206250000000000    5.206835937499999    5.206542968750000
     5.206542968750000    5.206835937499999    5.206689453125000
     5.206542968750000    5.206689453125000    5.206616210937500

    -0.038360501617149    0.139273355946131    0.022434458707975
    -0.038360501617149    0.022434458707975   -0.011823937247000
    -0.011823937247000    0.022434458707975    0.004114886114271
    -0.011823937247000    0.004114886114271   -0.004118333426015
    -0.004118333426015    0.004114886114271   -0.000071464033227
    -0.000071464033227    0.004114886114271    0.002003768028216
    -0.000071464033227    0.002003768028216    0.000961731374736
    -0.000071464033227    0.000961731374736    0.000444036509365
    -0.000071464033227    0.000444036509365    0.000186012938006
    -0.000071464033227    0.000186012938006    0.000057206250602
    -0.000071464033227    0.000057206250602   -0.000007145926390
    -0.000007145926390    0.000057206250602    0.000025025901417
    -0.000007145926390    0.000025025901417    0.000008938922581

     0.600000000000000    0.300000000000000
     0.300000000000000    0.150000000000000
     0.149999999999999    0.075000000000000
     0.074999999999999    0.037500000000000
     0.037500000000000    0.018750000000000
     0.018750000000000    0.009375000000000
     0.009374999999999    0.004687500000000
     0.004687500000000    0.002343750000000
     0.002343750000001    0.001171875000000
     0.001171875000000    0.000585937500000
     0.000585937499999    0.000292968750000
     0.000292968749999    0.000146484375000
     0.000146484375000    0.000073242187500

alfa =
     5.206689453125000
n =
    13
```

Fig. 6. Output for the Bisection method applied to solve (1)

Figure 7 show two plots indicating the error on the basis of number of iterations.

Bisection Method Analysis. What are the pros and cons of the Bisection method?

Since the Bisection method discards 50% of the current intervals at each step, it brackets the root much more quickly than the incremental search method does.

The pros are as follows:

- it is always convergent. Since the method brackets the root, the method is guaranteed to converge;
- as iterations are conducted, the interval gets halved. So one can guarantee the error in the solution of the equation.

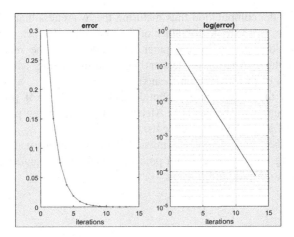

Fig. 7. Plot of the error as a function of the number of iterations for the Bisection method.

Since the Bisection method is like an incremental search, the following cons occur:

- it only finds roots where the function crosses the x axis. It cannot find roots where the function is tangent to the x axis;
- it can be fooled by singularities in the function;
- it cannot find complex roots of polynomials.

Analyzing the pros and cons, we can assert that the advantage of the Bisection method is that it is guaranteed to be convergent.

A disadvantage of the Bisection method is that it cannot detect multiple roots. In general, the Bisection method is used to get an initial rough approximation of the solution. Then faster converging methods are used to find the solution.

4 Conclusions

The computational approach utilized can be part of a more general didactical strategy and in no case can substitute for the teacher who has to choose and develop the didactical practice to support -step by step- the activity of the students, helping them to organize data and encourage creativity. Particularly, when software such as MATLAB is used, the teacher must plan the process of designing which starts from the defining step and she/he must structure the learning material in order to provide the students with tools to critically probe the concepts and to validate them with formal proofs in order and draw conclusions. These tools are important when searching for ways to work with students by simulation with mathematical models.

The computational approach brings students closer to the culture of analysis and numerical methods in an easy and understandable way because emphasis is given to the knowledge and to accompanying justifications. The basic pedagogy is that of making the students 'experience' rather than trying to provide them with all the concepts and results on the subject. The experience leads the students to a careful reading of the cultural aspects of the computation thus maturing the awareness that all these aspects are held together in the humble container of the algorithm. In fact, through the construction and implementation by computer the student has the real possibility of thinking in computational terms (understanding what estimating, analysing the accuracy and execution of that estimation means).

The algorithm then becomes a "cultural machine" because it produces objects, processes and cultural experiences [11]. Mechanically memorizing procedures is useless, students must learn the basic concepts.

Numerical mathematics is also experimental. Much can be learned simply by doing tests and observing how the calculation proceeds thanks to the algorithm as a stimulating and reflexive procedural criticism.

References

1. Aydin, E.: The use of computers in mathematics education: a paradigm shift from "computer assisted instruction" towards "student programming". Turk. Online J. Educ. Technol. TOJET **4**(2), 27–34 (2005)
2. Breed, E.A., Monteith, J.L. de K., Mentz, E.: Effective learning in computer programming: the role of learners' reflective thinking. In: Samways, B. (ed.) Proceedings 8th World Conference on Computers in Education – WCCE 2005, Western Cape, South Africa (2005)
3. Chapra, S.C., Canale, R.P.: Numerical Methods for Engineers. McGraw-Hill Higher Education, Boston (2010)
4. Costabile, F.A., Serpe, A.: Archimedes in secondary schools: a teaching proposal for the math curriculum. In: Paipetis, S., Ceccarelli, M. (eds.) The Genius of Archimedes–23 Centuries of Influence on Mathematics, Science and Engineering, vol. 11, pp. 479–491. Springer, Dordrecht (2010). https://doi.org/10.1007/978-90-481-9091-1_36
5. Costabile, F.A., Serpe, A.: Computer-based mathematics instructions with MATCOS: a pedagogical experiment. In: Handbook of Research on Didactic Strategies and Technologies for Education: Incorporating Advancements, pp. 724–738. IGI Global (2013). https://doi.org/10.4018/978-1-4666-2122-0.ch063
6. Cretchley, P., Harman, C., Ellerton, N., Fogarty, G.: MATLAB in early undergraduate mathematics: an investigation into the effects of scientific software on learning. Math. Educ. Res. J. **12**(3), 219–233 (2000). https://doi.org/10.1007/BF03217086
7. Daponte, P., Grimaldi, D., Molinaro, A., Sergeyev, Y.D.: Fast detection of the first zero-crossing in a measurement signal set. Measurement **19**(1), 29–39 (1996). https://doi.org/10.1016/S0263-2241(96)00059-0
8. Dubinsky, E., Tall, D.: Advanced mathematical thinking and the computer. In: Tall, D. (ed.) Advanced Mathematical Thinking. Mathematics Education Library, vol. 11, pp. 231–248. Springer, Dordrecht (2002). https://doi.org/10.1007/0-306-47203-1_14
9. Duval, R.: Representation, Vision and Visualization: Cognitive Functions in Mathematical Thinking. Basic Issues for Learning (1999)
10. Feurzeig, W., Papert, S., Bloom, M., Grant, R., Solomon, C.: Programming-language as a conceptual framework for teaching mathematics. Newslett. SIGCUE Outlook **4**(2), 13–17 (1970)
11. Finn, E.: What Algorithms Want: Imagination in the Age of Computing. MIT Press, Boca Raton (2017)

12. Frassia, M.G., Serpe, A.: Learning geometry through mathematical modelling: an example with GeoGebra. Turk. Online J. Educ. Technol. **2017**, 411–418 (2017). (November Special Issue INTE)
13. Gautschi, W.: Numerical Analysis. Springer, Heidelberg (2012). https://doi.org/10.1007/978-0-8176-8259-0
14. Goos, M., Galbraith, P., Renshaw, P., Geiger, V.: Perspectives on technology mediated learning in secondary school mathematics classrooms. J. Math. Behav. **22**(1), 73–89 (2003). https://doi.org/10.1016/S0732-3123(03)00005-1
15. Guide, Getting Started. "MATLAB® 7" (2009)
16. Hatfield, L.: Toward comprehensive instructional computing in mathematics. Comput. Math. Educ. 1–9 (1984)
17. Johnson, D.C. Educ. Inf. Technol. **5**, 201 (2000). https://doi.org/10.1023/A:1009658802970
18. Kurland, D.M., Pea, R.D., Clement, C., Mawby, R.: A study of the development of programming ability and thinking skills in high school students. J. Educ. Comput. Res. **2**(4), 429–458 (1986). https://doi.org/10.2190/BKML-B1QV-KDN4-8ULH
19. Liao, Y.K.C., Bright, G.W.: Effects of computer programming on cognitive outcomes: a meta-analysis. J. Educ. Comput. Res. **7**(3), 251–268 (1991). https://doi.org/10.1016/j.chb.2014.09.012
20. Lye, S.Y., Koh, J.H.L.: Review on teaching and learning of computational thinking through programming: what is next for K-12? Comput. Hum. Behav. **41**, 51–61 (2014). https://doi.org/10.1016/j.chb.2014.09.012
21. Nickerson, R.S.: Computer programming as a vehicle for teaching thinking skills. Thinking: J. Philos. Children **4**(3/4), 42–48 (1983). https://doi.org/10.5840/thinking19834310
22. Oprea, J.M.: Computer programming and mathematical thinking. J. Math. Behav. **7**, 175–190 (1988)
23. Papert, S.: Mindstorms: Children, Computers, and Powerful Ideas. Basic Books, Inc., New York (1980)
24. Pea, R.D., Kurland, D.M.: On the cognitive effects of learning computer programming. New Ideas Psychol. **2**(2), 137–168 (1984). https://doi.org/10.1016/0732-118X(84)90018-7
25. Robins, A., Rountree, J., Rountree, N.: Learning and teaching programming: a review and discussion. Comput. Sci. Educ. **13**(2), 137–172 (2003). https://doi.org/10.1076/csed.13.2.137.14200
26. Rushkoff, D.: Program or be Programmed: Ten Commands for a Digital Age. Or Books (2010)
27. Saeli, M., Perrenet, J., Jochems, W.M., Zwaneveld, B.: Teaching programming in secondary school: a pedagogical content knowledge perspective. Inform. Educ. **10**(1), 73–88 (2011). https://academic.microsoft.com/paper/1509940907
28. Serpe, A., Frassia, M.G.: Computer-based activity's development for probability education in high school. Turk. Online J. Educ. Technol. **2017**, 613–621 (2017). (October Special Issue INTE)
29. Serpe, A., Frassia, M.G.: Technology will solve student's probability misconceptions: integrating simulation, algorithms and programming. In: Dooley, T., Gueudet, G. (eds.) Proceedings of 10th Congress of the European-Society-for-Research-in-Mathematics-Education (CERME 10), pp. 828–835 (2017). http://www.mathematik.uni-dound.de/ieem/erme_temp/CERME10_Proceedings_final.pdf
30. Sfard, A., Leron, U.: Just give me a computer and i will move the earth: programming as a catalyst of a cultural revolution in the mathematics classroom. Int. J. Comput. Math. Learn. **1**(2), 189–195 (1996). https://doi.org/10.1007/BF00571078
31. Sonar, T.: Angewandte Mathematik Modellbildung und Informatik, Vieweg-Verlag, Braunschweig. Wiesbaden (2001). https://doi.org/10.1007/978-3-322-80225-5

Task Mathematical Modelling Design in a Dynamic Geometry Environment: Archimedean Spiral's Algorithm

Annarosa Serpe[1]([✉]) and Maria Giovanna Frassia[2]

[1] Department of Mathematics and Computer Science, University of Calabria,
Rende, CS, Italy
annarosa.serpe@unical.it
[2] IIS IPSIA-ITI "E. Aletti", Trebisacce, CS, Italy
frassia@mat.unical.it

Abstract. Over the last twenty years, several research studies have recognized that integrating, not simply adding, technology (Computer Algebra System - CAS, Dynamic Geometry Software - DGS, spreadsheets, programming environments, etc.) in the teaching of mathematics helps students develop essential understandings about the nature, use, and limits of the tool and promotes deeper understanding of the mathematical concepts involved. Moreover, the use of technology in the Mathematics curricula can be important in providing the essential support to make mathematical modelling a more accessible mathematical activity for students. This paper presents an example of how technology can play a pivotal role in providing support to explore, represent and resolve tasks of mathematical modelling in the classroom. Specifically, a mathematical modelling task design on the tracing of Archimedean spiral with use of a Dynamic Geometry Environment is shown. The aim is to emphasize the meaning and the semantic value of this rich field of study that combines tangible objects and practical mechanisms with abstract mathematics.

Keywords: Archimedes' spiral · Dynamic Geometry Environment · Mathematical modelling task

1 Introduction

Research on the use of technology in secondary Mathematics education has proliferated over the last 30 years and has demonstrated that the strategic use of technological tools such as graphing calculators, Dynamic Geometry Environment (DGE), programming environment and spreadsheets support students' mathematical thinking and discourse [3, 4, 6, 8, 16]. At the same time, awareness and interest in students' mathematical thinking, reasoning, and sense-making has increased [7], so much so that in the last twenty years many researchers have focused their research studies on how the use of technology can support students' mathematical thinking and reasoning [31].

Three themes emerge as researchers have examined mathematical thinking in the context of mathematical activity that occurs in technological environments [17]:

© Springer Nature Switzerland AG 2020
Y. D. Sergeyev and D. E. Kvasov (Eds.): NUMTA 2019, LNCS 11973, pp. 478–491, 2020.
https://doi.org/10.1007/978-3-030-39081-5_41

- the extent to which students develop tools for mathematical thinking and learning in these environments;
- the ways in which students engage in metacognitive activity as a result of mathematical activity in these environments;
- the level of generality of the mathematical thinking of students in these environments.

As regards the first theme, some technological instruments facilitate symbolic reasoning and/or conceptual understanding, but they can also inhibit mathematical thought. Some studies indeed highlight how symbolic manipulation skills do not automatically develop in a technology-based environment [34], but are the result of adequate integration of technology into classroom practice.

As far as the second theme is concerned, technological environments have two features that may enhance metacognitive activity: when used as tools they have the capacity to offload some of the routine work associated with mathematical activity leaving more time for reflection, and with their strict communication requirements they may help bring to consciousness mathematical ideas and procedures [17].

Regarding the third theme, some studies reported a higher level of generality in the thinking of students in the technology-based activity.

In the context of geometry, Hollebrand and colleagues in [17] observed that learners' generalizations were likely to be situated abstractions, not generalizing beyond the situation in which they were developed. Doerr and Pratt in [17, p. 428] noted 'the lack of evidence for any claim that the students reasoned better about real-world phenomena, even when they reasoned appropriately within the microworld' and that 'there is however little evidence that students can abstract beyond the modelling context'. The DGE were developed in the mid-1980s, in the thread of the powerful idea of 'direct manipulation' [11] to simulate ruler and compass constructions and assist/help in the precise design of geometric figures. A DGE is a computer microworld with Euclidean geometry as the embedded infrastructure. In this computational environment, a person can evoke geometrical figures and interact with them [18]. It is a virtual mathematical reality where abstract concepts and constructions can be visually reified. In particular, the traditional deductive logic and linguistic-based representation of geometrical knowledge can be re-interpreted, or even refined, in DGE as dynamic and process-based interactive 'motion picture' in real time [23]. The rapid evolution of Geometer's Sketchpad and Cabri Géomètre, the first two DGEs, highlighted the need for increased dynamism. Sketchpad, for example, was initially conceived as a program for drawing accurate static figures of Euclidean geometry. In the course of their development, the initial idea for the different DGEs was adapted to the emerging needs. The result was software able to build dynamic geometric shapes, in which points and segments could be dragged maintaining the properties that characterize the constructed figures. The relatively fast drag and drop operation (dragging) defined the future of DGE: the functional versatility and corresponding complexity of the operation were not anticipated, and have only gradually been dealt with [1]. Interest in DGE for learning geometry was evident right from the start: DGE through dragging offered new possibilities for the visualization and verification of the properties of geometric objects.

This has been recognized by the various educational areas that have integrated DGE in their curricula [20]. Direct manipulation, a special feature of DGE, allows the student to have a simultaneous response; according to [22], this simultaneity is a key element which can shorten the distance between experimental and theoretical Mathematics, or for switching between conjecture and formalization.

Laborde in [16] illustrates this potential for the teaching of geometry with an example on the "black box" situations:

> *In the black box situations, the students are given a diagram on the screen of the computer and they are asked questions about it. This kind of situation was used in our scenarios for introducing new trasformations. A point P and its image P' through an unknown transformation were given to the students. They could move P and observe the subsequent effect on P'. Students were asked to find the properties of the unknown transformation by means of this black box. In such a task, students must ask themselves questions about the transformation.*

This process emphasizes the students' responsibility in the formulation of questions concerning the transformation, but at the same time provides food for thought on the fact that this may well distract learners from the main goal, which is to identify geometric properties. On the other hand, before they can use the software properly, the students require a preliminary phase of instruction on the functions offered by DGE (including the dragging[1]); only at a later time should they experience the constraints and relations that restrict the movement of the points and of the figures in general [19].

This procedure adds useful meaning to the validation of the constructed figures.

In addition to the function of dragging, the DGE also have the following features/affordances:

- measuring (of lengths of segments, the amplitudes of angles, shapes of areas, …);
- tracing, place, animation (that let you see the evolution of models);
- The representation of functions and the investigation of their graph, at local or general level;
- The integration of different representation registers (such as the geometrical and analytical), that allows you to model problematic situations.

The potential of DGE is not just restricted to the field of geometry, but it can be exploited to introduce other notions of mathematics (e.g. analysis, probability, etc.).

Furthermore, the geometrical situations can also be associated with specific problems, in which the DGE help in the visualization of the dynamic aspects of geometrical objects [11]. In this prospective, the paper presents a mathematical modelling task design in a DGE. The paper is structured in four sections.

Section 2 covers the theoretical framework adopted. Section 3 covers all actions of task based on the tracing method – in Euclidean plane – of Archimedean spiral with the use of GeoGebra software. Section 4 covers the conclusions.

---

[1] Dragging in DGE is a powerful dynamic tool to acquire mathematical knowledge. It serves as a kind of interactive amplification tool for leaner to see global behavior via variation.

2 Theoretical Framework

Mathematics education is characterized by a high conceptual standard in regard to the development of central terms and through the construction of numerous algorithmic processes. In both areas, great importance can be attached to the use of digital mathematics tools. As mentioned above, the digital tools are not only a pedagogical medium for organizing processes in education, in particular they strengthen the activity of doing mathematics, such as experimenting, visualizing, applying, etc. [2, 33].

According to a well-considered use of technology, the tools used are and will remain cognitive tools because they help to represent and work on the individual problem (after input by the person that is working on them). Naturally, the work on this should be designed so that it supports the mental processes of the learners, who control the learning process, however, they should by no means be restricted. From the point of view of the discovery of mathematical interrelations, digital mathematics tools are of particular importance for example in simulations, understood as experimenting with models [15].

The International Commission on Mathematical Instruction (ICMI[2]) has promoted and continues to promote cultural debate in order to really support the students' development of mathematical modelling competences because relevant for their further education and for their subsequent professions. Several research studies recognize that the development of technology creates more opportunities for practicing mathematical modelling in the classroom [14, 32]. The use of technology can lead to the simplification of difficult and complex modeling operations, especially when solving, as [13] has shown (Fig. 1).

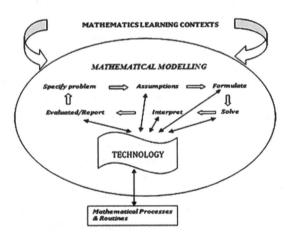

Fig. 1. Mathematical modelling including the use of technology.

---

[2] https://www.mathunion.org/activities/international-commission-mathematical-instruction-icmi.

In the teaching process, the use of technology allows the teacher not only to cope with traditional content using different methods, but also to explore and encounter new content. Especially, in the teaching and learning of Euclidean geometry, and solving problems related to geometry concepts, the DGE are appropriate tools to help enhances conceptual understanding [12, 28–30].

Duval [9] argued that DGE are superior to paper-and-pencil based (PPB) methods as they dissociate the "figure" from the "process of drawing". This allows students to understand the properties of the figure before it is sketched on the screen.

The use of DGE continually opens up new teaching perspectives in the teaching - learning of geometry because it enhances the constructive aspect without detracting from deductive accuracy, from clarity of hypotheses and related consequences pertaining to the discipline [1, 21]. Thanks to DGE the graphic-constructive phase, both prior to the acquisition of some concepts and geometrical properties, and subsequently as verification and/or further study, is not only enjoyable, but also greatly helps teaching, as it offers both visualization and exemplification and/or exploration.

On the other hand, according to Federigo Enriques [10]:

> [...] It does not help to develop with impeccable deduction series of theorems of Euclidean geometry, if you do not return to contemplate the constructed edifice, inviting the disciples to distinguish the truly significant geometric properties (for example: the sum of the angles of a triangle and Pythagoras' theorem) from those that have value only as links in the chain.

It is not superfluous to recall here that though geometry was created as modelling of the physical world around us, the Italian teaching tradition has followed the hypothetical-deductive Euclidean interpretation and has progressively emphasized its formal character, 'decontaminated' from the figural and constructive aspect, even in the most harmless of terms. In short, the surveyor's traditional tools (ruler, square ruler, compasses), retrieved and simulated by DGE, on the one hand facilitate geometrical intuitions, while on the other raise and stimulate interest and learners' imagination, enabling speculation, which is sometimes immediately verifiable, thanks to the simultaneous computer feedback [28, 29]. The connection between a drawing and a geometric object in everyday teaching practice is nearly always established through a process of approximation. This is based on the idea that with subsequent, better attempts the drawing can eventually achieve something close to the ideal figure.

The essential didactic value of the Euclidean frame has always been the perception of its nature as a comprehensive frame, which begins with the 'simple and evident' and progresses to the complex and 'non-evident'. The integrated tools offered by a DGE represent a valid aid along the way as they progress in the same way from what is predefined to what is made by the user. According to this perspective, the authors have chosen GeoGebra software because it is a free open-source DGE for mathematics teaching and learning. GeoGebra[3] combines features of DGE and CAS in a single, integrated, and easy to-use system. It is may be the bridge for the cognitive gap that hinders a student from carrying out a modelling task.

---

[3] GeoGebra was created by Markus Hohenwater and now has been translated into 40 languages. Users all over the world can freely download this software from the official GeoGebra website at http://www.geogebra.org.

However, it should also be noted that this software (and technology in general) should never replace the mathematics, much less the teacher; it should be viewed as a timely, and sometimes temporary, means of overcoming a difficulty.

3 Mathematical Modelling Task Design

The mathematical modelling task design aims to facilitate the understanding of the concrete-abstract relationship, in order to the acquisition of meaning of geometrical objects. The task implies the study of the geometrical method for tracing Archimedean spiral in Euclidean register using drawing tools such as a ruler, set square and compasses and later also with a computer as an educational tool. We chose to use Geo-Gebra because it is a constructive and creative activity, which reinforces the acquisition of concepts as well as abstraction skills.

The reason for the choice of the Archimedean spiral lies in the fact that the spiral is an ambiguous and double curve, on one part it gives the image of expansion and totality; a curve that, by rotating, always remain similar to itself, but at the same time it widens and extends to infinity, as if in rotating new parts are always being born from the centre to move around towards the periphery. However next to this clear and serene face, the spiral shows a second dark and disturbing one: the movement is transformed from an expansion to a continuous contraction that hypnotically pulls you into the centre.

Despite these contrasting images of it, the generation of spiral lines is very simple, a rare example of depth and at the same time of geometric evidence; two reasons that have led mathematicians to study their properties since ancient times, regardless of their scarce applications.

The first and simplest of the spiral lines is the one studied by Archimedes, which bears his name. Furthermore, the spirals are also at the base of the fractals [5, 24–27], which may constitute a further topic of further study in the direction of the theoretical framework outlined. Archimedean spiral is a curve of great charm and beauty. To date there is no simple and precise tool capable of tracing it and for this reason it is not easily used for practical purposes[4]. To draw it is necessary to identify a series of points on the plane that must be connected by a curve or simply, as more often happens, tracing the curve freehand. Although governed by a geometric principle[5], the Archimedean spiral cannot be drawn simply. Some points can be identified, but the task remains, of having to trace the form continuously. A substantial aid for tracing spiral-shaped curves comes from the polycentric family, that is from those curves defined by the appropriate combination of different arcs of circumference, arranged in such a way that there are no discontinuities in the connections between the various parts. To obtain this condition it is necessary that at the points of contact between two arcs they have a

---

[4] A concrete image of the Archimedean spiral is the grooves of a vinyl disc, equidistant from each other and separated by a very small constant distant.

[5] The mechanical generation of the spiral is described in a plane by a point that moves with a uniform motion along a straight line, while the line rotates in a uniform circular motion around a point.

common tangent, so that their rays belong to the same line. In this way the passages between the curves will be fluid and continuous.

The tracing method for the construction of the Archimedean spiral in the Euclidean plane represents a strategic phase from the teaching point of view because it helps the learner to perceive the complexity of the whole, which starts from 'simple and evident' and arrives at the 'complex and not evident'. The interpretation of data with GeoGebra enables us to turn the meaning of a representation into its construction through a model which shows and communicates geometrical synthesis. Mathematical modelling has expressive and evocative potential from the metric and formal point of view; the virtual tracing of a curve through simple steps adds value to the learning experience.

Table 1 summarizes the competencies that the mathematical modeling task design aims to achieve, with an indication of the levels that can be pursued.

Table 1. Levels of competencies of mathematical modelling task.

Levels	Interpretation of task	Task solution	Mathematical reasoning and computation
1st-level Initial	Failure to identify the significant information of the task	Lack of autonomy in solution strategies	Lack of control in the resolution process. Representation with software and analysis of mathematical objects with appropriate guidance
2nd-level Basic	Partial identification of significant information of the task	Partial autonomy in solution strategies	Lack of control in the resolution process. Representation with software and analysis of mathematical objects partially correct
3rd-level Intermediate	Identification of the significant information of the task	Autonomy in the solution strategies	Control in the resolution process. Correct representation of mathematical objects with software and analysis
4th-level Advanced	Identification and contextualization of the significant information of the task	Autonomy and creativity in the solution strategies	Rigor, coherence and logic in the resolution process. Correct and original representation of mathematical objects with software and analysis

Such a methodology avoids the traditional pitfalls of Mathematics classes, and gives students a true chance to improve their understanding.

Based on the theoretical framework described above, the mathematical modelling task design is shown in Fig. 2.

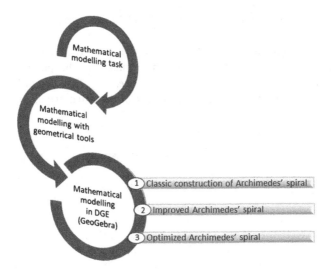

Fig. 2. Mathematical modelling task design.

The class is divided into groups, and each group gets the required material and tools (graph paper, ruler, compasses, protractor, pencils and rubbers); then the teacher sets the task for the classic geometric construction of Archimedean spiral (Interpretation and solution of the task without the technology). Then, the students do the construction process with paper and pencil and this is followed by use of the GeoGebra spreadsheet.

The last part of the mathematical modelling task design, when the student use GeoGebra – mathematical modelling computer-based environment – is divided in to three actions.

3.1   Action 1 – Classic Construction of Archimedes' Spiral

Classic construction of Archimedes' spiral through the "predefined objects" available in the GeoGebra toolbar: point, regular polygon, median point, half line, circumference and arc. This procedure will allow an analysis of the construction from a mathematical point of view.

Algorithm 1 – Archimedes' spiral. The steps are as follows:

1. Draw two points A and B;
2. Construct the square $ABCD$ with extreme side A and B;
3. Determine the median point (E, F, G, H) of the sides of the square;
4. Construct the rays (e, h, g, f) with origin the median point of the square and passing through one of the vertexes of the square and pertaining to the side on which the median point lies (anticlockwise direction);
5. Construct the circumference with centre A and passing through point B;
6. Determine the point of intersection I between the circumference and the ray with origin the median point of side AB and passing through A;

7. Draw the arc *p* with centre *A* and points *B* and *I* as extremes;
8. Repeat from 5 to 7:
   a. Construct the circumference *CC* with centre *A, B, C, D* (in this order) and radius the distance between the vertex of the square and the point of intersection between the old circumference and the rays *e, h, g, f* (in this order);
   b. Construct the arc with centre in the vertex of the square (Fig. 3).

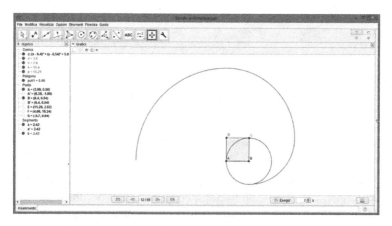

Fig. 3. Output of Archimedes' spiral (initial construction).

At this point the teacher tells the students that the spiral curve is called spiral of Archimedes[6] - the great mathematician from Syracuse - exhibited for the first time in his work *"On the Spirals"*, a treatise addressed to mathematician Dositeo of Alessandria. This treatise was much admired but little read, since it was considered as the most difficult of all of Archimedes' works. The spiral is defined as the flat place of a point which, starting from the end of a ray or half-line, moves uniformly along this radius while the radius in turn rotates uniformly around its end.

The study that Archimedes made of the spiral was set in the wake of research, typical of Greek mathematics, aimed at finding solutions to the three famous classic problems. Ancient testimonies of the spiral are found in some Minoan paintings dating back to 1650 BC., Although the Minoan were not aware of the properties of the curve, their representation was realized in an extraordinary way.

The use of history in mathematics education is important for experimenting method in order to obtain a full learning. Indeed, the history of mathematics offers the possibility to explore the development of mathematical knowledge.

---

[6] Archimedes, (born c. 287 BCE, Syracuse, Sicily [Italy]-died 212/211 BCE, Syracuse), the most-famous mathematician and inventor in ancient Greece.

3.2   Action 2 – The Improved Archimedes' Spiral

In the Archimedean spiral the execution of a turn requires the construction of four arcs traditionally constructed with ruler and compasses; so the need to simplify the repetition process about the construction of the connected arcs becomes apparent. As a result, the students must find a "geometrical strategy" for the solution of the problem.

In this case, GeoGebra represents a valid help as it allows for the creation of new computational tools in the spreadsheet interface, which can then be used as predefined objects. Students are then asked to design a new tool called *ArcSpiral* used for the construction of all the connected arcs so as to reduce the execution time of the previous algorithm. This action is very sensitive because the students have to face yet another abstraction leap. They have to plan the design of a new tool called *ArcSpiral* with the aim of the constructing the spiral of Archimedes starting from the construction of the first arc, which requires the identification of the initial (points *A, B*, median point *E* between *A* and *B* and the half line *EA*) and final (point *I* and arc *p*) objects.

Algorithm 2 – The improved Archimedes' spiral. After the first seven steps of algorithm 1, we have:

1. Create a new tool "*ArcSpiral*", having the points *A, B* and median point *E* between *A* and *B*, and the half line passing through *E* and *A* as initial objects, and points *I* and arc *p* as final objects. The tool thus created can be used by clicking on its icon;
2. Construct the connected spiral arcs, through the tool "*ArcSpiral*" created, with appropriate choice of points (Fig. 4).

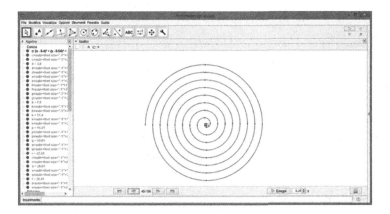

Fig. 4. Output of Archimedes' spiral (8 spires).

Refining the previous process of construction is relevant because the students are required to make a considerable abstractive leap: what is difficult is the identification of the construction, within the algorithm, which enables the repetition of an operation (construction of joined arcs) as long as a certain condition is true.

3.3   Action 3 – Optimized Archimedes' Spiral

Design a new tool "*ArcSpiral*" which contains a lower number of initial objects. Planning the "*ArcSpiral*" tool with only two initial objects (points *A* and *B*) and one final object (arc with centre *B* and points *A* and *A'* as extremes). This step has the aim of further perfecting the construction of the curve.

Such a highly educational process trains the students in the appreciation of the potential of the language of mathematics and at the same time offers them a reading awareness of theories.

Algorithm 3 – Optimized Archimedes' spiral. The steps are follows:

1. Draw two points *A* and *B*;
2. Draw point *A'*, obtained from the rotation of *A* with respect to centre *B* by an anticlockwise angle;
3. Construct the arc with centre *B* and points A and *A'* as extremes;
4. Create a new tool called "*Arcofspiral*", with points *A* and *B* as initial objects and as final objects the arc with centre *B* and *A* and *A'* as extremes; the tool created can be used by clicking on its icon (Fig. 5).

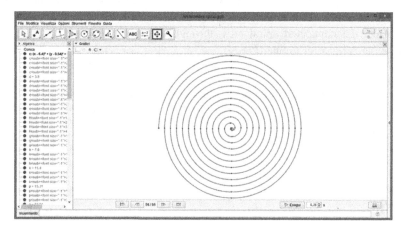

Fig. 5. Output of the optimized Archimedes' spiral (12 spires).

4   Conclusions

Today it is important to plan and experiment new modalities for the teaching of Euclidean geometry, bearing in mind the potential of new technological devices.

The mathematical modelling tasks design suggests important messages for mathematics education.

This practical contribution to mathematics education can prove that computer-based classroom activities can be effectively used in the teaching and learning environment. Firstly, the students can become familiar with DGE; secondly, geometric facts, figures, shapes and their proprieties with the constructions can be observed by using the software's features. Thus, the students can have the chance to verify the conditions by exploring and observing the geometric properties of the shapes with all sufficient conditions. This also can give opportunities to check and prove all features dynamically with the program itself. Consequently, the students have the chance to prove the terms and to observe construction conditions of geometric features for each case. Then, starting by simply drawing the geometric shapes and figures and providing all conditions for the construction requires students to consider all of the related facts and features together with associated geometric realities.

The implementation of the first construction algorithm for the curve, repeated several times, facilitates the understanding and use of the geometrical objects; however, at the same time the need to shorten the repetition sequence for the construction of the joined arcs emerges for the first time.

The use of GeoGebra, suitably exploited in teaching practice, favours the structuring of knowledge in meaningful networks thus maturing the students' skills. It offers an effective impact on mathematics education and has the potential to promote student-centred learning and active learning. However, the mathematical modelling task design is in no way limiting. The creative teacher can use the design as a springboard for new teaching initiatives which are instructive and engaging.

References

1. Arzarello, F., Olivero, F., Paola, D., Robutti, O.: A cognitive analysis of dragging practices in Cabri environments. ZDM **34**(3), 66–72 (2002). https://doi.org/10.1007/BF02655708
2. Barzel, B., Hußmann, T., Leuders, T.: Computer, internet & co *im Mathematik-Unterricht*. Scriptor-Cornelsen, Berlin (2005)
3. Bonanno, A., Camarca, M., Sapia, P., Serpe, A.: Archimedes and caustics: a twofold multimedia and experimental approach. In: Paipetis, S., Ceccarelli, M. (eds.) The Genius of Archimedes – 23 Centuries of Influence on Mathematics, Science and Engineering. History of Mechanism and Machine Science, vol. 11, pp. 45–56. Springer, Dordrecht (2010). https://doi.org/10.1007/978-90-481-9091-1_4
4. Burrill, G., Allison, J., Breaux, G., Kastberg, S., Leatham, K., Sanchez, W.: Handheld graphing technology in secondary mathematics: Research findings and implications for classroom practice. Retrieved from the Texas Instruments website (2002). http://education.ti.com/sites/UK/downloads/pdf/References/Done/Burrill,G.%20(2000).pdf
5. Caldarola, F.: The Sierpinski curve viewed by numerical computations with infinities and infinitesimals. Appl. Math. Comput. **318**, 321–328 (2018). https://doi.org/10.1016/j.amc.2017.06.024
6. Costabile, F.A., Sapia, P., Serpe, A.: Archimedes in secondary schools: a teaching proposal for the math curriculum. In: Paipetis, S., Ceccarelli, M. (eds.) The Genius of Archimedes – 23 Centuries of Influence on Mathematics, Science and Engineering. History of Mechanism and Machine Science, vol. 11, pp. 479–491. Springer, Dordrecht (2010). https://doi.org/10.1007/978-90-481-9091-1_36
7. Dick, T.P., Hollebrands, K.F.: Focus in high school mathematics: Technology to support reasoning and sense making, pp. xi-xvii. National Council of Teachers of Mathematics, Reston, VA (2011)

8. Drijvers, P., Ball, L., Barzel, B., Heid, M.K., Cao, Y., Maschietto, M.: Uses of Technology in Lower Secondary Mathematics Education. ITS. Springer, Cham (2016). https://doi.org/10.1007/978-3-319-33666-4

9. Duval, R.: Geometry from a cognitive point of view. In: Mammana, C., Villani, V. (eds.) Perspectives on the teaching of geometry for the 21st century, pp. 37–52 (1998)

10. Enriques, F.: Insegnamento dinamico. Periodico di Matematica Serie IV, vol. I, pp. 6–16 (1921)

11. Gueudet, G., Trouche, L.: Mathematics teacher education advanced methods: an example in dynamic geometry. ZDM **43**(3), 399–411 (2011). https://doi.org/10.1007/s11858-011-0313-x

12. Frassia, M.G., Serpe, A.: Learning geometry through mathematical modelling: an example with GeoGebra. Turk. Online J. Educ. Technol. **2017**(Nov Spec. Issue INTE), 411–418 (2017)

13. Galbraith, P., Goos, M., Renshaw, P., Geiger, V.: Technology enriched classrooms: some implications for teaching applications and modelling. In: Mathematical modelling in education and culture, pp. 111–125. Woodhead Publishing, Cambridge (2003). https://doi.org/10.1533/9780857099556.3.111

14. Rodríguez Gallegos, R., Quiroz Rivera, S.: Developing modelling competencies through the use of technology. In: Stillman, G.A., Blum, W., Biembengut, M.S. (eds.) Mathematical Modelling in Education Research and Practice. IPTLMM, pp. 443–452. Springer, Cham (2015). https://doi.org/10.1007/978-3-319-18272-8_37

15. Greefrath, G., Weigand, H.-G.: Simulieren: Mit Modellen experimentieren. Mathematik lehren **174**, 2–6 (2012)

16. Laborde, C.: The use of new technologies as a vehicle for restructuring teachers' mathematics. In: Lin, F.L., Cooney, T.J. (eds.) Making sense of mathematics teacher education, pp. 87–109. Springer, Dordrecht (2001). https://doi.org/10.1007/978-94-010-0828-0_5

17. Heid, M.K., Blume, G.W.: Research on technology and the teaching and learning of mathematics: Vol. 1. Research syntheses. Information Age, Charlotte (2008)

18. Hohenwarter, J., Hohenwarter, M., Lavicza, Z.: Introducing dynamic mathematics software to secondary school teachers: the case of GeoGebra. J. Comput. Math. Sci. Teach. **28**(2), 135–146 (2009). Retrieved 18 September 2019, from https://www.learntechlib.org/primary/p/30304/

19. Hollebrands, K.F.: The role of a dynamic software program for geometry in the strategies high school mathematics students employ. J. Res. Math. Educ. **38**(2), 164–192 (2007). https://doi.org/10.2307/30034955

20. Kotenkamp, U., Blessing, A.M., Dohrmann, C., Kreis, Y., Libbrecht, P., Mercat, C.: Interoperable interactive geometry for Europe-First technological and educational results and future challeges of the Intergeo project. In: Durand-Guerrier, V., Soury-Lavergne, S., Arzarello, F. (eds.) Proceeding of the Sixty European Conference on Research on Mathematics Education, pp. 1150–1160 (2010). http://www.inrp.fr/publications/edition-electronique/cerme6/wg7-11-kortenkamp.pdf. Accessed 18 September 2019

21. Leikin, R., Grossman, D.: Teachers modify geometry problems: from proof to investigation. Educ. Stud. Math. **82**, 515–531 (2013). https://doi.org/10.1007/s10649-012-9460-4

22. Leung, A.: Dynamic geometry and the theory of variation. In: Pateman, N.A., Dougherty, B. J., Zillox, J. (eds.) Proceedings of the 27th Conference of the International Group for the Psychology of Mathematics Education, Honolulu, USA, vol. 3, pp. 197–204 (2003). https://eric.ed.gov/?id=ED501009. Accessed 18 September 2019

23. Lopez-Real, F., Leung, A.: Dragging as a conceptual tool in dynamic geometry environments. Int. J. Math. Educ. Sci. Technol. **37**(6), 665–679 (2006). https://doi.org/10.1080/00207390600712539

24. Mandelbrot B.B.: Fractal Geometry of Nature. Freeman, San Francisco (1983). https://doi.org/10.1002/esp.3290080415
25. Pickover, C.A.: Mathematics and beauty: a sampling of spirals and 'Strange' spirals in science. Nat. Art. Leonardo **21**(2), 173–181 (1988). https://doi.org/10.5539/jmr.v3n3p3
26. Sergeyev, Y.D.: Blinking fractals and their quantitative analysis using infinite and infinitesimal numbers. Chaos, Solitons Fractals **33**(1), 50–75 (2007). https://doi.org/10.1016/j.chaos.2006.11.001
27. Sergeyev, Y.D.: Un semplice modo per trattare le grandezze infinite ed infinitesime. Matematica nella Società e nella Cultura: Rivista della Unione Matematica Italiana 8(1), 111–147 (2015)
28. Sergeyev, Y.D.: The exact (up to infinitesimals) infinite perimeter of the Koch snowflake and its finite area. Commun. Nonlinear Sci. Numer. Simul. **31**(1-3), 21–29 (2016). https://doi.org/10.1016/j.cnsns.2015.07.004
29. Serpe, A., Frassia, M.G.: legacy and influence in Mathematics and Physics with educational technology: a laboratory example. In: Magazù, S. (ed.) New Trends in Physics Education Research, pp. 77–96. Nova Science Publisher, New York (2018)
30. Serpe, A.: Geometry of design in high school - an example of teaching with Geogebra. In: GomezChova, L., LopezMartinez, A., CandelTorres, I. (eds.), INTED2018, Proceedings 12th International Technology, Education and Development Conference, Valencia, Spain, pp. 3477–3485 (2018). https://doi.org/10.21125/inted.2018.0668
31. Sherman, M.: The role of technology in supporting students' mathematical thinking: Extending the metaphors of amplifier and reorganizer. Contemporary Issues in Technology and Teacher Education, 14 (3), 220–246 (2014). https://www.learntechlib.org/primary/p/130321/. Accessed 18 September 2019
32. Siller, H.S., Greefrath, G.: Mathematical modelling in class regarding to technology. In: Durand-Guerrier, V., Soury-Lavergne, S., Arzarello, F. (eds.) Proceeding of the Sixty European Conference on Research on Mathematics Education, pp. 1150–1160 (2010). http://www.inrp.fr/editions/cerme6. Accessed 18 September 2019
33. Weigand, H.G., Weth, T.: Computer im Mathematikunterricht: Neue Wege zu alten Zielen. Spektrum, Heidelberg (2002)
34. Yerushalmy, M.: Student perceptions of aspects of algebraic function using multiple representation software. J. Comput. Assist. Learn. **7**, 42–57 (1991). https://doi.org/10.1111/j.1365-2729.1991.tb00223.x

Optimization and Management of Water Supply

Numerical Experimentations for a New Set of Local Indices of a Water Network

Marco Amos Bonora[1], Fabio Caldarola[2] , Joao Muranho[3], Joaquim Sousa[4] ,
and Mario Maiolo[1]([✉])

[1] Department of Environmental and Chemical Engineering (DIATIC),
Università della Calabria, Cubo 42/B, 87036 Arcavacata di Rende, CS, Italy
{marcoamos.bonora,mario.maiolo}@unical.it
[2] Department of Mathematics and Computer Science, Università della Calabria,
Cubo 31/B, 87036 Arcavacata di Rende, CS, Italy
caldorola@mat.unical.it
[3] Department of Computer Science, Universidade da Beira Interior,
Rua Marquês d'Ávila e Bolama, 6201-001 Covilhã, Portugal
jmuranho@ubi.pt
[4] Department of Civil Engineering, Instituto Politécnico de Coimbra,
Rua Pedro Nunes, 3030-199 Coimbra, Portugal
jjoseng@isec.pt

Abstract. Very recently, a new set of local performance indices has
been proposed for an urban water supply system together with a useful
mathematical model or, better, framework that organizes and provides
the tools to treat the complex of these local parameters varying from
node to node. In this paper, such indices are considered and examined
in relation to hydraulic software using Demand Driven Analysis (DDA)
or Pressure Driven Analysis (PDA). We investigate the needed hypotheses to obtain effective numerical simulations employing, in particular,
EPANET or WaterNetGen, and the concrete applicability to a real water
supply system known in literature as the KL network.

Keywords: Urban water supply systems · Performance indices ·
Mathematical modeling · EPANET 2.0.12 · WaterNetGen 1.0.0.942

1 Introduction

In the resolution of the hydraulic problem associated with drinking water distribution networks (WDNs), a hydraulic software model solves the continuity
equations in the junction nodes and the energy equations in the links. There
are two main resolutive approaches in literature: if the flow rate supplied in the
demand nodes is considered constant and defined upstream of the simulation,
the solver software will look for a solution that will guarantee that the flow and
load regime will meet the required supply in the nodes. In this case we speak of
Demand Driven Analysis (DDA). If instead the model foresees that the supply

© Springer Nature Switzerland AG 2020
Y. D. Sergeyev and D. E. Kvasov (Eds.): NUMTA 2019, LNCS 11973, pp. 495–505, 2020.
https://doi.org/10.1007/978-3-030-39081-5_42

may differ from the request in the nodes, depending on the pressure regime, it is a *Pressure Driven Analysis* (PDA).

The use of PDA models is more expensive in computational terms but presents results more representative of reality in the case of WDNs characterized by a poor regime of pressures; in the event that the pressure regime is sufficient to guarantee the supply in all the demand nodes there are no real advantages to using a PDA approach, which will provide very similar results if not identical to DDA models.

A Demand Driven approach is typical of software such as EPANET (see [23]). This software allows to model the hydraulic behavior of the water distribution network (WDN) and also to perform water quality simulations. EPANET is one of the most widespread software in WDNs simulation. On the other hand, a well-known software for PDA analysis is WaterNetGen, an EPANET extension developed by Muranho et al. (see [21, 22]).

The aim of this paper is to study in real contexts a new set of local performance indices recently developed by Caldarola and Maiolo in [1] (cf. also [2]) and, in particular, analysing and applying them to the WDN described by Kang and Lansey in [11]. In the following sections will be examined the hypotheses needed for a computational assessment and a practical application of these indices to the considered WDN, the results obtained by using DDA and PDA approaches, and the relation between them and some well-known indices as the resilience measures proposed and discussed in [5–8, 24, 25].

For similar indices concerning the vulnerability of infrastructures, the sustainability of water resources and various types of hydropotable risk, the reader can see [3, 4, 12–19].

2   Performance Indices

For a given water network we denote by n the number of its junction nodes, by r the number of tanks (or reservoirs), by q_i, $\hbar_i$ and $h_i = \hbar_i + z_i$ the discharge, the *pressure head* and the *piezometric head* at the i-th node, respectively, where z_i stands for the *elevation head*. Hence $p_i = \gamma \, q_i h_i$ represents the *delivered power* at the node i, where γ is the specific weight of water. We also use the notations q_i^*, $\hbar_i^*$, h_i^* and p_i^* to indicate the (minimal) project requests relative to the above defined quantities, as is usual in much current literature (see, e.g., [5–10, 25] and the references therein).

Inside a structured mathematical framework described in [1], the following *local indices* are proposed as "elementary building bricks" to construct (new) local and global indices for the needs of a WDN, and also useful to recover the well-known global indices ordinarily used in WDN analysis and implemented in many hydraulic simulation software:

$$q_i^s := \frac{q_i - q_i^*}{q_i^*} \qquad \textit{Local discharge surplus index,}$$

$$\hbar_i^s := \frac{\hbar_i - \hbar_i^*}{\hbar_i^*} \qquad \textit{Local pressure head surplus index,}$$

$$h_i^s := \frac{h_i - h_i^*}{h_i^*} \qquad \textit{Local piezometric head surplus index,}$$

$$p_i^s := \frac{p_i - p_i^*}{p_i^*} = \frac{q_i h_i - q_i^* h_i^*}{q_i^* h_i^*} \quad \textit{Local power surplus index,}$$

(1)

where $i = 1, 2, \ldots, n$ (see [1,2] for more details).

Example 1. An example of how it is possible to recover many well-known global indices of a WDN using (1) and the mathematical framework exposed in [1], is provided by the new formulation, given in (4) and (5), of the following two resilience indices

$$I_r = \frac{\sum_{i=1}^{n} q_i^* \left(h_i - h_i^* \right)}{\sum_{k=1}^{r} Q_k H_k - \sum_{i=1}^{n} q_i^* h_i^*} \tag{2}$$

and

$$I_R = \frac{\sum_{i=1}^{n} \left(q_i h_i - q_i^* h_i^* \right)}{\sum_{k=1}^{r} Q_k H_k - \sum_{i=1}^{n} q_i^* h_i^*}, \tag{3}$$

where Q_k and H_k are the discharge and the head, respectively, from the tank k. The former has been introduced by Todini in [25], while the second is a modified version of I_r used by Di Nardo et al. in [5–10].

The resilience indices I_r and I_R are written in [1,2] as follows

$$I_r = \frac{h^s \cdot p^*}{\gamma \left(Q \cdot H - q^* \cdot h^* \right)} \tag{4}$$

and

$$I_R = \frac{p^s \cdot p^*}{\gamma \left(Q \cdot H - q^* \cdot h^* \right)}, \tag{5}$$

where

- $Q := (Q_1, Q_2, \ldots, Q_r)$,
- $H := (H_1, H_2, \ldots, H_r)$,
- $q^* := (q_1^*, q_2^*, \ldots, q_n^*)$,
- $h^* := (h_1^*, h_2^*, \ldots, h_n^*)$,
- $p^* := (p_1^*, p_2^*, \ldots, p_n^*)$,
- $h^s := (h_1^s, h_2^s, \ldots, h_n^s)$,
- $p^s := (p_1^s, p_2^s, \ldots, p_n^s)$,

and "$\cdot$" denotes the *standard scalar product* between real vectors (of dimension n or r in our case).

For more details and examples the reader can see [1] and [2].

3 Hydraulic Solvers

In this work, two different approaches are studied using known software for the hydraulic modeling of WDNs. To obtain a solution, EPANET solves the continuity Eq. (6)(a) in each junction node and the energy law one, expressed in (6)(b), for each pipe connecting two nodes. The energy law links the headloss to the flow, depending on the pipe characteristics. Such equations are generally expressed as in the following form

$$
\begin{cases}
\sum_{j=1}^{n(i)} Q_{ij} - q_i = 0 & \text{for all } i = 1, 2, \ldots, n, & \text{(a)} \\
h_i - h_j = h_{ij} = R \cdot Q_{ij}^{\,e} + m_l \cdot Q_{ij}^{\,2}, & & \text{(b)}
\end{cases}
\tag{6}
$$

where q_i is the flow demand and h_i the nodal head at note i, n the total number of junction nodes, $n(i)$ the number of those linked to the node i, h_{ij} and Q_{ij} the headloss and the flow rate in the pipe between the linked nodes i and j respectively, R the resistance coefficient, e the flow exponent in the headloss formula (resistance law) and m_l a minor loss coefficient (see for example [23]).

EPANET uses a DDA approach, so the water demand at junction nodes is a known term for solving the continuity Eq. (6)(a). If the solver fails to find a combination of heads in the junction nodes and flows in the links that satisfy the total demand, it will stop the simulation without obtaining a full solution.

In order to simulate the network with a PDA approach, WaterNetGen is used. In addition to continuity equations on nodes (6)(a) and energy for links (6)(b), WaterNetGen makes changes to the EPANET solver in order to simulate a difference between supply and demand in case of insufficient pressure. This software adds a third equation (see [21, 22]), which expresses the water supply according to:

$$
q_i^{\text{avl}}(\hbar_i) = q_i^{\text{req}} \cdot
\begin{cases}
1 & \text{if } \hbar_i \geq \hbar_i^{\text{ref}}, \\
\left(\dfrac{\hbar_i - \hbar_i^{\min}}{\hbar_i^{\text{ref}} - \hbar_i^{\min}} \right)^{\alpha} & \text{if } \hbar_i^{\min} < \hbar_i < \hbar_i^{\text{ref}}, \\
0 & \text{if } \hbar_i \leq \hbar_i^{\min},
\end{cases}
\tag{7}
$$

where the measures explained below are referred to the node i of the network:

$\hbar_i$ = node pressure
q_i^{avl} = available water supply
q_i^{req} = water demand
$\hbar_i^{\text{ref}}$ = service pressure (necessary to completely satisfy the demand)
$\hbar_i^{\min}$ = minimum pressure below which there is no supply
α = exponent of the pressure-demand relationship.

The local surplus and the resilience indices recalled in Sect. 2 use some design conditions (q_i^*, $\hbar_i^*$ and h_i^*) as reference requests (see, e.g., [5–8, 24, 25]) that serve as comparison terms for the actual functioning parameters of the network (q_i, $\hbar_i$ and h_i, respectively).

A problem related to the use of these indices is the difficulty in defining and identifying those conditions. Scientific literature shows, in fact, that it is rare to have precise indications on design pressures and water demand. Some authors use a pressure value equal for the whole network [5–10, 25]) but it was not possible to identify publications that gave explicit indications on the design values of the water demand.

3.1 Demand Driven Analysis

In absence of indications on the design water demand, it seems that some authors have placed the design water demand equal to that assigned to the software as known terms. This means that, the design water demand is equal on all nodes to the actual water supply, that is

$$q_i = q_i^* \qquad \text{for all } i = 1, 2, \dots, n. \tag{8}$$

Then, it is immediate that the local discharge surplus index defined in (1) will be null everywhere,

$$q_i^s = \frac{q_i}{q_i^*} - 1 = 1 - 1 = 0 \qquad \text{for all } i = 1, 2, \dots, n,$$

and the local power surplus index will be equal to the local head surplus one:

$$p_i^s = \frac{p_i}{p_i^*} - 1 = \frac{q_i h_i}{q_i^* h_i^*} - 1 = \frac{h_i}{h_i^*} - 1 = h_i^s. \tag{9}$$

In particular, Eq. (8) implies that there cannot be surplus or deficit on water supply since the values will always be identical. No predictions or assumptions can be made on the sign of the local head and pressure head indices in (1) since it will solely depend on the regime of the water heads in the WDN.

The design water head (or pressure) assigned to the network will not affect the EPANET results, because these values will only affect the performance indices. Moreover, the resilience index I_r proposed by Todini and the resilience index I_R by Di Nardo et al., will in this case coincide:

$$
\begin{aligned}
I_R &= \frac{\sum\limits_{i=1}^{n} \left(q_i h_i - q_i^* h_i^* \right)}{\sum\limits_{k=1}^{r} Q_k H_k - \sum\limits_{i=1}^{n} q_i^* h_i^*} \\
&= \frac{\sum\limits_{i=1}^{n} q_i^* \left(h_i - h_i^* \right)}{\sum\limits_{k=1}^{r} Q_k H_k - \sum\limits_{i=1}^{n} q_i^* h_i^*} = I_r.
\end{aligned}
$$

It is also immediate to notice equality between the two formulations in (4) and (5), if we recall that the local head surplus index and the local power surplus index coincide (see (8)).

3.2   Pressure Driven Analysis

In PDA approach there is an effective difference between water demand and supply in the nodes. Taking into account the relation (7), the water demand and service pressure coincide with the design conditions used in the performance indices. The relation (7) becomes

$$q_i = q_i^* \cdot \begin{cases} 1 & \text{if } \hbar_i \geq \hbar_i^*, \\ \left(\dfrac{\hbar_i - \hbar_i^{\min}}{\hbar_i^* - \hbar_i^{\min}} \right)^{\alpha} & \text{if } \hbar_i^{\min} < \hbar_i < \hbar_i^*, \\ 0 & \text{if } \hbar_i \leq \hbar_i^{\min}. \end{cases} \qquad (10)$$

With this relation no longer worth (8), in fact, the supply may differ from the design demand and the local discharge surplus index in (1) can get non-zero values.

In particular, in WaterNetGen analysis, two situations can occur:

– the network is characterized by a good hydraulic head regime, so the behavior of PDAs is similar to that of DDAs;
– there are head deficits and the water supply is lower compared to the demand.

Note moreover that in the first scenario above, Eq. (8) remains still valid. For a head-deficient WDN we have instead

$$\begin{cases} q_i = q_i^* & \text{if } \hbar_i \geq \hbar_i^* \\ q_i < q_i^* & \text{if } \hbar_i < \hbar_i^* \end{cases},$$

whence

$$q_i^s \leq 0 \qquad \text{for all } i = 1, 2, \ldots, n.$$

This means that in the second scenario, the local discharge surplus index can only get non-positive values.

4   Application to the KL Network

The local surplus indices were tested in a practical application on a real network, well-known in literature. The used WDN is the network proposed by Kang and Lansey in [11] (*KL network*, for short) which consists of 935 nodes and 1274 pipes. In average condition the network has a consumption of 177 l/s (2808 gal/min) and the authors propose in [11] a peak factor of 1.75 to take into account hourly peaks. The authors also provide an indication of the minimum pressure in the network (design pressure) which is equal to 28 m (40 psi).

In this work a higher peak coefficient is used in order to establish a low-pressure regime and analyze the differences between the DDA and PDA approaches in the surplus index and in the resilience indices. The peak factor used is 2.5 (see [20, pag. 22]). The minimum pressure is 15 m and the design one is 28 m.

The local surplus indices are plotted for each node of the network for both analysis types and the results are shown in Figs. 1 and 2, respectively. A colorimetric scale that takes into account 3 intervals was used:

- Red: Deficit conditions;
- Yellow: Conditions close to the project requests;
- Green: Condition of surplus.

For the local discharge surplus index and the local pressure head surplus index the bands are:

$[-1; -0.1]$ Deficit conditions, for values that are less than 90% of the design values;

$[-0.1; 0.1]$ Design conditions, for values that do not differ more than 10% from the design values;

$[0.1; 1]$ Surplus conditions, for values that are greater than 110% of the design values.

For the local power surplus index and the local head surplus index, the bands are different because the presence of the geodetic elevation reduces the variability of the index, therefore they are:

$[-1; -0.01]$ Deficit conditions, for values that are less than 99% of the design values;

$[-0.01; 0.01]$ Design conditions, for values that do not differ more than 1% of the design values;

$[0.01; 1]$ Surplus conditions, for values that are greater than 101% of the design values.

In correspondence to Fig. 1, hence with the DDA approach, both resilience indices I_R and I_r coincide to the following value

$$I_R = I_r = 0.0532.$$

Instead, by using the PDA approach and hence referring to Fig. 2, we obtained

$$I_R = -1.46 \qquad \text{and} \qquad I_r = 0.448.$$

For the results interpretation of Figs. 1 and 2, it is clear that the multiplicative peak coefficient of the flows establishes in the network a condition characterized by a poor pressures regime. The increase of the elevation parameter, which grows further away from the reservoir (SE direction), contributes to the load losses along the pipelines, and this creates a deficit condition especially in the areas farthest from the reservoir. Recall that Fig. 1 shows the results obtained from the application of a solver that uses the DDA model and, as explained in Subsect. 3.1, the information provided by the local discharge surplus index q_i^s and the local power surplus index p_i^s (i.e., the first and the last index of the four defined in (1)) are not significant. The former, in fact, is everywhere equal to zero by the hypotheses made in (8) about the coincidence between nodal discharges q_i and minimal design discharge requests q_i^*. The reader can note that this agrees

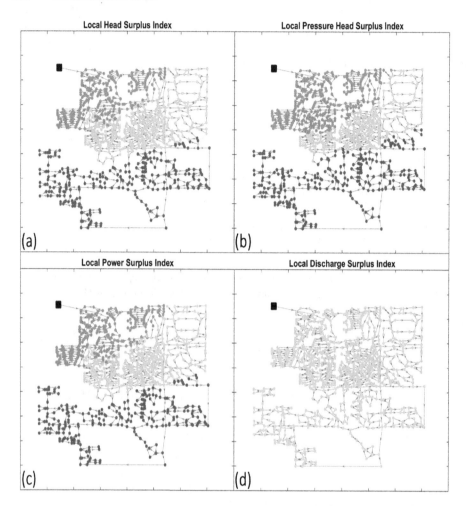

Fig. 1. The graphical display of local indices for the Kang and Lansey WDN, where the results are obtained from a DDA hydraulic simulation. Subfigures (a) and (b) show the local head and the local pressure head surplus index, respectively. Subfigures (c) and (d), instead, show the local power and the local discharge surplus index, respectively.

with the bottom right picture in Fig. 1 where all the junction nodes are colored in yellow. Similarly, the local power surplus index p_i^s coincides with the local head surplus index h_i^s (recall (9)) and this agrees, as well, with the same nodes coloration between the two left pictures Fig. 1.

The surplus indices relating to pressure and load give, instead, an immediate graphical information of the network status. Due to the distance from the reservoir, the nodes in the southern area of the network will be characterized by deficit conditions compared to the design ones.

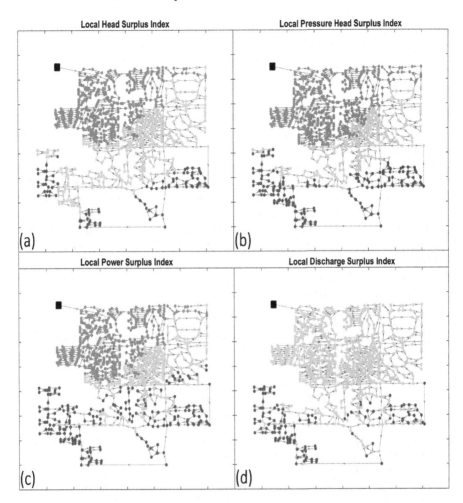

Fig. 2. The graphical display of local indices for the Kang and Lansey WDN, where the results are obtained from a PDA hydraulic simulation. The Subfigures (a), (b), (c) and (d) are in full correspondence with those of Fig. 1, for the same local indices.

On the other hand, the information provided by local indices, calculated for the result of a PDA model (see Fig. 2), with the same design conditions and peak coefficients, give different results. In this case, the local discharge surplus index q_i^s provides information on the nodes that, due to the pressure deficit, cannot guarantee that the supply meets the request. As highlighted in the Subsect. 3.2, there cannot be nodes in surplus. The shown condition of pressure and load deficit is less critical than the one obtained with the DDA model. It is clear that, being minor the supply than the previous case, there are lower flow rates, lower speeds and consequently the network contains larger load losses. Finally, the local power surplus index p_i^s provides a new set of information which, in this case, will not coincide with the ones provided by the local head surplus index.

5 Conclusions

The new set of local indices proposed by Caldarola and Maiolo in [1] are assessed for a WDN in addition to two well-known resilience ones. The mathematical framework described there allowed to simplify the automatic calculation of resilience indices. It was also possible to visualize the local surplus indices graphically and this approach allowed to have an immediate feedback on the state of the network. The assessment of these indices on a WDN leads moreover to a series of observations about their application limits. Both softwares that use DDA and PDA models were employed. The models hypotheses limit the possible results, sometimes preventing the achievement of indices representative of the WDN situation. The type of solution model used influences the value of the indices and the resilience assessment. The lack of a precise definition of design conditions is the main factor of uncertainty in the results.

Aknowledgments. This work is partially supported by the research projects *"origAMI, Original Advanced Metering Infrastructure"*, CUP J48C17000170006, POR Calabria FESR-FSE 2014–2020, and *"I-BEST"*, CUP B28I17000290008, PON "Innovazione e competitività" 2014/2020 MiSE Horizon 2020.

References

1. Caldarola, F., Maiolo, M.: Local indices within a mathematical framework for urban water distribution systems. Cogent Eng. **6**, 1643057 (2019). https://doi.org/10.1080/23311916.2019.1643057
2. Caldarola, F., Maiolo, M.: Algebraic tools and new local indices for water networks: some numerical examples. In: Sergeyev, Y., Kvasov, D. (eds.) NUMTA 2019. LNCS, vol. 11973, pp. 517–524. Springer, New York (2020)
3. Carini, M., Maiolo, M., Pantusa, D., Chiaravalloti, F., Capano, G.: Modelling and optimization of least-cost water distribution networks with multiple supply sources and users. Ric. Matem. **67**(2), 465–479 (2018). https://doi.org/10.1007/s11587-017-0343-y
4. Cervarolo, G., Mendicino, G., Senatore, A.: Re-modulating water allocation in a complex multi-reservoir system under current and climate change scenarios. Eur. Water **37**, 47–57 (2012)
5. Di Nardo, A., Di Natale, M.: A heuristic design support methodology based on graph theory for district metering of water supply networks. Eng. Optim. **12**, 193–211 (2011)
6. Di Nardo, A., Di Natale, M.: A design support metodology metodology for district metering of water supply networks. Water Distrib. Syst. Anal. **2010**, 870–887 (2012)
7. Di Nardo, A., et al.: Redundancy features of water distribution systems. Procedia Eng. **186**, 412–419 (2017)

8. Di Nardo, A., Di Natale, M., Santonastaso, G.: A comparison between different techniques for water network sectorization. Water Sci. Technol. Water Supply **14**, 961–970 (2014)
9. Di Nardo, A., Di Natale, M., Santonastaso, G., Tzatchkov, V., Alcocer-Yamanaka, V.: Water network sectorization based on graph theory and energy performance indices. J. Water Res. Planning Manag. **140**, 620–629 (2014)
10. Di Nardo, A., Di Natale, M., Santonastaso, G., Tzatchkov, V., Alcocer-Yamanaka, V.: Performance indices for water network partitioning and sectorization. Water Sci. Technol. Water Supply **15**, 499–509 (2015)
11. Kang, D., Lansey, K.: Revisiting optimal water-distribution system design: issues and a heuristic hierarchical approach. J. Water Res. Planning Manag. **138**, 208–217 (2012)
12. Maiolo, M., Carini, M., Capano, G., Piro, P.: Synthetic sustainability index (SSI) based on life cycle assessment approach of low impact development in the Mediterranean area. Cogent Eng. **4**, 1410272 (2017). https://doi.org/10.1080/23311916.2017.1410272
13. Maiolo, M., Martirano, G., Morrone, P., Pantusa, D.: Assessment criteria for a sustainable management of water resources. Water Pract. Technol. **1** (2006). https://doi.org/10.2166/wpt.2006.012
14. Maiolo, M., Mendicino, G., Pantusa, D., Senatore, A.: Optimization of drinking water distribution systems in relation to the effects of climate change. Water **9**, 803 (2017). https://doi.org/10.3390/w9100803
15. Maiolo, M., Pantusa, D.: A methodological proposal for the evaluation of potable water use risk. Water Pract. Technol. **10**, 152–163 (2015). https://doi.org/10.2166/wpt.2015.017
16. Maiolo, M., Pantusa, D.: An optimization procedure for the sustainable management of water resources. Water Sci. Technol. Water Supply **16**, 61–69 (2016). https://doi.org/10.2166/ws.2015.114
17. Maiolo, M., Pantusa, D.: Combined reuse of wastewater and desalination for the management of water systems in conditions of scarcity. Water Ecol. **72**, 116–126 (2017)
18. Maiolo, M., Pantusa, D.: Infrastructure vulnerability index of drinking water supply systems to possible terrorist attacks. Cogent Eng. **5**, 1456710 (2018). https://doi.org/10.1080/23311916.2018.1456710
19. Maiolo, M., Pantusa, D., Chiaravalloti, F., Carini, M., Capano, G., Procopio, A.: A new vulnerability measure for water distribution network. Water **10**, 1005 (2019). https://doi.org/10.3390/w10081005
20. Milano, V.: Acquedotti. Hoepli Editore, Milano (1996)
21. Muranho, J., Ferreira, A., Sousa, J., Gomes, A., Sa Marques, A.: Waternetgen - an epanet extension for automatic water distribution network models generation and pipe sizing. Water Sci. Technol. Water Supply **12**, 117–123 (2012)
22. Muranho, J., Ferreira, A., Sousa, J., Gomes, A., Sa Marques, A.: Pressure dependent demand and leakage modelling with an epanet extension - waternetgen. Procedia Eng. **89**, 632–639 (2014). 16th Conference on Water Distribution System Analysis, WDSA
23. Rossman, L.A.: Epanet 2: users manual (2000)
24. Shin, S., et al.: A systematic review of quantitative resilience measures for water infrastructure systems. Water **10**, 164 (2018)
25. Todini, E.: Looped water distribution networks design using a resilience index based on heuristic approach. Urban Water **12**, 115–122 (2000)

Performance Management of Demand and Pressure Driven Analysis in a Monitored Water Distribution Network

Marco Amos Bonora[1], Manuela Carini[1] (ID), Gilda Capano[1], Rocco Cotrona[1], Daniela Pantusa[1] (ID), Joaquim Sousa[2] (ID), and Mario Maiolo[1(✉)] (ID)

[1] Department of Environmental and Chemical Engineering (DIATIC), Università della Calabria, Cubo 42/B, 87036 Arcavacata di Rende, CS, Italy
{marcoamos.bonora,manuela.carini,gilda.capano,daniela.pantusa, mario.maiolo}@unical.it, roccocotrona@hotmail.it
[2] Department of Civil Engineering, Polytechnic Institute of Coimbra, Coimbra, Portugal
jjoseng@isec.pt

Abstract. A smart management of water distribution networks requires the infrastructure to be operated with high efficiency. For many years the hydraulic modelling of water distribution networks has been conditioned by the scarce availability of quality data but the technological advances contributed to overcome this drawback. The present work describes the research activity carried out about water distribution network modelling, focusing on model construction and calibration. For this purpose, Water-NetGen, an extension of the Epanet hydraulic simulation software, has been used. EPANET simulation model assumes that the required water demand is always fully satisfied regardless the pressure (Demand Driven Analysis - DDA), while WaterNetGen has a new solver assuming that the required water demand is only fully satisfied if the pressure conditions are adequate (Pressure Driven Analysis - PDA). A comparison between the software outputs is the starting point for a new method of allocating and distributing water demand and water losses along the network, leading to model results closer to the measurements obtained in the real network. The case study is the water distribution network of the municipality of Nicotera, in Southern Italy.

Keywords: Water distribution network (WDN) · Water management · Pressure Driven Analysis (PDA) · Demand Driven Analysis (DDA) · Calibration

1 Introduction

The water demand increase and the water scarcity require the use of management practices sensitive to the importance of water in human life. Consequently, all the aspects related to the proper allocation of available resources, to the evaluation of the climate change effects on water resources and schemes, to the use of

© Springer Nature Switzerland AG 2020
Y. D. Sergeyev and D. E. Kvasov (Eds.): NUMTA 2019, LNCS 11973, pp. 506–516, 2020.
https://doi.org/10.1007/978-3-030-39081-5_43

unconventional water resources, to the correct system design and operational efficiency, to the assess of vulnerability to natural and man-made disasters, are scientific topics of current interest [2, 6–8, 10, 18]. In this context, research activities
concerning the design and management of water distribution networks (WDN)
are of the utmost importance. The efficient management of water resources is
related to the hydraulic balances of the parameters which describe the network
behaviour. In the scientific literature there are many hydraulic-mathematical
models which allow evaluating the water distribution network efficiency. Among
these there are models based on the performance indices that play an important role, e.g. resilience, entropy and vulnerability [9, 13, 21]. These models allow
monitoring of the WDN correct functioning also for risk management. However,
the number and variability of the parameters which determine the risk in WDN
require the use of specific analytical methods, capable of quantifying it. In the
literature there are other models which allow increasing the knowledge of the
network, facilitating the planning, design and management phases. Simulation,
skeletonization, calibration and optimization models are very important in this
context, for the ability to show the basic behavioural structure of a WDN, allowing the implementation of improvement measures (optimization of characteristic parameters) and the WDN analysis in various scenarios. The improvement
of WDN management is frequently achieved by using simulation models, built
with the help of new software tools. In most cases, the model validity depends
on the ability to guarantee a distribution suited to the requests; therefore the
water demand is an essential parameter. Recently, due to the increasing need
of more realistic models, many of these incorporate special pressure-demand
relationships, that enable real prediction of the WDN behaviour [5]. In general,
simulation models may include a DDA or a PDA approach. In the first case
water demand is fixed while in the second it depends on the nodal pressures.
The traditional DDA approach can be used for planning, design and operation
of WDN working under normal conditions, while the PDA approach is used
in particular scenarios in which pressure conditions restrict water availability.
In the literature there are many models based on these approaches. The best-
known DDA application is presented in [17]. This is the open source software
Epanet, one of the most used tools for hydraulic simulation. The original node-
pipe equations, derived by [22], are solved by the Generalized Reduced Gradient
(GRG) method, which the same as the Newton-Raphson method. [23] presented
a unified framework for deriving simultaneous equations algorithms for WDN,
comprising all sets of equations solved by the Newton-Raphson method. The
Epanet software uses a DDA approach and so assumes that water demand can
always be met regardless the network pressures. However, if the network has
to operate in unusual operating conditions (excessive demands, pipe breaks or
fire scenarios, causing low pressures) it is shown that the DDA approach is not
able to correctly simulate operation [1]. In these situations, PDA simulation is
preferred, although often the analysis of pressure variability in a WDN can be
temporary and unpredictable [4]. Among the PDA approaches, [16] proposed a
methodology which takes into account the residual head versus outflow relation-

ship having no clamping of the outflow. [15] presented a pressure-driven method relying on a co-energy minimization problem formulation that do not requires to define any pressure-consumption relationship and does not rely on the detection of topological changes. This method permits better hydraulic predictions for network sections supplied by nodes with negative pressures. [20] developed a new head-discharge relationship which is based on volumetric demand (pressure independent) and head dependent demand, separately. [14] proposed a method based on the insertion of a sequence of devices at each water demand node (a General Purpose Valve, a fictitious junction, a reach with a check valve and a reservoir) to transform the DDA into a PDA. In this context it may be useful to have tools available to facilitate the implementation of a PDA approach also in Epanet software. There are many studies that deal with increasing its potential, also introducing applications which enable the PDA analysis. [3] proposed an Epanet extension to insert the pressure/demand functions into OOTEN toolkit, which aimed to simulate WDN under failures conditions. This study showed that when WDN work under abnormal pressure conditions the DDA approach fails and the PDA approach produces realistic results. Among these tools, WaterNet-Gen [11] plays an important role, it is an EPANET extension for the automatic generation and the sizing of synthetic models for WDN, particularly useful for the introduction of the PDA solver [19]. The extension, which can also be used to assess the technical performance of WDN [12] and analyse critical scenarios (abnormal pressure conditions), maintains the original user interface of EPANET to preserve the user experience, with the introduction of new functionalities particularly useful in contexts where it is necessary to modify elements from imported models, such as those obtained from CAD drawings [11]. An important characteristic is the ability to use this interface to split pipes (by automatically introducing intermediate nodes), change nodes to tanks/reservoirs, and easily insert pressure reduction valves to avoid or prevent the pressure exceeding the maximum value allowed [11]. Furthermore, WaterNetGen considers pressure independent demands (DDA) and three different types of pressure dependent demands (PDA): node consumption, pipe consumption and pipe losses. This software enables assigning the pipe consumption and losses through the use of pipe consumption and losses coefficients and respective patterns, very useful for large size WDN. In this paper, the Nicotera WDN (Southern Italy) is used to compare the calibration results from Epanet with those from WaterNetGen. The objective of this procedure is to define reliable data for a correct flow distribution, thus defining a new method of allocating and distributing demand and water losses along the network.

2   Case Study: Nicotera WDN

The Nicotera municipality ($38^{\circ}33'20''N, 15^{\circ}56'15''E$) is located in the Calabria region, in southern Italy (Fig. 1), with a territorial extension equal to 28.25 km^2.

The municipal water supply system, which must serve 6,353 inhabitants (the per capita water consumption is about 475 l/inhab/day), extends for about

Fig. 1. Nicotera municipality position.

51.6 km and includes 6 tanks, 4 wells and 3 reservoirs (Figs. 2 and 3). The Nicotera centre WDN, which is the subject of this work, is directly fed by 2 tanks (Piraino and Madonna della Scala). The average flow rate entering the Nicotera centre WDN from both tanks are as follows:

- out of the Piraino tank 27.8 l/s,
- out of the Madonna della Scala tank 0.4 l/s.

During the data collection, the flow rates coming from both tanks (2), the tank water levels (2) and some network pressures (3) and flows (3) were monitored. These measurements were conducted simultaneously and continuously for a period of two consecutive days, with an acquisition frequency equal to five minutes, using the following technologies:

- Flow - ultrasonic flow meters with transit time;
- pressure - LoLog Vista data logger with internal pressure sensors;
- level - LoLog Vista data logger with Radcom depth sensors.

The pressure meters were installed far from the tanks so that the measurements represent the general behaviour of the WDN. The water supply system model has 7 water sources (4 wells and 3 reservoirs) and comprises 6 District Metered Areas (DMAs), although this study is focused only on two of them (Nicotera centre): Nicotera Est and Nicotera Ovest. The model of these two

DMAs has 2 tanks (Piraino and Madonna della Scala), 186 nodes and 202 pipes with diameters in the range between 12 and 150 mm and 4 different materials (steel, cast iron, galvanized iron, HDPE).

3   Numerical Results and Discussions

The first simulation step, carried out using the Epanet software, required the pipe roughness coefficient, the allocation of water demand (with a spatial approach for each junction in the network, in this case calculated with Thiessen polygons obtained with QGIS), calculation of the water loss and its allocation for all junctions in the network, and 24 hours' demand patterns. This input data was used to simulate the Nicotera centre WDN behaviour for a 24 h period, between the 0:00 h of May 21st 2009 and 0.00 h of May 22nd 2009.

The following analysis showed great differences between the simulation results and the measurements (gaps bigger than 30% in the tank levels, 40% in pressures and 80% in flows). The chart in Fig. 4 presents some of the pressure results for node P2 from the first simulation. The average simulated pressure was equal to 26.07 m, while that observed was 30.15 m, and this was the best fit obtained in this simulation). To reduce the gap between the simulated and observed data, it was necessary to proceed with the calibration, which was carried out with WaterNetGen, an Epanet extension that has some interesting features to simplify the calibration work. The pressure and flow results obtained in the first simulation shown that it was necessary to correct the flow distribution in the network, namely increase it in the network served by the Piraino municipal tank (Nicotera East) and decrease them by the same amount for the network served by Madonna della Scala municipal tank (Nicotera West).

Firstly, demand coefficients were assigned to the pipes based on the building density observed in Google Earth. Secondly, water loss coefficients were assigned to the pipes in order to balance water loss in both networks and obtain good fittings with the flow measurements. Finally, roughness coefficients were adjusted to achieve good agreements with the pressure measurements. At the end of this procedure the simulation results were quite similar to the measurements (Figs. 5, 6, 7 and 8). Taking again node P2 as an example, the average simulated pressure is now equal to 30.41 m which compares with the observed value 30.15m, and the mean error is 0.424 m (1.4%). The biggest mean errors were: tank level - 0.035 m (0.7% Fig. 9), pressure - 1.253 m (4.0%) and flow - 0.436 l/s (8.4%). A considerable reduction of the gaps between the observed and simulated values is evident, confirming a good calibration. The results obtained show the advantage of using WaterNetGen in the calibration process, mainly due to a proper allocation of water demand to the nodes (water consumption and water losses assigned to the pipes). This issue, as previously mentioned, is essential to obtain good calibration results. In the case of WaterNetGen, the assignment of the nodal demands becomes quite simple by assigning to the pipes water demand and water loss coefficients and respective patterns. The pipe water demand and water loss are automatically assigned to the pipe end nodes using specific nodal demand categories for their representation [11].

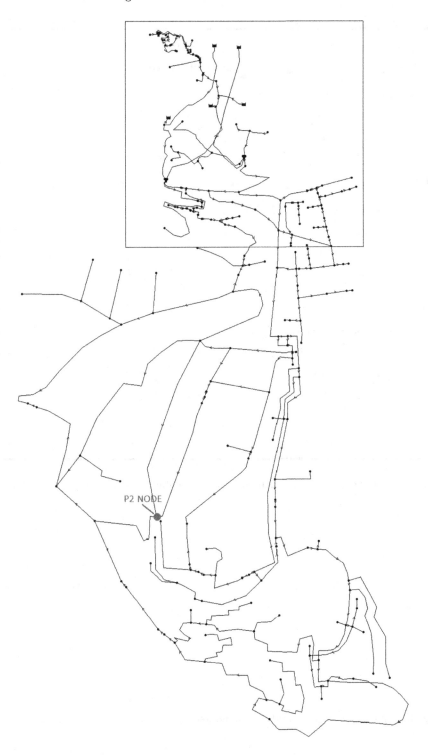

Fig. 2. (a) Nicotera municipality water distribution network scheme.

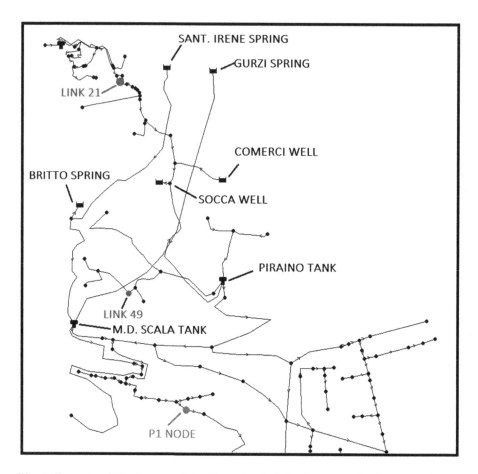

Fig. 3. Zoom-in of Nicotera municipality water distribution network scheme north part

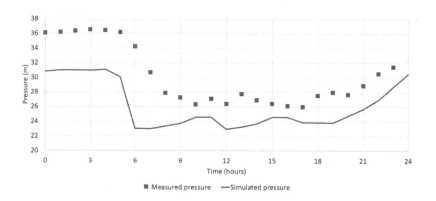

Fig. 4. Comparison of measured and simulated pressures for Node P2 obtained with Epanet.

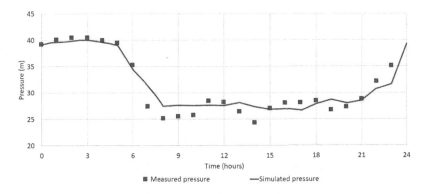

Fig. 5. Comparison of measured and simulated pressures for Node P1 obtained with Epanet.

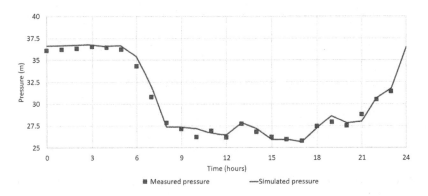

Fig. 6. Comparison of measured and simulated pressures for Node P2 obtained with WaterNetGen.

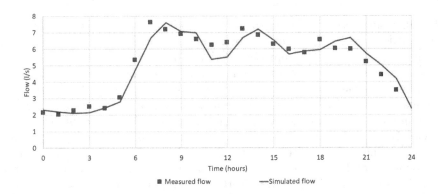

Fig. 7. Comparison of measured and simulated flows for Link 49 obtained with Water-NetGen.

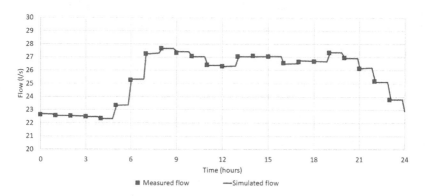

Fig. 8. Comparison of measured and simulated flows for Link 21 obtained with Water-NetGen.

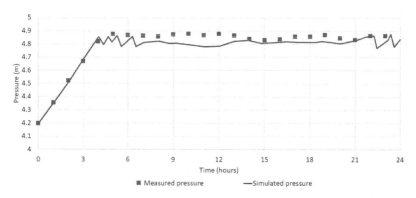

Fig. 9. Comparison of measurements and simulation results from WaterNetGen: Piraino Tank Level.

4 Conclusions

The DDA simulation, despite representing the traditional methodology, does not provide realistic results for networks under abnormal operating conditions. In this context, the PDA approach produces better results, especially in the implementation through software tools such as WaterNetGen, a useful Epanet extension. This software easily enables the assessment of the WDN performance, and analysis of critical scenarios, by applying a simple method to allocate and distribute water demand and losses along the network. The reliability of Water-NetGen is justified by obtaining simulation values similar to the measurements (which in some cases are perfect matches), resulting in a good calibration. This methodology applied to the real case of the Nicotera WDN showed how the EPANET simulator with the WaterNetGen extension is a useful tool to achieve accurate simulations. The calibration results provide useful data to define some new criteria for allocating and distributing water demand and losses along the network.

Aknowledgments. This work is partially supported by the Italian Regional Project (POR CALABRIA FESR 2014–2020): *"origAMI, Original Advanced Metering Infrastructure"* CUP J48C17000170006.

References

1. Ackley, J.R.L., Tanyimboh, T.T., Tahar, B., Templeman, A.B.: Head-driven analysis of water distribution systems. Water Softw. Syst. Theor. Appl. **1**(3), 183–192 (2001)
2. Barthel, R., Janisch, S., Nickel, D., Trifkovic, A., Hörhan, T.: Using the multiactor-approach in Glowa-Danube to simulate decisions for the water supply sector under conditions of global climate change. Water Resour. Manag. **23**(2), 239–275 (2010). https://doi.org/10.1007/s11269-009-9445-y
3. Cheung, P.B., Van Zyl, J.E., Reis, L.F.R.: Extension of EPANET for pressure driven demand modeling in water distribution system. Comput. Control Water Indus. **1**, 311–316 (2005)
4. Ciaponi, C., Creaco, E.: Comparison of pressure-driven formulations for WDN simulation. Water **10**(4), 523–537 (2018). https://doi.org/10.3390/w10040523
5. Giustolisi, O., Walski, T.M.: Demand components in water distribution network analysis. J. Water Resour. Plan. Manag. **138**(4), 356–367 (2011). https://doi.org/10.1061/(ASCE)WR.1943-5452.0000187
6. Maiolo, M., Pantusa, D.: Combined reuse of wastewater and desalination for the management of water systems in conditions of scarcity. Water Ecol. **4**(72), 116–126 (2017). https://doi.org/10.23968/23053488.2017.22.4.116-126
7. Maiolo, M., Pantusa, D.: Infrastructure Vulnerability Index of drinking water systems to terrorist attacks. Cogent Eng. **5**(1), 1456710 (2018). https://doi.org/10.1080/23311916.2018.1456710
8. Maiolo, M., Mendicino, G., Pantusa, D., Senatore, A.: Optimization of drinking water distribution systems in relation to the effects of climate change. Water **9**(10), 803–817 (2017). https://doi.org/10.3390/w9100803
9. Maiolo, M., Pantusa, D., Carini, M., Capano, G., Chiaravalloti, F., Procopio, A.: A new vulnerability measure for water distribution network. Water **10**(8), 1005 (2018). https://doi.org/10.3390/w10081005
10. Minville, M., Brissette, F., Krau, S., Leconte, R.: Behaviour and performance of a water resource system in Québec (Canada) under adapted operating policies in a climate change context. Water Resour. Manag. **24**, 1333–1352 (2010). https://doi.org/10.1007/s11269-009-9500-8
11. Muranho, J., Ferreira, A., Sousa, J., Gomes, A., Marques, A.S.: WaterNetGen: an EPANET extension for automatic water distribution network models generation and pipe sizing. Water Sci. Technol. Water Supply **12**(1), 117–123 (2012). https://doi.org/10.2166/ws.2011.121
12. Muranho, J., Ferreira, A., Sousa, J., Gomes, A., Marques, A.S.: Pressure-dependent demand and leakage modelling with an EPANET extension-WaterNetGen. Procedia Eng. **89**, 632–639 (2014). https://doi.org/10.1016/j.proeng.2014.11.488
13. Nachtnebel, H.P.: Irreversibility and Sustainability in Water Resources Systems. Risk, Reliability, Uncertainty, and Robustness of Water Resources Systems. International Hydrology Series. Cambridge University Press, Cambridge (2002). https://doi.org/10.1017/CBO9780511546006

14. Pacchin, E., Alvisi, S., Franchini, M.: Analysis of non-iterative methods and proposal of a new one for pressure-driven snapshot simulations with EPANET. Water Resour. Manag. **31**(1), 75–91 (2017). https://doi.org/10.1007/s11269-016-1511-7

15. Piller, O., Van Zyl, J.E.: Pressure-driven analysis of network sections supplied via high-lying nodes. In: Boxal, J., Maksimovic, C. (eds.) Proceedings of the Computing and Control in the Water Industry. Taylor and Francis Group, London (2009)

16. Reddy, L.S., Elango, K.: Analysis of water distribution networks with head-dependent outlets. Civil Eng. Syst. **6**(3), 102–110 (1989)

17. Rossman, L.A.: EPANET 2: users manual (2000)

18. Samani, H.M.V., Mottaghi, A.: Optimization of water distribution networks using integer linear programming. J. Hydraul. Eng. **132**(5), 501–509 (2006). https://doi.org/10.1061/(ASCE)0733-9429(2006)132:5(501)

19. Sousa, J., Muranho, J., Sá Marques, A., Gomes, R.: Optimal management of water distribution networks with simulated annealing: the C-Town problem. J. Water Resour. Plan. Manag. **142**(5), C4015010 (2015). https://doi.org/10.1061/(ASCE)WR.1943-5452.0000604

20. Tabesh, M., Shirzad, A., Arefkhani, V., Mani, A.: A comparative study between the modified and available demand driven based models for head driven analysis of water distribution networks. Urban Water J. **11**(3), 221–230 (2014). https://doi.org/10.1080/1573062X.2013.783084

21. Todini, E.: Looped water distribution networks design using a resilience index based heuristic approach. Urban Water **2**(2), 115–122 (2000). https://doi.org/10.1016/S1462-0758(00)00049-2

22. Todini, E., Pilati, S.: A gradient method for the solution of looped pipe networks. In: Coulbeck, B., Orr, C.H. (eds.) Computer Applications in Water Supply, vol. 1, pp. 1–20. Wiley, London (1988)

23. Todini, E., Rossman, L.A.: Unified framework for deriving simultaneous equations algorithms for water distribution networks. J. Hydraul. Eng. **139**(5), 511–526 (2013). https://doi.org/10.1061/(ASCE)HY.1943-7900.0000703

Algebraic Tools and New Local Indices for Water Networks: Some Numerical Examples

Fabio Caldarola[1] and Mario Maiolo[2]([⊠])

[1] Department of Mathematics and Computer Science, Università della Calabria,
Cubo 31/B, 87036 Arcavacata di Rende, CS, Italy
`caldarola@mat.unical.it`
[2] Department of Environmental and Chemical Engineering (DIATIC),
Università della Calabria, Cubo 42/B, 87036 Arcavacata di Rende, CS, Italy
`mario.maiolo@unical.it`

Abstract. Very recently, a new set of *local indices* for urban water networks has been proposed by the authors, within a mathematical framework which is unprecedented for this field, as far as we know. Such indices can be viewed as the "elementary bricks" that can be used to construct as many global (and local) indices as one needs or wants, where the glue, or mortar, is given by the mathematical tools of the aforementioned framework coming mostly from linear algebra and vector analysis. In this paper, after a brief description of the setting as explained above, we recover, through new formulations, some well-known global indicators like the *resilience index* I_r introduced by Todini. Then we also give some explicit numerical computations and examples, sometimes with the help of the hydraulic software EPANET 2.0.12.

Keywords: Mathematical modeling · Linear algebra · Urban water networks · Performance indices · EPANET 2.0.12

1 Introduction

In recent years many authors have introduced, sometimes with considerable success, a multitude of indices, especially of energetic-hydraulic nature (for example, indices of resilience, robustness, pressure, failure, flow deficit, mechanical redundancy, balance, reliability, entropy, etc.) to characterize and summarize in a single parameter some of the most important peculiar characteristics of a complex water network (for instance, a recent review of 21 different resilience measures was given last year in paper [20]). Therefore these indices, which are expressly designed to be of global nature, do not adapt very well to local analysis even applying them to a small portion of the network: a small portion of a network, in fact, is not the same as a small independent network.

In [3] the authors propose a new set of local indicators within a mathematical framework which is also unprecedented, as far as we know, for hydraulic-engineering purposes. Such indices, besides providing the basis for a local analysis

© Springer Nature Switzerland AG 2020
Y. D. Sergeyev and D. E. Kvasov (Eds.): NUMTA 2019, LNCS 11973, pp. 517–524, 2020.
https://doi.org/10.1007/978-3-030-39081-5_44

of the water distribution network (WDN), can be seen as the "elementary bricks" with which, by means of the mathematical tools offered by the aforementioned framework, which act as glue or mortar for the bricks, one can construct as many global (and local) indices as one needs or wants, for the study of the considered WDN. Moreover, he can also recover many already known global indicators, often even giving a deeper structural interpretation of the same. This possibility is explicitly illustrated in [3] in several cases, giving *per se* a strong automatic validation to the new proposed machinery.

In this paper we first give a brief description of the *local indices* introduced in [3], then we examine their building relations with some resilience indices like that proposed by Todini in [21], or the one used by Di Nardo et al. in [5–10], or others (see Sect. 2). Section 3 instead is devoted to numerical computations and explicit examples. In particular, with the help of the hydraulic software EPANET 2.0.12, we will examine some complementary cases of those considered in [3] for the prototypical looped system known as the *two-loop network* (TLN) and we will computing explicitly the family of local indices, deriving also from them the global resilience measures mentioned above. Finally, in Sect. 4, we will remove the ubiquitous assumption of a uniform minimal design pressure on the network, and we will consider analogous pipes calibrations of those of the previous section. Once the calculations are done again, we will briefly compare the obtained results.

2   The Local Performance Indices for a WDN

As noticed in the Introduction, the contemporary trend of designing and developing indices of a global nature for water distribution networks (WDNs) is strongly growing and involves a number of aspects of WDNs. But not only; multiple lines of research have introduced a wide range of similar indices and measures concerning as well the vulnerability of infrastructures, the sustainability of water resources, various types of hydropotable risk, etc. (the reader can see, for instance, [4, 12–19] and the references therein). On the contrary, instead, indices, measures and parameters of local nature are really very little present in the literature of this field.

To introduce local indices for a WDN we need a certain amount of notations. Let $\mathscr{N}$ be a given WDN and we denote by n the number of its junction nodes, by m the number of pipes in $\mathscr{N}$ connecting two nodes, by r the number of reservoirs, by H_k and Q_k the head and the discharge, respectively, outgoing from the k-th reservoir, $k = 1, 2, \ldots, r$, by z_i, q_i and $\hbar_i$, the *elevation height*, the *discharge* and the *pressure head*, respectively, at the node i, where $i = 1, 2, \ldots, n$. Then we also pose $h_i = z_i + \hbar_i$ for the *piezometric* or *hydraulic head* and $p_i = \gamma\, q_i h_i$ for the power delivered at the node i, where $\gamma = \rho g$ denotes, as usual, the specific weight of water. We moreover use the widespread star notation, i.e., q_i^*, $\hbar_i^*$, $h_i^* = z_i + \hbar_i^*$ and $p_i^* = \gamma\, q_i^* h_i^*$, to indicate the minimal requests or the design conditions relative to the above defined quantities (see [3, 5–10, 21] and many references mentioned there).

In [3] the following local performance indices are defined and proposed as elementary factors to perform a local-global analysis on the WDN:

$$q_i^s := \frac{q_i - q_i^*}{q_i^*} \qquad \textit{Local discharge surplus index,}$$

$$\hbar_i^s := \frac{\hbar_i - \hbar_i^*}{\hbar_i^*} \qquad \textit{Local pressure head surplus index,}$$

$$h_i^s := \frac{h_i - h_i^*}{h_i^*} \qquad \textit{Local piezometric head surplus index,} \tag{1}$$

$$p_i^s := \frac{p_i - p_i^*}{p_i^*} = \frac{q_i h_i - q_i^* h_i^*}{q_i^* h_i^*} \quad \textit{Local power surplus index,}$$

where $i = 1, 2, \ldots, n$. We moreover collect such local indices in vectors obtaining the following *local surplus vectors*

$$\boldsymbol{q}^s := (q_1^s, q_2^s, \ldots, q_n^s) \quad \textit{Local discharge surplus vector,}$$

$$\boldsymbol{\hbar}^s := (\hbar_1^s, \hbar_2^s, \ldots, \hbar_n^s) \quad \textit{Local pressure head surplus vector,}$$

$$\boldsymbol{h}^s := (h_1^s, h_2^s, \ldots, h_n^s) \quad \textit{Local piezometric head surplus vector,} \tag{2}$$

$$\boldsymbol{p}^s := (p_1^s, p_2^s, \ldots, p_n^s) \quad \textit{Local power surplus vector.}$$

Finally let D_j be the diameter of the j-th pipe, $j = 1, 2, \ldots, m$, and we also pose $\boldsymbol{D} := (D_1, D_2, \ldots, D_m)$, $\boldsymbol{Q} := (Q_1, Q_2, \ldots, Q_r)$, $\boldsymbol{H} := (H_1, H_2, \ldots, H_r)$, $\boldsymbol{q}^* := (q_1^*, q_2^*, \ldots, q_n^*)$, $\boldsymbol{h}^* := (h_1^*, h_2^*, \ldots, h_n^*)$ and $\boldsymbol{p}^* := (p_1^*, p_2^*, \ldots, p_n^*)$, to have all the data in vectorial form.

Example 1. The well-known resilience index introduced by E. Todini in [21] is defined as

$$I_r = I_r(\mathcal{N}) := \frac{\sum\limits_{i=1}^{n} q_i^* (h_i - h_i^*)}{\sum\limits_{k=1}^{r} Q_k H_k - \sum\limits_{i=1}^{n} q_i^* h_i^*} \tag{3}$$

and is used and implemented in various hydraulic software. Using local vectors and some elementary mathematical tools, (3) can be written as

$$I_r = \frac{\boldsymbol{h}^s \cdot \boldsymbol{p}^*}{\gamma (\boldsymbol{Q} \cdot \boldsymbol{H} - \boldsymbol{q}^* \cdot \boldsymbol{h}^*)} \tag{4}$$

where "$\cdot$" denotes the standard scalar product between real vectors (of dimension n or r in this case).

Example 2. The resilience index used by Di Nardo et al. in [5–10] is defined as follows

$$I_R = I_R(\mathcal{N}) := \frac{\sum\limits_{i=1}^{n} (q_i h_i - q_i^* h_i^*)}{\sum\limits_{k=1}^{r} Q_k H_k - \sum\limits_{i=1}^{n} q_i^* h_i^*}, \tag{5}$$

and using (2) and our mathematical framework we have

$$I_R = \frac{p^s \cdot p^*}{\gamma (Q \cdot H - q^* \cdot h^*)}. \tag{6}$$

Meaning, applications and, for instance, some advantages of the new formulas (4) and (6), are discussed in [3] and also in [2]. Here we just notice as the linear algebra language allows an easy and effective implementation in many engineering software and numerical computing systems as, for example, MATLAB[1].

Remark 1. Among the local indices defined in (1) it is easy to find relations as the following

$$p_i^s = q_i^s h_i^s + q_i^s + h_i^s,$$

which yields

$$p^s = q^s \circ h^s + q^s + h^s, \tag{7}$$

where "$\circ$" denotes the *Hadamard product* defined entrywise between vectors and matrices (see [3] or [11]).

For future use we recall here another simple algebraic tool (for several others see [3]); for an n-tuple $x = (x_1, x_2, \ldots, x_n)$ belonging to $\mathbb{R}^n$, the *1-norm*, or *taxicab norm* $\| \cdot \|_1$ is defined as

$$\|x\|_1 = \|(x_1, x_2, \ldots, x_n)\|_1 := \sum_{i=1}^{n} |x_i|. \tag{8}$$

3   Local Indices Applied to the TLN

The very simple TLN, after its appearance in [1], has become very widespread in literature for various types of examples, both theoretical and computational. It is represented in Fig. 1 where each of the 8 pipes is 1000 m long, the altimetric nodal data are given by

$$z = (150\,\text{m}, 160\,\text{m}, 155\,\text{m}, 150\,\text{m}, 165\,\text{m}, 160\,\text{m}),$$

and the reservoir R has $Q = 1120\,\text{m}^3/\text{h}$ and $H = 210\,\text{m}$ (coincident with its geodetic height).

Example 3. If we take

$$\begin{aligned} q^* = q &= (100\,\text{m}^3/\text{h}, 100\,\text{m}^3/\text{h}, 120\,\text{m}^3/\text{h}, 270\,\text{m}^3/\text{h}, 330\,\text{m}^3/\text{h}, 200\,\text{m}^3/\text{h}) \\ &= (1, 1, 1.2, 2.7, 3.3, 2) \cdot 10^2\,\text{m}^3/\text{h}, \\ h^* &= (30\,\text{m}, 30\,\text{m}, 30\,\text{m}, 30\,\text{m}, 30\,\text{m}, 30\,\text{m}) \end{aligned} \tag{9}$$

as in [21], we immediately obtain

$$\begin{aligned} h^* &= (180\,\text{m}, 190\,\text{m}, 185\,\text{m}, 180\,\text{m}, 195\,\text{m}, 190\,\text{m}), \\ p^* &= (18, 19, 22.2, 48.6, 64.35, 38) \cdot 10^3\,\gamma\,\text{m}^4/\text{h}, \end{aligned} \tag{10}$$

---

[1] The linear algebra foundation, in this case, is clear even from the name: MATrix LABoratory.

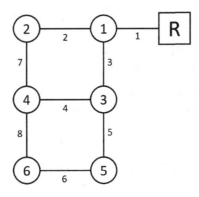

Fig. 1. A schematic representation of the TLN.

where $\gamma = \rho g = 9810\,\mathrm{N/m^3}$ is the specific weight of water. By using the Hazen-Williams formula with a coefficient of 130 and the software EPANET 2.0.12 with the following input diameters (in inches)

$$\boldsymbol{D} \; = \; (D_1, \ldots, D_8) \; = \; (20, 14, 14, 6, 14, 1, 14, 10) \tag{11}$$

(see [21, Table 3, Sol. B]), the pressure output is

$$\hslash \; = \; (55.96\,\mathrm{m}, 40.39\,\mathrm{m}, 45.82\,\mathrm{m}, 46.63\,\mathrm{m}, 33.43\,\mathrm{m}, 31.79\,\mathrm{m}), \tag{12}$$

hence, consequently,

$$h \; = \; (205.96\,\mathrm{m}, 200.39\,\mathrm{m}, 200.82\,\mathrm{m}, 196.63\,\mathrm{m}, 198.43\,\mathrm{m}, 191.79\,\mathrm{m}). \tag{13}$$

We therefore obtain the following explicit values for the local surplus vectors (2)

$$
\begin{aligned}
\boldsymbol{q}^s \; &= \; (\text{since } \boldsymbol{q} = \boldsymbol{q}^*) \; = \; (0,0,0,0,0,0), \\
\hslash^s \; &\approx \; (0.8653, 0.3463, 0.5273, 0.5543, 0.1143, 0.0597), \\
h^s \; &\approx \; (0.1442, 0.0547, 0.0855, 0.0924, 0.0176, 0.0094), \\
\boldsymbol{p}^s \; &= \; (\text{using } (7) \text{ and } \boldsymbol{q}^s = \boldsymbol{0}) \; = \; h^s.
\end{aligned}
\tag{14}
$$

Thus we can compute the resilience index I_r using (14) and (4), (9), (10) as follows[2]

$$
\begin{aligned}
I_r \; &= \; \frac{h^s \cdot p^*}{\gamma(Q \cdot H - q^* \cdot h^*)} \\
&\approx \; \frac{(0.1442,0.0547,0.0855,0.0924,0.0176,0.0094) \cdot (18,19,22.2,48.6,64.35,38) \cdot 10^3\,\gamma\,\mathrm{m^4/h}}{\gamma\,[1120\,(\mathrm{m^3/h}) \cdot 210\,\mathrm{m} - ((1,1,1.2,2.7,3.3,2)10^2\,(\mathrm{m^3/h})) \cdot (18,19,18.5,18,19.5,19)10\,\mathrm{m}]} \\
&= \; \frac{11\,513.4}{25\,050} \; \approx \; 0.45962.
\end{aligned}
$$

The index I_R coincides with I_r as well: see (4), (6), and the last equation in (14).

---

[2] Our value for I_r is very close to 0.47, the one computed in [21, Tab. 3].

Example 4. A different choice of the diameters in (11) affects all the local vectors except q^s which continues to be null. If, for example, we alter just the diameter of the eighth pipe adding $2\,\mathrm{in.}$, i.e. replacing $D_8 = 10$ with $D_8 = 12$ in (11) (see [21, Table 3, Sol. D]), then, by using the software EPANET 2.0.12, we obtain

$$\hbar = (55.96\,\mathrm{m}, 40.38\,\mathrm{m}, 45.83\,\mathrm{m}, 46.63\,\mathrm{m}, 33.43\,\mathrm{m}, 34.63\,\mathrm{m}),$$
$$h = (205.96\,\mathrm{m}, 200.38\,\mathrm{m}, 200.83\,\mathrm{m}, 196.63\,\mathrm{m}, 198.43\,\mathrm{m}, 194.63\,\mathrm{m}),$$

and the following slight differences in the local surplus vectors defined in (2)

$$\hbar^s \approx (0.8653, 0.346, 0.5277, 0.5543, 0.1143, 0.1543),$$
$$h^s = p^s \approx (0.1442, 0.0546, 0.0856, 0.0924, 0.0176, 0.0244). \tag{15}$$

Using (4), (9), (10) and (15), we then compute[3]

$$I_r = \frac{h^s \cdot p^*}{\gamma(Q \cdot H - q^* \cdot h^*)} = \frac{12\,081.6}{25\,050} \approx 0.48223.$$

4   The TLN with *Non*-uniform Minimal Design Pressure

Almost all WDNs in literature have uniform design conditions, when they are present, for all the junction node. Here we have not the space to develop several or new examples, hence we consider the cases studied in the previous section but adding some alterations on the minimal design pressure request, i.e. on the vector $\hbar^*$ in such a way as to make its components no longer constant.

Example 5. Referring to the conditions of Example 3, we replace the vector $\hbar^*$ with the following

$$\hbar^* = (30\,\mathrm{m}, 31\,\mathrm{m}, 30\,\mathrm{m}, 24\,\mathrm{m}, 33\,\mathrm{m}, 32\,\mathrm{m}),$$

hence

$$h^* = (180\,\mathrm{m}, 191\,\mathrm{m}, 185\,\mathrm{m}, 174\,\mathrm{m}, 198\,\mathrm{m}, 192\,\mathrm{m}). \tag{16}$$

This change affects the local head surplus vectors (and the local power one) as follows

$$\hbar^s \approx (0.8653, 0.3029, 0.5273, 0.9429, 0.013, -0.0066),$$
$$h^s = p^s \approx (0.1442, 0.0492, 0.0856, 0.1301, 0.0022, -0.0011), \tag{17}$$

and using (4), (16) and (17), we compute the resilience indices obtaining

$$I_R = I_r = \frac{h^s \cdot p^*}{\gamma(Q \cdot H - q^* \cdot h^*)} = \frac{11\,643.4}{25\,180} \approx 0.46241. \tag{18}$$

Comparing Example 3 and 5 is very interesting and it is enlightening about the limits of global indices and the need for a local-global analysis of WDNs.

For brevity, let $\mathcal{N}_3$ be the network in Example 3 and $\mathcal{N}_5$ the one in Example 5 ($\mathcal{N}_3$ and $\mathcal{N}_5$ differ just for the design pressure minimal condition in the junction

---

[3] The result, this time, agrees perfectly with 0.48 calculated in [21, Tab. 3].

nodes 2,4,5,6). Note, first of all, that $\mathcal{N}_5$ is slightly more resilient than $\mathcal{N}_3$; more precisely

$$I_r(\mathcal{N}_5) \approx 1.006\,\% \text{ more than } I_r(\mathcal{N}_3).$$

But $\mathcal{N}_5$ certainly has many more criticalities[4] than the other: we just highlight the two major ones. First of all, it has a pressure deficit in node 6 (note the minus sign in the last component of $\hbar^s$ in (17)). Then, as it can be shown using EPANET 2.0.12, any failure, even very small at any point of the network, causes a pressure deficit at node 5 as well, long before a pressure deficit occurs in the network $\mathcal{N}_3$. But it is also very important to note that the greater fragility of $\mathcal{N}_5$ is not due to a higher request for minimum pressure in total, in fact it is identical for both the networks:

$$\|\hbar^*(\mathcal{N}_3)\|_1 \;=\; \|\hbar^*(\mathcal{N}_5)\|_1 \;=\; 180\,\mathrm{m} \tag{19}$$

(see (8) for the definition of the 1-norm $\|\cdot\|_1$). We therefore conclude that the greater vulnerability of $\mathcal{N}_5$, despite its resilience index, is mainly due to the worst distribution between the nodes of the same total pressure design request of $180\,m$ as shown by (19). Such a worst distribution could be immediately noted by comparing the local pressure surplus vectors $\hbar^s(\mathcal{N}_3)$ and $\hbar^s(\mathcal{N}_5)$: for instance, just note that in the node 4 of $\mathcal{N}_5$ we have a pressure surplus of 94,29 % (i.e. about twice the design request) *vs* the 55,43 % for the same node of $\mathcal{N}_3$.

Aknowledgments. This work is partially supported by the research projects *"origAMI, Original Advanced Metering Infrastructure"*, CUP J48C17000170006, POR Calabria FESR-FSE 2014–2020, and *"I-BEST"*, CUP B28I17000290008, PON "Innovazione e competitività" 2014/2020 MiSE Horizon 2020.

References

1. Alperovits, E., Shamir, U.: Design of optimal water distribution systems. Water Resour. Res. **13**, 885–900 (1977)
2. Bonora, M., Caldarola, F., Muranho, J., Sousa, J., Maiolo, M.: Numerical experimentations for a new set of local indices of a water network. In: Sergeyev, Y. D. and Kvasov, D. E. (eds.) Proceedings of the 3rd International Conference "Numerical Computations: Theory and Algorithms". Lecture Notes in Computer Science. vol. 11973. Springer, New York (2020). https://doi.org/10.1007/978-3-030-39081-5_42
3. Caldarola, F., Maiolo, M.: Local indices within a mathematical framework for urban water distribution systems. Cogent Eng. **6**, 1643057 (2019). https://doi.org/10.1080/23311916.2019.1643057
4. Carini, M., Maiolo, M., Pantusa, D., Chiaravalloti, F., Capano, G.: Modelling and optimization of least-cost water distribution networks with multiple supply sources and users. Ricerche di Matematica **67**(2), 465–479 (2018)

---

[4] Being the two WDNs so similar and being dimension and complexity so small, a "criticality" of one WDN with respect to the other must obviously be understood in an appropriate sense...

5. Di Nardo, A., Di Natale, M.: A heuristic design support methodology based on graph theory for district metering of water supply networks. Eng. Optim. **12**, 193–211 (2011)
6. Di Nardo, A., Di Natale, M.: A design support metodology metodology for district metering of water supply networks. Water Distrib. Syst. Anal. **2010**, 870–887 (2012)
7. Di Nardo, A., et al.: Redundancy features of water distribution systems. Proc. Eng. **186**, 412–419 (2017)
8. Di Nardo, A., Di Natale, M., Santonastaso, G.: A comparison between different techniques for water network sectorization. Water Sci. Technol.: Water Supply **14**, 961–970 (2014)
9. Di Nardo, A., Di Natale, M., Santonastaso, G., Tzatchkov, V., Alcocer-Yamanaka, V.: Water network sectorization based on graph theory and energy performance indices. J. Water Resour. Plann. Manage. **140**, 620–629 (2014)
10. Di Nardo, A., Di Natale, M., Santonastaso, G., Tzatchkov, V., Alcocer-Yamanaka, V.: Performance indices for water network partitioning and sectorization. Water Sci. Technol.: Water Supply **15**, 499–509 (2015)
11. Horn, R., Johnson, C.: Matrix Analysis. Cambridge University Press, Cambridge (2012)
12. Maiolo, M., Carini, M., Capano, G., Piro, P.: Synthetic sustainability index (SSI) based on life cycle assessment approach of low impact development in the Mediterranean area. Cogent Eng. **4**, 1410272 (2017)
13. Maiolo, M., Martirano, G., Morrone, P., Pantusa, D.: Assessment criteria for a sustainable management of water resources. Water Pract. Technol. 1(1): wpt2006012 (2006). https://doi.org/10.2166/wpt.2006.012
14. Maiolo, M., Mendicino, G., Pantusa, D., Senatore, A.: Optimization of drinking water distribution systems in relation to the effects of climate change. Water **9**, 803 (2017). https://doi.org/10.3390/w9100803
15. Maiolo, M., Pantusa, D.: A methodological proposal for the evaluation of potable water use risk. Water Pract. Technol. **10**, 152–163 (2015)
16. Maiolo, M., Pantusa, D.: An optimization procedure for the sustainable management of water resources. Water Sci. Technol.: Water Supply **16**, 61–69 (2016). https://doi.org/10.2166/ws.2015.114
17. Maiolo, M., Pantusa, D.: Combined reuse of wastewater and desalination for the management of water systems in conditions of scarcity. Water Ecol. **72**, 116–126 (2017)
18. Maiolo, M., Pantusa, D.: Infrastructure vulnerability index of drinking water supply systems to possible terrorist attacks. Cogent Eng. **5**, 1456710 (2018)
19. Maiolo, M., Pantusa, D., Chiaravalloti, F., Carini, M., Capano, G., Procopio, A.: A new vulnerability measure for water distribution network. Water **10**, 1005 (2019)
20. Shin, S., et al.: A systematic review of quantitative resilience measures for water infrastructure systems. Water **10**, 164 (2018)
21. Todini, E.: Looped water distribution networks design using a resilience index based on heuristic approach. Urban Water **12**, 115–122 (2000)

Identification of Contamination Potential Source (ICPS): A Topological Approach for the Optimal Recognition of Sensitive Nodes in a Water Distribution Network

Gilda Capano, Marco Amos Bonora, Manuela Carini⬤, and Mario Maiolo⁽⊠⁾⬤

Department of Environmental and Chemical Engineering,
University of Calabria, Rende, Italy
{gilda.capano,marcoamos.bonora,manuela.carini,mario.maiolo}@unical.it

Abstract. The correct management of urban water networks have to be supported by monitoring and estimating water quality. The infrastructure maintenance status and the possibility of a prevention plan availability influence the potential risk of contamination. In this context, the Contamination Source Identification (CSI) models aim to identify the contamination source starting from the concentration values referring to the nodes. This paper proposes a methodology based on Dynamics of Network Pollution (DNP). The DNP approach, linked to the pollution matrix and the incidence matrix, allows a topological analysis on the network structure in order to identify the nodes and paths most sensitive to contamination, namely those that favor a more critical diffusion of the introduced contaminant. The procedure is proposed with the aim of optimally identifying the potential contamination points. By simulating the contamination of a synthetic network, using a bottom-up approach, an optimized procedure is defined to trace back to the chosen node as the most probable contamination source.

Keywords: Water quality · Contamination sources · Graph theory

1 Introduction

The sustainable management of water resources requires an efficient control of the distribution systems performance to guarantee an adequate supply to the users. A measure of non-achievement of qualitative and quantitative standards is associated with the assessment of the potable water risk, which is particularly useful for the careful planning of infrastructural and management interventions. The potable water risk evaluation is complex because it depends on many factors that are sometimes difficult to estimate, for example, the source pollution, obstructions or dysfunctions, the water quality alterations and water losses [18]. The evaluation of the water quality alterations due to accidental or intentional events (natural or artificial) is of equal complexity. In fact, especially in recent

© Springer Nature Switzerland AG 2020
Y. D. Sergeyev and D. E. Kvasov (Eds.): NUMTA 2019, LNCS 11973, pp. 525–536, 2020.
https://doi.org/10.1007/978-3-030-39081-5_45

years, the water systems vulnerability is considered a priority, for the importance dedicated to the critical infrastructures [2,21]. The water quality in the drinking water networks, their distribution, utilization, discharge and purification, aimed at reuse in the agricultural and industrial field, conditions typical processes of the circular economy [19,20]. For these aspects, it is necessary to equip the water systems with the monitoring of qualitative and quantitative parameters using increasingly reliable instruments, based on real-time control, in order to facilitate forecasting and risk prevention operations [17]. Being able to continuously acquire reliable data on electro-filters placed in the network is a very ambitious result also depending on the correct location of the water quality sensors in the strategic network points. In this regard, in the literature, there are different modeling approaches based on the correct positioning of the sensors in the network [15,25]. This observation confirms the attention to this topic. A correct survey of the location of the measuring instruments has to take into account the objective (easy identification of the contamination source) allowing the contaminant tracing. The contamination source identifying is a priority problem. In scientific literature this problem has been widely discussed and interpreted using various methodologies, which are generally called Contamination Source Identification (CSI) [2]. These models provide an adequate calculation of three parameters: the location of the contamination source, the pollute concentration and the intrusion time, through different modeling approaches. [9] propose a simulation-optimization method for complexes water distribution system, which does not focus on a topological view. This method is based on an optimal predictor-corrector algorithm to locate the sources and their release histories. The optimization approach is used to define the similarity between the simulated and measured output response data at monitoring points. [24] characterize of the contaminant source with an optimization approach using a genetic algorithm linked to the EPANET simulation software. The sources characterization is based on the three different sensors types, assuming that contamination intrusions are associated with a single location. [5] propose a methodology for identifying the contamination source through an optimization problem using the water fraction matrix concept. [4] propose a Bayesian belief network (BBN) method which comparison the sensors data with other simulation of contamination scenarios. The approach presented clarified how the uncertainties on the mass and the position of the source influence the probability of sensors detecting. [28] use a methodological approach based on the probability of interaction between pollutant concentration data and injection duration with the feedback provided by consumers. [16] propose an integrated simulation-optimization procedure with a logistic regression and a local improvement method to accelerate convergence. This method is based on the pre and post screening technique with the aim of accelerating convergence by reducing the investigation field. [13], using Artificial Neural Networks (ANN), have developed a methodology to identify the position of release of contaminants in the network. The water systems sensitivity, in relation to the risk of the resource alteration, can detect erroneous dosages of the reagents which are spilled into drinking water: among these, chlorine has a role of particular scientific interest and in the management practice. Chlorine, which is often used as a water disinfectant, can be used as an indicator of the gradual deterioration of the water quality. In fact, the chlorine concentration released into the network by the treatment plant is progressively reduced due to the reactions with the bacterial component. For this reason, it is important to estimate the correct content of residual chlorine in the water network nodes. The analysis models of chlorine decay in the network are many and focus their attention on the characteristics of the decay reaction kinetics. In principle, most of these models refer to [26], on which based the criteria set for quality simulations in Epanet software. For these models, the chlorine decay occurs due to the reactions that are generated within the mass flow and to the reactions that are activated along the pipe wall. For this reason, the scientific interest focuses on the estimation of the Bulk decay coefficient (kb) and the Wall decay coefficient (kw). The methods of calculating these parameters can be classification in direct methods [1,22,27] which are based on the application of specific formulations, and indirect methods [6,14,23] which mainly use calibration techniques. In this work, an expeditious methodology is proposed for an easy identification of sensitive nodes within an urban distribution network based on the Dynamics of Network Pollution (DNP). The DNP summarizes a series of essential aspects in such assessments as the concentration of the pollutants in the single node, topology and the number of times a node appears in the potential pollutant paths in the network (node occurrence). This last aspect represents an essential element for the evaluation, because the most sensitive node (the node to which, in the event of contamination, a worse pollution dynamic is associated) will be identified in the list of nodes with the greatest number of occurrences. One of the salient aspects of the proposed methodology is based on the predictive capacity of the contamination source. The DNP represents a tool to obtain, information on the potential location of meters (for this reason it can influence the definition of monitoring costs). For this reason, the proposed methodological scheme is not inventible but is bound to the logic of the detailed setting in the following sections. The methodology is applied in the Kang and Lansey network (KL) [11] which, according to a black box logic is contaminated providing a chlorine spill outside the node to which, in the event of contamination, a worse pollution dynamic is associated limits (0.2 mg/l).

2   Materials and Methods

The DNP is a useful tool to carry out a preliminary screening on the contaminant diffusion in the network, contributing to the identification of sensitive paths and nodes, starting from topological information. On an analytical level the DNP is defined by the following matrix product:

$$DNP = IM \ x \ PM \tag{1}$$

where: IM indicates the incidence matrix, with dimension (n, t), with $n = $ nodes number and $t = $ pipelines number connecting nodes (arcs), and PM indicates the pollution matrix, with size $(t, 1)$. DNP, having dimensions $(n, 1)$, is a vector that, starting from the concentration values in the pipelines, provides an estimate of the pollutants concentration in the nodes. The DNP vector refers to a single time instant and for this reason, for the time discretization k, it is necessary to define a DNP for each k-th time sampling chosen. Regarding the relation (1), PM is a vector that contains the average concentration values in the pipelines, in a generic instant To + dt after the contamination occurred at in To time. This quantity is weighed with respect to the water volume passed through the generic pipeline and for this reason, it can be interpreted as a mass flow rate. This vector is descriptive of a brief contamination scenario, defined by a criterion based on the node sensitivity. The contamination scenario is defined in the hypothesis of a chlorine overdose and is determined by the setting of the parameters of the chlorine reaction kinetics. In the specific case, using the [26] approach, a first order reaction is identified for the bulk coefficient, using a kb value of $-0.55d^{-1}$ and neglecting, instead, the mass transport linked to the interaction with the pipe walls. PM, therefore, contains values of the pollulant concentration in terms of mg/l. The matrix differs from similar literature cases [5, 12] because it does not provide binary information, but identifies the pollutant quantity in the node, defining the contribution of each node to the dynamic of contaminant diffusion.

2.1   Identification of Contamination Potential Sources

The definition of contamination potential sources in the KL network is aimed at studying the sensitivity of the nodes involved in the pollution dynamics. In order to carry out this investigation, it is necessary that the network be considered as a weighted oriented graph (see [8])

$$N = (J, P)$$

where

J: the set of n vertices;
P: the set of m edges or arcs;
ρ_{ij}: the non-negative weight associated to the arc ij if $ij \in P$
and $\rho_{ij} = 0$ if $ij \notin P$ (i.e., if i, j are disconnected vertices in N).

More in detail, the graph of the water network association considers the demand nodes and the reservoirs/tanks as Junction and the pipeline, together with the longitudinal elements (Valves and Pumps), as arches. The arcs direction in the directed graph is defined by orientation of the pipes through the flow in conditions of average flow (HYP). The arches weight is defined as the inverse of the pipeline volume. This choice is linked to an application requirement, because it is wanted to select the path with the highest pollutant volume by an algorithm that identifies the paths with minimum weight:

$$\rho_{ij} = \frac{1}{W_{ij}} \tag{2}$$

$$W_{ij} = \frac{\pi (D_{ij})^2}{4} L_{ij} \tag{3}$$

Where:
ρ_{ij} weight of the ij pipe $[m^3]$
W_{ij} volume of the ij pipe $[m^3]$
L_{ij} length of the ij pipe $[m^3]$
D_{ij} diameter of the ij pipe $[m^3]$

However, as a preface to the definition of the procedure the following definitions are useful. Given an oriented graph:

$$G = (V, E) \tag{4}$$

with $v \in V$ is a G vertex, the In-Degree of v in G is equal to the number of incident arcs in v, while the Out-Degree of v in G is equal to the number of incident arcs from v. Notations:

$$indeg(v) \; o \; indeg_G \; (v) \qquad outdeg(v) \; o \; outdeg_G \; (v) \tag{5}$$

From a purely computational point of view, the implementation of the DNP estimation algorithm is based on the identification of the Source and Sink nodes [3,10] in the weighted oriented graph describing the network. Specifically, in a directed graph $G = (V, E)$ a node $v \in V$, which has only arcs starting from it, is called Source:

$$indeg_G \; (v) = 0 \qquad outdeg_G \; (v) > 0 \tag{6}$$

While, a node $v \in V$ that has only incident arcs in it, is called Sink:

$$indeg_G \; (v) > 0 \qquad outdeg_G \; (v) = 0 \tag{7}$$

This definition highlights a hierarchical criterion between nodes, based on the definition of dominator. In a oriented graph $G = (V, E)$ taken the nodes $d, \in V$, it can say that d dominates n if every path that starts from the entry node to reach n node, has to pass through d. Notation:

$$d \; dom \; n \tag{8}$$

$$d \; dom \; n \; con \; d \neq n \tag{9}$$

To identify the most sensitive nodes within the network an automatic procedure has been implemented that develops in two phases. In the first phase the nodes to be discarded are identified, according to the following criterion: all the nodes that are "directly dependent" from the reservoir will be discarded. A node is said to be directly dependent on the reservoir nodes if it is Strictly Dominated by one of them or if it is Strictly Dominated by a node dominated by the reservoir. More in detail it can hypothesize the following synthetic schematization

– source nodes are placed in a list;
– source nodes will be classified: (i) strictly Dominated by the nodes in the list; (ii) with *indeg* = 1; (iii) that are not terminal nodes.

These nodes are defined as "directly dependent" on the reservoir and are placed in the list of excluded nodes. The "directly dependent" nodes identified in the previous phase are added to the list. The procedure is repeated identifying the nodes Strictly Dependent from other nodes which are Strictly Dependent on the reservoir. In the second phase, a minimum path search algorithm is applied, using the inverse of the volume to identify the paths with maximum water volume in the pipes. Considering that the weights are always positive, the Dijkstra algorithm is used [7]. This is applied to the chosen network, interpreted as an oriented graph, and the minimum weight path that connects, if possible, each Source node to each Sink node is identified. From the list of paths with maximum volume the number of occurrences of each node is obtained. The occurrences number is the number of times a node appears in the paths found. The importance linked to the quantification of node occurrences depends on the objective of the work, which, in a network composed of n nodes with different topological hierarchical levels of participation in the pollution dynamics, wants to identify the single most significant nodes. These nodes are the nodes that appear many times, that is those present in the more sensitive paths with maximum volume. In fact, the most sensitive node or path will be identified in the nodes list with the greatest occurrences number. The procedure is applied to KL network (Fig. 1), which has 935 nodes and 1274 pipes. The average total demand is 177 l/s. The network is simulated under peak conditions with the total consumption is 336 l/s.

3   Numerical Results

In order to identify the most sensitive node in the KL network, classified as a potential source of contamination, it is necessary to discuss the results obtained for individual application phases. The first part of the procedure, aimed at carrying out a preliminary screening of the nodes hierarchy, allows to eliminate the obligatory passage nodes (the source nodes and the nodes Strictly Dependent on the reservoir) and the terminal nodes (Fig. 2a). Then proceed with the calculation of the occurrences from the paths list with maximum volume (Fig. 2b). The Fig. 2b has an important role, because it makes clear the criterion for identifying the interesting node for contamination. This node will be chosen among those that have the highest number of filtered occurrences. The total occurrences (or simply occurrences) are obtained by summing the number of times that the node appears in all paths. The filtered occurrences instead discriminate the reservoir node and do not count the occurrences of the "obligatory passage nodes" in the single Dijkstra application, that is those excluded in the first selection phase.

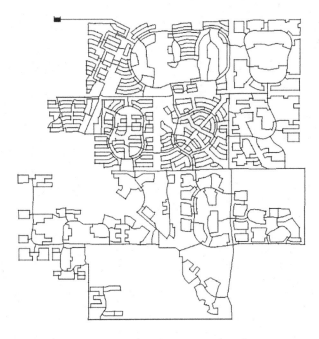

Fig. 1. Kang and Lansey network.

The histogram shows that the most critical situation is associated with nodes 249 and 250, which have a 24 value for a filtered occurrence. According to the criterion identified, the choice between the two might seem equivalent, however, in the network topology it is clear that 249 is upstream of 250, so it will be chosen as a "node to be polluted". The hydraulic simulation is carried out on a permanent motion setting, forecasting an impulsive input of chlorine in the chosen node, lasting 30 min. The pollutant diffusion is shown in Fig. 3, where it has been chosen to show the trend of the chlorine concentration.

Now it is possible to proceed with the DNP calculation. The DNP values are calculated for each node and for each time interval. For reasons of synthesis and clarity, only the DNP values associated with the nodes identified in Fig. 2a will be shown for a period of 8 hours. It is important to specify that the DNP vector takes positive and negative values. The DNP values are indicative of the pollutant balance in the single node. Negative values indicate the pollutant mass leaving the node and vice versa the positive values. In the case of the path identified in Fig. 2a, the trend of the DNP can be summarized by the following graph. The trends of the curves oscillate according to the pipes number and their characteristics, which converge in the single node, determining the contaminant dilution more or less fastly. The information that can be obtained from the resulting DNP values are of two types: the peaks values, both positive and negative, provide quantitative information that allows an indication of the decay degree and mixing of the pollutant. Their position and shape, on the

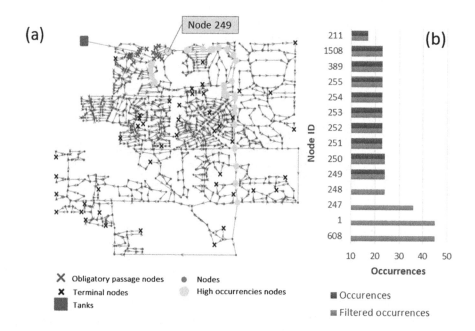

Fig. 2. (a) The occurrences location in the KL network. The red crosses indicate the obligatory passage nodes, the black crosses the terminal nodes, the yellow circles indicate the nodes with the maximum filtered occurrences and the green circle indicates the chosen node (ID 249). (b) Histogram of total and filtered occurrences of the nodes in the most sensitive path. (Color figure online)

other hand, provides an indication of the network pollution dynamics. As an example, Fig. 5 shows the DNP graph for some particularly significant nodes. As mentioned above, with the DNP there is information on the pollution dynamics in the network, which for some nodes is slow and lasts over time, while for others it is impulsive. Through the graphic representation of the DNP it can therefore have immediate feedback on the behavior of some nodes. Some nodes, among those closest to the immission area, show an impulsive trend, indicative of a temporally rapid and highly concentrated pollution dynamics, while, other downstream nodes are distinguished by a trend in which the peak less high, lasts over time. The first situation indicates the pollution periods with high concentrations with minor duration, while, in the second case, the "polluted" period continues for longer times. From the evaluation of the results it is clear that the interesting node cannot be one, since two different pollution dynamics have been identified. These nodes are the 253, with the maximum DNP value (impulsive behavior) and the node 256, which corresponds to a contamination dynamic more durable in overtime (Fig. 4).

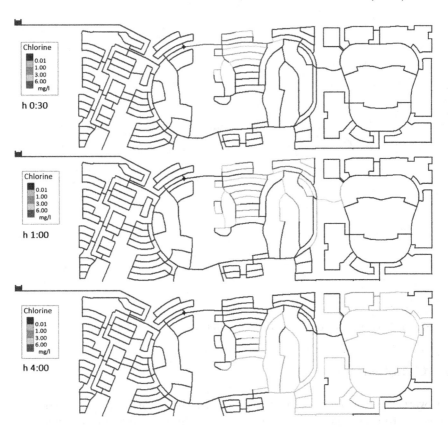

Fig. 3. Network contamination status. (a) situation after 30 min, (b) situation after 1 h and (c) situation after 4 h.

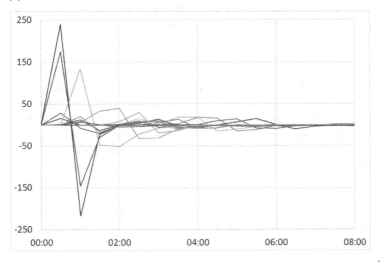

Fig. 4. DNP values for the nodes along the path shown (highlighted in yellow) in Fig. 2. (Color figure online)

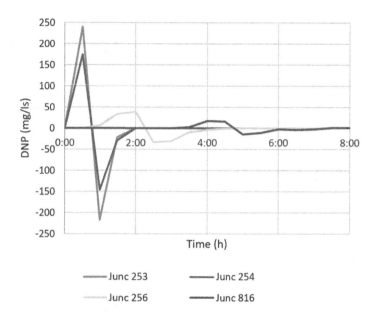

Fig. 5. DNP trends for the most significant nodes, useful for peak identifying.

4   Conclusions

The water quality of a drinking water network can be easily compromised during transport to users due to the intrusion of a pollutant through tanks, nodes and broken pipes. In this work, an expeditious procedure is proposed for the identification of one or more sensitive nodes in a network, which are the nodes that can determine a worse contaminant distribution if contaminated. The method is based on the DNP calculation, a vector that for each node and for each chosen time sampling, provides an indicator of the pollutant quantity that switches. DNP is based on a quick procedure that summarizes hydraulic and topological evaluations of the network. The importance linked to the topology determines the application conditions of the method, in fact, it is not possible to define a priori and absolutely the necessary measurement points because this information is linked to the topological characteristics of the network chosen as a case study. The importance linked to hydraulic evaluations, on the other hand, does not constrain the study to the analysis of chlorine concentration alone, but the interest can be directed towards other characteristic parameters (PH, temperature), considering that their variability defines the conditions of decay of the chlorine. The case study is well suited to experiment with the validity of the proposed methodology, as it is characterized by a good number of loops and therefore by a high degree of contaminant dilution. The particularity and the interest of the procedure depend on the ability to manage a problem with unknown input parameters and evaluations on partially deductible outputs.

Aknowledgments. This work is partially supported by the Italian Regional Project (POR CALABRIA FESR 2014–2020): *"origAMI, Original Advanced Metering Infrastructure"* CUP J48C17000170006.

References

1. Al-Jasser, A.O.: Chlorine decay in drinking-water transmission and distribution systems: pipe service age effect. Water Res. **41**(2), 387–396 (2007). https://doi.org/10.1016/j.watres.2006.08.032
2. Adedoja, O.S., Hamam, Y., Khalaf, B., Sadiku, R.: Towards development of an optimization model to identify contamination source in a water distribution network. Water **10**(5), 579–606 (2018). https://doi.org/10.3390/w10050579
3. Borowski, E.J., Borwein, J.M.: The HarperCollins Dictionary of Mathematics. HarperCollins, New York (USA) (1991)
4. Dawsey, W.J., Minsker, B.S., VanBlaricum, V.L.: Bayesian belief networks to integrate monitoring evidence of water distribution system contamination. J. Water Resour. Plann. Manage. **132**(4), 234–241 (2006). https://doi.org/10.1061/(ASCE)0733-9496(2006)132:4(234)
5. Di Cristo, C.D., Leopardi, A.: Pollution source identification of accidental contamination in water distribution networks. J. Water Resour. Plann. Manage. **134**(2), 197–202 (2008). https://doi.org/10.1061/(ASCE)0733-9496(2008)134:2(197)
6. Digiano, F.A., Zhang, W.: Pipe section reactor to evaluate chlorine-wall reaction. J.-Am. Water Works Assoc. **7**(1), 74–85 (2005)
7. Dijkstra, E.W.: A note on two problems in connexion with graphs. Numerische Mathematik **1**(1), 269–271 (1959)
8. Di Nardo, A., Di Natale, M.: A heuristic design support methodology based on graph theory for district metering of water supply networks. Eng. Optim. **43**(2), 193–211 (2011). https://doi.org/10.1080/03052151003789858
9. Guan, J., Aral, M.M., Maslia, M.L., Grayman, W.M.: Identification of contaminant sources in water distribution systems using simulation-optimization method: case study. J. Water Resour. Plann. Manage. **132**(4), 252–262 (2006). https://doi.org/10.1061/(ASCE)0733-9496(2006)132:4(252)
10. Harary, F.: Graph Theory. Reading. Addison-Wesley, Boston (USA) (1994)
11. Kang, D., Lansey, K.: Revisiting optimal water-distribution system design: issues and a heuristic hierarchical approach. J. Water Resour. Plann. Manage. **138**(3), 208–217 (2012). https://doi.org/10.1061/(ASCE)WR.1943-5452.0000165
12. Kessler, A., Ostfeld, A., Sinai, G.: Detecting accidental contaminations in municipal water networks. J. Water Resour. Plann. Manage. **124**(4), 192–198 (1998). https://doi.org/10.1061/(ASCE)0733-9496(1998)124:4(192)
13. Kim, M., Choi, C.Y., Gerba, C.P.: Source tracking of microbial intrusion in water systems using artificial neural networks. Water Res. **42**(4–5), 1308–1314 (2008). https://doi.org/10.1016/j.watres.2007.09.032
14. Kim, H., Kim, S., Koo, J.: Modelling chlorine decay in a pilot scale water distribution system subjected to transient. In: Civil and Environmental Engineering, Pusan National University (Korea) and Environmental Engineering, University of Seoul (2015) https://doi.org/10.1016/j.proeng.2015.08.89
15. Kim, S.H., Aral, M.M., Eun, Y., Park, J.J., Park, C.: Impact of sensor measurement error on sensor positioning in water quality monitoring networks. Stoch. Environ. Res. Risk Assess. **31**(3), 743–756 (2017). https://doi.org/10.1007/s00477-016-1210-1

16. Liu, L., Zechman, E.M., Mahinthakumar, G., Ranji Ranjithan, S.: Identifying contaminant sources for water distribution systems using a hybrid method. Civil Eng. Environ. Syst. **29**(2), 123–136 (2012). https://doi.org/10.1080/10286608.2012.663360
17. Maiolo, M., Carini, M., Capano, G., Pantusa, D., Iusi, M.: Trends in metering potable water. Water Pract. Technol. **14**(1.1), 1–9 (2019). https://doi.org/10.2166/wpt.2018.120
18. Maiolo, M., Pantusa, D.: A methodological proposal for the evaluation of potable water use risk. Water Pract. Technol. **10**(1), 152–163 (2015). https://doi.org/10.2166/wpt.2015.017
19. Maiolo, M., Pantusa, D.: Combined reuse of wastewater and desalination for the management of water systems in conditions of scarcity. Water Ecol. **4**(72), 116–126 (2017). https://doi.org/10.23968/2305-3488.2017.22.4.116-126
20. Maiolo, M., Pantusa, D.: A proposal for multiple reuse of urban wastewater. J. Water Reuse Desalin. **8**(4), 468–478 (2018a). https://doi.org/10.2166/wrd.2017.144
21. Maiolo, M., Pantusa, D.: Infrastructure Vulnerability Index of drinking water systems to terrorist attacks. Cogent Eng. **5**(1), 1456710 (2018b). https://doi.org/10.1080/23311916.2018.1456710
22. Mostafa, N.G., Minerva, E., Halim, H.A.: Simulation of Chlorine Decay in Water Distribution Networks. Public Works Department, Faculty of Engineering, Cairo (Egypt), Using EPANET. Case Study - Sanitary and Environmental Engineering Division (2013)
23. Nagatani, T., et al.: Residual chlorine decay simulation in water distribution system. In: The International Symposium on Water Supply Technology, Yokohama (Japan) (2008)
24. Preis, A., Ostfeld, A.: Genetic algorithm for contaminant source characterization using imperfect sensors. Civil Eng. Environ. Syst. **25**(1), 29–39 (2008). https://doi.org/10.1080/10286600701695471
25. Rashid, B., Rehmani, M.H.: Applications of wireless sensor networks for urban areas: a survey. J. Netw. Comput. Appl. **60**, 192–219 (2016)
26. Rossman, L.A., Clark, R.M., Grayman, W.M.: Modeling chlorine residuals in drinking water distribution systems. J. Environ. Eng. **120**(4), 803–820 (1994)
27. Rossman, L.A.: EPANET 2: Users Manual (2000)
28. Tao, T., Huang, H.D., Xin, K.L., Liu, S.M.: Identification of contamination source in water distribution network based on consumer complaints. J. Central S. Univ. Technol. **19**, 1600–1609 (2012)

Seeking for a Trade-Off Between Accuracy and Timeliness in Meteo-Hydrological Modeling Chains

Luca Furnari[(✉)] [iD], Alfonso Senatore [iD], and Giuseppe Mendicino [iD]

DIATIC, University of Calabria, Rende, CS, Italy
{luca.furnari,alfonso.senatore,giuseppe.mendicino}@unical.it

Abstract. The level of detail achieved by operational General Circulation Models (e.g., the HRES 9 km resolution forecast recently launched by the ECMWF) raises questions about the most appropriate use of Limited Area Models, which provide for further dynamical downscaling of the weather variables. The two main objectives targeted in hydrometeorological forecasts, i.e. accuracy and timeliness, are to some extent conflicting. Accuracy and precision of a forecast can be evaluated by proper statistical indices based on observations, while timeliness mainly depends on the spatial resolution of the grid and the computational resources used. In this research, several experiments are set up applying the Advanced Research Weather Research and Forecasting (WRF-ARW) Model to a weather event occurred in Southern Italy in 2018. Forecast accuracy is evaluated both for the HRES ECMWF output and that provided by WRF dynamical downscaling at different resolutions. Furthermore, timeliness of the forecast is assessed adding to the time needed for GCM output availability the time needed for Limited Area simulations at different resolutions and using varying core numbers. The research provides useful insights for the operational forecast in the study area, highlighting the level of detail required and the current weaknesses hindering correct forecast of the hydrological impact of extreme weather events.

Keywords: Hydro-meteorological modeling · Warning lead time · Weather model resolution

1 Introduction

Numerical weather prediction (NWP) is usually based on modeling chains where forecasts are dynamically downscaled from a General Circulation Model (GCM) coarse grid (of the order of $\sim 10^1$ km) to the desired resolution ($\sim 10^0$ km). GCM forecasts provide the initial and boundary conditions to the Limited Area Models (LAMs), which adopt finer grids to improve the accuracy. Downscaled fields of meteorological variables can be used then for many purposes, e.g. simply like triggers for activating warning procedures or as input data for hydrological

© Springer Nature Switzerland AG 2020
Y. D. Sergeyev and D. E. Kvasov (Eds.): NUMTA 2019, LNCS 11973, pp. 537–544, 2020.
https://doi.org/10.1007/978-3-030-39081-5_46

models, in order to predict the ground impact of the meteorological forcings (e.g., [4,10]).

Though very widely used, the dynamical downscaling approach is continuously subject to very detailed analysis in order to understand its real benefits in terms of forecasting accuracy (e.g., [5,7,9]), despite the greater computational burden required and the longer calculation times, which can be an important issue in operational contexts. This research question is particularly timely if the recent amazing improvements of the operational GCMs are considered, whose current spatial resolutions are comparable to that of LAMs run operationally only a few years ago. Discussion within the scientific community about the real improvement given by GCMs' increased resolution [13] is providing positive feedback [8]. A very important novelty from an operational point of view is that on March 2016 the European Centre for Medium-Range Weather Forecasts (ECMWF) released a new version of its operative GCM forecast model, the Integrated Forecasting System - High Resolution (IFS-HRES), where the horizontal resolution of the deterministic forecasts has been increased from 16 km to 9 km.

This work presents a preliminary analysis over a single case study about which output resolution of an NWP modeling chain could be the optimal trade-off between the accuracy of the forecasting and the computational cost. The selected event affected the northeastern side of the Calabrian peninsula (southern Italy) during summer of 2018. From a forecasting point of view, it is very challenging, due to the deep convective activity leading to intense and highly localized rainfall. Nevertheless Calabria, due to its complex orography and its position in the middle of the Mediterranean sea, is often interested by this type of high energy events, which are being subjected to several studies (e.g., [2,11]).

The paper is structured as follows: next Sect. 2 briefly describes the event analyzed and the modeling strategy adopted, Sect. 3 analyzes the results of simulations both in terms of accuracy and computational performances and finally in the conclusions (Sect. 4) the main findings of the research are summarized and briefly discussed.

2   Data and Methods

2.1   Description of the Event

On the early afternoon of 20 August 2018 a highly localized and intense event hit the northeastern part of the Calabria region (southern Italy). Specifically, the event affected the Raganello Creek catchment, a small Mediterranean catchment characterized by a complex and irregular orography in its upper/mountain area, which amplified the convective characteristics of the event. Unfortunately, 10 casualties occurred nearby the village of Civita (downstream a catchment area of about $100 \, km^2$), despite the local flood warning agency (the Centro Funzionale Multirischi) emitted a 2 out of 4 level warning message.

One important feature of this event is that the regional rain gauge network was not able to detect significant rainfall measures, as Fig. 1b shows. The main source publicly released about the real extent and intensity of the event is the technical report of the event by the Centro Funzionale Multirischi [3], where a static image of the Surface Rainfall Total (SRT) derived from the national radar network is shown Fig. 2a. This SRT image, showing the accumulated three-hour rainfall from 10 to 13 UTC of 20 August 2018, highlights that only a small area of the catchment, not covered by rain gauges, was interested by rain. This static image will be also used as comparison with the output of the meteorological model.

2.2 Modeling Strategy

The Limited Area Model (LAM) adopted in this study is the Weather Research and Forecasting model, in the Advanced Research Weather core (WRF-ARW, [12]), version 4.1. WRF and its libraries, such as NetCDF, HDF5, zlib and others, were compiled using Intel Compiler version 12.1.4.

Two different domain configurations were used in this work. The first (called C1 hereafter) is the same used in [1], with two nested (one-way) grids, the outermost (D01) covering the central Mediterranean basin at 6 km horizontal resolution with 312 × 342 cells and the innermost (D02) focusing on the Calabrian Peninsula with 2 km resolution (200 × 200 cells) (Fig. 1a). Both computational domains extend over 44 vertical layers. In order to guarantee numerical stability of the model, the time step size was set equal to 36 s and to 12 s, respectively for D01 and D02.

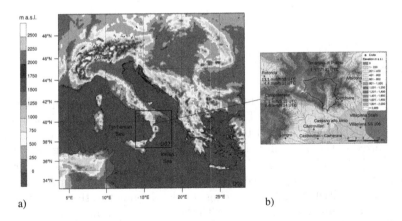

a)   b)

Fig. 1. (a) Overview of the two grids nested used in configuration C1, (b) zoom on the Raganello Creek catchment closed at Civita (red point), showing rain gauges measurements, orography and river network. (adapted from [1]) (Color figure online)

The second domain configuration (called C2 hereafter) adopts only one domain, whose geographic coverage is approximately the same as the D02 of C1 configuration, with 3 km grid spacing (140 × 140 grid points). Also, this domain configuration extends over 44 vertical layers. Time step size was fixed to 18 s.

All simulations forecasted 24 h, starting at 00 UTC of 20 August 2018, in order to guarantee enough spin up time before the occurrence of the event. The ECMWF Integrated Forecasting System (IFS) in its high-resolution deterministic forecast version (HRES - 9 km horizontal resolution) was used to provide initial and boundary conditions every 3 h. The ECMWF products, even if referred to 00 UTC, are available starting from 6 UTC, due to the time needed by GCM to complete 24 h of simulation. Physical schemes adopted for both configurations are summarised in Table 1. Cumulus parameterization adopted was the Tiedtke cumulus scheme, turned on only on D01 domain in the C1 configuration, whose horizontal resolution is greater than 5 km.

All simulations were performed on the 4-socket/16-core AMD Opteron 62xx class CPU 2.2 GHz provided by the ReCaS infrastructure (for further details please refer to [6]).

Table 1. WRF physical parameterization.

Component	Scheme adopted
Microphysics	New Thompson
Planet boundary layer	Mellor-Yamada-Janjic scheme
Shortware radiation	Goddard
Longwawe radiation	Rapid Radiative Transfer Model (RRTM)
Land surface model	Unified NOAH

3   Results

Simulations results are shown in Figs. 2b–d and are directly comparable with the SRT radar image (Fig. 2a). The radar image shows a large area in the northern part of Calabria interested by medium/high precipitation intensity, with a peak localized in the proximity of the northern boundary of the Raganello Creek catchment. Both radar and simulations show a kind of "C" shape for the precipitation pattern and highlight the effect of surrounding mountain ranges, acting as obstacles and contributing to enhancing the development of the convective cell. The rainfall amount clearly increases with the resolution. Lower rainfall amounts in the ECMWF forecast and in configuration C2 are connected to the coarser resolutions that both flatten rainfall values over larger cell areas and smooth the complex orographic features of the region.

In agreement with [1], WRF simulations delayed about one-hour the rainfall peak observed by the radar, with the highest simulated rainfall intensity occurring between 13 and 14 UTC (this is the main reason for the discrepancies between observations and simulations in Fig. 2). Configuration C1 forecasted about 20 mm averaged rainfall over the catchment between 10 and 14 UTC; for the same time, configuration C2 and ECMWF forecasts simulated only about

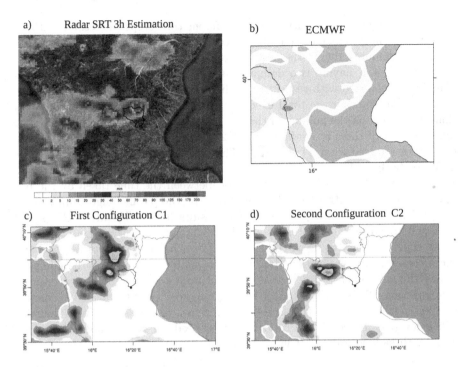

Fig. 2. (a) Radar SRT estimate between 10 and 13 UTC, the black line represents the border of the catchment and the red point the outlet at Civita; (b) ECMWF 9 km accumulated rainfall forecast for the same time; (c) WRF 3-h accumulated precipitation provided by configuration C1, the black line represents the catchment borders, the black dot Civita, the grey lines are the regional administrative borders, (d) same as c, but for configuration C2. (Color figure online)

2 mm and 3 mm average precipitation over the catchment, respectively. Specifically, the drier configuration C2 concentrated rainfall over smaller areas than C1, thus missing the event over the Raganello Creek catchment and being not eligible to perform further hydrological simulations. However, it was able to simulate medium/high rainfall amounts not far from the western catchment borders, with a peak value of about 72 mm in three hours. With configuration C1 the peak amount of 79 mm was located north of the catchment.

Both configurations C1 and C2 proved to be able to forecast enough high precipitation amounts to trigger warning procedures, even though the level of accuracy concerning the correct localization of the event was different.

The computational burden was evaluated through an analysis of the computational scalability of the system, using 1, 2, 4, 8, 16, 32 and 64 threads respectively. Execution times were calculated using the default WRF output configuration, which produces relatively small output files (WRF allows choosing several sets of variables, depending on the aims of the users, which could easily generate very large output files and not negligible writing times). Of course, the processing time is very sensitive to the different time steps (increasing with the resolution), the number of cells in the computational domain(s) and also the number of domains, due to the nesting procedure. Therefore, it is expected that configuration C1, having two domains and higher resolution in the innermost domain, is penalized with respect to configuration C2.

Figure 3a, showing the total execution times of the two different configurations depending on the number of threads used, clearly highlights that, given the same number of threads, configuration C2 is always faster than C1, with execution times reduced from 60% to 70%. The optimal number of threads resulted in 32 for both configurations, with 83 and 23 min, respectively for C1 and C2. Speed-up (Fig. 3b), calculated as the ratio between the time taken by the best available serial algorithm and the time taken by the parallel algorithm running in different threads, also provides better results for configuration C2 (speed-up equal to 16 with 32 threads) and highlights, for both configurations, the reduced performance using 64 threads. The latter result is most probably due to the increasing communication time between distinct threads, overhanging and hiding the gain given by the greater computing power. The higher speed-up of configuration C2 also means that it is more able to exploit the available computing power. This outcome is connected to the lack of nesting between domains. Efficiency (calculated as the ratio between speed up and number of threads) with 32 threads is equal to 0.36 for C1 and 0.49 for C2.

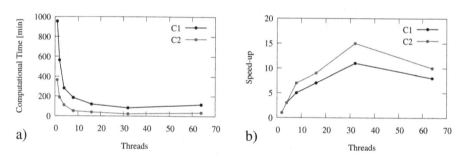

Fig. 3. (a) Total execution time achieved by WRF with configurations C1 (black line) and C2 (red line); (b) Speed-up obtained by WRF with configurations C1 (black line) and C2 (red line) (Color figure online)

4   Conclusions

The paper presented a case study investigating the optimal configuration of the NWP modeling chain, balancing accuracy and timeliness. The results showed that: (1) notwithstanding the relatively high resolution of the GCM, dynamical downscaling is required to reproduce reliably the high convective event; (2) both configurations C1 and C2 of the LAM are capable to reproduce the main features of the event, forecasting enough rainfall amount to trigger warning procedures for a larger area than the Raganello Creek catchment, but (3) lower resolution configuration C2 almost completely misses the localization of the event over the catchment, preventing any further meaningful hydrological simulation, and (4) higher resolution configuration C1 takes about 1 more hour, optimizing the available computational resources, to provide its results, therefore reducing the forecast lead time.

With the conditions analyzed in this specific case study, it would have been possible to exploit all the 64 available threads running simultaneously both configurations C1 and C2, having right away a general framework and then more detailed results. Of course, it is a specific solution for this case, envisaging however a more complex (and most probably needed for this kind of events) ensemble approach. With different domain extents or different architecture of the computing infrastructure (e.g., higher CPU clock speed), alternative solutions could have been found. Also, the number of case studies analyzed needs to be highly increased to get more general findings. Nevertheless, as a preliminary indication, the analysis performed suggests that, fixed specific lead time, in topographically complex areas it is desirable to provide forecasts with the highest space resolution as possible.

Acknowledgements. L. Furnari acknowledges for the program "POR Calabria FSE/FESR 2014/2020 - Mobilitá internazionale di Dottorandi e Assegni di ricerca/Ricercatori di Tipo A" Actions 10.5.6 and 10.5.12. We thank the "Centro Funzionale Multirischi" of the Calabrian Regional Agency for the Protection of the Environment, for providing the observed precipitation data. We, also, would like to acknowledge the INFN group at University of Calabria for giving the opportunity to use their cluster.

References

1. Avolio, E., Cavalcanti, O., Furnari, L., Senatore, A., Mendicino, G.: Brief communication: preliminary hydro-meteorological analysis of the flash flood of 20 august 2018 in Raganello Gorge, Southern Italy. Nat. Hazards Earth Syst. Sci. **19**(8), 1619–1627 (2019)
2. Avolio, E., Federico, S.: WRF simulations for a heavy rainfall event in southern Italy: verification and sensitivity tests. Atmos. Res. **209**, 14–35 (2018)
3. CFM: Evento meteopluviometrico del 20 agosto 2018 - torrente raganello. Technical report, Centro Funzionale Multirischi della Calabria (2018)
4. Davolio, S., Miglietta, M.M., Diomede, T., Marsigli, C., Morgillo, A., Moscatello, A.: A meteo-hydrological prediction system based on a multi-model approach for precipitation forecasting. Nat. Hazards Earth Syst. Sci. **8**(1), 143–159 (2008)
5. Di Luca, A., Argüeso, D., Evans, J.P., de Elía, R., Laprise, R.: Quantifying the overall added value of dynamical downscaling and the contribution from different spatial scales. J. Geophys. Res. Atmos. **121**(4), 1575–1590 (2016)
6. Guarracino, N., Lavorini, V., Tarasio, A., Tassi, E.: The ReCaS project cosenza infrastructure, pp. 43–55 (2017). Chapter 5

7. Hong, S.Y., Kanamitsu, M.: Dynamical downscaling: fundamental issues from an NWP point of view and recommendations. Asia Pac. J. Atmos. Sci. **50**(1), 83–104 (2014)

8. Neumann, P., et al.: Assessing the scales in numerical weather and climate predictions: will exascale be the rescue? Philos. Trans. Roy. Soc. A Math. Phys. Eng. Sci. **377**(2142) (2019). 20180148

9. Rummukainen, M.: Added value in regional climate modeling. Wiley Interdisc. Rev. Clim. Change **7**(1), 145–159 (2016)

10. Senatore, A., Mendicino, G., Gochis, D.J., Yu, W., Yates, D.N., Kunstmann, H.: Fully coupled atmosphere-hydrology simulations for the central mediterranean: impact of enhanced hydrological parameterization for short and long time scales. J. Adv. Model. Earth Syst. **7**(4), 1693–1715 (2015)

11. Senatore, A., Mendicino, G., Knoche, H.R., Kunstmann, H.: Sensitivity of modeled precipitation to sea surface temperature in regions with complex topography and coastlines: a case study for the mediterranean. J. Hydrometeorol. **15**(6), 2370–2396 (2014)

12. Skamarock, W.C., et al.: A description of the advanced research WRF version 3. NCAR technical note-475+str (2008)

13. Wedi, N.P.: Increasing horizontal resolution in numerical weather prediction and climate simulations: illusion or panacea? Philos. Trans. Roy. Soc. A Math. Phys. Eng. Sci. **372**(2018), 20130289 (2014)

Optimization Model for Water Distribution Network Planning in a Realistic Orographic Framework

Mario Maiolo[1]([✉])[iD], Joaquim Sousa[4][iD], Manuela Carini[1][iD],
Francesco Chiaravalloti[2][iD], Marco Amos Bonora[1], Gilda Capano[1],
and Daniela Pantusa[3][iD]

[1] Department of Environmental and Chemical Engineering (DIATIC),
Università della Calabria, Cubo 42/B, 87036 Arcavacata di Rende, CS, Italy
{mario.maiolo,manuela.carini,marcoamos.bonora,gilda.capano}@unical.it
[2] Research Institute for Geo-Hydrological Protection (IRPI), CNR,
87036 Rende, CS, Italy
francesco.chiaravalloti@irpi.cnr.it
[3] Innovation Engineering Department, University of Salento, Lecce, Italy
daniela.pantusa@unisalento.it
[4] Department of Civil Engineering, Polytechnic Institute of Coimbra,
Coimbra, Portugal
jjoseng@isec.pt

Abstract. Defining criteria for correct distribution of water resource is a common engineering problem. Stringent regulations on environmental impacts underline the need for sustainable management and planning of this resource usage, which is sensitive to many parameters. Optimization models are often used to deal with these problems, identifying the optimal configuration of a Water Distribution Network (WDN) in terms of minimizing an appropriate function propotional to he construction cost of the WDN. Generally, this cost function increases as the distance between the source-user connection increases, therefore in minimum cost optimization models is important to identify minimum source-user paths compatible with the orography. In this direction, the methodology presented in the present work proposes a useful approach to find minimum-length paths on surfaces, which moreover respect suitable hydraulic constraints and are therefore representative of reliable gravity water pipelines. The application of the approach is presented in a real case in Calabria.

Keywords: Optimization model · Water planning model · Graph theory

1 Introduction

Due to the possible presence of multiple water sources and different types of conflicting water users, one of the most challenging problems in water resources

© Springer Nature Switzerland AG 2020
Y. D. Sergeyev and D. E. Kvasov (Eds.): NUMTA 2019, LNCS 11973, pp. 545–556, 2020.
https://doi.org/10.1007/978-3-030-39081-5_47

management is the optimal allocation of water resources with respect the water demand and availability [4,11]. An useful approach to deal with this problem is to identify the optimal configuration of a Water Distribution Network (WDN)[3,6] in terms of minimizing an appropriate function which is proportional to the construction cost of the WDN. In particular, in [2,8] the authors propose a minimum-cost optimization model that determines an idealised water distribution system providing the optimal allocations of water resources among different sources and users. The cost of the WDN, which defines the optimization problem, is in the form:

$$C \propto \sum_{i,j} f(Q_{i,j}, L_{i,j}, h_{i,j}) \tag{1}$$

where $Q_{i,j}$ is the flow rate between the source i and the user j, $L_{i,j}$ is the source-user distance and $h_{i,j}$ the altitude difference (piezometric head difference between i and j). Beyond the details of the expression (1), it is obvious that the cost of a pipeline increases as its length increases. Therefore, it is reasonable to consider minimum-length path joining i and j. A critical approximation in [2,8] is the calculation of $L_{i,j}$ as Euclidean distance between the points i nd j.

With this approach the proposed procedure is expeditious, because only positions of source and destinations are needed in order to completely define the geometry of the problem, but the resulting optimal minimum-cost WDN is less representative from a physically feasible hydraulic infrastructure.

An improvement in this sense is obtained by taking into account the orography of the territory and by calculating $L_{i,j}$ as the length of the shortest path between i and j, lying on the orography.

The present work describes a methodology that allows to identify the shortest path between two point on a topographic surface which moreover is compatible with appropriate hydraulic criteria, making it possible the use of such path as a realistic trace for a gravity water pipeline.

Finding shortest paths (geodesic) on surfaces is a challenging problem in computational geometry, with important application in robotics, path planning, texture mapping, computer graphics [9].

A possible approach to this problem consists in converting the geometric problem to a graph problem and find approximate solutions [1,7,10]. In effect, the surface under examination can be transformed into a graph and procedures, e.g. Dijkstra's algorithm [5], capable to identify minimum paths joining two nodes of the graph can be used.

A 2-dimensional surface S embedded in $\mathbb{R}^3$ can be represented with a polyhedral surface, i.e. a set of polygonal faces that constitutes a piecewise-flat approximation of the surface. Two polygons do not intersect, except at a common point v (vertex) or an a common edge e. Sampling a sufficiently large number of points on the orographic surface,its spatial structure is preserved.

The sets of vertices (or nodes) $V = \{v_1, v_2, ..., v_n\}$ and edges (or arcs) $E = \{e_1, e_2, ..., e_m\}$ actually constitute a graph. If the weight of each arc is set equal to the euclidean distance of the nodes it joins, then the shortest path between two points on the surface will be approximated by the shortest path between the corresponding nodes in the graph.

The degree of novelty of the present work lies in identifying minimum-length paths that respect hydraulic limitations allowing them to constitute possible layouts for gravity water pipelines. For example,paths joying source-user nodes should preferably avoid the uphill sections in order to guarantee an adequate hydraulic load along the entire route.

2   Methodology

2.1   Polyhedral Surface Construction

The presented methodology use a digital terrain elevation model in raster format as a database. The raster format is a matrix structure that represents a rectangular grid of pixels. This structure can be used to store topographic information and is common-use format in GIS (Geographic Information System) software. In a DTM (Digital Terrain Model) raster, the type of information stored is the altimetry. Each pixel of the raster corresponds to the mean elevation of the area covered by that pixel, to each of them will be associated a pair of coordinates of a geodetic datum. The raster images used in this work consist of square pixels of equal size. The topographic surface is modelled as a weighted oriented graph:

$$G = (V, E) \tag{2}$$

with n nodes and m links with non-negative weight ϵ_{ij}. The nodes are placed in the center of the DTM cells. Each node will get a triplet of coordinates (x, y, z) from the cells in which is placed. To identify the nodes, in addition to matrix position indices of the corresponding cell, it can be useful to have single index k identifying the kth node. For example, the index map $k \to (i(k), j(k))$ can be easily written out for the lexicographical ordering, according to which the nodes are numbered by proceeding from left to right, from top to the bottom. In the usual approach to approximate the geodesics on a surface with the minimal paths on a graph, the weights ϵ_{ij} of the arcs are set equal to the Euclidean distances between the corresponding nodes i and j, and each arc can be travelled indifferently in both directions

$$\epsilon_{ij} = \epsilon_{ji} \tag{3}$$

for each corresponding arcs $\in E$. In this work, in order to take into account the hydraulic constraints and penalize the uphill sections of the path, different weights are assigned to uphill and downhill arches; more precisely, the weight of the upward arc is increased:

$$\epsilon_{ij} > \epsilon_{ji} \tag{4}$$

if $z_i > z_j$, where z is the altitude of the nodes. Both a directed and an undirected graph can be used to model a topographic surface. Since it will be necessary to take into account whether if an arc is rising or descending, a direct graph is chosen. As regards the linking criterion, there are eight possible connections along the directions: North, South, East, West, North-West, South-West, North-East, South-East (Fig. 2). These connections can be automatically generated whether the position of the cell in the matrix (i, j) or the node index k is used. The weight of a link is placed equal to the Euclidean distance between cell centers. For a DTM with square cells it will depend only on the type of connection (straight or diagonal) and on the elevation difference between starting and ending node. The distance between two adjacent nodes is:

$$d_{xyz} = \sqrt{dx^2 + dy^2 + dz^2} \tag{5}$$

For a DTM with square cells :

$$dx = dy = \Delta l \tag{6}$$

Hence, if d_{xy} indicates the horizontal distance between nodes, Eq. (5) can be rewritten as:

$$d_{xyz} = \sqrt{d_{xy}^2 + dz^2} \tag{7}$$

where $d_{xy} = \Delta l$ for N, S, E, W links, and $d_{xy} = \sqrt{2}\Delta l$ for NW, SE, NE, SW links. A dummy length is added in order to differently weight uphill and downhill links:

$$d_{xyz} = \sqrt{d_{xy}^2 + dz^2} + \text{Pen} \cdot dz \tag{8}$$

Fig. 1. Digital terrain model in raster format.

where $Pen \geq 0$ represents the penalty for the ascending traits of the path: for uphill links, the penalty is positive, for others is null. If Pen is equal to zero,

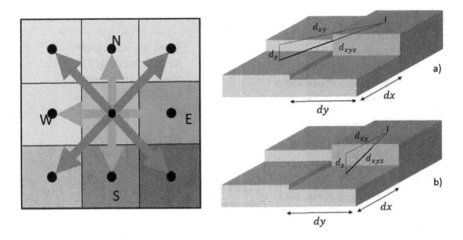

Fig. 2. Link generation method

the procedure gives the simple geodesic path. In order to reduce the extension of the graph and reduce the calculus time, all nodes with a higher elevation than the source node are excluded from the calculation. It is possible to exclude these nodes because the shortest paths that will represent the route of the supply pipes have to work only by gravity. Once the constrained geodesic path has been identified, it will be necessary to verify that this is usable as a path of a gravity supply pipe. Hence, a hydraulic check is implemented that takes into account the slope and the burial of the pipeline.

2.2 Hydraulic Check

To find out if the elevation trace of supplying pipe can work, it is necessary to solve the hydraulic problem related to it. The novelty of this work is the insertion of a procedure that influences the automatic choice of the curves that represent the supply paths taking into account their purpose. The motion formulas related to the problem, use the geometrical and hydraulic characteristics of the pipe. If the piezo metric line is always higher than the topographic surface, the pipeline can work only with gravity. The hydraulic check of a geodesic takes place immediately after finding the curve. In this phase, there is no information about the flows that will pass through the pipes and neither on diameters and roughness. A complete hydraulic check is impossible. The check that will be carried out imposes a limit slope, which must be ensured. In the preliminary design phases of supply pipes the engineering practice uses a constant slope for the piezometric line. This is done to consider a constant headloss along the pipe. The constant slope line will follow the geodesic curvilinear abscissa. If the line intersects the topographic surface then the verification is not satisfied. Given a tridimensional curve defined by a set of point triplets: (x_i, y_i, z_i), then the line will follow the planar trace of the line defined by (x_i, y_i)

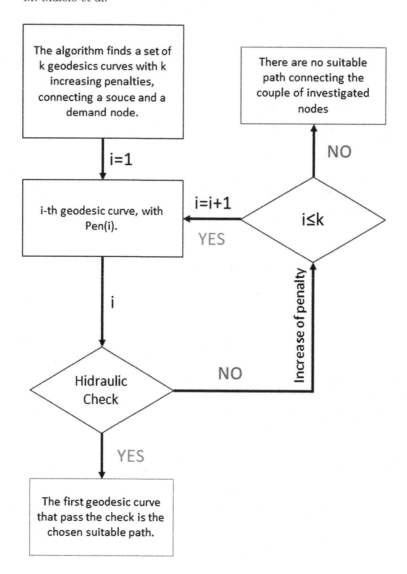

Fig. 3. Hydraulic check decision algorithm with an increasing-penalty geodesic curve set.

$$s_i = \sum_{j=2}^{i} \sqrt{(x_{j-1} - x_j)^2 + (y_{j-1} - y_j)^2} \qquad (9)$$

Given

$$E : \text{ possible excavation depth}$$
$$slope_{tg} : \text{ target slope}$$

the line will start from (x_1, y_1, z_1) with the following equation:

$$\begin{cases} x_i \\ y_i \\ z_i{}^{tg} = z_1 + s_i \cdot slope_{tg} + E \end{cases} \tag{10}$$

It is worth to note that the hydraulic check is a process that can be carried out only after the constrained geodesic is identified. It would be necessary an iterative calculation methodology that increases the penalty until the hydraulic verification is satisfied. This algorithm is too expensive in computational resources, as it requires the calculus of the Eq. (1), for each links of the graph. In this work 5 values of increasing penalities were considered:

$$Pen = \{0, 5, 10, 50, 100\}; \tag{11}$$

and the hydraulic control is carried out starting from a simple geodesic ($Pen = 0$) and then increasing the penalty until the corresponding path does not respect the imposed hydraulic limitation (Fig. 3).

3   Application

The constrained geodesics have been searched for the DTM showed in Fig. 1. An altimetry model of an area in the city of Rende (Italy, Calabria) was used, in which possible source points and demand points were identified. 5 source nodes and 4 demand nodes have been placed on this area. The source node ($S5$) is on a high ground, the demand node on a flat area. On the straight path that connects

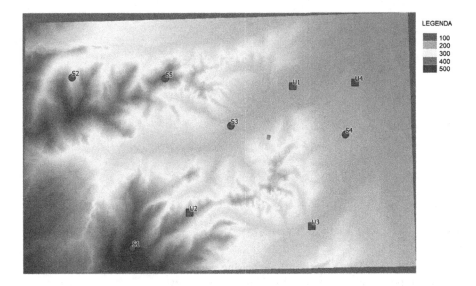

Fig. 4. DTM of the city of Rende (Italy), position of the source and demande nodes

a)

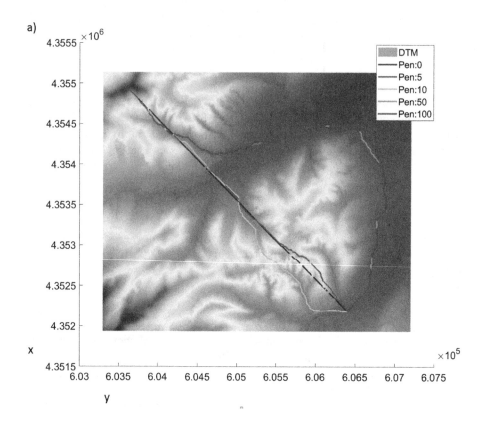

b)

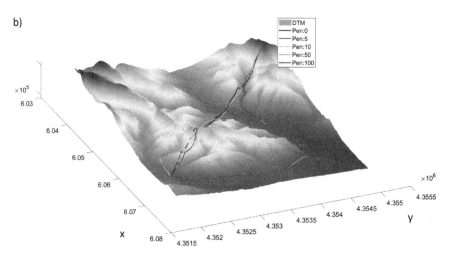

Fig. 5. set of 5 geodesic curves with increasing penalties found from a source point to destination point. (a) Plan view. (b) Tridimensional view

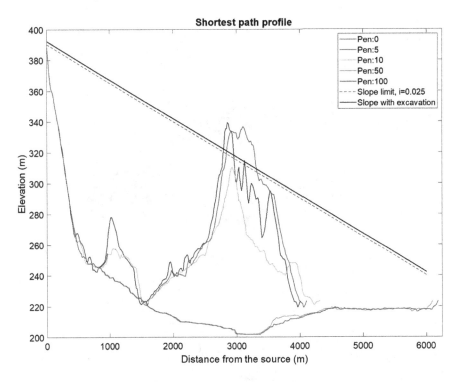

Fig. 6. Section view of the 5 geodesic curves with the hydraulic check line

points, there is another high ground lower than the first one. For each of the considered 5 penalties corresponding minimum-path are obtained (Fig. 7). For null penalties, the output is a simple geodesic, which in plan view is an almost straight line. As the penalty increases, the resulting constrained geodesic tries to avoid the uphill areas by moving away from the geodesic path, making it longer. As long as the starting point is at a height higher than the end point, there is a constrained geodesic connecting them. The hydraulic control allows choosing a suitable curve for the path of a supply pipe. Setting the possible excavation depth and a target slope:

$$E = 2\,m$$
$$slope_{tg} = 0.025 \tag{12}$$

The simple geodesic and the lowest penalty does not pass the check (see Fig. 6). The results for the entire set of user nodes and source nodes are shown below. The information obtained is the length of the path that connects the couples of points. The length values of the constrained geodesics are compared with the Euclidean distances. It is immediate to notice some things: For the Euclidean distance, the only condition for the existence of a connection is that the source node is at a higher altitude than the destination one. In constrained geodesics, this condition is necessary but not sufficient. This reduces the number of possible connections. The use of (constrained or simple) geodesics implies an

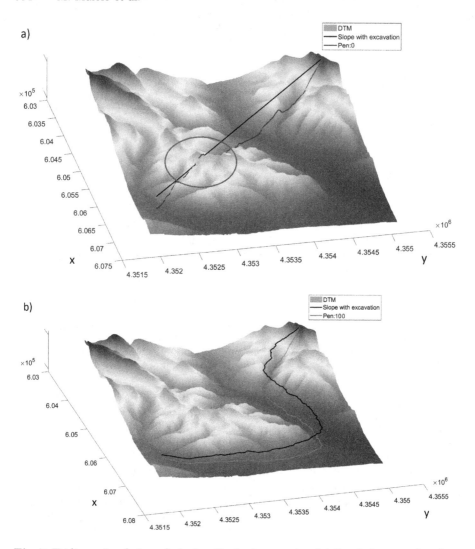

Fig. 7. Tridimensional view of a hydraulic check examples. (a) Geodesic curve that does not pass the check, highlighted the point where the check line intersects the surface. (b) Constrained geodesic that pass the check.

increase in length. The greater is the penalty used, the greater is the length of the optimal curve. The application of the hydraulic check, in addition to leading to the choice of routes at higher penalty, therefore longer, introduces the possibility that a path between a source node with a higher altitude and a destination with a lower altitude may not exist. By removing a certain number of links from the set of usable links in the optimization models, the final solution, in some cases, could differ much from that obtainable with methods that estimate the distance without take into account the hydraulics aspects. In the application carried out, as shown in Table 1, the use of a very high target slope led to the elimination of 7 links, and to the increase in the distance of one of them (Figs. 4 and 5).

Table 1. Table showing the connection matrix containing the distance between source and users nodes. The Table matrix assess the increase in distance due to the use of geodesic curves compared to the Euclidean ones.

Connection matrix: Constrained Geodesic length [m]				Connection matrix: Euclidean line [m]			
U1	U2	U3	U4	U1	U2	U3	U4
S1 4351.6	1386.6	3620.7	5562.1	S1 4292.9	1279.1	3446.0	5241.1
S2				S2 4211.3		5338.0	5401.3
S3				S3 1386.7			2497.2
S4			1035.5	S4 1351.9			977.7
S5 2510.4		4316.1	3682.2	S5 2425.3	2527.7	3904.4	3612.2
Geodesic penalty information				Length increase using constrained geodesics			
U1	U2	U3	U4	U1	U2	U3	U4
S1 Pen 0	Pen 0	Pen 0	Pen 0	S1 1.37%	8.40%	5.07%	6.13%
S2 HC	TP	HC	HC	S2			
S3 HC	TP	TP	HC	S3			
S4 HC	TP	TP	Pen 0	S4			5.91%
S5 Pen 0	HC	Pen 0	Pen 0	S5 3.51%		10.54%	1.94%

HC: Hydraulic check not passed.
TP: Impossible path because the source node is lower than the demand node.

4   Conclusion

The present work describes a methodology aimed at identifying the minimum paths that follows the topological surface to connect a source node with a destination node. The novelty of the work consists in the use of an approach that provides distances and paths in a more representative way than reality with respect to Euclidean distances and geodetic curves, and also allows to take into account hydraulic constraints in the identification of the paths. The use of this methodology in the context of Water Distribution Systems optimization models allows to obtain more detailed and realistic solutions for the subsequent design phase. The application carried out to a real surface model has allowed to make a comparison between the Euclidean distances and the lengths of the constrained paths obtained with the proposed approach, showing how the distances that take into account the topographic surface are longer of the Euclidean ones. The use of Euclidean distances therefore leads to underestimating quantities of materials and excavations necessary for the realization of the optimization configuration identified by the model.

Acknowledgments. Research supported by the Italian Regional Project (POR CALABRIA FESR 2014-2020): origAMI original Advanced Metering Infrastructure [J48C17000170006].

References

1. Balasubramanian, M., Polimeni, J.R., Schwartz, E.L.: Exact geodesics and shortest paths on polyhedral surfaces. IEEE Trans. Pattern Anal. Mach. Intell. **31**(6), 1006–1016 (2008). https://doi.org/10.1109/TPAMI.2008.213

2. Carini, M., Maiolo, M., Pantusa, D., Chiaravalloti, F., Capano, G.: Modelling and optimization of least-cost water distribution networks with multiple supply sources and users. Ricerche Mat. **67**(2), 465–479 (2018)

3. Coelho, A.C., Labadie, J.W., Fontane, D.G.: Multicriteria decision support system for regionalization of integrated water resources management. Water Resour. Manag. **26**(5), 1325–1346 (2012). https://doi.org/10.1007/s11269-011-9961-4

4. Davijani, M.H., Banihabib, M.E., Anvar, A.N., Hashemi, S.R.: Multi-objective optimization model for the allocation of water resources in arid regions based on the maximization of socioeconomic efficiency. Water Resour. Manag. **30**(3), 927–946 (2016). https://doi.org/10.1007/s11269-015-1200-y

5. Dijkstra, E.W.: A note on two problems in connexion with graphs. Numer. Math. **1**(1), 269–271 (1959)

6. Hsu, N.S., Cheng, K.W.: Network flow optimization model for basin-scale water supply planning. J. Water Resour. Plan. Manag. **128**(2), 102–112 (2002). https://doi.org/10.1061/(ASCE)0733-9496(2002)128:2(102)

7. Kanai, T., Suzuki, H.: Approximate shortest path on a polyhedral surface based on selective refinement of the discrete graph and its applications. In: Proceedings Geometric Modeling and Processing 2000. Theory and Applications, pp. 241–250. IEEE (2000). https://doi.org/10.1109/GMAP.2000.838256

8. Maiolo, M., Pantusa, D.: An optimization procedure for the sustainable management of water resources. Water Supply **16**(1), 61–69 (2015). https://doi.org/10.2166/ws.2015.114

9. Porazilova, A.: The Geodesic Shortest Path (2007)

10. Surazhsky, V., Surazhsky, T., Kirsanov, D., Gortler, S.J., Hoppe, H.: Fast exact and approximate geodesics on meshes. ACM Trans. Graphics (TOG) **24**(3), 553–560 (2005). https://doi.org/10.1145/1073204.1073228

11. Zhanping W., Juncang T.: Optimal allocation of regional water resources based on genetic algorithms. J. Converg. Inf. Technol. (JCIT) **7**(13) (2012). https://doi.org/10.4156/jcit.vol7.issue13.51

Scenario Optimization of Complex Water Supply Systems for Energy Saving and Drought-Risk Management

Jacopo Napolitano[✉] and Giovanni M. Sechi

University of Cagliari, Via Marengo 2, Cagliari, Italy
{jacopo.napolitano, sechi}@unica.it

Abstract. The management of complex water supply systems needs a close attention to economic aspects concerning high costs related to energy requirements in water transfers. Specifically, the optimization of activation schedules of water pumping plants is an important issue, especially managing emergency and costly water transfers under drought-risk. In such optimization context under uncertainty conditions, it is crucial to assure simultaneously energy savings and water shortage risk alleviating measures. The model formulation needs to highlight these requirements duality to guarantee an adequate water demand fulfillment respecting an energy saving policy. The proposed modeling approach has been developed using a two stages scenario optimization in order to consider a cost-risk balance, and to achieve simultaneously energy and operative costs minimization assuring an adequate water demand fulfillment for users. The optimization algorithm has been implemented using GAMS interfaced with CPLEX solvers. An application of the proposed optimization approach has been tested considering a water supply system located in a drought-prone area in North-West Sardinia (Italy). By applying this optimization procedure, a robust strategy in pumping activation was obtained for this real case water supply system.

Keywords: Scenario analysis · Energy optimization · Water management

1 Introduction

The management optimization of complex water supply systems, aimed to the energy saving, is an interesting and actual research topic [2, 10, 16]. Problems pertaining to water system management policies and specifically concerning the effectiveness of emergency and costly water transfers activation to alleviate droughts, are faced with different methodological approaches [8, 9, 12]. Solving these optimization problems frequently leads to complex computational models: their solution needs efficient approaches to deal with many uncertainties which arise modeling real systems and trying to achieve optimal decision rules, in order to provide robust solutions to the water resource system's Authorities [17].

In the water resource modeling field, problems affected by uncertainty have been treated implementing several computational solutions, especially with application of

© Springer Nature Switzerland AG 2020
Y. D. Sergeyev and D. E. Kvasov (Eds.): NUMTA 2019, LNCS 11973, pp. 557–569, 2020.
https://doi.org/10.1007/978-3-030-39081-5_48

stochastic dynamic programming to multi-reservoir systems. Under scarcity conditions, the system reliability evaluation is intimately related to a quantitative evaluation of future water availability and the opportunity to provide water through the activation of emergency and costly water transfers. Hence, the water system optimization problem needs to deal with uncertainties particularly in treating the effectiveness of emergency measures activation to face droughts.

On the occasion of drought occurrences, water managers must be able to manage this criticality, alleviating the effect of shortages on water users. A useful solution could be to define some emergency policies, supplying additional water to demand centers. Modern water supply systems could count on the presence of emergency sources with enough capacity for these critical occurrences. Frequently, the activation of these emergency transfers requires additional costs, such as for activation of pumping schedules. Optimal strategies on this costly transfer activation is a hard decision problem: it is conditioned by uncertainties, mainly due to future demands behavior and hydrological inputs that are normally characterized by high variability during the time horizon normally considered in modeling the systems.

According some authors [9, 11, 14], the uncertainty could be described through different scenarios, which occurrence follows a probability distribution. In case of previous pessimistic forecasts, uncertainty and variability can generate water excess or subsequent spilling from water storage reservoirs, causing losses and therefore resulting as regrets costs [6]. On the other hand, the definition of emergency policies in reservoirs management could consider early warning measures, taking advantages of lower energy prices in some time-periods and achieving economic savings also related to avoid the occurrence of water shortages.

The hereafter described research aims to develop an optimization under uncertainty modeling approach, in order to deal with water resources management problems especially referring to the definition of optimal activation rules for emergency activation of pumping stations in drought conditions. Therefore, this study aims to define a cost-risk trade-off considering the minimization of water shortage damages and the pumping operative costs, under different hydrological and demand scenarios occurrence possibilities. The expected results should be able to provide the water system's authority with a strategic information, defining optimal rules, and specifically optimal activation reservoir-storage triggers for water pumping stations.

The formulation of the related optimization model needs to highlight this duality: to guarantee water demands fulfillment respecting an energy saving policy.

Optimal rules in order to minimize the emergency transfer's costs will be defined through a cost/risk balanced management and formulating a multistage scenario optimization model. The proposed modeling approach has applied to a real case concerning a multi-reservoir and multi-user water supply system in a drought-prone area, located in the North-West Sardinia (Italy) region, characterized by South-Mediterranean climate.

2   Scenario Optimization for a Water Supply System Management

Scenario analysis approach considers that all future events can be described through a set of different and statistically independent scenarios [14]. A single scenario describes a possible sequence in the realization of some sets of uncertain data along the analyzed time horizon. Considering all together the structure of the scenarios temporal evolution, it is possible to obtain a robust decision policy, minimizing the risk of wrong future decisions.

This modelling approach can be represented as a tree-graph (Fig. 1), according to appropriate aggregation rules, called congruity constraints. Some scenarios sharing a common initial portion of data can be considered partially aggregated with the same decision variables for the aggregated part, taking into account possible evolutions in the subsequence of different behaviors.

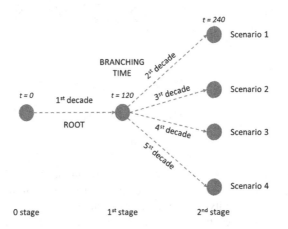

Fig. 1. Scenario-tree aggregation

In order to perform correctly the scenario aggregation, must be defined stages and branching-times as represented in Fig. 1. A branching-time identifies the time-period in which some scenarios, that are identical up to that period, begin to be different.

The root of the scenario-tree corresponds to the time at which decisions (common to all scenarios) have been taken, while the leaves represent the state of the system in the last stage. Each path from the root to a leaf identifies a possible sequence of occurrences along the entire time horizon.

Therefore, each possible scenario corresponds with a dynamic multi-period graph [11], following a particular sequence of decisions. E.g., in the water resource management, it is possible to consider a hydrological series of water inflows or a sequence of management decisions related to a reservoir [16].

2.1   Trigger Values Optimization Model and Cost-Risk Trade Off

Dealing with uncertainty, in order to find a balanced water system management criterion between alternatives represented through the scenario tree and to improve the decision-making process robustness, a cost-risk balancing optimization approach has been used.

This model formulation tries to attain to a robust decision policy minimizing the risk to assume wrong and harmful decisions in the system management. Therefore, the water system's authority should be able to define a reliable decision system, minimizing the energetic and management system costs and reducing the deficit hardships for users.

Considering the entire set $g \in G$ of scenarios, the objective function (1) minimizes the energy and management costs, optimizing the configuration of the flows x^g along the network. Moreover, it minimizes the weighted distance between a single scenario activation threshold $\widehat{S}^g$ and the optimal barycentric value S^b.

$$\underset{x^g,\widehat{S}^g,S^b}{\text{Minimize}}(1-\lambda)\sum_{g \in G}p^g c^g x^g + \lambda \sum_{g \in G}\sum_{i \in P}p^g\left[w^g\left(\widehat{S}_i^g - S_i^b\right)^2\right] \tag{1}$$

subject to

$$A^g x^g = b^g \qquad \forall g \in G \tag{2}$$

$$l^g \leq x^g \leq u^g \qquad \forall g \in G \tag{3}$$

$$x^* \in \Phi \tag{4}$$

All decision variables and data are scenario dependent, hence the index g. A weight p^g can be assigned to each scenario characterizing its relative importance. Weights could represent the probability of occurrence of each scenario.

The vector c^g describes the unit cost of different activities like delivery cost, opportunity cost related to unsatisfied demand, opportunity cost of spilled water, energy cost and so on. The set of standardized equality constraints A^g describes the relationships between storage, usage, spill, and exchange of water at different nodes and in subsequent time periods. The RHS values b^g are given from scenario occurrences and are related to data of inflows and demands. The lower and upper bounds l^g and u^g are defined by structural and policy constraints on operating the system. All constraints (2–4) are collected from all scenarios and must be considered in the aggregated model. The additional set of constraints (4) are called non-anticipative constraints and $x^* \in \Phi$ represents the congruity constraints derived by the scenario aggregation rules [11].

In general terms, the first part of the objective function (1) can be defined as a cost function and it tries to look for the system flows configuration that allows minimizing the costs supported during the water system management. The second part can be considered as a risk function, it is quadratic and it tries to minimize the quadratic weighted distance between the barycentric value and each single scenario trigger value. The weight w^g is the cost related to the risk occurrences of each scenario $g \in G$.

In this way, giving a weighted value to both terms of the objective function, we can find a solution of the cost-risk balancing problem.

The relationship between cost function and risk function is regulated by the parameter λ called weight factor, which can vary between 0 and 1. Intermediate values of λ provide different tradeoffs between costs and risks.

In order to guarantee a correct operation of pumping stations, the model (1–4) should be completed introducing a new set of constraints (5–9).

$$-h_i^g BM \leq \sum_{j=1}^{R_i} xv_j^g - \widehat{S}_i^g \leq (1 - h_i^g) BM \quad \forall i \in P, \forall g \in G \tag{5}$$

$$x_i^g = (1 - h_i^g) P_i \quad \forall i \in P, \forall g \in G \tag{6}$$

$$x_i^g = (1 - h_i^g) T_k \quad \forall k \in K, \forall i \in P, \forall g \in G \tag{7}$$

$$\widehat{S}_i^g < \sum_{j=1}^{R_i} K_j \quad \forall i \in P, \forall g \in G \tag{8}$$

$$S_i^b < \sum_{j=1}^{R_i} K_j \quad \forall i \in P, \forall g \in G \tag{9}$$

Where:

$i \in P\{1, \ldots, n_P\}$ Pumping stations in the system;
$k \in K\{1, \ldots, n_K\}$ Diversion dams in the system;
$j \in R\{1, \ldots, n_R\}$ Reservoirs in the system;
$h_i^g \in \{0, 1\}$ Binary variable

The activation of pump stations is supposed to be dependent on the stored volume levels in reservoirs that could supply the downstream demand nodes by gravity or, anyway, without recurring to emergency and costly water transfer activation. Therefore, to model the pump activation, a binary variable h_i^g to each i-th pump station should be assigned. This variable represents the on/off condition for a single pump station as it can assume one or zero value.

In the optimization model h_i^g is dependent on the sum of the stored levels xv_j^g in the j-th reservoirs supplying water by gravity, according the activation dependences shown for the real case in Table 5. Therefore, constraints (5) allows the i-th pump station activation if the sum of the stored volume in reservoir j is under the threshold value $\widehat{S}_i^g$. In this constraint, the parameter BM is a large scalar.

Constraint (6) guarantees that, in the case of activation of the i-th pump station, the flow along the pumping arc starting from the i-th station will be equal to its capacity P.

If the pump station i-th is located downstream to a transshipment node k, the constraint (7) assures that, in case of activation, the flow along the arc will be equal to the potential water volumes withdrawal from this node $k \in K$ [15], which should be lower to the pump capacity P.

Constraints (8–9) impose an upper bound on the activation storage levels of the i-th pumping station equal to the sum of the reference reservoir's capacity K_j.

The adopted optimization procedure, defining trigger values in the pumps activation, is summarized in the flowchart shown in Fig. 2.

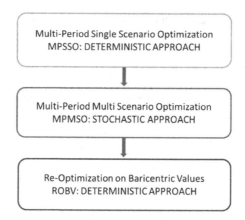

Fig. 2. Scenario-optimization: main modelling steps

In a first step of the analysis, it is possible to work using a single scenario optimization in order to calibrate the main parameters of the process (spilling costs, storage benefit and so on). In a second phase, the model examines the set of different scenarios. Once evaluated optimal values using scenario analysis, a re-optimization phase could be performed in order to verify the reliability of the water system and to obtain the network's flows in single scenario configurations. Through this last phase is possible to reach the sensitivity analysis and to verify output caused by assumptions about the adopted parameters.

3   Case Study: North-West Sardinia Water Supply System

Considering a water management problem, the graph representation approach [3] has been considered as an efficient support for the mathematical modeling [1, 5]. According to this, $J = (N, L)$ are sets satisfying $L \subseteq [N]^2$, where the elements of N are the nodes of the graph J, while the elements of L are its arcs. In the common water system notation, nodes can represent groundwater, sources, reservoirs, demands, etc. Arcs represent the connections between nodes, where water could flow.

This approach allows drafting a complex water system problem through a simple flow network on a graph. In the single period, we can represent the physical system and the static situation by a direct network called basic graph, as reported in Fig. 3.

This kind of analysis could be extended to a wide time-horizon T, assuming a time step (month) t. By replicating the basic graph for each period of the time horizon, it is possible to generate a dynamic multi-period network [11].

3.1   Water System's Features

An application of the proposed optimization procedure has been tested considering the draft of a real water supply system located in a drought-prone area in North-West Sardinia (Italy).

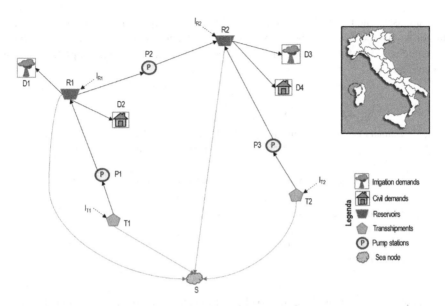

Fig. 3. North-West Sardinia water supply system

The main water sources are provided by 2 artificial reservoirs and 2 diversion dams. Their main features are reported in Tables 1 and 2.

The evaluation of potentiality in the water supply arises from historically hydro-logical inflows evaluated from 1922 to 1992 considering the values reported in the Sardinia Region Water Plan [13].

Table 1. Reservoir's features

Code	Reservoir	Capacity[10^6 m^3]	Spilling cost [€/m^3]
R1	Temo Rocca Doria	70	0.001
R2	Bidighinzu	10.9	0.001

Reservoirs are represented in the sketch by system's storage nodes. Diversion dams do not have a large storage volume: therefore, only a partial incoming flow can be diverted to demand centers or to larger capacity reservoirs.

Table 2. Diversion dams

Code	Transshipment	Spilling cost [€/m^3]
T1	Cumone - Badu Crabolu	0.001
T2	Ponte Valenti - Calambru	0.001

Water demands have been grouped in four centers, according to two different users: civil and irrigation. As shown in Table 3, an annual volume of demand, rate and deficit costs are associated to each demand center. Deficit costs quantify the possible damages supported by the users in the case of shortage occurrences. These costs have been evaluated starting from the water annual rates for unit of volume applied by each stakeholder.

Table 3. Demand centers

Code	Demand center	Demand [10^6 m^3/year]	Rate cost [€/m^3]	Planned deficit cost [€/m^3]	Unplanned deficit cost [€/m^3]
D1	Irrigation Cuga	15.36	0.006	0.06	0.6
D2	Civil Temo	6.93	0.025	0.25	2.5
D3	Irrigation Valle Giunchi	0.95	0.006	0.06	0.6
D4	Civil Bidighinzu	15.56	0.025	0.25	2.5

Deficits can be categorized in two classes: planned and unplanned. Planned deficits can be forecasted and communicated in advance to water users, while the unplanned ones arise when the realization of water inflows follow hydrological scenarios of unpredictable scarcity, affecting and harming several users. In order to take into account these costs in the mathematical model, in case of shortage occurrences, the initial 15% of the monthly demand not satisfied will be compute as planned deficit while remaining surplus of shortages should be considered as unplanned deficits.

Table 4. Pump stations' features

Code	Pumping station	Capacity [10^6 m^3/month]	Pumping cost [€/m^3]
P1	Padria	7.88	0.071
P2	Rocca Doria	1.66	0.196
P3	Su Tulis	3.15	0.087

Pump stations reported in Table 4 allow demand centers to be supplied with an increased economic burden, namely incurring energy costs in addition to the ordinary management.

Table 5. Pump stations' activation dependences

Reservoir → Pump station ↓	R1	R2
P1	1	0
P2	0	1
P3	0	1

Modeling the system, threshold levels for pumps activation refer to the stored volume in reservoirs that supply the downstream demand nodes. These functional dependencies are reported in the Table 5, where 1 means dependence, while 0 independence for pump activation.

3.2 Optimization of the Pumping Schedules Thresholds

For each pump station, two season's activation thresholds should be identified, in order to define a barycentric seasonal value. These trigger values refer specifically to the dry (in Mediterranean countries months from April to September) and wet (months from October to March) hydrological semesters.

By applying the scenario optimization process, a robust decision strategy for seasonal trigger values of pumps activation was retrieved. Reference scenarios were settled considering the regional hydrological database [13] modified in order to take into account climatological trends observed during the last decades in this region. Starting from the river-runoff historical database of 50 years horizon, scenarios of different criticism are extracted from the observed data: each one has been of 20 years, defined at monthly step. Inside the 240 monthly periods the branching time is located at the 120th period. As shown in Fig. 1, the resulting scenario-tree has been composed. Therefore, the historical database has been used in order to generate 4 hydrological scenarios characterized by the same length, organized with a common root of 10 years and the following data diversified in the following 10 years by climate criticism.

The scenario analysis approach needs a complex mathematical formulation in terms of model dimensions and number of variables and constraints considered. Moreover, the mathematical model described in the Eqs. (1–4) should be solved through a Mixed Integer Quadratically Constraints Programming. Therefore, in order to afford efficaciously this problem, guaranteeing a reliable solution with a reasonable computational time, it has been implemented by the software GAMS [4] calling CPLEX [7] Branch and Cut algorithm as optimization solver.

The optimized activation thresholds reported in Table 6 have been evaluated considering this scenario-tree and solving the optimization model (1–4) using a λ value equal to 0.5, then assuring an equal weight to cost and risk elements in the objective function.

Table 6. Optimized activation thresholds

Activation threshold [10^6 m^3]	S1		S2		S3	
	Wet	Dry	Wet	Dry	Wet	Dry
	60.22	67.13	1.04	3.26	7.64	8.9

According the scenario analysis, these optimized values are barycentric among all hydrological scenarios, and they will guarantee a compromise among different water resource availabilities. These barycentric values have been adopted during the re-optimization phase, where the whole process has been developed assigning these thresholds as fixed parameters.

Table 7 reports the average annual costs evaluated by an economic post-processor taking into account deficit and pumping costs. Only the real costs of the system, related to energy consumption and drought occurrences have been considered during this analysis.

Table 7. Economical post-processor

Mean costs	10^6 €/year
Pumping	1.82
Deficit	0.19
Total	2.01

Results show that almost the total cost amount is due to the energy costs, while only the 10% is caused by deficit occurrences. It highlights that the cost-risk balancing approach has been able to assure simultaneously energy costs minimization and a reduction of possible damages caused by water shortages.

Figures 4, 5 and 6 show the percentage of water deficit along the considered four scenarios for the affected demand centers. As expected, deficits affect especially irrigation users in order to preserve the civil demands, which represent the water system's major priority users.

This representation highlights planned and unplanned deficit occurrences, indeed: planned deficit are included under the 15%, while unplanned deficits across this threshold.

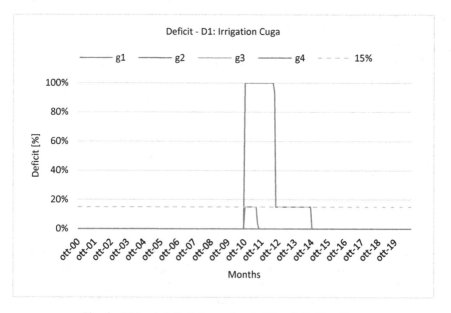

Fig. 4. Water deficit at demand node D1 – Irrigation Cuga

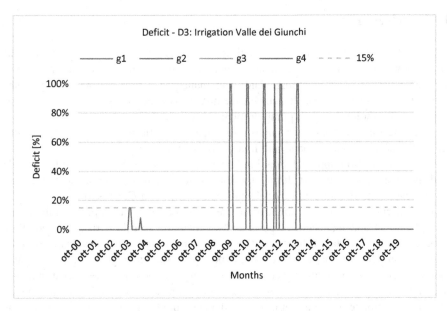

Fig. 5. Water deficit at demand node D3 – Irrigation Valle dei Giunchi

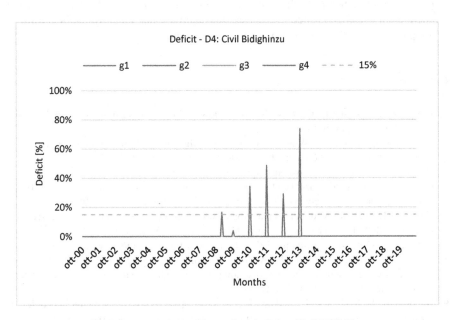

Fig. 6. Water deficit at demand node D4 – Civil Bidighinzu

4   Conclusions

The scenario-optimization approach confirmed its potentiality when applied to the real case of a water resource management problem and specifically optimizing the activation schedules of water pumping plants and managing emergency and costly water transfers under drought-risk. It allowed identifying two barycentric seasonal optimal activation thresholds for each pumping plant located in the water system.

These results are obtained using a two stage stochastic programming taking into account expected water demand and hydrological series.

This optimization approach was implemented using GAMS that could be considered as an excellent support during the model development. The software allowed writing easily the optimization models and interfacing with CPLEX solvers.

Considering the real case application, the cost-risk balancing approach minimized the operational and management costs and contextually restricted risks and conflicts between users in shortage conditions. Costs and penalties have been evaluated in a re-optimization phase through an economic post-processor taking into account water shortage penalties and pumping costs and assuring a trade-off between cost and risk elements. For the considered water scheme, the proposed methodologies guarantees almost the complete fulfilment of the water demand: unplanned deficits still remain but only for few periods of the considered time horizon.

Using a more adherent to the reality simulation approach, the effectiveness of the obtained results has been tested interacting with the regional water system's Authority by comparison with the occurred management behavior.

References

1. Ahuja, R., Magnanti, T., Orlin, J.: Network Flows: Theory, Algorithms, and Applications. Prentice Hall, Englewood Cliffs (1993)
2. D'Ambrosio, C., Lodi, A., Wiese, S., Bragalli, C.: Mathematical programming techniques in water network optimization. Eur. J. Oper. Res. **243**(3), 774–788 (2015)
3. Diestel, R.: Graph Theory, 3rd edn. Springer, New York (2005)
4. GAMS: A user's guide. GAMS Development Corporation. Washington DC, USA (2008)
5. Jensen, P., Barnes, J.: Network Flow Programming. Wiley, New York (1980)
6. Kang, D., Lansey, K.: Multiperiod planning of water supply infrastructure based on scenario analysis. ASCE J. Water Resour. Plann. Manag. **140**, 40–54 (2014)
7. IBM: Cplex Optimization Studio (2017). http://www-03.ibm.com/software
8. Lerma, N., Paredes-Arquiola, J., Andreu, J., Solera, A., Sechi, G.M.: Assessment of evolutionary algorithms for optimal operating rules design in real water resource systems. Environ. Model Softw. **69**, 425–436 (2015)
9. Napolitano, J., Sechi, G.M., Zuddas, P.: Scenario optimization of pumping schedules in complex water supply system considering a cost-risk balancing approach. Water Resour. Manag. **30**, 5231–5246 (2016)
10. Nault, J., Papa, F.: Lifecycle assessment of a water distribution system pump. ASCE J. Water Resour. Plann. Manag. **141**(12), A4015–004 (2015)
11. Pallottino, S., Sechi, G.M., Zuddas, P.: A DSS for water resource management under uncertainty by scenario analysis. Water Resour. Manag. **28**(12), 3975–3987 (2014)

12. Pasha, M.F.K., Lansey, K.: Strategies to develop warm solutions for real-time pump scheduling for water distribution systems. Environ. Model Softw. **20**, 1031–1042 (2004)
13. RAS: Piano stralcio di bacino regionale per l'utilizzazione delle risorse idriche. Regione autonoma della Sardegna, Italy (2006)
14. Rockafellar, R.T., Wets, R.J.B.: Scenario and policy aggregation in optimization under uncertainty. Math. Oper. Res. **16**, 119–147 (1991)
15. Sassu, E., Zucca, R., Sechi, G.M.: Calibration of regional flow-duration curves evaluating water resource withdrawal from diversion dam. In: Garrote, L., Tsakiris, G., Tsihrintzis, V. A., Vangelis, H., Tigkas, D. (eds.) Managing Water Resources for a Sustainable Future, Proceedings of the 11th World Congress of EWRA on Water Resource and Environment, Madrid, 25–29 June 2019 (2019)
16. Sechi, G.M., Gaivoronski, A.A., Napolitano, J.: Optimizing pumping activation in multi-reservoir water supply systems under uncertainty with stochastic quasi-gradient methods. Water Resour. Manag. **33**(2), 1881–1895 (2019)
17. Sechi, G.M., Sulis, A.: Water system management through a mixed optimization-simulation approach. ASCE J. Water Resour. Plann. Manag. **135**, 160–170 (2009)

Optimizing Rainwater Harvesting Systems for Non-potable Water Uses and Surface Runoff Mitigation

Stefania Anna Palermo[1]([✉]) [iD] , Vito Cataldo Talarico[1] [iD] ,
and Behrouz Pirouz[2] [iD]

[1] Department of Civil Engineering, University of Calabria, Rende, CS, Italy
{stefania.palermo,vitocataldo.talarico}@unical.it
[2] Department of Mechanical, Energy and Management Engineering,
University of Calabria, Rende, CS, Italy
behrouz.pirouz@unical.it

Abstract. Rainwater harvesting systems represent sustainable solutions that meet the challenges of water saving and surface runoff mitigation. The collected rainwater can be re-used for several purposes such as irrigation of green roofs and garden, flushing toilets, etc. Optimizing the water usage in each such use is a significant goal. To achieve this goal, we have considered TOPSIS (Technique for Order Preference by Similarity to Ideal Solution) and Rough Set method as Multi-Objective Optimization approaches by analyzing different case studies. TOPSIS was used to compare algorithms and evaluate the performance of alternatives, while Rough Set method was applied as a machine learning method to optimize rainwater-harvesting systems. Results by Rough Set method provided a baseline for decision-making and the minimal decision algorithm were obtained as six rules. In addition, The TOPSIS method ranked all case studies, and because we used several correlated attributes, the findings are more accurate from other simple ranking method. Therefore, the numerical optimization of rainwater harvesting systems will improve the knowledge from previous studies in the field, and provide an additional tool to identify the optimal rainwater reuse in order to save water and reduce the surface runoff discharged into the sewer system.

Keywords: Rainwater harvesting · Water supply · Flood risk mitigation

1 Introduction

There are many benefits in Rainwater harvesting (RWH) systems mainly water saving for non-potable water uses and surface runoff mitigation. Moreover, the collected rainwater can be re-used for several purposes including green roofs and garden, flushing toilets, etc. In previous studies, the optimization of rainwater harvesting systems was mostly limited to optimum size of the tankers according to hydrological and hydraulic analysis and in some cases combined with economic analysis.

However, the design of RWH systems depends on many elements and even optimizing different water usages is significant. Therefore, in this paper, Multi-Objective Optimization approaches have been applied, and ranking methods such as TOPSIS

© Springer Nature Switzerland AG 2020
Y. D. Sergeyev and D. E. Kvasov (Eds.): NUMTA 2019, LNCS 11973, pp. 570–582, 2020.
https://doi.org/10.1007/978-3-030-39081-5_49

(Technique for Order Preference by Similarity to Ideal Solution) have been considered to compare algorithms and evaluate the performance of alternatives to reach the ideal solution. Moreover, the attributes analysis such as Rough Set method has been used in analysis of vague description of decisions.

1.1 Rainwater Harvesting (RWH) Systems

The combined effect of global urbanization, climate change and water scarcity, requires a transition towards a sustainable, smart and resilient urban water management. In this regard, nowadays the sustainability concept is a basilar element for scientific, technical and socio-economic discussion. Therefore, the implementation of decentralized stormwater controls systems, as LID (Low Impact Development) systems, represents a promising strategy to achieve several benefits at multiple scales [1, 2].

Among these techniques, Rainwater Harvesting (RWH), considered an ancient practice used all over the world to meet the water demand, is now supported by many countries as a suitable solution to limit potable water demand, reduce frequency, peaks and volumes of stormwater runoff at the source, and participate in the restoration of natural hydrological cycle [3–6].

The principal component of a conventional RWH system is the rainwater tank which temporally stores the water from a capturing surface, normally the building roof or others impervious surfaces closely to the building. In a single-family building, above-ground tank, named "rain barrels", are generally used for irrigation and runoff control, while, in the case of multi-story building, above or below-ground concrete cisterns are implemented. In addition, a system consisting of gutters and downspouts lead the runoff from the collecting surface to the tank, while a dedicated piping network is needed for rainwater reuse. One or more pumps can be used to assure the pressure head for different usages, while other devices as first flus diverters, debris screen, and filters are generally implemented for water quality control. These information and more specific detail regarding this technique can be found in several studies [4, 7, 8].

Recent advances have showed the possibility for real-time monitoring and control of these systems in order to increase their efficiency in terms of reduction of urban flooding or combined sewer overflows [9] and optimize the rainwater reuse.

Harvested rainwater can be considered a renewable water source that is perfect for different non-potable water uses, as toilet flushing, laundry, car washing, terrace cleaning, private garden irrigation and green roof irrigation [7, 10–15].

In addition, as source control technology distributed at urban catchment scale, these systems are suitable to reduce stormwater runoff volume. In this regard, several studies have evaluated also the hydrological efficiency of RWH in terms of reduction of the runoff volume and peak discharged to the sewer system [5, 7, 11, 14, 16].

Several studies have been carried out to show the RWH efficiency for water saving and runoff mitigation, as study of Campisano and Modica [11] showed that the performance depends of site-specific factors, such as roof type and surface, precipitation regime, demand usage, tank size, number of people in the household, etc.

Based on a deeper literature review, in this paper some factors, here after called "attributes" have been selected and considered in a mathematical optimization of RWH.

2 Methodology

In the current study, the Rough Set method applied as a machine learning method to optimize rainwater-harvesting systems. The process is reviewed in details and the result is achieved with analysis of different case studies.

2.1 Rough Set Theory

The Rough Set theory is attributes analysis based on data or knowledge about a decision. The results of analysis can provide clear rules for similar decisions [17–19]. The Rough Set rules in a given approximation space apr = (U, A) can be calculated as follow:

The lower approximations of X in A $\qquad \overline{apr}(A) = \{x \mid x \in U, U/ind(A) \subset X\}$ (1)

The upper approximations of X in A $\qquad \underline{apr}(A) = \{x \mid x \in U, U/ind(A) \cap X \neq \varphi\}$ (2)

$$U/ind(a) = \{(x_i, x_j) \in U \times U, f(x_i, a) = f(x_j, a), \forall a \in A\} \qquad (3)$$

$$\text{Boundary is BN}(A) = \overline{apr}(A) - \underline{apr}(A) \qquad (4)$$

$$\text{The reduct} = \text{minimal set of attributes B} \subseteq A \text{ such that } r_B(U) = r_A(U) \qquad (5)$$

The quality of approximation of U by B $\qquad r_B(U) = \dfrac{\sum card(\underline{B}(X_i))}{card(U)}$ (6)

$$\text{Decision rule is } \varphi \Rightarrow \theta \qquad (7)$$

Where:

U is a set,
A is attributes of the set,
φ is the conjunction of elementary conditions,
θ is the disjunction of elementary decisions.

2.2 TOPSIS Method

TOPSIS (Technique for Order Preference by Similarity to Ideal Solution) method developed by Hwang and Yoon in 1981 is a method to solve the ranking and comparing decisions [20–22] and can be applied to wide range of multi-attribute decision making with several attributes [22–24]. The ranking in this method can be done for matrix $(n_{ij})m \times n$ according to the similarity to ideal solution as follow:

Normalized decision matrix: $N = n_{ij} = \dfrac{a_{ij}}{\sqrt{\sum_i^m = 1 a_{ij}^2}}, i = 1, 2, \ldots, m, j = 1, 2, \ldots, n$ (8)

Weighted normalized decision matrix: $V = N \times Wn \times n$ (9)

Determining the solutions:

Ideal solution: $A^+ = \left\{ \left\langle max_i \left(a_{ij} | j \in J_- \right) \right\rangle, \left\langle min_i \left(a_{ij} | j \in J_+ \right) \right\rangle \right\}$ (10)

Negative-ideal solutions: $A^- = \left\{ \left\langle min_i \left(a_{ij} | j \in J_- \right) \right\rangle, \left\langle max_i \left(a_{ij} | j \in J_+ \right) \right\rangle \right\}$ (11)

$J_+ = \{ j = q, 2, \ldots, n | j \}$ Associated with positive impact criteria
$J_- = \{ j = q, 2, \ldots, n | j \}$ Associated with negative impact criteria
Determining the distances from the solutions:

Distance from ideal solution: $d_i^+ = \sqrt{\sum_{j=1}^n \left(v_{ij} - v_j^+ \right)}, \quad i = 1, 2, \ldots, m$ (12)

Distance from negative-ideal solution: $d_i^- = \sqrt{\sum_{j=1}^n \left(v_{ij} - v_j^+ \right)}, \quad i = 1, 2, \ldots, m$ (13)

Ranking in order to the highest closeness to the negative-ideal condition:

$$CL_i^* = \frac{d_i^-}{d_i^- + d_i^+}, \quad 0 \leq CL_i^* \leq 1 \, \& \, i = 1, 2, \ldots, m$$ (14)

2.3 Case Studies

To carry the analysis by rough set and TOPSIS a set of data is required such as case studies. The selection of the case studies has been done in a way that considers the main possible attributes/factors confronting in rainwater harvesting systems. The first and second case studies (CS1–CS2) are taken from the study carried out by Herrmann and Schimida [7] that considers the development and performance of rainwater utilization systems in Germany, and specifically in Bochum where the mean annual precipitation is 787 mm. More in detail, the data considered there for CS1, are related to a one-family house (with an effective roof area of 150 m², 4 persons, combined demand of 160 l/d, a storage volume of 6 m³ and an additional retention volume of 15 m³, with a covering efficiency of 98%). The CS2 case refers to a multi-story building (CS2) (with an effective roof area of 320 m², 24 persons, toilet flushing demand of 480 l/d, a storage volume of 14 m³ and an additional retention volume of 35 m³).

A study carried out by Domènech and Saurí [13] was considered to select the third (CS3) and the fourth (CS4) case study. Both case studies are in Sant Cugant del Vallès – Spain. More in detail, CS3 is a single-family house with a rooftop catchment area of

107 m², 3 residents, a toilet and laundry usage demand of 27 LCD and 16 LCD, respectively. While CS4 refers to a multi-family building with a rooftop catchment area of 625 m², 42 residents, a toilet and laundry usage demand of 30 LCD and 16 LCD, respectively. For CS3, we considered a model scenario in which a tank of 13 m³ can meet 80% of the combined demand of toilet flushing and laundry; while for CS4 the model scenario is a tank of 31 m³ covering 59.5% of the combined demand of toilet flushing and laundry.

Case studies CS5, CS6 and CS7 are considered from the study of Palla et al. [5]. These three case studies are located in Genoa (Italy) with a mean annual precipitation of 1340 mm. More in detail, CS5 is a 4-flat house with 16 inhabitants, a roof area of 420 m², an annual toilet flushing demand of 233.6 m³/y and a tank capacity of 14 m³. CS6 is a 6-flat house with 24 inhabitants, a roof area of 420 m², an annual toilet flushing demand of 350.4 m³/y and a tank capacity of 21 m³. CS7 is a condominium with 32 inhabitants, a roof area of 680 m2, an annual toilet flushing demand of 467.2 m³/y and a tank capacity of 28 m³. For the three case studies, the modeling results show a water saving efficiency of 0.83, 0.79 and 0.76 for CS5, CS6 and CS7, respectively.

The CS8 case study considers the values of the example of application found in Campisano and Modica [25], where a 4 people residential house with a daily toilet flushing demand of 0.168 m³, a roof area of 186 m², daily precipitation of 0.0018 m and a size tank of 2.93 m³, achieving a water saving of 67%, was considered.

The CS9 case study refers to a real case study at University of Calabria in Southern Italy [15], where a tank of 1.5 m³ is located at the base of an university building to collect the water for an experimental full-scale green roof implementation and the water is reused to irrigate the same green roof in the dry period. Finally, three hypothetical cases, CS10, CS11 and CS12, have been considered to evaluate remain factors under different conditions. Specifically, CS10 represent the hypothetical implementation of RWH systems in the old town of Cosenza, CS11 in the old town of Matera, and CS12 in the new area of Quattromiglia, in the town of Rende, respectively.

Table 1. Case studies

Locations of case studies	Case study
Bochum – Germany [7]	CS1
Bochum – Germany [7]	CS2
Sant Cugant del Vallès – Spain [13]	CS3
Sant Cugant del Vallès – Spain [13]	CS4
Genoa – Italy [5]	CS5
Genoa – Italy [5]	CS6
Genoa – Italy [5]	CS7
Sicily – Italy [25]	CS8
University of Calabria (Rende) – Italy [15]	CS9
Old town of Cosenza – Italy	CS10
Old town of Matera – Italy	CS11
New Area of Quattromiglia (Rende) – Italy	CS12

3   Results

3.1   Application of Rough Set Theory in Optimizing Rainwater-Harvesting Systems

In real projects there is an enormous quantity of data that may be considered and this makes hard the decision making process. In Rough Set method, all data should be categorized. In this regard, the correlated RWH attributes must be determined. All the information about the case studies in form of determined attributes, classification of attributes and decision level for each of them should be provided. According to the data gathered in Table 1, the main RWH attributes have been determined and are presented in Table 2. The attributes have been classified based on 3 classes which denote the suitability conditions for decisions and are high (H), medium (M) and low (L).

Table 2. Conditional attributes for ranking decisions of selected case studies

Conditional attributes	Classification of individual situations	Decision
(a) Building type	1 - One-family building/one-office with garden	H
	2 - One-family building/one-office without garden	M
	3 - Multi-family building/multi offices with garden	
	4 - Multi-family building/multi offices without garden	L
(b) Roof type	1 - Slope roof with tiles, corrugated plastic, plastic or metal sheets	H
	2 - Flat roof covered with plastic or metal sheet	
	3 - Flat roof with concrete or asphalt slabs	
	4 - Flat roofs with gravel	M
	5 - Extensive green roof	
	6 - Intensive green roof	L
(c) Roof Size (collecting area)	1 - Big surface capture (>250)	H
	2 - Average surface capture (100–250)	M
	3 - Small surface capture (<100)	L
(d) The age of the building	1 - New building	H
	2 - Average age building	M
	3 - Old building	L
(e) Average Annual Precipitation	1 - >700 or <300	H
	2 - 300 to 700	M
(f) Number of building residents (based on demand)	1 - 1 to 4	H
	2 - 5 to 20	M
	3 - more 20	L

(*continued*)

Table 2. *(continued)*

Conditional attributes	Classification of individual situations	Decision
(g) Density of city (based on the location of the barrels)	1 - Low Density	H
	2 - Medium Density	M
	3 - High Density	L
(h) Type of urban area	1 - New urban area	H
	2 - Average age urban area	M
	3 - Old urban area	L
(i) Demand usage (m^3/y)	1 - combined usage	H
	2 - one usage (toilet flushing or garden irrigation)	M
	3 - laundry	L
	4 - terrace cleaning	
	5 - car washing	
(j) Tank size	1 - Big Tank (>20 m^3)	H
	2 - Medium Tank (6–20 m^3)	M
	3 - Low Tank (<6 m^3)	L
(k) Economic	1 - Very Economic	H
	2 - Partly Economic	M
	3 - Expensive	L

According to 11 attributes and classes, the selected case studies have been ranked from 1 to 3 and the results are presented in Table 3. For instance, in the first case study (CS1), since the conditional attribute (a) that is "Building type" is "One-family building", the highest rank, i.e. 3, has been selected. Since the table represents the correlation between the case studies and conditional attributes it is named "decision rules".

Table 3. Data Inspection for analysis of site selection decision ranking

Case study	Conditional attributes											Decision level
	a	b	c	d	e	f	g	h	i	j	k	
CS1	3	3	2	2	3	3	1	2	3	1	2	M
CS2	1	3	3	2	3	1	1	2	2	1	2	M
CS3	3	3	2	2	2	3	1	2	3	2	1	H
CS4	2	3	3	2	2	1	1	2	3	3	1	H
CS5	2	3	3	2	3	2	1	2	2	2	2	M
CS6	2	3	3	2	3	1	1	2	2	3	2	H
CS7	2	3	3	2	3	1	1	2	2	3	2	H
CS8	2	3	2	2	2	3	2	2	2	1	3	M
CS9	2	2	2	2	3	1	2	2	2	1	3	H
CS10	1	3	2	1	3	2	1	1	2	2	1	L
CS11	1	3	1	1	3	2	1	1	3	3	1	L
CS12	1	3	3	3	3	1	1	3	2	3	2	H

All the decisions and attributes have been checked to find out the existence of non-deterministic rules that means that for case studies of similar attributes decisions are different. The number of non-deterministic rules in Table 3 was zero. Therefore, the number of conditional attributes is sufficient for determining the decisions. The found reduction in the data is presented in Table 4.

Table 4. The founded reduction

Raw	Reduction	Raw	Reduction	Raw	Reduction
1	{a, b, d, j}	13:	{a, b, i, j}	25:	{a, f, j}
2	{a, c, e, j}	14:	{a, b, j, k}	26:	{d, e, g, j}
3	{b, c, e, j}	15:	{a, b, h, j}	27:	{c, f, j}
4:	{a, b, c, j}	16:	{b, c, h, j}	28:	{b, h, i, j}
5:	{b, d, e, j}	17:	{a, e, h, j}	29:	{e, g, h, j}
6:	{a, d, e, j}	18:	{b, e, h, j}	30:	{f, g, h, j}
7:	{b, c, d, j}	19:	{c, e, i, j}	31:	{b, c, i, j}
8:	{a, d, e, f}	20:	{a, e, i, j}	32:	{b, d, j, k}
9:	{a, e, f, h}	21:	{d, f, g, j}	33:	{b, h, j, k}
10:	{b, d, f, j}	22:	{b, d, i, j}	34:	{e, j, k}
11:	{a, d, f, k}	23:	{c, e, g, j}	35:	{f, j, k}
12:	{a, f, h, k}	24:	{b, f, h, j}		

After deriving the reducts, the decision rules can be achieved by overlaying the determined reducts on the data. A decision table free of contradiction and determining a minimal decision algorithm can be achieved after elimination of all non-deterministic rules that was zero in this study. The contradictions have been analyzed based on the conditional attributes and the decisions in selected case studies. Moreover, if the attributes do not cause any contradiction they can be removed. In order to check the impact of an attribute on the result, the attributes can be removed one by one. For example, if the conditional attributes (a), (b), (c), and (d) be removed, the decision rules of case studies 1 and 2 might be contradictory that means the decision levels of these two case studies are subordinate to the mentioned conditional attributes. In this regard, and after elimination of all removable conditional attribute or classes, the minimal decision algorithm has been obtained and is presented in Table 5.

Table 5. Minimal decision algorithm

Rules	
Rule 1	$(d = 1) \Rightarrow (\text{Decision} = L)$
Rule 2	$(b = 3) \,\&\, (j = 1) \Rightarrow (\text{Decision} = M)$
Rule 3	$(a = 2) \,\&\, (j = 2) \Rightarrow (\text{Decision} = M)$
Rule 4	$(f = 1) \,\&\, (j = 3) \Rightarrow (\text{Decision} = H)$
Rule 5	$(b = 2) \Rightarrow (\text{Decision} = H)$
Rule 6	$(a = 3) \,\&\, (e = 2) \Rightarrow (\text{Decision} = H)$

The validation of the rules is presented in Tables 6 and 7. It must be mentioned that, since the selected case studies are only 12, the accuracy of the rules might not be high. To be able to extend the result of the method to other similar case studies in RWH systems, more field data might be required.

Table 6. Confusion matrix (sum over 10 passes)

	1	2	3	None
1	2	0	0	0
2	0	2	2	0
3	1	4	1	0

Table 7. Average accuracy [%]

	Correct	Incorrect	None
Total	40.00 ±43.59	60.00 ± 43.59	0.00 ± 0.00
1	20.00 ± 40.00	0.00 ± 0.00	0.00 ± 0.00
2	15.00 ± 32.02	15.00 ± 32.02	0.00 ± 0.00
3	5.00 ± 15.00	45.00 ± 47.17	0.00 ± 0.00

3.2   Application of TOPSIS in Ranking of Rainwater-Harvesting Systems

In this section, the TOPSIS method has been used to rank the selected case studies and the results are compared with those of a simple ranking. The results of simple ranking are presented in Tables 8 and 9 and those obtained by TOPSIS method in Tables 10, 11 and 12.

Table 8. Values of each attribute for each case study

Case study	Attributes										
	a	b	c	d	e	f	g	h	i	j	k
CS1	3	3	150	2	787	4	1	2	58.4	6	2
CS2	1	3	320	2	787	24	1	2	175.2	14	2
CS3	3	3	107	2	612	3	1	2	47.1	13	1
CS4	2	3	625	2	612	42	1	2	705.2	31	1
CS5	2	3	420	2	1086	16	1	2	233.6	14	2
CS6	2	3	420	2	1086	24	1	2	350.4	21	2
CS7	2	3	680	2	1086	32	1	2	467.2	28	2
CS8	2	3	186	2	657	4	2	2	61.3	2.93	3

Table 9. Simple ranking of the factors for each case study

Case study	Rank of attributes											Sum of ranking	Final rank
	a	b	c	d	e	f	g	h	i	j	k		
CS1	1	1	6	1	2	2	4	1	7	6	2	33	6
CS2	3	1	4	1	2	4	4	1	5	4	2	23	4
CS3	1	1	7	1	4	1	1	1	8	5	3	24	5
CS4	2	1	2	1	4	6	1	1	1	1	3	18	2
CS5	2	1	3	1	1	3	3	1	4	4	2	19	3
CS6	2	1	3	1	1	4	3	1	3	3	2	18	2
CS7	2	1	1	1	1	5	3	1	2	2	2	17	1
CS8	2	1	5	1	3	2	2	1	6	7	1	23	4

Table 10. TOPSIS matrix without scale (Normalized)

Case study	Attributes										
	a	b	c	d	e	f	g	h	i	j	k
CS1	0.48	0.35	0.13	0.35	0.32	0.11	0.46	0.35	0.06	0.11	0.36
CS2	0.16	0.35	0.27	0.35	0.32	0.26	0.46	0.35	0.18	0.26	0.36
CS3	0.48	0.35	0.09	0.35	0.25	0.25	0.03	0.35	0.05	0.25	0.18
CS4	0.32	0.35	0.53	0.35	0.25	0.59	0.03	0.35	0.73	0.59	0.18
CS5	0.32	0.35	0.36	0.35	0.44	0.26	0.44	0.35	0.24	0.26	0.36
CS6	0.32	0.35	0.36	0.35	0.44	0.40	0.44	0.35	0.36	0.40	0.36
CS7	0.32	0.35	0.58	0.35	0.44	0.53	0.44	0.35	0.48	0.53	0.36
CS8	0.32	0.35	0.16	0.35	0.27	0.06	0.04	0.35	0.06	0.06	0.54

Table 11. TOPSIS matrix without scale and equal weighted

Case study	Attributes										
	a	b	c	d	e	f	g	h	i	j	k
CS1	0.044	0.032	0.012	0.032	0.029	0.010	0.041	0.032	0.005	0.010	0.033
CS2	0.015	0.032	0.025	0.032	0.029	0.024	0.041	0.032	0.016	0.024	0.033
CS3	0.044	0.032	0.008	0.032	0.023	0.022	0.003	0.032	0.004	0.022	0.016
CS4	0.029	0.032	0.048	0.032	0.023	0.053	0.003	0.032	0.066	0.053	0.016
CS5	0.029	0.032	0.033	0.032	0.040	0.024	0.040	0.032	0.022	0.024	0.033
CS6	0.029	0.032	0.033	0.032	0.040	0.036	0.040	0.032	0.033	0.036	0.033
CS7	0.029	0.032	0.053	0.032	0.040	0.048	0.040	0.032	0.044	0.048	0.033
CS8	0.029	0.032	0.014	0.032	0.024	0.005	0.003	0.032	0.006	0.005	0.049
V+	**0.044**	**0.032**	**0.053**	**0.032**	**0.040**	**0.053**	**0.003**	**0.032**	**0.066**	**0.053**	**0.049**
V−	**0.015**	**0.032**	**0.008**	**0.032**	**0.023**	**0.005**	**0.041**	**0.032**	**0.004**	**0.005**	**0.016**

Table 12. Ranking in TOPSIS based on higher CL and comparison with simple ranking

Case study	d+	d−	CL	Rank in TOPSIS	Simple rank method	Decision level
CS4	0.040	0.109	0.731	1	2	H
CS7	0.049	0.090	0.645	2	1	H
CS6	0.063	0.064	0.504	**3**	3	**H**
CS5	0.077	0.049	0.389	**4**	4	**M**
CS3	0.095	0.054	0.363	5	7	H
CS8	0.101	0.053	0.342	**6**	6	**M**
CS2	0.088	0.038	0.302	7	5	M
CS1	0.105	0.035	0.250	**8**	8	**M**

Despite the fact that in some case studies the difference in ranking methods are minor since in TOPSIS all correlated attributes and the differences among the values are taken into consideration the results could be more accurate.

4   Conclusions

There are many benefits in Rainwater harvesting (RWH) systems mainly water saving for non-potable water uses and surface runoff mitigation. Moreover, the collected rainwater can be re-used for several purposes including green roofs and garden, flushing toilets, etc. Our analysis showed that, in previous studies, some important factors in the analysis and the feasibility of the RWH systems have been neglected and the optimization of RWH systems mostly is limited to optimize the size of the tankers according to hydrological and hydraulic analysis and in some cases, this is combined with an economic analysis. In this paper, multi-objective optimization approaches have been considered for comparing algorithms and evaluating the performance of alternatives to identify the ideal solution. For this, a limited set of data extracted from several case studies has been used. The selection of the case studies has been made considering the main possible attributes/factors confronting in rainwater harvesting systems. The results show that the Rough Set method is a suitable way for analysis of RWH systems and the outcomes can be useful in decision making by decreasing the uncertainties, reducing the cost, and increasing the efficiency. According to the results, TOPSIS ranking method showed good agreement with the decision levels in the case studies. This may be due to the consideration of all correlated attributes and of the differences between the values of this ranking method. In conclusion, the numerical optimization of RWH systems may improve previous studies in the field. Moreover, the Rough Set and TOPSIS methods could be applied as a useful approach in rainwater harvesting systems investigations and provide an additional tool to identify the optimal system and the best site.

Acknowledgements. The study was co-funded by the "Innovative Building Envelope through Smart Technology (I-Best)" Project funded by the Italian National Operational Program "Enterprise and Competitiveness" 2014–2020 ERDF – I AXIS "Innovation" - Action 1.1.3 – "Support for the economic enhancement of innovation through experimentation and the adoption of innovative solutions in processes, products and organizational formulas, as well as through the financing of the industrialization of research results".

References

1. Palermo, S.A., Zischg, J., Sitzenfrei, R., Rauch, W., Piro, P.: Parameter sensitivity of a microscale hydrodynamic model. In: Mannina, G. (ed.) UDM 2018. GET, pp. 982–987. Springer, Cham (2019). https://doi.org/10.1007/978-3-319-99867-1_169
2. Maiolo, M., Pantusa, D.: Sustainable water management index, SWaM_Index. Cogent Eng. **6**(1), 1603817 (2019). https://doi.org/10.1080/23311916.2019.1603817
3. Christian Amos, C., Rahman, A., Mwangi Gathenya, J.: Economic analysis and feasibility of rainwater harvesting systems in urban and peri-urban environments: a review of the global situation with a special focus on Australia and Kenya. Water **8**(4), 149 (2016). https://doi.org/10.3390/w8040149
4. Campisano, A., et al.: Urban rainwater harvesting systems: research, implementation and future perspectives. Water Res. **115**, 195–209 (2017). https://doi.org/10.1016/j.watres.2017.02.056
5. Palla, A., Gnecco, I., La Barbera, P.: The impact of domestic rainwater harvesting systems in storm water runoff mitigation at the urban block scale. J. Environ. Manag. **191**, 297–305 (2017). https://doi.org/10.1016/j.jenvman.2017.01.025
6. Petrucci, G., Deroubaix, J.F., De Gouvello, B., Deutsch, J.C., Bompard, P., Tassin, B.: Rainwater harvesting to control stormwater runoff in suburban areas. An experimental case-study. Urban Water J. **9**(1), 45–55 (2012). https://doi.org/10.1080/1573062X.2011.633610
7. Herrmann, T., Schmida, U.: Rainwater utilisation in Germany: efficiency, dimensioning, hydraulic and environmental aspects. Urban Water **1**(4), 307–316 (2000). https://doi.org/10.1016/S1462-0758(00)00024-8
8. GhaffarianHoseini, A., Tookey, J., GhaffarianHoseini, A., Yusoff, S.M., Hassan, N.B.: State of the art of rainwater harvesting systems towards promoting green built environments: a review. Desalin. Water Treat. **57**(1), 95–104 (2016). https://doi.org/10.1080/19443994.2015.1021097
9. Oberascher, M., Zischg, J., Palermo, S.A., Kinzel, C., Rauch, W., Sitzenfrei, R.: Smart rain barrels: advanced LID management through measurement and control. In: Mannina, G. (ed.) UDM 2018. GET, pp. 777–782. Springer, Cham (2019). https://doi.org/10.1007/978-3-319-99867-1_134
10. Li, Z., Boyle, F., Reynolds, A.: Rainwater harvesting and greywater treatment systems for domestic application in Ireland. Desalination **260**(1–3), 1–8 (2010). https://doi.org/10.1016/j.desal.2010.05.035
11. Campisano, A., Modica, C.: Rainwater harvesting as source control option to reduce roof runoff peaks to downstream drainage systems. J. Hydroinform. **18**(1), 23–32 (2016). https://doi.org/10.2166/hydro.2015.133
12. Jones, M.P., Hunt, W.F.: Performance of rainwater harvesting systems in the southeastern United States. Resour. Conserv. Recycl. **54**(10), 623–629 (2010). https://doi.org/10.1016/j.resconrec.2009.11.002
13. Domènech, L., Saurí, D.: A comparative appraisal of the use of rainwater harvesting in single and multi-family buildings of the Metropolitan Area of Barcelona (Spain): social experience, drinking water savings and economic costs. J. Clean. Prod. **19**(6–7), 598–608 (2011). https://doi.org/10.1016/j.jclepro.2010.11.010
14. Cipolla, S.S., Altobelli, M., Maglionico, M.: Decentralized water management: rainwater harvesting, greywater reuse and green roofs within the GST4Water project. In: Multidisciplinary Digital Publishing Institute Proceedings, vol. 2, no. 11, p. 673 (2018). https://doi.org/10.3390/proceedings2110673

15. Piro, P., Turco, M., Palermo, S.A., Principato, F., Brunetti, G.: A comprehensive approach to stormwater management problems in the next generation drainage networks. In: Cicirelli, F., Guerrieri, A., Mastroianni, C., Spezzano, G., Vinci, A. (eds.) The Internet of Things for Smart Urban Ecosystems. IT, pp. 275–304. Springer, Cham (2019). https://doi.org/10.1007/978-3-319-96550-5_12

16. Becciu, G., Raimondi, A., Dresti, C.: Semi-probabilistic design of rainwater tanks: a case study in Northern Italy. Urban Water J. **15**(3), 192–199 (2018). https://doi.org/10.1080/1573062X.2016.1148177

17. Pawlak, Z.: Rough set theory and its applications to data analysis. Cybern. Syst. **29**(7), 661–688 (1998). https://doi.org/10.1080/019697298125470

18. Arabani, M., Sasanian, S., Farmand, Y., Pirouz, M.: Rough-set theory in solving road pavement management problems (Case Study: Ahwaz-Shush Highway). Comput. Res. Prog. Appl. Sci. Eng. (CRPASE) **3**(2), 62–70 (2017)

19. Arabani, M., Pirouz, M., Pirouz, B.: Optimization of geotechnical studies using basic set theory. In: 1st Conference of Civil and Development, Zibakenar, Iran (2012)

20. Hwang, C.L., Yoon, K.P.: Multiple Attributes Decision-Making Methods and Applications. Springer, Berlin (1981). https://doi.org/10.1007/978-3-642-48318-9

21. Balioti, V., Tzimopoulos, C., Evangelides, C.: Multi-criteria decision making using TOPSIS method under fuzzy environment. Appl. Spillway Sel. Proc. **2**, 637 (2018). https://doi.org/10.3390/proceedings2110637

22. Krohling, R.A., Pacheco, A.G.: A-TOPSIS an approach based on TOPSIS for ranking evolutionary algorithms. Procedia Comput. Sci. **55**, 308–317 (2015). https://doi.org/10.1016/j.procs.2015.07.054

23. Haghshenas, S.S., Neshaei, M.A.L., Pourkazem, P., Haghshenas, S.S.: The risk assessment of dam construction projects using fuzzy TOPSIS (case study: Alavian Earth Dam). Civil Eng. J. **2**(4), 158–167 (2016). https://doi.org/10.28991/cej-2016-00000022

24. Haghshenas, S.S., Mikaeil, R., Haghshenas, S.S., Naghadehi, M.Z., Moghadam, P.S.: Fuzzy and classical MCDM techniques to rank the slope stabilization methods in a rock-fill reservoir dam. Civil Eng. J. **3**(6), 382–394 (2017). https://doi.org/10.28991/cej-2017-00000099

25. Campisano, A., Modica, C.: Optimal sizing of storage tanks for domestic rainwater harvesting in Sicily. Resour. Conserv. Recycl. **63**, 9–16 (2012). https://doi.org/10.1016/j.resconrec.2012.03.007

New Mathematical Optimization Approaches for LID Systems

Behrouz Pirouz[1]([✉]) [ID], Stefania Anna Palermo[2] [ID],
Michele Turco[3] [ID], and Patrizia Piro[2] [ID]

[1] Department of Mechanical, Energy and Management Engineering,
University of Calabria, Rende, CS, Italy
behrouz.pirouz@unical.it
[2] Department of Civil Engineering, University of Calabria, Rende, CS, Italy
{stefania.palermo,patrizia.piro}@unical.it
[3] Department of Environmental and Chemical Engineering,
University of Calabria, Rende, Italy
michele.turco@unical.it

Abstract. Urbanization affects ecosystem health and downstream communities by changing the natural flow regime. In this context, Low Impact Development (LID) systems are important tools in sustainable development. There are many aspects in design and operation of LID systems and the choice of the selected LID and its location in the basin can affect the results. In this regard, the Mathematical Optimization Approaches can be an ideal method to optimize LIDs use. Here we consider the application of TOPSIS (Technique for Order Preference by Similarity to Ideal Solution) and Rough Set theory (multiple attributes decision-making method). An advantage of using the Rough Set method in LID systems is that the selected decisions are explicit, and the method is not limited by restrictive assumptions. This new mathematical optimization approach for LID systems improves previous studies on this subject. Moreover, it provides an additional tool for the analysis of essential attributes to select and optimize the best LID system for a project.

Keywords: Optimization · LID · Rough Set Theory · TOPSIS method

1 Introduction

Low Impact Development (LID) systems are important tools in sustainable development. There are different types of LID practices such as green roofs, green wall, bioretention cell, permeable pavements, rainwater harvesting systems, etc. In the design and operation of LID systems, many components must be considered. When choosing and designing the best LID practices many factors can affect their efficiency in terms of flooding risk mitigation, water quality improvement, water saving, urban heat island reduction, air pollution decreasing. Previous studies are generally limited to focus the design of a type of LID based on a determined scenario and location. However, these elements are not fixed.

In this research, the application of mathematical optimization approaches by TOPSIS ranking method and attributes analysis by Rough Set in evaluation of alternative decisions is described.

Y. D. Sergeyev and D. E. Kvasov (Eds.): NUMTA 2019, LNCS 11973, pp. 583–595, 2020.
https://doi.org/10.1007/978-3-030-39081-5_50

1.1   LID Systems

In the last decades, the combined effect of urbanization and climate change produced several environmental adverse impacts as flooding risk, water and air pollution, water scarcity, and urban heat island effect [1–5].

Specifically, from a hydrologic point of view, land use changes led to significant alterations of the natural hydrological cycle with a reduction of infiltration and evapotranspiration rates, a decreasing of groundwater recharge and baseflow, and a growth of surface runoff. This phenomenon makes cities vulnerable to local flooding and combined sewer overflows (CSOs) [6, 7].

In response to these issues, a novel approach, which considers the implementation of sustainable nature-based stormwater management strategies, is necessary [8].

In this regard, several strategies, known as Low Impact Development (LID) systems [9], have gained increasing popularity. These techniques are small-scale units, which provide several benefits at multiple scales [10–13].

LID systems consist of a series of facilities - such as green roofs, green wall, rain gardens, bioretention cells, permeable pavements, infiltration trenches, rain water harvesting systems, and so on - whose purpose is to infiltrate, filter, store, evaporate, and detain runoff close to its source [14–16].

Therefore, their implementation in an urban area can create a complex system of temporal storage facilities allowing an innovative urban stormwater management approach [17].

In recent years, the implementation of these sustainable solutions has attracted widespread interest from researchers, urban planners, and city managers researchers [18] and several studies have been devoted to LID design, benefits, and simulation models of their behavior.

From these studies, as pointed out by Eckart et al. [15], it appears that: (i) the appropriate design of LID is affected by the site location (in terms of soil type/conditions, plants, rainfall regime, land use types and other meteorological and hydrological properties); (ii) the optimal selection and placement of LID is one of the most crucial factors to consider to achieve the maximum efficiency at the minimum cost.

The implementation of these systems can improve the sustainable development and mitigate several other environmental impacts in addition to urban flooding risk, such as water and air pollution, water scarcity, urban heat island effect. Therefore, a large number of factors have to be considered during the design process and for the choose of the site location.

After a deeper analysis of different studies carried out on LID solutions, several factors affecting their efficiency at multiple scale have been identified and are here considered here for optimizing their use.

2   Methodology

In this paper, we discuss the application of TOPSIS (Technique for Order Preference by Similarity to Ideal Solution) and Rough Set theory (multiple attributes decision-making method) in optimization of LIDs.

2.1 Rough Set Theory

Low Impact Development approaches are Multi-Objectives and there are many uncertainties in the selection of these objectives [19]. The Rough Set method, introduced by Pawlak [20] in 1982, can be used as an excellent mathematical tool in analysis of vague description of decisions such as quality of data, means variation or uncertainty that follows from information. The rough sets philosophy is based on assumption that, for every objects a certain information (data, knowledge) exists, that can be expressed as attributes [21].

With respect to the available data, the objects with the same description are indiscernible and a set of indiscernible objects, named elementary set, can be provided to build knowledge about a real system. Deal with quantitative or qualitative data depends on the input information and before the analysis, it is necessary to remove the irregularities. With respect to the output data, the relevance of particular attributes and their subsets to the quality of approximation can be acquired [22].

In this regard, the induction-based approach can provide clear rules for decision-makers (DMs) in the form of "if…, then…". The concept of approximation space in rough set method can be described in a given approximation space as follows:

$$apr = (U, A) \tag{1}$$

U is a finite and non-empty set and A is set of attributes in the given space. Based on the approximation space, the lower and upper approximations of a set can be defined. Let X be a subset of U and the upper and lower approximation of X in A are:

$$\overline{apr}(A) = \{x | x \in U, U/\text{ind}(A) \subset X\} \tag{2}$$

$$\underline{apr}(A) = \{x | x \in U, U/\text{ind}(A) \cap X \neq \varphi\} \tag{3}$$

where:

$$U/\text{ind}(a) = \{(x_i, x_j) \in U \times U, f(x_i, a) = f(x_j, a), \forall a \in A\} \tag{4}$$

Equation (2), that is the best upper approximation of X in A, means the minimum composed set in A containing X, and Eq. (3), that is the best lower approximation, means the maximum composed set in A contained in X. The graphical illustration of approximations in the rough set method is shown in Fig. 1.

The boundary represent as:

$$BN(A) = \overline{apr}(A) - \underline{apr}(A) \tag{5}$$

The reducts and decision rules can be defined as follows:

The reduct RED (B), is a minimal set of attributes $B \subseteq A$ such that rB (U) = rA (U). rB (U) indicates the quality of approximation of U by B.

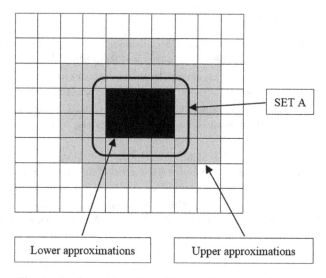

Fig. 1. Graphical illustration of the rough set approximations

The equation is:

$$r_B(\mathrm{U}) = \frac{\sum card(\underline{B}(X_i))}{card(U)} \tag{6}$$

After providing the result of reducts, the decision rules can be derived by using the overlaying of the reducts on the information systems. An expressed decision rule can be as follow:

$$\varphi \Rightarrow \theta \tag{7}$$

where:

φ is the conjunction of elementary conditions,
$\Rightarrow$ Represents indicates
θ represents the disjunction of elementary decisions.

2.2   TOPSIS Method

TOPSIS (Technique for Order Preference by Similarity to Ideal Solution) is a method developed by Hwang and Yoon in 1981 to solve the ranking and compare problems [23]. The ranking in this method is made according to the similarity to ideal solution [24, 25]. The TOPSIS method can be applied to a wide range of multi-attribute decision making with several attributes [26–28]. The graphical illustration of the TOPSIS methodology is presented in Fig. 2.

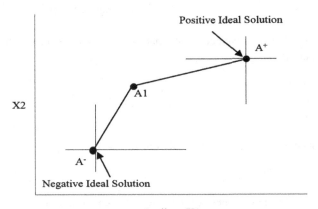

Fig. 2. Graphical illustration of the TOPSIS methodology, (A+ represent the ideal point, A− represent the negative-ideal point)

The ranking by TOPSIS is carried out through seven steps:
Step 1: Create the matrix

$$(n_{ij})m \times n \tag{8}$$

Step 2: Construct the normalized decision matrix

$$N = n_{ij} = \frac{a_{ij}}{\sqrt{\sum_{i}^{m} = 1 a_{ij}^2}}, i = 1, 2, \ldots, m \quad \& \quad j = 1, 2, \ldots, n \tag{9}$$

Step 3: Construct the weighted normalized decision matrix

$$V = N \times Wn \times n \tag{10}$$

Step 4: Determine the solutions (the ideal and negative-ideal solutions)

$$\text{The ideal solution } A^+ = \left\{ \left\langle max_i \left(a_{ij} | j \in J_- \right) \right\rangle, \left\langle min_i \left(a_{ij} | j \in J_+ \right) \right\rangle \right\} \tag{11}$$

$$\text{The negative−ideal solutions } A^- = \left\{ \left\langle min_i \left(a_{ij} | j \in J_- \right) \right\rangle, \left\langle max_i \left(a_{ij} | j \in J_+ \right) \right\rangle \right\} \tag{12}$$

where,

$$J_+ = \{j = q, 2, \ldots, n|j\} \text{ Associated with positive impact criteria}$$
$$J_- = \{j = q, 2, \ldots, n|j\} \text{ Associated with negative impact criteria}$$

Step 5: Determine the distance of alternatives v_{ij} from the ideal solution and the negative-ideal solutions

The distance from ideal solution: $d_i^+ = \sqrt{\sum_{j=1}^{n} \left(v_{ij} - v_j^+ \right)}, \quad i = 1, 2, \ldots, m$ (13)

The distance from negative$-$ideal solution : $d_i^- = \sqrt{\sum_{j=1}^{n} \left(v_{ij} - v_j^+ \right)},$ (14)
$i = 1, 2, \ldots, m$

Step 6: Calculate the closeness to the negative-ideal condition, CL^*

$$CL_i^* = \frac{d_i^-}{d_i^- + d_i^+}, \quad 0 \le CL_i^* \le 1 \,\&\, i = 1, 2, \ldots, m \tag{15}$$

where,

$CL_i^* = 1$ *if* The solution has the best condition and means the highest rank

$CL_i^* = 0$ *if* The solution has the worst condition and means the lowest rank

Step 7: Rank in order to the highest CL^*.

2.3 Case Studies

In order to perform the analysis of LID practices by Rough Set and TOPSIS a set of data is required. For this purpose, sites have been chosen according to different conditions and considering the main factors in the LIDs selection. The selected site along with hypothesis are presented in Table 1.

Table 1. Site selection hypothesis to rank the attributes

Site	Description
S1	An high urbanized area (78% impervious surfaces), with extremely wet condition, average age, presenting High Flooding Risk, Low Water Scarcity, High Water Pollution, Medium Heat Island Effect, High Air Pollution, supposing to replace 50% of impervious area with LID (permeable pavement and green roof); and that the inhabitants are supported to implement them, thus partially economic
S2	An urbanized area (65% of impervious surfaces), with moderately wet condition, old area, presenting High Flooding Risk, Medium Water Scarcity, High Water Pollution, High Heat Island Effect, High Air Pollution, where it is not possible to implement LID; not economic
S3	A peri-urban area (45% of impervious surfaces), with extremely dry condition, new area, presenting Low Flooding Risk, High Water Scarcity, Medium Water Pollution, Low Heat Island Effect, Medium Air Pollution, where it is possible to implement combined usage of LID (rainwater harvesting and biofiltration trench) for a percentage of 65%; high economic
S4	A peri-urban area (40% of impervious surfaces), with moderate climate condition, average age, presenting Medium Flooding Risk, Medium Water Scarcity, Low Water Pollution, Low Heat Island Effect, Medium Air Pollution, where it is possible to implement combined usage of LID (rainwater harvesting and green wall) for a percentage of 40%; partially economic
S5	A normal urban area (70% of impervious surfaces), with extremely dry climate condition, average age, presenting Low Flooding Risk, High Water Scarcity, Low Water Pollution, High Heat Island Effect, High Air Pollution, where it is possible to implement combined usage of LID (rainwater harvesting, green roof and green wall) for a percentage of 50%; partially economic
S6	A rural area (20% of impervious surface), with extremely dry climate condition, average area, presenting s Low Flooding Risk, High Water Scarcity, Low Water Pollution, Low Heat Island Effect, Low Air Pollution, where it would be better to implement rainwater harvesting for a percentage of 70%; partially economic

As it can be recognized, in Table 1. The main factors considered to define the case studies include climate condition, urbanization level, age of site, flood risk, water scarcity, water and air pollution, Heat Island Effect, LID implementation percentage and economical condition.

3   Results

3.1   Application of Rough Set Theory in Optimizing LIDs

In Rough Set method, at the first stage all factors must be categorized in form of attributes that are classified. This, and the decision level for each of them, can be done according to previous standards, papers or experts. We identified 12 conditional decision attributes for LIDs that are presented in Tables 2 and 3.

Table 2. Conditional decision attributes in selected sites, part 1

Conditional attributes	Classification	Decision
(a) Type of area	1- Urban	H
	2- Peri-urban	M
	3- Rural	L
(b) Climate condition based on precipitation	1- Extremely wet or extremely dry	H
	2- Moderately wet or moderately dry	M
(c) Age of area	1- New area	H
	2- Average age area	M
	3- Old area	L
(d) Impervious surfaces of the selected area	1- >75% of area	H
	2- 50%–75% of area	M
	3- 25%–50% of area	L
	4- <25% of area	N
(e) Flooding risk	1- High risk	H
	2- Medium risk	M
	3- Low risk	L
(f) Water scarcity	1- High water scarcity	H
	2- Medium water scarcity	M
	3- Low water scarcity	L
(g) Water pollution	1- High water pollution	H
	2- Medium water pollution	M
	3- Low water pollution	L

Table 3. Conditional decision attributes in selected sites, part 2

Conditional attributes	Classification	Decision
(h) Urban heat island effect	1- High heat island effect	H
	2- Medium heat island effect	M
	3- Low heat island effect	L
(i) Air pollution	1- High air pollution	H
	2- Medium air pollution	M
	3- Low air pollution	L
(j) LID percentage implementation	1- >60% of area	H
	2- 30%–60% of area	M
	3- <30% of area	L
(k) LID usage	1- Combined implementation (more than 1 LID)	H
	2- Only one type implementation	M
	3- No implementation	N
(l) Economic	1- Economic	H
	2- Partially economic	M
	3- Not economic	L

As it is clear from Tables 2 and 3 the conditional attributes have been categorized in at most four classes with high (H), medium (M), low (L) and no (N) suitability conditions for decisions. As next step, the selected sites presented in Table 1 have been ranked according to the attributes of Tables 2 and 3. The result is presented in Table 4. In Table 4 the ranks are based on the conditional attribute in the site and for ranks from 0 to 4. For example, in site 1 (S1), the conditional attribute (a) that is "Type of Area" is "Urban", and therefore the highest rank (3) has been selected.

Table 4. Ranking of decisions attributes in selected sites

Sites	Conditional attributes												Decision levels
	a	b	c	d	e	f	g	h	i	j	k	l	
S1	3	3	2	3	3	1	3	2	3	2	3	2	H
S2	3	2	1	2	3	2	3	3	3	1	0	1	L
S3	2	3	3	1	1	3	2	1	2	3	3	3	H
S4	2	2	2	1	2	2	1	1	2	2	3	2	M
S5	3	3	2	2	1	3	1	3	3	2	3	2	H
S6	1	3	2	0	1	3	1	1	1	3	2	2	M

3.2   Determining Minimal Decision Algorithm

The finding of a minimal decision algorithm can be achieved by analysis of decision rules in all sites and finding non-deterministic rules or sites, thus assigning different decision levels for different sites under the same class for every conditional attribute. In this regard, at the next stage, the contradiction between data can be checked according to the conditional attributes ranks and the correlated decisions and some of the attributes can be removed if this does not cause any contradiction. For this purpose, the conditional attribute can be eliminated one by one to check the role of that attribute in the result. For instance, if the conditional attributes (a), (b), and (c) were removed, the decision rules of sites S1 and S2 might be contradictory to each other, which means that the decision levels of these two sites are subordinate to one of the conditional attributes. Finally, and after checking all of the combinations, the minimal decision algorithm can be extracted that will be the main rules for each level of decisions. The determined rules according to the attributes data of Table 4 and the six case studies of Table 1 are as follows:

$$\text{Rule 1} : (c = 1) \Rightarrow (D = L)$$
$$\text{Rule 2} : (g = 1) \& (h = 1) \Rightarrow (D = M)$$
$$\text{Rule 3} : (b = 3) \& (k = 3) \Rightarrow (D = H)$$

Thus, by the determined rules it is possible to make a decision with minimum attributes. However, these rules have been generated by the decision on the selected case studies. Therefore, by increasing the number of the case studies, the accuracy of the rules will increase and after checking the validation and accuracy of the rules, it will be possible to extend the rules for other sites.

3.3   Application of TOPSIS in Selection of LID Practices

In this section, the application of TOPSIS method in ranking the selected sites for LIDs is presented. To provide the final ranking, the data of Tables 2 and 3 that are the decision attributes for the selected sites have been used. The result of the process as explained in the methodology section are presented in Tables 5, 6 and 7.

Table 5. Attributes matrix without scale

Case study	Attributes											
	a	b	c	d	e	f	g	h	i	j	k	l
S1	0.50	0.45	0.39	0.69	0.60	0.17	0.60	0.40	0.50	0.36	0.47	0.54
S2	0.50	0.30	0.20	0.46	0.60	0.33	0.60	0.60	0.50	0.18	0.00	0.18
S3	0.33	0.45	0.59	0.23	0.20	0.50	0.40	0.20	0.33	0.54	0.47	0.54
S4	0.33	0.30	0.39	0.23	0.40	0.33	0.20	0.20	0.33	0.36	0.47	0.36
S5	0.50	0.45	0.39	0.46	0.20	0.50	0.20	0.60	0.50	0.36	0.47	0.36
S6	0.17	0.45	0.39	0.00	0.20	0.50	0.20	0.20	0.17	0.54	0.32	0.36

Table 6. Attributes matrix without scale and with equal weight

Case study	Attributes											
	a	b	c	d	e	f	g	h	i	j	k	l
S1	0.04	0.04	0.03	0.06	0.05	0.01	0.05	0.03	0.04	0.03	0.04	0.04
S2	0.04	0.03	0.02	0.04	0.05	0.03	0.05	0.05	0.04	0.01	0.00	0.01
S3	0.03	0.04	0.05	0.02	0.02	0.04	0.03	0.02	0.03	0.04	0.04	0.04
S4	0.03	0.03	0.03	0.02	0.03	0.03	0.02	0.02	0.03	0.03	0.04	0.03
S5	0.04	0.04	0.03	0.04	0.02	0.04	0.02	0.05	0.04	0.03	0.04	0.03
S6	0.01	0.04	0.03	0.00	0.02	0.04	0.02	0.02	0.01	0.04	0.03	0.03

Table 7. Final ranking in based on higher CL and comparison with initial decision level

Case study	d+	d−	CL	Rank by TOPSIS	The initial decision level
S1	0.039	0.102	0.722	1	H
S5	0.057	0.086	0.598	2	M
S3	0.066	0.080	0.548	3	H
S2	0.072	0.081	0.530	4	H
S4	0.074	0.059	0.446	5	L
S6	0.094	0.055	0.369	6	M

The results of ranking by TOPSIS in Table 7 and the comparisons with the initial decision levels represent that in some sites the results are the same such as S1, S3, S2 and S4. However, the decision levels in S6 was M but it is at the end of TOPSIs ranking. This might be based on the consideration of all correlated factors at the same time and more exact.

4 Conclusions

The analysis showed that in design and operation of the LID systems, many components can be considered and that in choosing the best LID practices and implementation percentage many factors can affect the results. In previous studies, generally, the attentions was limited to design a type of LID based on determined scenarios and for a selected site that both are not fixed elements and might need to be optimized.

The results of this application of mathematical optimization approaches by TOPSIS ranking method and attributes analysis by Rough Set in evaluation of alternative decisions confirm the advantage of using these methods. The rules provided by Rough Set method can improve the designing decisions. The generated decisions are explicit, and the results are not limited to restrictive assumptions. With consideration of more case studies, more stringent decision rules can be achieved. Moreover, the final ranks of TOPSIS shows the advantages in compared with simple ranking method.

In conclusion, the new presented mathematical optimization approaches can improve the previous studies about LIDs. They provide an additional tool for engineers in analysis of essential attributes to select and optimize the best LID system for a project and accordingly define the scenarios and hydrologic or hydraulic modeling. This means that the presented methods would provide a baseline for decision-making and would increase the efficiency of the systems and decrease the project cost by preventing uncertainties.

Acknowledgements. The study was co-funded by the Italian Operational Project (PON)—Research and Competitiveness for the convergence regions 2007/2013—I Axis "Support to structural changes" operative objective 4.1.1.1. "Scientific-technological generators of transformation processes of the productive system and creation of new sectors" Action II: "Interventions to support industrial research".

References

1. Zhang, D.L., Shou, Y.X., Dickerson, R.R.: Upstream urbanization exacerbates urban heat island effects. Geophys. Res. Lett. **36**(24), 1–5 (2009)
2. Haase, D.: Effects of urbanisation on the water balance–A long-term trajectory. Environ. Impact Assess. Rev. **29**(4), 211–219 (2009)
3. Jacob, D.J., Winner, D.A.: Effect of climate change on air quality. Atmos. Environ. **43**(1), 51–63 (2009)
4. Piro, P., et al.: Flood risk mitigation in a Mediterranean urban area: the case study of Rossano Scalo (CS – Calabria, Italy). In: Mannina, G. (ed.) UDM 2018. GET, pp. 339–343. Springer, Cham (2019). https://doi.org/10.1007/978-3-319-99867-1_57

5. Miller, J.D., Hutchins, M.: The impacts of urbanisation and climate change on urban flooding and urban water quality: a review of the evidence concerning the United Kingdom. J. Hydrol.: Reg. Stud. **12**, 345–362 (2017)

6. Piro, P., Turco, M., Palermo, S.A., Principato, F., Brunetti, G.: A comprehensive approach to stormwater management problems in the next generation drainage networks. In: Cicirelli, F., Guerrieri, A., Mastroianni, C., Spezzano, G., Vinci, A. (eds.) The Internet of Things for Smart Urban Ecosystems. IT, pp. 275–304. Springer, Cham (2019). https://doi.org/10.1007/978-3-319-96550-5_12

7. Raimondi, A., Becciu, G.: On pre-filling probability of flood control detention facilities. Urban Water J. **12**, 344–351 (2015)

8. Zischg, J., Rogers, B., Gunn, A., Rauch, W., Sitzenfrei, R.: Future trajectories of urban drainage systems: a simple exploratory modeling approach for assessing socio-technical transitions. Sci. Total Environ. **651**, 1709–1719 (2019)

9. Fletcher, T.D., et al.: SUDS, LID, BMPs, WSUD and more–The evolution and application of terminology surrounding urban drainage. Urban Water J. **12**(7), 525–542 (2015)

10. Razzaghmanesh, M., Beecham, S., Salemi, T.: The role of green roofs in mitigating Urban Heat Island effects in the metropolitan area of Adelaide, South Australia. Urban For. Urban Green. **15**, 89–102 (2016)

11. Maiolo, M., Carini, M., Capano, G., Piro, P.: Synthetic sustainability index (SSI) based on life cycle assessment approach of low impact development in the Mediterranean area. Cogent Eng. **4**(1), 1410272 (2017)

12. Zahmatkesh, Z., Burian, S.J., Karamouz, M., Tavakol-Davani, H., Goharian, E.: Low-impact development practices to mitigate climate change effects on urban stormwater runoff: case study of New York City. J. Irrig. Drain. Eng. **141**(1), 04014043 (2014)

13. Jia, H., Yao, H., Shaw, L.Y.: Advances in LID BMPs research and practice for urban runoff control in China. Front. Environ. Sci. Eng. **7**(5), 709–720 (2013)

14. Piro, P., Carbone, M., Morimanno, F., Palermo, S.A.: Simple flowmeter device for LID systems: from laboratory procedure to full-scale implementation. Flow Meas. Instrum. **65**, 240–249 (2019)

15. Turco, M., Brunetti, G., Carbone, M., Piro, P.: Modelling the hydraulic behaviour of permeable pavements through a reservoir element model. In: International Multidisciplinary Scientific Geo Conference: SGEM: Surveying Geology & mining Ecology Management, vol. 18, pp. 507–514 (2018)

16. Eckart, K., McPhee, Z., Bolisetti, T.: Performance and implementation of low impact development–a review. Sci. Total Environ. **607**, 413–432 (2017)

17. Palermo, S.A., Zischg, J., Sitzenfrei, R., Rauch, W., Piro, P.: Parameter sensitivity of a microscale hydrodynamic model. In: Mannina, G. (ed.) UDM 2018. GET, pp. 982–987. Springer, Cham (2019). https://doi.org/10.1007/978-3-319-99867-1_169

18. Jia, H., et al.: Field monitoring of a LID-BMP treatment train system in China. Environ. Monit. Assess. **187**(6), 373 (2015)

19. Zhang, G., Hamlett, J.M., Reed, P., Tang, Y.: Multi-objective optimization of low impact development designs in an urbanizing watershed. Open J. Optim. **2**, 95–108 (2013)

20. Pawlak, Z.: Rough set theory and its applications. J. Telecommun. Inf. Technol. **3**, 7–10 (2002)

21. Arabani, M., Sasanian, S., Farmand, Y., Pirouz, M.: Rough-set theory in solving road pavement management problems (case study: Ahwaz-Shush Highway). Comput. Res. Prog. Appl. Sci. Eng. (CRPASE) **3**(2), 62–70 (2017)

22. Arabani, M., Pirouz, M., Pirouz, B.: Geotechnical investigation optimization using rough set theory. In: 9th International Congress on Civil Engineering (9ICCE), Isfahan, Iran (2012)

23. Hwang, C.L., Yoon, K.P.: Multiple Attributes Decision-Making Methods and Applications. Springer, Berlin (1981). https://doi.org/10.1007/978-3-642-48318-9

24. Haghshenas, S.S., Neshaei, M.A.L., Pourkazem, P., Haghshenas, S.S.: The risk assessment of dam construction projects using fuzzy TOPSIS (case study: Alavian Earth Dam). Civ. Eng. J. 2(4), 158–167 (2016)

25. Balioti, V., Tzimopoulos, C., Evangelides, C.: Multi-criteria decision making using TOPSIS method under fuzzy environment, application in spillway selection. In: Multidisciplinary Digital Publishing Institute Proceedings, vol. 2, p. 637 (2018)

26. İç, Y.T.: A TOPSIS based design of experiment approach to assess company ranking. Appl. Math. Comput. 227, 630–647 (2014)

27. Krohling, R.A., Pacheco, A.G.: A-TOPSIS an approach based on TOPSIS for ranking evolutionary algorithms. Procedia Comput. Sci. 55, 308–317 (2015)

28. Haghshenas, S.S., Mikaeil, R., Haghshenas, S.S., Naghadehi, M.Z., Moghadam, P.S.: Fuzzy and classical MCDM techniques to rank the slope stabilization methods in a rock-fill reservoir dam. Civ. Eng. J. 3(6), 382–394 (2017)

Evaluation of an Integrated Seasonal Forecast System for Agricultural Water Management in Mediterranean Regions

Alfonso Senatore[1]([✉]) [iD], Domenico Fuoco[1], Antonella Sanna[2] [iD],
Andrea Borrelli[2] [iD], Giuseppe Mendicino[1] [iD], and Silvio Gualdi[2] [iD]

[1] University of Calabria, Rende, CS, Italy
alfonso.senatore@unical.it
[2] Euro-Mediterranean Center on Climate Change, Bologna, Italy
antonella.sanna@cmcc.it

Abstract. The Euro-Mediterranean Center on Climate Change (CM-CC) seasonal forecasting system, based on the global coupled model CMCC-CM, performs seasonal forecasts every month, producing several ensemble integrations conducted for the following 6 months. In this study, a performance evaluation of the skills of this system is performed in two neighbouring Mediterranean medium-small size catchments located in Southern Italy, the Crati river and the Coscile river, whose hydrological cycles are particularly important for agricultural purposes.

Initially, the performance of the system is evaluated comparing observed and simulated precipitation and temperature anomalies in the irrigation periods of the years 2011–2017. Forecasts issued on April 1st (i.e., at the beginning of the irrigation period) are evaluated, considering two lead times (first and second trimester). Afterward, the seasonal forecasts are integrated into a complete meteo-hydrological system. Precipitation and temperature provided by the global model are ingested in the spatially distributed and physically based In-STRHyM (Intermediate Space-Time Resolution Hydrological Model) model, which analyzes the hydrological impact of the seasonal forecasts.

Though the predicted precipitation and temperature anomalies are not highly correlated with observations, the integrated seasonal forecast for the hydrological variables provides significant correlations between observed and predicted anomalies, especially concerning mean discharge (>0.65). Overall, the system showed to provide useful insights for agricultural water management in the study area.

Keywords: Meteo-hydrological system · Seasonal forecast · Agricultural water management

1 Introduction

Filling the so-called subseasonal to seasonal (S2S) forecast gap is a challenging issue for both weather and climate science communities. Nevertheless, improv-

Y. D. Sergeyev and D. E. Kvasov (Eds.): NUMTA 2019, LNCS 11973, pp. 596–603, 2020.
https://doi.org/10.1007/978-3-030-39081-5_51

ing meteorological forecast skill up to leads out to one or more months portends invaluable social and economic benefits, particularly in the field of water resources management. Therefore, several initiatives have been running in the last years at the international level, like the joint World Weather Research Program (WWRP) and World Climate Research Program (WCRP) S2S Project. Furthermore, since more than a decade many operational centers produce routine dynamical seasonal forecasts [1]. For example, in the US the North American Multi-Model Ensemble (NMME) real-time forecasts are incorporated as part of National Oceanic and Atmospheric Administration's (NOAA) operational forecast suite. In Europe, several operational forecast systems contribute to the Copernicus Climate Change Service (C3S). Among them, the Euro-Mediterranean Center on Climate Change (CMCC) seasonal forecast system [2] performs seasonal forecasts every month, producing a number of ensemble integrations conducted with the coupled model for the following 6 months. The skill of the CMCC forecast system has been thoroughly tested from decadal [3] to seasonal lead times, evaluating different aspects like the impact of initial conditions [4] or the capability to represent the extratropical low-frequency variability [5,6].

The usefulness of seasonal forecasts for practical applications is particularly highlighted when they are coupled to impact models, dealing with various aspects of the terrestrial hydrological cycle. This kind of integrated systems can provide timely, clear and useful information to support policy decisions, with obvious socio-economic benefits.

Unlike many meteorological variables (e.g., precipitation), time series data of hydrological variables like river discharge or aquifer levels are subjected to a much higher autocorrelation. The hydrologic persistence is a well known phenomenon [7] and has been widely exploited for predicting catchments seasonal behavior only with statistical tools, not necessarily relying on weather forecasts (e.g., [8]). Nevertheless, recent advances in seasonal forecasts have allowed statistical models to be profitably combined with seasonal forecasts (hybrid statistical-dynamical forecasts) and, in several cases, to directly couple seasonal models to cropping system models for predicting crop yield [9] or to hydrological models for streamflow forecast (e.g., [10,11]), aimed either at overall water resources management or at specific objectives, e.g., snow accumulation forecast, reservoir operations or inland waterway transport management.

In this study, the performance evaluation of a forecasting chain based on the CMCC seasonal forecast system and the spatially distributed, physically based In-STRHyM (Intermediate Space-Time Resolution Hydrological Model) model [12] was performed in two neighbouring Mediterranean medium-small size catchments located in Southern Italy, where water management (especially in the irrigation summer period) is related to remarkable agricultural activities. The evaluation concerned the irrigation periods of the years 2011–2017, for which the seasonal forecasts are integrated in a complete meteo-hydrological system aimed at providing useful indications for agricultural water management in the area.

Next Sect. 2 deals with the description of the study area, the seasonal forecast system, the hydrological model and the methodology followed for the experiment setup, while Sect. 3 briefly summarizes the preliminary results. Finally, conclusions (Sect. 4) will sketch out future steps for improving the forecast skill of the integrated system.

2   Data and Methods

2.1   Study Area

The study area includes two catchments in Calabria (southern Italy), namely the Crati river catchment closed at the Santa Sofia gauge station and the Coscile river catchment closed at the Cammarata gauge station (1281 km^2 and 405 km^2, respectively; Fig. 1). These catchments are both sub-basins of the Crati river basin, the biggest in Calabria, and cover about 70% of its extent. Their hydrological cycles are particularly important for agricultural purposes, because they feed the Sibari plain, the main agricultural area of a region which is often subject to drought [13].

The mean altitude of the Crati river catchment is 672 m a.s.l. (ranging from 1856 m a.s.l. on the Sila plateau to 67 m a.s.l.), while for the Coscile river catchment is 698 m a.s.l. (from 2248 m a.s.l. on the Pollino Massif to 82 m a.s.l.). The whole Crati river basin has a Mediterranean sub-humid climate. The real-time monitoring network managed by the "Centro Funzionale Multirischi" of the Calabrian Regional Agency for the Protection of the Environment provides about 15 temperature and 20 precipitation recording stations within the catchments. In the analyzed period 2011–2017, the mean annual accumulated precipitation values for the Crati river catchment and the Coscile river catchment were 1196 mm and 1097 mm (64.2% and 64.1% occurring between October and March), and the mean annual temperatures were 13.0 °C and 13.3 °C, respectively.

2.2   The CMCC Seasonal Forecasting System

The CMCC-SPS3 is the Seasonal Prediction System developed at CMCC to perform seasonal forecasts operationally. It is based on the CMCC-CM2 coupled model, which consists of several independent model components, simulating at the same time the Earth's atmosphere, ocean, land, river routing and sea-ice.

In the SPS implementation, the number of vertical levels is increased from 30 to 46 in order to better resolve (up to 0.3 hPa) the stratosphere, which recent evidences show to be of greater importance in the seasonal forecast framework [14,15]. Horizontal resolution is about 1×1 degree for the atmosphere and land, 0.5×0.5 for river routing and 0.25×0.25 for ocean and sea-ice.

In order to take account of the uncertainty associated with the initial conditions, 50 perturbations of the initial state are built. Perturbations are generated by combining the initial states of the three main components: the atmosphere (10 perturbations of the initial state), the ocean (8 perturbations) and the land surface (3 perturbations). Out of the possible 240, 50 unique combinations are randomly chosen to initialize the CMCC-SPS3 each month.

The complete system has been run in retrospective forecast way over the 24-year-period 1993–2016 (hindcast period), with a slightly reduced ensemble population of 40 members. The results of the hindcasts were used to construct a reference climatology. Forecasts are presented in terms of anomalies with respect to the reference climatology of the considered season (to take into account also model climate drift). Because the model systematic error (that is the bias with respect to observed climate) is one of the main sources of uncertainties, removing the climatology allows analysing seasonal tendencies, where SPS exhibits positive skills in reproducing the observed variability.

The CMCC-SPS3 contributes to C3S climate service, together with other four providers (ECMWF, DWD, MeteoFrance and MetOffice), in the multi-model seasonal forecast system and it is also a recognized WMO Global Producing Center for Long-Range Forecasts. More details on model initialization, performances, skills and shortcomings can be found in [2].

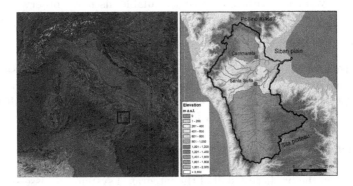

Fig. 1. Study area. The right panel zooms the framed area in the left panel: the whole Crati river basin is shown (black contours), together with the Crati river catchment closed at the Santa Sofia gauge station (yellow) and the Coscile river catchment closed at the Cammarata gauge station (orange) (Color figure online)

2.3   The In-STRHyM Hydrological Model

The original distributed hydrological model In-STRHyM [12] has been rewritten in Python and integrated in the QGIS environment as a suitable tool both for analysis over long periods, such as climate change scenarios, and for operational monitoring and subseasonal to seasonal forecasts of water resource availability (drought risk). In-STRHyM estimates the hydrological balance using a physically-based modeling approach that explicitly simulates the hydrological processes of canopy interception, snow accumulation and melting, actual evapotranspiration, root zone infiltration and percolation towards groundwater, runoff generation due to either infiltration or saturation excess, surface and subsurface

(1D) runoff routing and baseflow (2D) routing. Being typically applied with a time resolution of 1 day and a spatial resolution of 1 km^2, In-STRHyM allows the distributed estimate of the main components of the hydrological balance together with the discharge in selected outlets.

By default the model input is totally distributed, even if in the absence of data some parameters can be assumed constant in the catchment. For the validation of the In-STRHyM model in Calabria, the parameters relating to the topographical features, the soil hydraulic properties, the depth of the root zone and the soil uses were spatially distributed. Regarding the meteorological input, daily accumulated precipitation and mean, minimum and maximum temperature grids are required (in the In-STRHyM model version used in this study, a simplified method has been adopted for estimating the reference evapotranspiration [16]). Furthermore, for the purpose of actual evapotranspiration estimate, remote sensing maps (MODIS) are used providing space-time distributed information of the NDVI (Normalized Difference Vegetation Index) and LAI (Leaf Area Index) vegetation indices. The model was calibrated on the Crati river catchment from October 1961 to December 1966 (Nash-Sutcliffe coefficient E^2 = 0.83), then validated with more recent available discharge observations [12].

2.4   Experiment Setup

The performance evaluation concerned seven consecutive irrigation seasons (6 months, from April to September) from 2011 to 2017. The 40-member ensemble hindcasts (from 2011 to 2016) and forecast (2017) issued by CMCC-SPS3 on April 1st of each of these years for the following 6 months were evaluated comparing observed and simulated precipitation and temperature anomalies. In this preliminary evaluation, the reference period for calculating the anomalies is 2011–2016, both for observations and hindcasts/forecast.

Afterwards, the predictive skill of the integrated system was assessed against the hydrological impact. The In-STRHyM model was first run for the two catchments from October 1st 2009 (thus ensuring more than one year of spin-up time) to September 30th 2017, using as meteorological input the daily precipitation and temperature grids achieved by the spatial interpolation of the available observations. Then, the state of the hydrological model on April 1st of each of the years 2011–2017 was imposed as the initial condition for the hydrological forecast in the same year. The meteorological driver for the following six months was obtained perturbing the mean daily grids of observed precipitation and temperatures with the anomalies (i.e., differences from the average values) calculated at the monthly time scale for the 40 members of the ensemble forecast. Finally, the anomalies of the main output variables of the hydrological model calculated using the observed meteorological input were compared with those achieved through the ensemble seasonal forecasts.

3   Preliminary Results

The seasonal forecasts carried out by CMCC-SPS3 do not exhibit significant skill with the investigated meteorological variables. The correlation between observed three-month accumulated precipitation anomalies and the corresponding average values of the predicted anomalies during the 7 analyzed years is rather low for both lead 0 (i.e., months of April, May and June - AMJ, Fig. 2a) and lead 1 (i.e., months of July, August and September - JAS, Fig. 2b). Similar results were achieved comparing temperatures (Figs. 2c,d). Specifically, the model did not seem able to correctly predict the intense summer drought occurred in 2017 in the analyzed area (and in most of southern Italy). These results are affected, at least partially, by the too short reference period used for calculating anomalies and, mainly, by the reduced extent of the analyzed area (for the analysis of the meteorological variables 11 grid points were used).

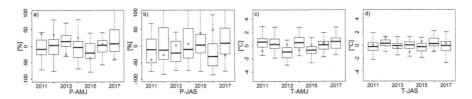

Fig. 2. Precipitation (P) and mean temperature (Tmed) anomalies achieved using observations and SPS predictions for the Crati river catchment. In the box-and-whiskers plots, the first and third quantiles are represented by the lower and upper edges of each box, respectively, while the horizontal thin bar in the box is the median value. The tip of the upper (lower) whisker represents the maximum (minimum) value

Hydrological variables provided more promising results, thanks also to their higher persistence. Figures 3 and 4 show the monthly evolution of the observed and predicted actual evapotranspiration (ET) and mean discharge (Q_{med}) for the Crati river catchment in the summer period (from April to September) from 2011 to 2017, highlighting a general slight underestimation of predictions. However, to evaluate the system performance, in agreement with the SPS approach, it is worth it to focus on the observed and predicted anomalies. ET largely depends on predicted temperature and is relatively unaffected by the catchment "memory": in the AMJ period, the correlation between observed anomalies and the medians of the predicted anomalies is −0.21, but almost in all cases the deviations from zero are very low, such as the variances of the forecasts. In JAS, the variance of the predicted anomalies is higher, but the median is better correlated with observations (+0.61).

On the other hand, Q_{med} is a much more persistent variable and the predicted precipitation, especially in the dry summer period, can only relatively affect mean streamflow characteristics. For the Crati river catchment, anomalies correlation is high both in AMJ (+0.66) and, quite surprisingly, in JAS (+0.95).

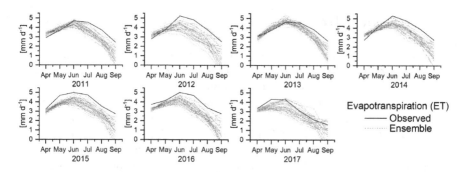

Fig. 3. Monthly averaged daily ET evolution for each year from 2011 to 2017 achieved using observed and predicted meteorological drivers for the Crati river catchment

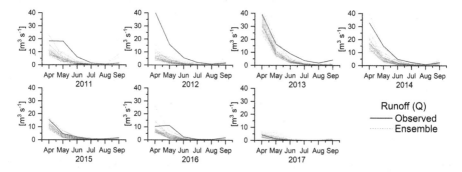

Fig. 4. Same as Fig. 3, but for Q_{med}

Such promising results highlight the potential of the proposed approach for water resources seasonal planning and management, and pave the way for further more detailed analyses.

4   Conclusions

An integrated meteo-hydrological seasonal forecast system in a Mediterranean region is described in this paper, together with a preliminary evaluation of its performance. First results highlight both weak points and interesting insights. Further investigation is required, future work will be directed towards several aspects, e.g.: improving both the reference and application periods for calculating anomalies; enlarging the ensemble members and the models involved; testing dynamical downscaling of the seasonal forecasts, possibly with two-way coupled atmospheric-hydrological models [17]. These analyses will be addressed to exploit at its best the potential of the seasonal forecast systems, to reduce forecast uncertainty and to concur to the long-term objective of developing valuable tools for supporting water resources management.

References

1. National Academies of Sciences: Engineering, and Medicine: Next Generation Earth System Prediction: Strategies for Subseasonal to Seasonal Forecasts. The National Academies Press, Washington, DC (2016)
2. Sanna, A., Borrelli, A., Athanasiadis, P., et al.: The CMCC Seasonal Prediction System. CMCC Research Paper 285 (2017)
3. Bellucci, A., Haarsma, R., et al.: An assessment of a multi-model ensemble of decadal climate predictions. Clim. Dyn. **44**, 2787–2806 (2015)
4. Materia, S., Borrelli, A., Bellucci, A., et al.: Impact of atmosphere and land surface initial conditions on seasonal forecasts of global surface temperature. J. Clim. **27**(24), 9253–9271 (2014)
5. Athanasiadis, P.J., et al.: The representation of atmospheric blocking and the associated low-frequency variability in two seasonal prediction systems. J. Clim. **27**(24), 9082–9100 (2014)
6. Athanasiadis, P.J., et al.: A multisystem view of wintertime NAO seasonal predictions. J. Clim. **30**(4), 1461–1475 (2017)
7. Koutsoyiannis, D.: Hydrologic persistence and the hurst phenomenon. In: Lehr, J.H., Keeley, J. (eds.) Water Encyclopedia. Surface and Agricultural Water, vol. 4. Wiley, New York (2005)
8. Mendicino, G., Senatore, A., Versace, P.: A Groundwater Resource Index (GRI) for drought monitoring and forecasting in a Mediterranean climate. J. Hydrol. **357**(3–4), 282–302 (2008)
9. Jha, P.A., Athanasiadis, P., Gualdi, S., et al.: Using daily data from seasonal forecasts in dynamic crop models for yield prediction: a case study for rice in Nepal's Terai. Agr. Forest Meteorol. **265**, 349–358 (2019)
10. Candogan Yossef, N., van Beek, R., Weerts, A., Winsemius, H., Bierkens, M.F.P.: Skill of a global forecasting system in seasonal ensemble streamflow prediction. Hydrol. Earth Syst. Sci. **21**, 4103–4114 (2017)
11. Bell, V.A., Davies, H.N., Kay, A.L., Brookshaw, A., Scaife, A.A.: A national-scale seasonal hydrological forecast system: development and evaluation over Britain. Hydrol. Earth Syst. Sci. **21**, 4681–4691 (2017)
12. Senatore, A., Mendicino, G., Smiatek, G., Kunstmann, H.: Regional climate change projections and hydrological impact analysis for a Mediterranean basin in Southern Italy. J. Hydrol. **399**, 70–92 (2011)
13. Maiolo, M., Mendicino, G., Pantusa, D., Senatore, A.: Optimization of drinking water distribution systems in relation to the effects of climate change. Water **9**(10), 83 (2017)
14. Marshall, A.G., Scaife, A.A.: Impact of the QBO on surface winter climate. J. Geophys. Res.-Atmos. **114**, D18110 (2009)
15. Doblas-Reyes, F.J., García-Serrano, J., Lienert, F., Biescas, A.P., Rodrigues, L.R.L.: Seasonal climate predictability and forecasting: status and prospects. WIREs Clim. Change **4**(4), 245–268 (2013)
16. Mendicino, G., Senatore, A.: Regionalization of the hargreaves coefficient for the assessment of distributed reference evapotranspiration in southern Italy. J. Irrig. Drainage Eng. **139**(5), 349–362 (2013)
17. Senatore, A., Mendicino, G., Gochis, D.J., Yu, W., Yates, D.N., Kunstmann, H.: Fully coupled atmosphere-hydrology simulations for the central Mediterranean: Iimpact of enhanced hydrological parameterization for short- and long-timescales. JAMES **7**(4), 1693–1715 (2015)

Optimization of Submarine Outfalls with a Multiport Diffuser Design

Salvatore Sinopoli, Marco Amos Bonora, Gilda Capano, Manuela Carini[ID],
Daniela Pantusa[ID], and Mario Maiolo[✉][ID]

Department of Environmental and Chemical Engineering (DIATIC), Università della
Calabria, Cubo 42/B, 87036 Arcavacata di Rende, CS, Italy
{salvatore.sinopoli,marcoamos.bonora,gilda.capano,manuela.carini,
daniela.pantusa,mario.maiolo}@unical.it

Abstract. Immission of civil sewage into sea is realized to complete
the onshore depurative process or to take out the already purified wastewater from the bathing area, ensuring a good perceived seawater quality.
Anyhow the compliance of the pollutant concentrations limits is necessary to ensure safe bathing. The design of submarine pipes is usually
completed contemplating a diffuser with a series of ports for the reparti-
tion of the wastewater discharge. The real process of pollutants diffusion
into the sea, simulated with complex diffusion-dispersion models in a
motion-field dependent from environmental conditions and drift speeds,
affect the submarine pipe design. A design optimization procedure has
been realized for the marine outfall pipe-diffuser system using a simplified zone model, subjected to a sensitivity analysis on the characteristic
parameter. The method is shown using an example project for the submarine outfall at service for the sewage treatment plant of Belvedere
Marittimo, on the southern Tyrrhenian Sea in Italy.

Keywords: Sea water quality · Submarine outfalls · Optimization
design · Multiport diffuser

1 Introduction

The increase in water demand and the need for environmental protection require
proper management of wastewater which is a scientific and engineering topic of
current interest [12,14,15,19]. Regarding marine outfall systems, they have long
been used for the discharge of industrial and domestic effluent as a means of
increasing its dilution and improving the assimilative capacity of, the receiving
environment [5]. The submarine outfall is a hydraulic structure that has the
purpose of discharge in the receiving marine water body the wastewater after
the wastewater treatment plant. This structure consists of onshore headwork, a
feeder pipeline, and a diffuser section [9]. A pumping station can installed into
the onshore headwork if the effluent discharge cannot take place by gravity, and
the most efficient diffuser design is a set of ports whereby effluent diffuses into

© Springer Nature Switzerland AG 2020
Y. D. Sergeyev and D. E. Kvasov (Eds.): NUMTA 2019, LNCS 11973, pp. 604–618, 2020.
https://doi.org/10.1007/978-3-030-39081-5_52

the submarine environment [10,18]. The characteristics of the effluent must be established taking into account volume, flow variations, and constituents concentrations. These concentrations are established by current regulations. The design of the submarine outfall system must keep into consideration hydraulic aspects, impact on the natural environment and economic criteria. The most efficient hydraulic design is influenced by a uniform flow distribution through all the ports and by the initial dilution, taking into account the necessity of avoiding saline intrusion into the outfall and preventing scour and sedimentation in the surrounding area [5].

Although these systems are mature and in use for many years, advancements in computational power increase, numerical algorithms, and optimization models allow for improved outfalls design by optimizing their performance.

In this context, this paper describes an optimization procedure that has been realized for the submarine outfall pipe-diffuser system using a simplified zone model, subjected to a sensitivity analysis on the characteristic parameter.

The construction of a submarine outfall system is a problem to be faced in different ways. If on the one hand there is the need to respect the environmental parameters, there is at the same time the need to design a diffusion system that can ensure the discharge of the wastewater below the average sea level, and thus guarantee the correct dilution. Furthermore, sizing must be carried out also considering the economic criteria, given that the construction is not that of a classic underground pipe. The proposed optimization procedure, therefore, tries to take into account all these decision-making aspects, thus seeking a solution that can satisfy all three of the aforesaid requirements.

Is then proposed a design criterion that first of all checks the functioning of the hydraulic sizing, and therefore provides the design parameters of the pipeline to a dilution model of the pollutant, used to verifies the diffuser's capacity to operate under the environmental aspect. From the optimization point of view, the chosen pipeline will be the one that reaches the two previous results with the lowest cost in terms of materials.

The optimization procedure and the sensitivity analysis carried out are described below, applying it to the case study of the submarine outfall of Belvedere Marittimo, in the south of Italy.

2   Materials and Methods

2.1   Hydraulic Dimensioning

The hydraulic dimensioning of this type of works is very complex. If the only criterion to be respected in the feeder pipeline is velocity (not exceeding a few m/s, and not much lower than 1 m/s to avoid sedimentation problems), for the section of the diffuser the correct sizing is necessary to respect the hydraulic operation and also to allow the complete formation of the buoyant jets.

To avoid the saline intrusion into the ports the densimetric Froude number must be greater than 1, and for the full mouth operation, it is necessary that the ratio between total ports area and pipe area is about 0.5:

$$\frac{\sum A_{ports}}{\sum A_{Pipe}} = 0,5 \tag{1}$$

Since their introduction in the 1950s, marine outfalls with diffusers have been prone to saline intrusion. Saline intrusion is likely to reduce the efficiency of the outfall and to inhibit the discharge of wastewater through long sea outfalls. Operational difficulties arising from saline intrusion in multi-port sea outfalls are mainly related to seawater corrosion of the pipework and reduced dilution. The problem associated with sea water intrusion into sea outfalls has been high-lighted in a number of papers and over the years several studies on physical and numerical models have been carried out on this topic and its implications [20, 28, 29]. Some studies have instead focused on measures to ensure rapid dilu-tion of effluent with sea water in a wide range of prevailing conditions. Possible design for pre-diluition devices have been studied and experiments have been carried out to test the efficiency of the devices in increasing diluition in different water depths [1].

The additional design criterion to be respected is to try to have a constant flow rate for all the diffuser ports; this criterion is difficult to respect as the last diffuser mouths are reached by a lower flow rate and are subject to a greater head. By accepting the variability of the flow rates, the hydraulic design criterion becomes that of identifying the head necessary for the operation of each mouth of the diffuser.

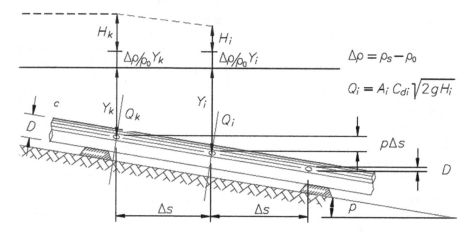

Fig. 1. Hydraulic scheme of the diffuser.

The provided criterion solves the problem taking into account the head losses located in the individual ports of the diffuser, and that starting from the one upstream can guarantee a practically constant outflow. The procedure then starts from the upstream mouth and considers a pipe with a fixed diameter Dc, along which a number of circular ports of equal and fixed diameter D are made, and placed at equal distance between them Δs, so that the turbulences produced by the single jet does not affect on the nearest ones.

The discharge with density ρ_0 coming out of the $i-th$ mouth is regurgitated under a head Y_i of sea water with density ρ, and can be expressed as:

$$Q_i = A_i \, C_{di} \, \sqrt{(2gHi\,)} \qquad (2)$$

where C_{di} is an experimental outflow coefficient that can be derived from the curve obtained by Olivotti [24] as a function of the ratio between the kinetic head on the diffuser upstream of the port and the total head on the port itself (Fig. 2).

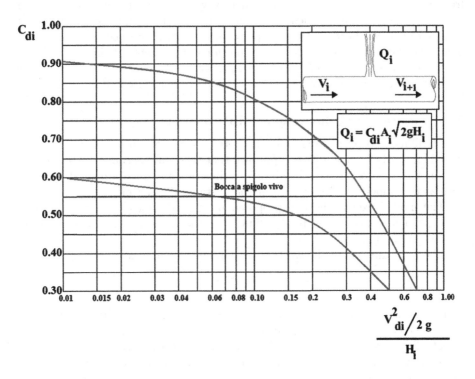

Fig. 2. Hydraulic scheme of the diffuser.

The calculation method consists in fixing a head value on the first mouth, using the value C_{di} obtained from the graph and identifying with the Eq. (2) the value of flow rate spill.

At this point it is possible to calculate the head losses distributed in the section of pipe necessary to reach the next port, placed at a distance Δs, considering that the initial flow must be subtracted from the outflow to the first port:

$$Q_{k-i} = Q_{initial} - \sum Q_{ports} \qquad (3)$$

The value of the head on the next mouth is expressed as:

$$H_k = H_i - (J_i \, \Delta s \,) - \left(\frac{\Delta \rho}{\rho_0} \, \rho\right) \, \Delta \tag{4}$$

The subsequent iterations will then lead to determining the outflow from each mouth, stopping the sizing when the outflow will be approximately equal to that coming from the treatment plant.

However, the total flow rate may not be coincident, due to the arbitrariness of the head chosen on the first port, and therefore does not allow a real solution.

Leaving the parameters of the pipeline unchanged (diameter and material, which also determine the head losses) and the size of the ports, the distance from the coast can be varied, consequently varying the depth of the diffuser (taken into account like a linear function of sea-bed slope and distance from the shoreline) and the head on the first port, until the flow rates coincide, and thus obtaining the correct hydraulic operation, for the diameter of the pipe Dc and for the diameter of the ports D both initially fixed, and thus identifying the number of diffuser ports that must be provided to dispose the flow rate.

2.2   Verification of Environmental Parameters

The fulfillment of environmental parameters is a procedure based on a zone model [4, 21, 24, 27], a simplified approach that takes into account both the engineering parameters related to hydraulic verification and environmental variables such as the velocity of the current or the bacterial decay time.

The model used is the one proposed by [21, 24] which takes into account three phenomena that lead to the dilution of the wastewater, and therefore to the reduction of concentration.

The total dilution of the wastewater can be calculated as:

$$S_c = \frac{C_0}{C_t} \tag{5}$$

with C_0 is the concentration of wastewater coming from the treatment plant and C_t is the concentration near the coast.

The first contribution is given by the so-called Initial Dilution, Si, a process extensively studied over the years both theoretically and experimentally [2, 3, 10, 17, 24].

According to these studies the concentration of pollutant assumes a Gaussian distribution along the axis of the jet, and the phenomenon of Initial Dilution Si can be expressed as a function of two dimensionless parameters $\frac{Y}{D}$ e F, where F is the densimetric Froude number, D is the diameter of the ports and Y is the distance between the jet release point and the average sea surface.

In the present model the formulation of Cederwall [8] is used, which has already been successfully tested in Italy [22, 26]:

$$S_i = 0.54 \, F \left(0.38 \, \frac{Y}{D} \, \frac{1}{F} + 0.66\right)^{\frac{3}{5}} \tag{6}$$

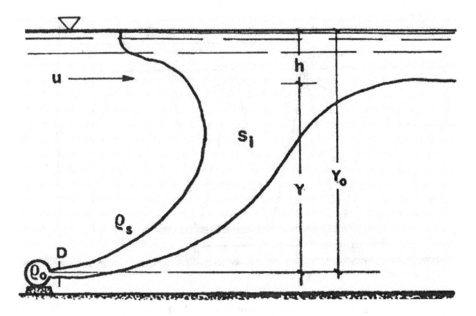

Fig. 3. Schematic representation of the buoyant jet in a quiet environment.

To apply the model to diffusers with multiple ports, the possible mutual interference between two adjacent jets must be taken into account. According to [3] it is possible to approximate the width of the jet as a function of its height, provided that the number of densimetric Froude is in the range 1–20. For this reason, it is possible to impose that the distance Δs between two ports is equal to $\dfrac{Y}{3}$ (Fig. 4), in which Y is again evaluated like the product of sea-bed slope and distance from the shoreline.

The second contribution to the reduction of concentration is given by the Subsequent Dilution, SS, related to the advective-diffusive phenomena and dependent on the velocity of the current, u. Also, in this case, the assumption is that the plume can be represented by a Gaussian type distribution, and with a complete vertical mixing [13] (Fig. 5).

Considering a plane flow, and velocity of the advancement of the plume practically constant to depth, it is possible to assume a constant horizontal dispersion coefficient according to Pearson's law:

$$\in = 0.01 \cdot \left(\frac{b}{10}\right)^{1.333333} \tag{7}$$

where b is the total width of the diffusor.

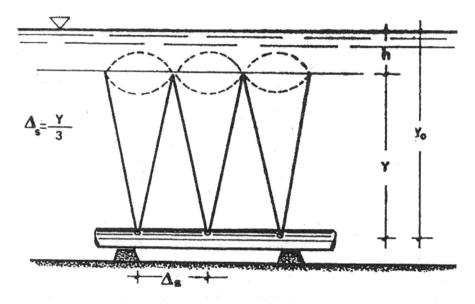

Fig. 4. Schematic representation of the buoyant jet.

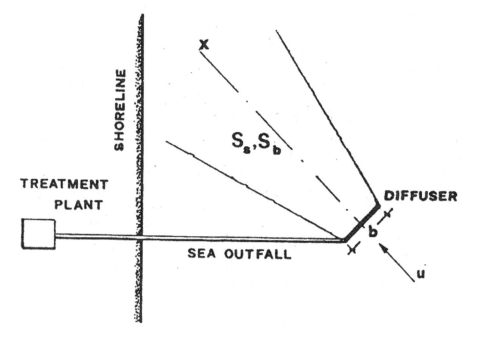

Fig. 5. Schematic representation of the far-field dispersion process.

Considering these assumptions the Subsequent Dilution can be calculated with the formulation proposed by [6,7] as:

$$S_s = \left[erf \sqrt{\frac{1.5}{\left(\left(1 + 0.67\beta\frac{x}{b}\right)^3 - 1 \right)}} \right]^{-1} \tag{8}$$

where β is a dimensionless parameter that recalls $\in$, and it is evaluated as:

$$\beta = 12 \frac{\in}{ub} \tag{9}$$

where u is the average velocity of the current incident on the diffuser.

The third dilution contribution comes instead from the bacterial decay S_b, due to both physical and chemical phenomena, and which can therefore be described by the well-known Chick's law, in which the concentration variation is a function of the drift period, and a parameter, k_e, related to reaction kinetics:

$$\frac{dC}{dt} = -k_e \cdot C \tag{10}$$

By integrating this equation:

$$S_c = \frac{C_0}{Ct} = S_i \cdot S_s \cdot S_b \tag{11}$$

Consequently:

$$S_b = \frac{C_0}{S_i \cdot S_s \cdot C_t} = 10^{\frac{k_e \cdot t}{2.3}} \tag{12}$$

where considering:

$$\frac{2.3}{k_e} = T_{90} \tag{13}$$

finally obtain:

$$S_b = 10^{\frac{t}{T_{90}}} \tag{14}$$

where T_{90} is the time necessary for the disappearance of 90% of the bacteria, and t is the ratio between the distance x that separates the diffuser from the coastline and the average velocity of the current u.

Once the three dilution values are known, it is, therefore, possible to calculate the total dilution value and then evaluate the concentration of coliforms near the coastline as:

$$C_t = \frac{C_0}{S_c} = \frac{C_0}{S_i \cdot S_s \cdot S_b} \tag{15}$$

2.3 Optimization Procedure

Due to the non-continuous nature of the variables relating to the diameters of the pipeline and the ports, the solution to the problem cannot be evaluated with a simple minimum law.

It has been necessary to identify a logical procedure that hallows to evaluate all the possible solutions, first from the hydraulic point of view, and subsequently from the environmental point of view.

To do this, starting on the base offered by [22] once a pipe diameter has been set, a calculation code has been implemented that sets a diameter for the pipe and varies the diameters of the ports, like in example scheme in Fig. 6.

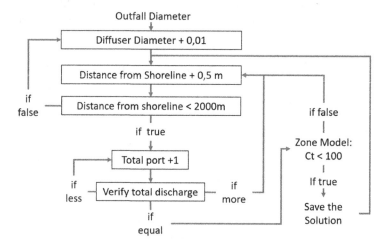

Fig. 6. Schematic representation of the algorithm logical process.

At each iteration, once the pair of diameters has been set, the program proceeds by setting a distance value from the coast. Based on the diameter and distance, the pipe head losses are evaluated, and the flow rate by the diffuser is checked.

If the flow is less than that coming from the treatment plant, the number of ports is increased and the check is repeated. If the flow rate is too large, the distance to the coast is increased and the flow rate check is repeated starting from the minimum number of ports. If, on the other hand, the flow rate is less than a tolerance value equal to that coming from the plant, an environmental check is performed.

If the environmental verification is not passed, the distance from the coast is increased and the number of ports is searched. If the environmental verification is exceeded, the solution is saved, and then, in this case, too the value of the distance from the coast is increased.

When the maximum value allowed for the distance from the coast has been reached, it means that all the possible solutions for the diameter of the previously set port have been evaluated and therefore it is increased.

The algorithm then proceeds with the new pair of diameters, sharing the minimum distance from the coast to evaluate other possible solutions.

3   Application and Results

3.1   Cost Analysis

An application of the optimization procedure has been carried out for the real case of the Belvedere Marittimo submarine outfall, in southern Italy. The pipe is a 600 m long HDPE, with a nominal diameter of 355 mm. The treatment plant is equipped with a pumping station which, together with the difference in height of the loading tank respect to the sea surface, provides a head value at the beginning of the pipeline equal to about 19 m. Regarding the environmental verification, a reduction up to the regulatory values has been considered, with a coliform concentration value of $5.5 \cdot 10^6$ [CFU/100 ml], conveyed by a flow rate of 0.125 m3/s.

To verify the validity of the construction of this structure, a series of diameters have been evaluated based on the velocity in the pipeline. To this aim, considering that velocity lower than 0.8 m/s are not valid for sedimentation problems, the DN280, 315, 355, 400, 450 and 500 have been selected.

Obviously for each diameter different solutions have been found, but the one with the least length, and therefore the lowest cost in terms of materials, has been considered the most valid solution.

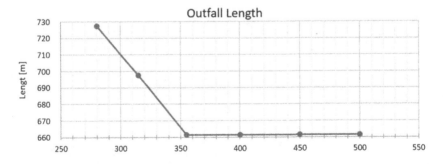

Fig. 7. Comparison of minimum outfall length for each diameter.

As shown in Fig. 7, the system reaches the minimum necessary development in terms of length for the DN355 with a value of 661.3 m, remaining unchanged for higher measurements, and increasing the path for smaller diameters up to 727.2 m of DN280.

All the diffusers corresponding to the shorter overall length have only two ports, and even in this case the minimum for the DN355 is reached.

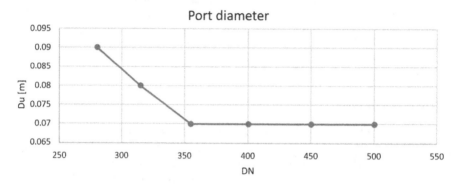

Fig. 8. Comparison of necessary port dimension for each diameter.

$$C_m = 785.94 \; DN^{1.9083} \tag{16}$$

The evaluation of the structure convenience has been made based on the cost per meter formula (16) for HDPE, already used by [11] for estimating the costs of the systems, through which it is possible to identify the cost per meter for every single diameter. Multiplying the values obtained for the respective lengths leads to the cost in materials of the system.

The economically most convenient pipe to be realized is that with the DN280 with €50,356.43, while the pipeline with DN355 has a cost equal to 1.43 times the most convenient with a total cost of €72025.89 (Fig. 9).

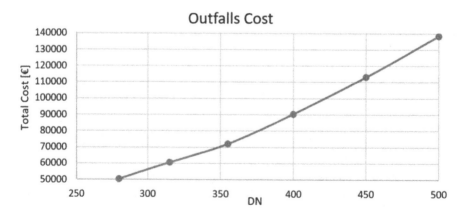

Fig. 9. Comparison of pipe cost for each diameter.

3.2   Sensitivity Analysis

The proposed optimization model contains design values, such as the diameters, the lengths, the capacity, which can be chosen by the designer, also through a procedure like the one followed so far. There are, however, two parameters that are never completely certain, the T_{90} bacterial decay time, which during the day can vary from 2 to more than 20 hours, and the average velocity of the current, which in the stretch of sea in question can vary greatly in direction and module [23, 25].

For this reason, a sensitivity analysis based on the fraction perturbation method [16] has been carried out, to understand how the variations of these two parameters can influence the dilution capacity.

Two systems have been analyzed, the one with DN280 and the one with DN355, since they represent the economic optimum and the system that actually exists, and to make the graphs more readable, the two parameters have been made dimensionless.

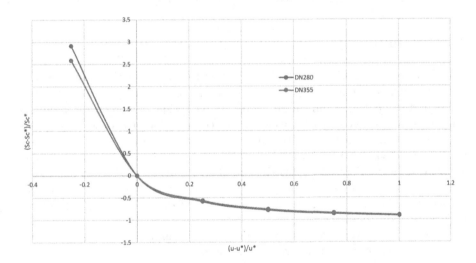

Fig. 10. Sensitivity analysis for u.

The analysis of the velocity graphs shows a significant variation in the dilution capacity, and the two plants undergo the same. In the same way, the response to changes in T_{90} significantly affects the dilution capacity. Compared to the velocity of the current the percentage variations are more moderate, but even in this case, there are no particular differences between the two plants.

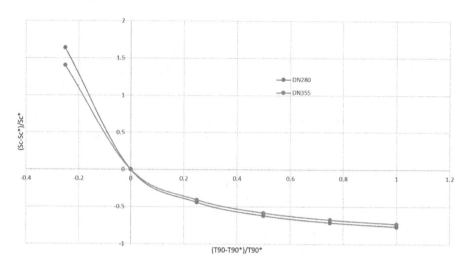

Fig. 11. Sensitivity analysis for $T90$.

4   Conclusions

Submarine outfalls are mature hydraulic structures, but scientific advances lead to the possibility of improving and optimizing the performance of these systems.

In particular, the optimization aspect of design and maintenance is important, as is the economic optimization. In this context, the present work proposes an optimization procedure for the pipe-diffuser system using a simplified zone model subjected to a sensitivity analysis of the characteristic parameters. The application to the Belvedere Marittimo submarine outfalls case study allowed to evaluate the feasibility of the proposed procedure while the sensitivity analysis carried out allowed to understand the influence on the dilution capacity of the variations of two parameters.

The results showed that the procedure is feasible and expeditious and allows to optimize the design of this hydraulic structures.

Aknowledgments. Research supported by the Italian Regional Project (POR CALABRIA FESR 2014–2020): "TEMAR" [J68C17000150006].

References

1. Agg, A.R., White, W.R.: Devices for the pre-dilution of sewage at submerged outfalls. In: Proceedings of the Institution of Civil Engineers, vol. 57, no. 1, part 2, pp. 1–20 (1974)
2. Abraham, G.: Jet diffusion in stagnant ambient fluid. Delft Hydraulics Laboratory **29** (1963)
3. Abraham, G.: Horizontal jets in stagnant fluid of other density. J. Hyrdraulics Div. **91**(4), 139–154 (1965)

4. Benfratello, G.: Scarichi sottomarini: aspetti idraulici e fenomeni di diffusione. In: Proceedings Immissione di acque reflue in mare. A.I.I., Ischia, Italy (1989)
5. Botelho, D.A., et al.: Marine Outfall Systems: Current Trends. Research and Challenges. IWA Specialist Group (2016)
6. Brooks, N.H.: Diffusion of sewage effluent in ocean current. In: Proceedings of the First International Conference on Waste Disposal in the Marine Environment. Pergamon Press (1959)
7. Brooks, N.H.: Conceptual design of submarine outfalls - jet diffusion. In: Water Resources Engineering Educational Series, Program VII, Pollution of Coastal and Estuarine Waters (1970)
8. Cederwall, K.: Gross parameter solution of jets and plumes. J. Hydraulics Div. 101(HY5), 489–509 (1975)
9. Charlton, J.A.: Developments in Hydraulic Engineerings. Sea Outfalls 3(3), 73–128 (1985)
10. Fisher, H.B., Imberger, J., List, E.J., Koh, R.C.Y., Brooks, N.H.: Mixing in Inland and Coastal Waters. Academic Press, New York (1979)
11. Forrest, S., Rustum, R.: Design optimisation of marine wastewater outfalls. Int. J. Sustainable Water Environ. Syst. 8(1), 7–12 (2016)
12. Hernández-Sancho, F., Lamizana-Diallo, B., Mateo-Sagasta, J., Qadir, M.: Economic valuation of wastewater - the cost of action and the cost of no action-United Nations Environment Programme (2015). ISBN: 978-92-807-3474-4
13. Holly, F.M.: Sea outfalls. In: Development in Hydraulic Engineering, vol. 3, no. 1, pp. 1–37. Elsevier Applied Science Publisher, London (1985)
14. Maiolo, M., Pantusa, D.: A proposal for multiple reuse of urban wastewater. J. Water Reuse Desalination 8(4), 468–478 (2017)
15. Maiolo, M., Pantusa, D.: Combined reuse of wastewater and desalination for the management of water systems in conditions of scarcity. Water Ecol. 4(72), 116–126 (2017)
16. Mc Cuen, R.H.: The role of sensitivity analisys in hydrologic modeling. J. Hidrology 18, 37–53 (1973)
17. Occhipinti, A.G.: Bacterial disappearance experiments in Brazilian coastal waters. In: Proceedings of Immissione di acque reflue in mare. A.I.I., Ischia, Italy (1989)
18. Rahm, S.L., Cederwall, K.: Submarine disposal of sewage. In Proceedings on the XI Congress IAHR, Leningrad (1965)
19. Rodriguez-Garcia, G., Molinos-Senante, M., Hospido, A., Hernandey-Sancho, F., Moreira, M.T., Feijoo, G.: Environmental and economic profile of six typologies of wastewater treatment plants. Water Res. 45, 5997–6010 (2011)
20. Shannon, N.R., Mackinnon, P.A., Hamill, G.: Modelling of saline intrusion in marine outfalls. In: Proceedings of the Institution of Civil Engineers. Maritime Engineering, vol. 158, no. MA2, pp. 47–58 (2005)
21. Veltri, P., Maiolo, M.: Un modello di zona per il trattamento marino dei liquami. In: Proceedings on the XXII Corso di Aggiornamento in Tecniche per la Difesa dall'Inquinamento, pp. 417–451. Bios, Cosenza (1991)
22. Veltri, P., Maiolo, M.: Un criterio ambientale per il calcolo degli scarichi dei liquami in mare. In: Proceedings on the XXIII Hydraulics and Hydraulic infrastructures, Firenze (1992)
23. Veltri, P., Maiolo, M.: Environmental aspects in the use of sea outfalls: a sensitivity analysis. In: Marina Technology. Computational Mechanics Publications (1992)
24. Veltri, P., Maiolo, M.: Modelli Matematici di dispersione dei liquami in Mare e Indagini su campo. Ingegneria Sanitaria Ambientale (1993)

25. Veltri, P., Morosini, A.F., Maiolo, M., Carratelli, E.P.: Analysis of Sea Current and Diffusion Data at a Submarine Sewage Outfall, pp. 169–179. Computational Mechanics Publications, Southampton (1997)

26. Vigliani, P.G., Sclavi, B., Olivotti, R., Visconti, A., Tartaglia, G.F.: Indagine sperimentale sulla diluizione iniziale dei liquami versati a mare da un diffusore, vol. 3, pp. 129–140. Ingegneria Sanitaria (1980)

27. Viviani, G.: Il progetto delle condotte sottomarine: una semplificazione nell'uso dei modelli complete, vol. 18, pp. 327–339. Ingegneria Ambientale (1989)

28. Wei, W., Zhong-min, Y.: Evaluation of three-dimensional numerical model for saline intrusion and purging in sewage outfalls. J. Hydrodynamics Ser. B **19**(1), 48–53 (2007)

29. Wilkinson, D.L.: Avoidance of seawater intrusion into ports of ocean outfalls. J. Hydraulic Eng. **114**(2), 218–228 (1988)

Author Index

Printed in the United States
By Bookmasters